Metapopulation
BIOLOGY
Ecology, Genetics, and Evolution

Metapopulation
BIOLOGY
Ecology, Genetics, and Evolution

Edited by

Ilkka Hanski

Department of Ecology and Systematics
University of Helsinki
FIN-00014 Helsinki, Finland

Michael E. Gilpin

Department of Biology
University of California, San Diego
La Jolla, California

ACADEMIC PRESS
San Diego   London   Boston   New York   Sydney   Tokyo   Toronto

Front cover photograph: The landscape of forests, fields, and lakes
in Åland, Finland, with the network of small meadows used by the
Glanville fritillary butterfly (*Melitaea cinxia*) shown as red and
white dots. The red dots represent meadows with a local population
in 1993 (see Chapter 4 by Hanski, this volume). Reproduced with permission
from the National Land Survey of Finland.

This book is printed on acid-free paper. ∞

Academic Press
a division of Harcourt Brace & Company
525 B Street, Suite 1900, San Diego, California 92101-4495, USA
http://www.apnet.com

Academic Press Limited
24-28 Oval Road, London NW1 7DX, UK
http://www.hbuk.co.uk/ap/

Library of Congress Cataloging-in-Publication Data

Metapopulation biology : ecology, genetics, and evolution / edited by
  Ilkka Hanski, Michael E. Gilpin.
          p.   cm.
    Includes index.
    ISBN 0-12-323445-X (alk. paper) CASE   ISBN 0-12-323446-8 (alk. paper) PAPER
    1. Population biology.   I. Hanski, Ilkka.   II. Gilpin, Michael
  E., date.
  QH352.M47   1996
  574.5'248--dc20   96-28242
                                                         CIP

PRINTED IN THE UNITED STATES OF AMERICA
      98 99   00   01   QW   9   8   7   6   5   4   3   2

Contents

PART **I** CONCEPTUAL FOUNDATIONS

1 The Metapopulation Approach, Its History, Conceptual Domain, and Application to Conservation

Ilkka Hanski and Daniel Simberloff

6   Two-Species Metapopulation Models

Sean Nee, Robert M. May, and Michael P. Hassell

7   From Metapopulation Dynamics to Community Structure: Some Consequences of Spatial Heterogeneity

Robert D. Holt

8   Genetic Effective Size of a Metapopulation

Philip W. Hedrick and Michael E. Gilpin

9   The Evolution of Metapopulations

N. H. Barton and Michael C. Whitlock

PART **III** METAPOPULATION PROCESSES

PART IV CASE STUDIES

16 Tritrophic Metapopulation Dynamics: A Case Study of Ragwort, the Cinnabar Moth, and the Parasitoid *Cotesia popularis*

Ed van der Meijden and Catharina A. M. van der Veen-van Wijk

17 Spatially Correlated Dynamics in a Pika Metapopulation

Andrew T. Smith and Michael E. Gilpin

18 A Case Study of Genetic Structure in a Plant Metapopulation

Barbara E. Giles and Jérôme Goudet

Contributors

Numbers in parentheses indicate the pages on which the authors' contributions begin.

N. H. Barton (183) Institute of Cell, Animal, and Population Biology, University of Edinburgh, Edinburgh EH9 3JT, Scotland

Patrick Foley (215) Department of Biological Sciences, California State University, Sacramento, Sacramento, California 95819

Steven A. Frank (325) Department of Ecology and Evolutionary Biology, University of California, Irvine, Irvine, California 92717

Barbara E. Giles (429) Department of Genetics, Umeå University, S-901 87 Umeå, Sweden

Michael E. Gilpin (165, 407) Department of Biology, University of California, San Diego, La Jolla, California 92093

Jérôme Goudet (429) Institut de Zoologie et d'Ecologie Animale, Université de Lausanne, CH-1015 Lausanne, Switzerland

Pierre-Henri Gouyon (293) Evolution et Systématique des Végétaux, , Université Paris-Sud, 91405 Orsay Cedex, France

Mats Gyllenberg (93) Department of Mathematics, University of Turku, FIN-20014 Turku, Finland

Ilkka Hanski (5, 69, 93, 359) Department of Ecology and Systematics, Division of Population Biology, University of Helsinki, FIN-00014 Helsinki, Finland

Susan Harrison (27) Division of Environmental Studies and Center for Population Biology, University of California, Davis, Davis, California 95616

Michael P. Hassell (123) Department of Biology, Imperial College at Silwood Park, Ascot, Berkshire SL5 7PY, United Kingdom

Alan Hastings (93) Division of Environmental Studies and Institute of Theoretical Dynamics, University of California, Davis, Davis, California 95616

Philip W. Hedrick (165) Department of Zoology, Arizona State University, Tempe, Arizona 85287

Robert D. Holt (149) Department of Systematics and Ecology, Natural History Museum, The University of Kansas, Lawrence, Kansas 66045

Rolf A. Ims (247) Department of Biology, Division of Zoology, University of Oslo, N-0316 Oslo 3, Norway

Veronica A. Johnson (267) Program in Ecology, Evolution and Conservation Biology, University of Nevada, Reno, Nevada 89512

Robert M. May (123) Department of Zoology, Oxford OX1 3PS, United Kingdom

Sean Nee (123) Department of Zoology, Oxford OX1 3PS, United Kingdom

Isabelle Olivieri (293) Institut des Sciences de l'Evolution, Université Montpellier 2, 34095 Montpellier cedex 05, France

Daniel Simberloff (5) Department of Biological Sciences, Florida State University, Tallahassee, Florida 32306

Andrew T. Smith (407) Department of Zoology, Arizona State University, Tempe, Arizona 85287

Peter B. Stacey (267) Program in Ecology, Evolution and Conservation Biology, University of Nevada, Reno, Nevada 89512

Mark L. Taper (267) Department of Biology, Montana State University, Bozeman, Montana 59717

Andrew D. Taylor (27) Department of Zoology, University of Hawaii, Honolulu, Hawaii 96822

Chris D. Thomas (359) Department of Biology, University of Leeds, Leeds LS2 9JT, United Kingdom

Ed van der Meijden (387) Institute of Evolutionary and Ecological Sciences, Leiden University, 2300 RA Leiden, The Netherlands

Catharina A. M. van der Veen-van Wijk (387) Institute of Evolutionary and Ecological Sciences, Leiden University, 2300 RA Leiden, The Netherlands

Michael C. Whitlock (183) Department of Zoology, University of British Columbia, Vancouver, British Columbia, Canada V6T 1Z4

John A. Wiens (43) Department of Biology and Graduate Degree Program in Ecology, Colorado State University, Fort Collins, Colorado 80521

Nigel G. Yoccoz (247) Laboratoire de Biométrie, Génétique et Biologie des Populations, URA CNRS 2055, Université Claude Bernard, F-69622 Villeurbanne Cedex, France; and Department of Biology, Division of Zoology, University of Oslo, N-0316 Oslo 3, Norway

Preface

In the past few years, the metapopulation concept has become widely and firmly established both in population biology and in conservation. The number of papers on metapopulations is growing exponentially, with a doubling time of less than two years. The metapopulation concept is beginning to appear in textbooks, and the metapopulation theory has to a large extent replaced the dynamic theory of island biogeography in conservation biology. As observed by *Science* magazine, metapopulation approaches are now "all the rage."

Our previous book, *Metapopulation Dynamics: Empirical and Theoretical Investigations* (Gilpin and Hanski, Academic Press, 1991), brought together a range of viewpoints and ecological models bearing on spatially fragmented populations. The book sold out rapidly, leaving an unmet demand. In considering the need for a new book on the same subject, we had a choice between an updated second edition and an entirely new volume. We chose the second alternative for two reasons. First, the field of metapopulation biology has advanced considerably, with a vigorous interplay among theory, models, and field studies, and we wanted to reflect the depth and the breadth of this growth in the new volume. Second, we wanted to shift some emphases and expand along new lines of inquiry. The first volume was biased toward a conceptual analysis of metapopulation ecology. In this volume, we cover more thoroughly both empirical studies and more advanced theories, and we have now included more information pertaining to genetics and evolution. The rapid progress that has occurred in field studies is evident on the covers of the two volumes. Whereas the cover of the previous book depicted a metapopulation of a hypothetical butterfly species, *Euphydryas macintoshus* G., the cover of this volume illustrates the fragmented landscape of a real butterfly metapopulation (the Glanville fritillary, *Melitaea cinxia*).

This volume consists of solicited chapters from selected authors working in the general area of metapopulation biology. We are pleased that everyone whom we asked to contribute did contribute. Several chapters in this volume are primarily empirical, while others are highly theoretical. Some chapters practically ignore ecology; many more ignore genetics and evolution. However, a few chapters describe both theory and empirical results, and others cover both ecology and genetics, a trend that we hope will become more prominent in the near future. Not all readers will equally appreciate every chapter, but we would be very disappointed (and truly surprised) if most readers would not be better informed and indeed stimulated by most of the chapters. The scope of the chapters in this volume represents our attempt to sketch the general limits of metapopulation biology. We hope that this volume will disseminate ideas, results, and conclusions across the customary academic confines.

One recent trend that we have noted is a definite broadening of the meaning of the term "metapopulation." In the first volume, population turnover, local extinctions and colonizations, was considered to be the key and practically indispensable feature of metapopulation dynamics. In this interpretation we followed the conceptual guidance of the root of all metapopulation models, the Levins model. Although we continue to think that metapopulation dynamics in this narrow sense forms the hard conceptual core of this area of population biology, it is now time to accept that a broader perspective is needed. This volume promotes such a view. Inevitably, an expansion of the metapopulation concept will attract applications to an even wider range of situations than we can foresee, and some of these applications will not turn out to be productive. During a period of rapid growth, excesses may occur and limits may be crossed. This is the time-honored process by which the scientific community tests the applicability of any worthwhile idea or model.

We thank Chuck Crumly of Academic Press for encouraging us to edit this second volume and for all his assistance during the process. The following colleagues greatly helped us in reviewing individual chapters: Milo Adkinson, Richard Barnes, Nick Barton, Jan Bengtsson, Jim Berkson, Ian Billick, Ted Case, Diane Debinski, Torbjörn Ebenhard, Gordon Fox, Andy Hansen, Alan Hastings, Phil Hedrick, Anthony Ives, Tad Kawecki, Joshua Kohn, Russ Lande, Simon Levin, Sean Nee, Isabelle Olivieri, Trevor Price, Jonathan Silvertown, Dan Simberloff, Monte Slatkin, Peter Stacey, Mark Taper, Chris Thomas, Rick Walker, Christian Wissel, and Greg Witteman. We also thank Pia Vikman for her secretarial contribution to the project. Chuck Crumly and Deborah Moses of Academic Press have been a pleasure to work with.

Ilkka Hanski
Michael E. Gilpin

I

CONCEPTUAL FOUNDATIONS

The three chapters in this section explore the scope of the metapopulation concept and its applications. Hanski and Simberloff sketch the history of metapopulation studies and the range of approaches, both theoretical and empirical, that have been used in single-species studies. Harrison and Taylor assess critically the pertinence of the metapopulation approach to field studies and expand their review to multispecies situations. Wiens more directly connects the metapopulation concept to the complexities of real landscapes. Hanski and Simberloff outline in some detail the use (and misuse) of the metapopulation concept in conservation, where an apparent paradigm shift has occurred from the dynamic theory of island biogeography to the metapopulation theories.

The gradual unfolding and evolution of the metapopulation concept from the pioneering studies of Sewall Wright, Andrewartha and Birch, Huffaker, Den Boer, Ehrlich, Gadgil, and Levins have been narrated previously and are summarized here by Hanski and Simberloff and by Harrison and Taylor. Harrison and Taylor make the interesting point that the origin of the metapopulation idea is different in single-species and in multispecies studies. Single-species studies have tended to emphasize the benefits of migration in leading to the establishment of new populations and thereby compensating for extinctions in small habitat patches. In the multispecies metapopulation scenarios, the key issue has been the locally unstable interaction among competitors and between a prey and its predator. Habitat fragmentation can be beneficial in creating the possibility for asynchronuous fluctuations, which can enhance regional stability. A high rate of migration may eliminate such asynchrony, and is hence potentially harmful for regional persistence in multispecies metapopulations. Multispecies metapopulation theory is further discussed by Nee, May, and Hassell and by Holt in Part II.

In the predecessor of this volume, *Metapopulation Dynamics: Empirical and Theoretical Investigations* (Gilpin and Hanski, 1991), metapopulation dynamics was seen to imply significant turnover of local populations, local extinctions, and colonizations. This notion follows directly from Levins's original concept of a metapopulation as a population of populations, analogous to a population of individuals with finite lifetimes. This narrow classical view of metapopulations has now become superceded by a broader view, where any assemblage of discrete local populations with migration among them is considered to be a metapopulation, regardless of the rate of population turnover. (In a nonequilibrium metapopulation declining toward extinction even among-population migration is not a necessary criterion, but a system with no turnover and no migration would not classify as a metapopulation.) There are important questions to be asked about the role of migration in (local) population dynamics, and these questions are most naturally asked in a metapopulation (regional) context. Metapopulation dynamics in the narrow sense, with significant population turnover, is of course included in metapopulation dynamics in the broad sense.

The realization that natural populations exemplify many kinds of spatial population structures has stimulated a terminology, originally due to Susan Harrison, and including entries such as patchy populations (not really metapopulations), classical (Levins) metapopulations, mainland–island metapopulations, source–sink metapopulations, and nonequilibrium metapopulations. These concepts and types of metapopulation structures are discussed by Hanski and Simberloff and by Harrison and Taylor. The danger here is that too much emphasis is given to classification, definition of ideal types, which in itself does not guarantee any better understanding of the ecology, genetics, and evolution of metapopulations. What matters is what works. Does the "metapopulation approach" help answer important questions about spatially structured populations? Does it provide us with scientific insight to the problems in which we are interested? All this being said, there still is a need to be concerned with the type of spatial structure of populations in any empirical study and in an application of the metapopulation concept and models to real populations. One should avoid the temptation of pigeonholing every population with some form of patchiness as a "metapopulation," as Harrison and Taylor warn. In the worst case, this may obscure what is important and draw attention to elements that are less critical. Unfortunately, there are no easy answers here; one simply has to know the species and one has to understand the interactions between the populations and their environment.

Metapopulation biology may be a multifaceted subject, but there is one common element that characterizes this approach to population biology. The metapopulation approach is based on the notion that space is not only discrete but that there is a binary distinction between suitable and unsuitable habitat types. If this does not fit one's idea of a particular environment, one is probably better off in using some approach other than the metapopulation approach. An important reason for the appeal of the metapopulation concept comes from our subjective conviction that natural lansdscapes truly are, for many species, patchworks of one or several habitat types.

Though the metapopulation view of nature is complex enough, it appears to be hopelessly simplified in comparison of how landscape ecologists view reality. Wiens in his chapter lists four components that characterize landscape ecology: variation in patch quality, variation in the quality of the surrounding environment, boundary effects, and how the landscape affects patch connectivity. Wiens is correct in suggesting that most of these elements are by and large missing from metapopulation models, which are typically focused on idealized habitat patches in a featureless landscape. Recent studies of Andrén and Green (cited by Wiens) appear to suggest that where the suitable habitat fragments for some species cover only a relatively small fraction of total area (let us call these LC landscapes, for low coverage), patch area and isolation effects tend to be significant; but where much of the area is covered by more or less suitable habitat (HC landscapes, for high coverage), other factors, such as exactly how individuals move in a complex landscape, begin to dominate. Now, it so happens that the classical metapopulation concept implicitly assumes a LC landscape, hence the tradition of representing habitat patches as dots on maps, rather than drawing them as realistic habitat fragments. There appears to be a real difference between the two traditions here, as they have been largely concerned with either LC landscapes (metapopulation ecology) or HC landscapes (landscape ecology). As Wiens stresses, it is imperative for the practical application of both metapopulation biology and landscape ecology in conservation and planning that more common ground is established by developing appropriate theory and designing appropriate field studies.

Some necessary constituents of a more unified approach seem relatively easy to achieve. For instance, it should not be too difficult to correct among-patch distances by taking into account how the features of the intervening landscape affect individual movement behavior. On the other hand, when considering HC landscapes, patch models are likely to be inadequate anyway. Metapopulation theory may well remain a useful practical tool for LC landscapes, with relatively small and isolated fragments of suitable habitat, but the "reserve mentality" that this approach implies should give away, as Wiens argues, to "mosaic management" of the environment in HC landscapes. Today, we do not yet have a conceptual and practical synthesis of metapopulation biology and landscape ecology, but no doubt the time will come when we will.

1

The Metapopulation Approach, Its History, Conceptual Domain, and Application to Conservation

Ilkka Hanski   *Daniel Simberloff*

I. INTRODUCTION

At no period in the history of ecology has the spatial structure of populations and communities been entirely ignored, but the role that space plays in forming ecological patterns and in molding processes has been viewed very differently in different times (McIntosh, 1991). In the 1960s and 1970s, theoretical ecology was largely focused on issues other than spatial dynamics (May, 1976a), with notable exceptions (MacArthur and Wilson, 1967), and field ecologists tended to follow suit. Today, space is in the forefront and space is introduced in various ways into all fields of ecology and population biology more generally. Whether one is interested in processes occurring at the level of genes, individuals, populations, or communities, spatial structure is widely seen as a vital ingredient of better and more powerful theories, and good empirical work involving space is seen as a great challenge (Kareiva, 1990).

Five years ago, before the publication of the predecessor of this volume (*Metapopulation Dynamics: Empirical and Theoretical Investigations,* Gilpin and Hanski, 1991), the metapopulation concept was new to most biologists. Since then, literature on metapopulations has grown exponentially, with a doubling time of less than 2 years (Fig. 1). The metapopulation concept has by now been firmly established in population biology and beyond; we review and analyze in this

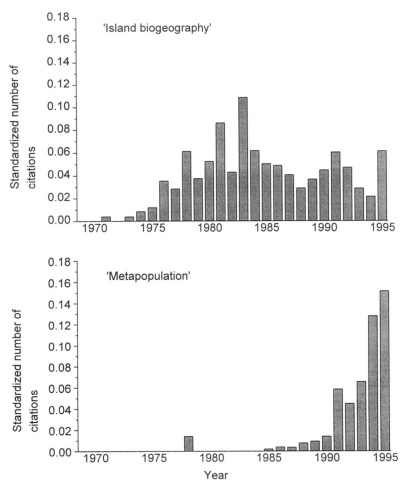

FIGURE 1   Numbers of citations to the key words "island biogeography" and "metapopulation" in the BIOSIS data base in 1970–1995, standardized by the respective total number of papers in the data base.

chapter the spread of the metapopulation concept to conservation biology and applications.

What is the metapopulation approach? A more complete explication is given below, but in a nutshell the two key premises in this approach to population biology are that populations are spatially structured into assemblages of local breeding populations and that migration among the local populations has some effect on local dynamics, including the possibility of population reestablishment following extinction. These premises contrast with those of standard models of demography, population growth, genetics, and community interaction that assume a panmictic population structure, with all individuals equally likely to interact with any others. Population biology has made productive use of such models for at least 100 years, but today there is a distinct need to account for the position of individuals and populations in space. This need has arisen from the intrinsic development of population biology as a science, but the trend has clearly been strengthened by the demand for professional advice on environmental issues typically involving space.

In the past few years, metapopulation studies have shed new light on such phenomena as patterns of distribution and population turnover dynamics in fragmented landscapes (Hanski, this volume; Harrison and Taylor, this volume; Thomas and Hanski, this volume; Smith and Gilpin, this volume; van der Meijden and van der Veen-van Wijk, this volume), landscape ecology (Wiens, this volume) and community structure (Holt, this volume), population viability and time to extinction (Gyllenberg *et al.*, this volume; Foley, this volume), coexistence of competing species, and of prey and their natural enemies (Nee *et al.*, this volume), evolution of migration rate and other life-history traits (Olivieri and Gouyon, this volume), ecological consequences of migration (Ims and Yoccoz, this volume; Stacey and Taper, this volume), unexpectedly high levels of inbreeding and low heterozygosity in natural populations (Hedrick and Gilpin, this volume), patterns of genetic differentiation (Giles and Goudet, this volume), adaptation (Barton and Whitlock, this volume), and coevolutionary processes (Frank, this volume). As is apparent from the citations, these developments are well represented in the chapters in this volume, which provide an excellent entree to the literature at large.

There are many advantages of a metapopulation approach, but success may also breed problems. As in any scientific field experiencing rapid growth, there is the danger of blurring of concepts. There is the temptation to view any system with any kind of patchiness at any spatial or even temporal scale as a "metapopulation." Harrison (1991, 1994b; Harrison and Taylor, this volume) cautions us about this tendency. Anticipating the kind of verbal entropy that has enveloped many terms in population biology, Hanski and Gilpin sketched in the 1991 volume the meaning of the term "metapopulation," highlighting issues of scale, hierarchy, and a requirement for some population turnover. We feel a need to dwell on the same issues in this chapter, and we provide a revised succinct glossary of the commonly used terms in the literature. First, however, let us examine briefly the history of the metapopulation concept.

II. BRIEF HISTORY OF METAPOPULATION STUDIES

The metapopulation concept has a pedigree dating back to the early part of this century, but until recently this tradition played only a minor and episodic role in the intellectual advance of population biology. For a long time, the prevailing view was one emphasizing persistence and stability of local populations, or as McIntosh (1991) put it, "the great tradition of balance of nature, going back to antiquity, imputed to nature homogeneity, constancy, or equilibrium and abhored thoughts of extinction and randomness."

In evolutionary biology, Sewall Wright (1931, 1940) had the insight that evolution might proceed rapidly in spatially structured populations, especially if there are local extinctions and recolonizations. Wright's shifting balance theory has remained an intriguing, imperfectly understood, and little tested model ever since (Barton and Whitlock, this volume). Wright's work may have stimulated interest in spatially structured populations in the first half of this century, represented for instance by studies of Boycott (1930), Diver (1938), and Lamotte (1951) on ecology and genetics of snail populations (for a more thorough discussion, see Hanski, 1996a). Pioneering quantitative studies in epidemiology (Ross, 1909, Kermack and McKendrick, 1927; see Anderson and May, 1991; Nee *et al.*, this volume) are now seen as closely linked conceptually and theoretically to metapopulation studies, but that connection remained without comment until recently (May, 1991; Lawton *et al.*, 1994; Nee, 1994).

The ecological implications of the metapopulation concept were not considered before 1954, when Andrewartha and Birch published their distinguished text on animal ecology. Drawing on their wide experience from insect population ecology, Andrewartha and Birch found the "dogma of density-dependent factors" unacceptable. They emphasized wild oscillations of populations, documented frequent local extinctions, but also recognized the possibility of reestablishment of populations at vacated localities. In brief, Andrewartha and Birch (1954) advocated the view that local population extinction was a common phenomenon: "spots that are occupied today may become vacant tomorrow and reoccupied next week or next year" (Andrewartha and Birch, 1954, p.87). However, why did their ideas fail to gain wider acceptance? We believe the reason is their nearly categorical rejection of the concept of density-dependent population regulation. The Andrewartha and Birch notion about population dynamics in space was largely ignored and eventually forgotten. The incipient metapopulation concept nonetheless had a quiet existence in the 1950s and 1960s, in works of Huffaker (1958), den Boer (1968), Ehrlich and Raven (1969), Gadgil (1971), and undoubtedly a few others. The MacArthur and Wilson (1963, 1967) dynamic theory of island biogeography has much in common with the metapopulation concept, even if MacArthur and Wilson were primarily concerned with multispecies communities, as we discuss below.

The term "metapopulation" was introduced in the works of Richard Levins in 1969 (1969a) and 1970. The word itself suggests a population of populations, with colonization and extinction of local populations in a metapopulation likened to births and deaths of individuals in a local population (hence the emphasis on population turnover in "classical" metapopulation studies). Levins's work marks the beginning of contemporary metapopulation biology. It is puzzling, though, that the early lead that Levins provided was followed by a period of nearly 20 years of recess (Fig. 1). We return to the possible reasons for this delay below, in the section on metapopulations and conservation biology.

III. CONCEPTUAL DOMAIN AND METAPOPULATION APPROACHES

A fundamental assumption of the original metapopulation concept (Levins, 1969a) is that space is discrete and that it is possible and useful to distinguish between habitat patches that are suitable for the focal species and the rest of the environment, often called the matrix. In this respect the metapopulation approach is closely akin to the dynamic theory of island biogeography (MacArthur and Wilson, 1967) but differs from landscape ecology (Wiens, this volume). The metapopulation concept also presumes that the habitat patches are large enough to accommodate panmictic local populations, but not larger. Other fields of ecology are concerned with spatial patchiness, but either at a smaller (foraging theories; Krebs and Davies, 1984) or at a larger scale (e.g., much of landscape ecology: Forman and Godron, 1986; GAP analyses: Scott *et al.*, 1991; geographical ecology: Ricklefs and Schluter, 1993) than the scale of (panmictic) local populations. The concept of an ideal metapopulation a la Levins includes three other simplifying assumptions: habitat patches have equal areas and isolation, local populations in the metapopulation have entirely independent (uncorrelated) dynamics, and the exchange rate of individuals among local populations is so low that migration has no real effect on local dynamics in the existing populations: local dynamics occur on a fast time scale in comparison with metapopulation dynamics.

No real metapopulation completely satisfies all these requirements. However, the more specific assumptions, such as equal patch areas and isolation, can be relaxed without need for a major conceptual amendment. This is not unlike how the population concept is used in population biology: no real population completely satisfies all the criteria of an ideal, closed and panmictic, population. What really matters is the notion of discrete local breeding populations connected by migration. We suggest that if this assumption cannot be defended, some other approach should be used instead of the metapopulation approach; and conversely, the more distinct and smaller the local breeding populations are, the more useful the metapopulation approach is likely to be. Hanski and Gilpin (1991) used population turnover, local extinctions and colonizations, as the hallmark of true metapopulations. By this definition, the mainland–island systems studied in the dynamic theory of island biogeography and in recent metapopulation models (Gotelli, 1991; Hanski and Gyllenberg, 1993) would not count as metapopulations. Following the current usage of the term, we now include mainland–island structures among other metapopulation structures.

It has been suggested that "much" migration among local populations makes the metapopulation approach less useful (Harrison, 1994b). While it is true that the classical concept (Levins, 1969a) implicitly assumes a low migration rate, so low that migration plays no role in the dynamics of existing local populations, more recent theoretical (Hassell *et al.*, 1991a, 1994, Gyllenberg and Hanski, 1992; Nee *et al.*, this volume) and empirical work (Hanski *et al.*, 1995a,b) has made good use of the metapopulation concept even when some tens of percents of individuals per generation leave their natal patch. An important issue here is the spatial scale of migration. Theoretical studies suggest that a low rate of long-distance migration has often about the same consequences as a high rate of short-distance migration (Nachman, 1991). Clearly, if migration rate is very high, say > 50%, and if migration distances are not limited, a metapopulation approach is unlikely to be helpful. The fundamental criterion, however, is whether or not the metapopulation approach is useful in elucidating the questions in which we happen to be interested, not whether migration rate is high or low. From the perspective of traditional population biology, the question is whether the implicit assumption that migration makes no difference to the dynamics of the focal population is a useful approximation or not.

Our remarks have been directed at the population ecological properties of metapopulations. Genetic and evolutionary consequences of these metapopulation structures enlarge the biological domain of the metapopulation concept as described by Olivieri and Gouyon (this volume) and Barton and Whitlock (this volume).

In the previous metapopulation book, Hanski and Gilpin (1991) defined a set of key metapopulation terms in the hope of promoting a more uniform terminology. We repeat this exercise here, with a revised and expanded list of terms (Table I). This list is largely self-explanatory, but a few comments are warranted. The source-sink concept continues to cause confusion in the literature. Pulliam (1988) defined sources and sinks on the basis of whether emigration exceeds immigration, or vice versa, at equilibrium. This definition is useful for population genetic purposes, in emphasizing asymmetry in gene flow, which may have important consequences for genetic structure and adaptation (Barton and Whitlock, this volume; Giles and Goudet, this volume). The definition given in Table I, which is based on the expected population growth rate at low density, in the absence of intraspecific density dependence, may often be preferable for ecological purposes. In the latter case, sinks are populations that would go extinct in the absence of immigration (by Pulliam's definition, a sink population may decline to a low but positive equilibrium value in the absence of immigration; Watkinson and Sutherland, 1995). A third and potentially misleading sense in which the source–sink concept is often used is for a mixture of small and large habitat patches. Populations in small patches typically have a high risk of extinction, but they are not necessarily "sinks" in the sense of Pulliam (1988) or Table I; small populations have a high risk of stochastic extinction, even if the expected growth rate at low density and the expected equilibrium population size are positive (Foley, this volume).

A. Modeling Approaches

The traditional approach to population biology assumes spatially unstructured populations. Modeling approaches to spatially structured populations can be divided conveniently into two classes, based on whether the model deals with

TABLE I Metapopulation Terminology [a]

Term	Synonyms and definition
Patch	Synonyms: Habitat patch, (habitat) island, (population) site, locality Definition: A continuous area of space with all necessary resources for the persistence of a local population and separated by unsuitable habitat from other patches (at any given time, a patch may be occupied or empty)
Local population	Synonyms: Population, subpopulation, deme Definition: Set of individuals that live in the same habitat patch and therefore interact with each other; most naturally applied to "populations" living in such small patches that all individuals practically share a common environment
Metapopulation	Synonyms: Composite population, assemblage (of populations) [population (when "local populations" are called "subpopulations")] Definition: Set of local populations within some larger area, where typically migration from one local population to at least some other patches is possible (but see nonequilibrium metapopulation)
Metapopulation structure	Synonyms: Metapopulation type Definition: Network of habitat patches which is occupied by a metapopulation and which has a certain distribution of patch areas and interpatch migration rates
Levins metapopulation	Synonyms: Classical metapopulation Definition: Metapopulation structure assumed in the Levins model: a large network of similar small patches, with local dynamics occurring at a much faster time scale than metapopulation dynamics; in a broader sense used for systems in which all local populations, even if they may differ in size, have a significant risk of extinction
Mainland–island metapopulation	Synonyms: Boorman–Levitt metapopulation Definition: System of habitat patches (islands) located within dispersal distance from a very large habitat patch (mainland) where the local population never goes extinct (hence mainland–island metapopulations do not go extinct)
Source–sink metapopulation	Definition: Metapopulation in which there are patches in which the population growth rate at low density and in the absence of immigration is negative (sinks) and patches in which the growth rate at low density is positive (sources)
Nonequilibrium metapopulation	Definition: Metapopulation in which (long-term) extinction rate exceeds colonization rate or vice versa; an extreme case is where local populations are located so far from each other that there is no migration between them and hence no possibility for recolonization
Turnover	Synonyms: Colonization–extinction events (or dynamics) Definition: Extinction of local populations and establishment of new local populations in empty habitat patches by migrants from existing local populations
Metapopulation persistence time	Synonyms: Expected life-time of a metapopulation Definition: The length of time until all local populations in a metapopulation have gone extinct

(continues)

TABLE I *(continued)*

Term	Synonyms and definition
Patch model	Synonyms: Occupancy model, presence/absence model Definition: A metapopulation model in which local population size is ignored and the number (or fraction) of occupied habitat patches is modeled
Levins model	Definition: The model presented by Levins (1969a; see Hanski, this volume)
Structured metapopulation model	Definition: A model in which the distribution of local population sizes is modeled
Incidence function model	Definition: A model of the stationary probabilities (incidences) of patches being occupied, generally assumed to be functions of the sizes and isolations of the patches
Spatially implicit metapopulation model	Synonyms: Island model Definition: Model in which all local populations are equally connected; patch models and structured metapopulation models are spatially implicit models
Spatially explicit metapopulation model	Synonyms: Lattice (grid) model, cellular automata model, stepping-stone model Definition: Model in which migration is distance-dependent, often restricted to the nearest habitat patches; the patches are typically identical cells on a regular grid, and only presence or absence of the species in a cell is considered (the model is called a coupled map lattice model if population size in a patch is a continuous variable)
Spatially realistic metapopulation model	Synonyms: Spatially explicit model (note that we make a distinction between spatially explicit and spatially realistic models) Definition: Model that assigns particular areas, spatial locations, and possibly other attributes to habitat patches, in agreement with real patch networks; spatially realistic models include simulation models and the incidence function model

a Modified from Hanski and Gilpin, 1991, and Hanski, 1996a.

interactions among two conspecific populations connected by migration, or with interactions among many local populations. The former approach is useful when the focus of the study is specifically on the effect of migration on local dynamics and one is willing to assume that populations are so effectively regulated that extinctions do not occur (Levin, 1974; Holt, 1985; Gyllenberg *et al.*, 1993). In metapopulation studies in the narrow sense, when there is population turnover, it is necessary to resort to modeling approaches assuming many habitat patches and local populations. Among these approaches, we distinguish between spatially implicit, spatially explicit, and spatially realistic approaches (Hanski, 1994c).

1. Spatially Implicit Approaches

Truly significant insights are often based on a critical simplification of what at first appears a hopelessly complex problem. The model that Levins (1969a,

1970) constructed to caricature metapopulation dynamics is an excellent example. Instead of attempting to extend a model of a single population to many populations connected by migration, Levins modeled the changes in the number of such populations, effectively ignoring what happens in each one of them and where in space they happen to be located (Hanski, this volume). For the latter reason, the Levins model and other related patch models (Table I) are spatially implicit; the habitat patches and local populations are discrete (and are generally assumed to have independent dynamics), but they are assumed to be all equally connected to each other. In spite of this simplifying assumption, which can be generally defended only for metapopulations close to steady state and with no strong spatial aggregation, the patch models allow us to analyze many interesting questions about metapopulation dynamics, starting with the conditions of metapopulation persistence in a balance between local extinctions and colonizations. The advantage of the spatially implicit approach is that it greatly facilitates the mathematical and conceptual analysis; the disadvantage is that it can be used to study only a subset of all interesting questions.

Thinking about the restrictive assumptions of the Levins model and other patch models, ecologists have asked what happens when local dynamics are included in the metapopulation model. What happens when the habitat patches are of different sizes and when the local populations have different extinction probabilities? What if migration rate is high enough to "rescue" local populations before extinction? What are the consequences of real spatial locations of local populations? What if extinction events are correlated over the entire metapopulation? What if there is spatial asymmetry and source and sink populations? Some of these questions have been explored in the context of spatially implicit models (Pulliam, 1988; Harrison and Quinn, 1989; Hanski and Gyllenberg, 1993; Gyllenberg et al., this volume), but it comes as no surprise that at some point we have to turn to models that incorporate specific information on the spatial locations of populations. Incidentally, most analyses of metapopulation genetics (Barton and Whitlock, this volume; Hedrick and Gilpin, this volume) have been based on the Levins model, which is essentially equivalent to what population geneticists call the "island model." As in ecology, there is an increasing need to add space in a more explicit manner to metapopulation genetic models.

2. Spatially Explicit Approaches

Under the rubric of spatially explicit approaches are several related modeling frameworks, such as cellular automata models (Caswell and Etter, 1993), interacting particle systems (Durrett, 1989), and coupled map lattice models (Hassell et al., 1991a). These modeling approaches assume that "local populations" are arranged as cells on a regular grid (lattice), with population sizes modeled as either discrete or continuous variables. The key feature that distinguishes spatially explicit approaches from spatially implicit approaches is localized interactions: populations are assumed to interact only with populations in the nearby "cells." Localized interactions can have profound dynamic consequences, such as very long times before the metapopulation settles to a steady state (Hastings and Higgins, 1994) and spatially chaotic dynamics (Hassell *et al.*, 1991a; Nee *et al.*, this volume). The disadvantage is that the state of the metapopulation cannot be described simply by the fraction of cells occupied; an entire vector of presences and absences is needed. Such models require considerable computation. An advantage is that, since each cell on the grid has a constant area and constant spacing, the mathematical rules that govern local behavior are the same from cell to cell, and it is easy to write a computer program to model the dynamics.

Lattice-based models and raster-based GIS descriptions in landscape ecology share the same format of representing space. Thus it is possible to develop complex models that blur the distinction between spatially explicit and spatially realistic models (below). From raster-based description of habitat suitability, one can aggregate "cells" into patches on which local populations may exist, thus reverting to a patch-based metapopulation model for a dynamic analysis (Burgman *et al.*, 1993; Akçakaya, 1994).

3. Spatially Realistic Models

Spatially realistic models allow one to include in the model the specific geometry of particular patch networks: how many patches are there in the network, how large are they, and where exactly are they located? Including all this information in the model is necessary if one is interested in making specific quantitative predictions about the dynamics of real metapopulations. For instance, if we want to assess the likely consequences of destroying some particular patches in a patch network, we need a spatially realistic model. For obvious reasons, the spatially realistic approach is closely linked with empirical field studies.

The incidence function (IF) model (Hanski, 1994a,b, this volume) is perhaps the simplest spatially realistic metapopulation model. The IF model is conceptually related to the Levins model, but with the following critical differences: there is a finite number of habitat patches, and hence the model is stochastic in contrast to the deterministic Levins model; the patches are allowed to differ in area, which is assumed to affect local extinction probabilities; and the patches have specific spatial locations, which affect their probabilities of recolonization. Alternative spatially realistic approaches are based on extensive simulation of many local populations connected by migration (Hanski and Thomas, 1994; Akçakaya, 1994). Several generic models of this type are already available (Akçakaya, 1994; Sjögren Gulve and Ray, 1996). Not surprisingly, meaningful application of these models assumes much data. The extreme approach is to simulate the birth, movements, reproduction, and death of individuals (DeAngelis and Gross, 1992), but this approach, which can be used for any population structure, does not really take advantage of the metapopulation concept. An individually based modeling approach may nonetheless provide valuable insight into key processes affecting metapopulation dynamics, such as migration among populations (Kindvall, 1995).

B. Empirical Approaches

In a standard ecological metapopulation study, a key initial task is to make a practical distinction between habitat and nonhabitat and to delimit the suitable habitat patches in the study area. Suitable habitat may be defined subjectively or with the help of statistical methods (Lawton and Woodroffe, 1991). An experimental approach may be used to test the accuracy of an existing habitat classification: experimental introductions to empty habitat should succeed (Harrison, 1989; Oates and Warren, 1990; Thomas, 1992; Massot et al., 1994); introductions to nonhabitat should fail. Metapopulation studies focused on assemblages of extinction-prone local populations typically proceed to record the presence or absence of the focal species in the habitat patches and then to analyze the effects of various environmental factors on patch occupancy (Verboom et al., 1991b; Thomas and Harrison, 1992; Hanski et al., 1995a,b) and on the rates of extinction and colonization (Sjögren, 1991; Eber and Brandl, 1994; Hanski et al., 1995b). Other field studies have been concerned with more permanent local populations, but ones whose dynamics are significantly affected by migration (Stacey and Taper, this volume). Experimental studies have attempted to demonstrate the predicted temporal stability of local populations in a metapopulation as opposed to that in isolated local populations (Murdoch et al., 1996; Harrison and Taylor, this volume).

Landscape ecology (Forman and Godron, 1986; Turner, 1989; Wiens, this volume) and metapopulation ecology share a common focus on space and patchiness. The difference is primarily in the complex mosaic structure of real landscapes that is the object of landscape ecology (Wiens, this volume). In contrast, metapopulation studies typically assume that the patches which are used by the focal species are of the same type, though this assumption is made primarily for the sake of keeping the models reasonably simple (see Holt, this volume, for metapopulation models with two patch types). Empirical research in landscape ecology has been reluctant to use the population dynamic theory that metapopulation ecology purports to provide, even if in a rudimentary form, and consequently the two fields have developed largely independently. One trend that is beginning to change this situation is the use of GIS-based landscape descriptions in generic metapopulation simulation models (Akçakaya, 1994). Today, ecologists have access to huge data bases of digitized information about landscape structure, and the imminent arrival of low-cost global positioning systems will greatly facilitate further empirical research in this area.

It should come as no surprise that the bulk of current empirical research that is conceptually related to the metapopulation notion is conducted in conservation biology. We therefore devote the rest of this chapter to a more thorough scrutiny of the past and present links between metapopulation biology and conservation biology.

IV. METAPOPULATIONS AND CONSERVATION BIOLOGY

Conservation biology changed dramatically, beginning ca. 1975, from a heavy emphasis on habitat relationships of individual species to a focus on refuge design, guided by the dynamic theory of island biogeography and the genetic deterioration owing to drift and inbreeding (Simberloff, 1988). The two halves of this "new conservation biology" did not fit together well, as the former dealt with species richness of communities, while the latter aimed at the population level. Currently, a replacement of the island biogeographic component of conservation biology by metapopulation thinking is providing a more comfortable fit. Although the incorporation of metapopulation models into conservation biology has spurred important insights, it has also led to some misfocused proposals.

A. The Rise and Fall of the Theory of Island Biogeography

The theory of island biogeography (MacArthur and Wilson, 1963, 1967) quickly attracted much attention from ecologists (Fig. 1) by using simple mathematics to focus on an easily obtained statistic (species richness) and depicting a dynamic nature that is nonetheless readily understood because it is divided into small units, namely real or habitat islands (Simberloff, 1974, 1978a). The theory posits species richness on each island as a dynamic equilibrium maintained by continuing immigration of all species, balanced by ongoing local extinctions on the island, primarily owing to demographic and genetic stochasticity.

Clearly the island biogeographic theory shares key underpinnings with metapopulation models — the division of nature into discrete entities, with movement of individuals among relatively unstable local populations. There is also an apparent difference — island biogeographic theory treats communities, not individual species. Its key statistic is species richness. However, some island biogeographic models are formally composites of models for individual species, with the community-wide immigration and extinction rates being simply sums of the respective species-specific rates (Simberloff, 1969, 1983; Gilpin and Diamond, 1981). In these latter models, the underlying conception of what is happening in nature is just a mainland–island version of Levins's metapopulation concept (Hanski, this volume). However, even a species-based model of this type, and even one with migration from several sources, is still focused on a single island and on questions about the number of species and immigration and extinction rates on that island. A metapopulation model, even one in which the different sites and local populations are not modeled explicitly, focuses on the entire metapopulation of one or two species, using statistics such as the number of sites occupied.

In both types of models, an element of arbitrariness is just how much movement there is between sites for the models to remain useful. For metapopulations, the bone of contention is whether the movement is so frequent that one is dealing with a single population rather than a metapopulation, even if that population may be so large that individuals are likely to interact with their neighbors only (Harrison, 1991). For island biogeographic models, the argument is whether groups of conspecific individuals on sets of islands are separate populations or just transient parts of one widely ranging population (Smith, 1975; Simberloff, 1976). The latter argument has recrudesced recently with important conservation implications. Pimm *et al.* (1988), extrapolating from records of breeding birds on small British islands, suggested guidelines for how many individuals should constitute propagules for reintroduction efforts based on the body size of the species. Aside from problems with the statistics of extrapolation (Tracy and George, 1992), Haila and Hanski (1993) saw a more fundamental flaw: the assumption that the birds on each island constituted a population, and their disappearance an extinction. In their view, all these birds are parts of wide-ranging, large populations and their disappearances from specific islands within the range are not population phenomena. Diamond and Pimm (1993) retorted that the birds of each island could be viewed as a population within a metapopulation.

Within about a decade (Fig. 1), the theory of island biogeography came to dominate much of conservation biology, with a series of nearly simultaneous papers (Terborgh, 1974, 1975; Diamond, 1975a; Wilson and Willis, 1975) all advocating a set of "rules" of refuge design ostensibly based on the theory (Fig. 2). The rules each suggest a refuge configuration that would maximize species richness, and the papers describing the rules apparently stemmed from lectures given by E. O. Willis beginning in 1971 (Willis, 1984). Although some of the rules in fact were not based on the theory (references in Simberloff, 1988); they became popular in conservation circles, particularly after their publication in 1980 in the first synthetic plan for dealing with a perceived disastrous wave of extinctions, World Conservation Strategy, jointly authored by the International Union for the Conservation of Nature and Natural Resources, the United Nations, and the World Wildlife Fund. With this imprimatur, it is unsurprising that these rules, and the theory that supposedly supported them, became the governing paradigm in conservation biology, reproduced in textbooks and published in newspapers. The dominance of the island biogeographic paradigm was so strong that even studies that today would be seen as metapopulation research were published as island biogeographic studies, with no mention of the term "metapopulation" (e.g., Fritz, 1979).

It was noted early that most ecological publications citing island biogeographic theory simply interpreted a species-area relationship in terms of the theory, when alternative explanations were also possible (Simberloff, 1974), and that there was little empirical evidence for continuing local extinctions of the sort envisioned by the theory (Lynch and Johnson, 1974; Simberloff, 1974). Further, as noted by Smith (1975) and Simberloff (1976), by defining the comings to and goings from local sites of individuals within continuous populations as "immigration" and "extinction," one could almost always claim that extinctions and colonizations were occurring, even if the theory really envisioned most recruitment to local populations as being by *in situ* reproduction rather than immigration.

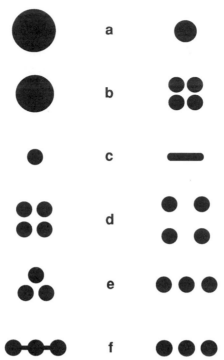

FIGURE 2   The "island biogeographic" rules for refuge design (after Wilson and Willis, 1975; International Union for the Conservation of Nature and Natural Resources, 1980). For each rule, the design on the left is seen as superior to the alternative on the right.

Nevertheless, ecologists on the whole tended to view the theory favorably until around 1980, when doubt about the existence of widespread local extinction became pervasive (Gilbert, 1980; Schoener and Spiller, 1987b; Williamson, 1989). The prevailing view now is that, in most systems, "turnover involves a subset of fugitive populations, with many others, mostly much larger, being permanent" (Schoener and Spiller, 1987b). The decline in citations of "island biogeography" (Fig. 1) reflects the declining faith in the theory. Though it is no longer seen as a model for much of nature, island biogeographic theory provided a theoretical perspective from which to view a number of patterns, such as the species–area relationship (Haila and Jarvinen, 1982).

The key conservation legacies of the dynamic theory of island biogeography were (1) the metaphor of a refuge as an island or spaceship, (2) interest in the fragility of the biota of individual refuges and causes of this fragility (Soulé and Simberloff, 1986; Simberloff, 1994a), and (3) the rules of refuge design (Fig. 2). The recognition that some of the rules, including the most widely debated one (SLOSS, single large or several small; Fig. 2b) are not related to the theory (Soulé and Simberloff, 1986, and references therein), lessened conservation interest in the theory, while documented exceptions to some of the rules, including SLOSS, led to their fall from status of conventional wisdom. For example, the third edition of one of the most widely used introductory ecology textbooks, *Ecology* (Krebs, 1985, p. 559), reprinted the figure of the rules as popularized by the IUCN and described them as flowing from island biogeographic theory. The fourth edition (Krebs, 1994) omits the figure, makes no mention of the rules, and cites the criticism of the theory as "true but trivial" by Williamson (1989).

B. Paradigm Shift

The waning of the theory of island biogeography as a dominant conservation paradigm in the late 1980s coincided with the burgeoning interest among biologists in the metapopulation concept (Fig. 1). As does Hanski (1989), Merriam (1991, p. 134) explicitly claims a paradigm shift: "Metapopulation models have largely replaced equilibrium island biogeography as a way of thinking about terrestrial habitat islands, fragmented habitats and heterogeneous terrestrial environments in general . . . " For other conservation biologists, the shift is tacit and consists simply of an assertion that nature is structured as metapopulations followed by discussion of what actions are required to preserve metapopulations (e.g., Noss, 1993). Perhaps most telling is *The Diversity of Life,* by Wilson (1992), a founder of the theory of island biogeography, in which species are typically seen as structured as metapopulations and the consequences of this structure for conservation are explored.

The causes of the shift are many. One must be the growing ecological literature, described above, doubting the verisimilitude of island biogeographic theory. However, scientific data and the weakness of a prevailing paradigm alone are unlikely to precipitate a paradigm shift (Kuhn, 1970; Haila, 1988), and we must seek other prevailing currents. It is worth recalling that, fundamentally, the theory of island biogeography can be construed as just a multispecies version of an analogous metapopulation theory, so it is hard to imagine objective scientific reasons for accepting one while rejecting the other.

One possible explanation is a shift among conservation biologists and ecologists from the conception of nature as an equilibrium world to that of a nonequilibrium one (Wiens, 1977, 1984; Chesson and Case, 1986). Island biogeographic theory is dynamic, of course, but the emphasis is on equilibrium species richness, hence the nickname, "equilibrium theory," and even the underlying immigration and extinction rates are seen as constant. Though metapopulation theories are not any more, or less, "equilibrium" theories than the theory of island biogeography, the emphasis in the latter on equilibrium species richness and in the former on population turnover may have created the sense of a conflict between an equilibrium and a nonequilibrium theory. The key point, of course, is that in both theories there is no equilibrium at the population level. However, the modus operandi of the island biogeographic theory is to ignore the changes in the presences and absences of individual species and to focus on the equilibrium pattern of species richnesses; this theory is spatially implicit in our taxonomy. Yet two growing interests in conservation are spatially explicit models, to a large extent fostered by an increase in spatial data and the use of GIS, and maintenance of species that are destined to be locally ephemeral, such as fugitive species and early successional ones. The metapopulation theory fits well with these interests. Indeed, a critical difference between the models of MacArthur and Wilson (1967) and Levins (1969a) is the presence of a permanent mainland population in the former but not in the latter.

In addition to island biogeographic theory, the other main component of the "new conservation biology" is population genetics, particularly the study of drift and inbreeding in small populations (Simberloff, 1988). This research tended to shift the focus of conservation biologists from communities to species and populations. Ecological aspects of conservation began also to be seen in terms of populations rather than species—the role of demographic and environmental stochasticity in setting minimum viable population sizes is the prime example (Simberloff, 1988). Again, a focus on populations rather than on communities is bound to make island biogeographic theory seem less relevant.

Finally, metapopulation models rescued small sites from their devaluation by island biogeographic theory. The main ecological data interpreted in terms of island biogeographic theory were simply species–area relationships, showing that, all other things being equal, large sites tend to have more species than small ones. The first rule of refuge design (Fig. 2a) expresses this relationship as a mandate for conservation planners. The rules, and the theory, were widely used to argue that large refuges are needed and the elevated extinction rates in small ones will inevitably render them depauperate (e.g., Diamond, 1972; Soulé et al., 1979). Indeed, to the extent that environmental stochasticity and catastrophies extinguish small populations, mathematical modeling suggested that even populations in enormous refuges, the size of the largest national parks in the United States, would be subject to collapse.

Conservationists eventually recognized that astute opponents could turn this emphasis on inviable small populations against conservation. For example, the refuge system of the small nation of Israel consists of some 200 reserves, many of which are very small. These are protected and managed to various degrees by the Nature Conservation Authority, and the Authority was under great pressure during the 1980s to abandon some small refuges, not because specific research showed declining populations within them but because island biogeographic theory, codified in the refuge design rules, shows that they will inevitably lose species (R. Ortal, personal communication, 1984).

This threat from island biogeographic theory to the maintenance of small reserves was forestalled in several ways. The species–area relationship was shown to have such wide confidence limits that an assertion of imminent faunal collapse could not be sustained (Boecklen and Simberloff, 1986). Some populations that had persisted as very small populations for millennia were adduced as cautions against taking the theory too literally (e.g., Walter, 1990). However, the main salvation of small sites was the shift by conservationists to the metapopulation paradigm. In the Levins model, at least, small sites containing small populations were the only homes of a species and thus the proper locus of conservation concern. The model even suggests that a certain number of unoccupied sites is required for metapopulation persistence (Lande, 1988a; Hanski, this volume), thus relieving beleaguered conservation biologists from having to justify a refuge for a given species by confirmed residence. A famous example in which local extinction rates are high and a supply of suitable empty sites is necessary is *Pedicularis furbishae,* the Furbish lousewort (Menges, 1990).

In sum, from a conservation standpoint, it is not surprising that citations of metapopulation studies increase exactly when those of island biogeography decline (Fig. 1). These trends represent a paradigm shift.

C. Use and Misuse of the Metapopulation Concept in Conservation Biology

Hanski and Gilpin (1991) observed that "metapopulation ideas have recently become the vogue in conservation biology," and numerous conservation strategies are explicitly based on metapopulation models (references in Harrison, 1994b). The general effect has been to draw attention to landscapes and networks, as opposed to individual reserves in isolation, for which the island metaphor of island biogeographic theory is appropriate. This is a salutary development. Even if there were no significant interactions among populations in different refuges, it would be good to have multiple refuges simply as insurance against local catastrophes (Soulé and Simberloff, 1986). Doak and Mills (1994) and Harrison (1994b) argue that, even if most species are not structured as Levins-type metapopulations in nature, the rise of the metapopulation paradigm has served and continues to serve a useful function by forcing conservation biologists to gather data that are important to effective conservation strategies of individual species—movement rates from site to site, relative reproduction and mortality rates at different sites, and the like. This is precisely the view of Haila and Järvinen (1982) for island biogeographic theory.

The problems arise, for metapopulation models as for island biogeographic theory, when it is assumed without empirical evidence that all species, or all species in a large class, conform to some particular model (Doak and Mills, 1994; Harrison, 1994b). If a conservation strategy is based on metapopulation dynamics that do not exist, it can misfire. Thus, for example, Murphy *et al.* (1990) suggested that small-bodied, short-lived species with high reproductive rates and high habitat specificity typically constitute Levins metapopulations, but there are simply insufficient data to evaluate this claim (Harrison, 1991, 1994b). To focus automatically on metapopulation dynamics for such species would not constitute effective science. No wide-ranging generalizations are yet possible, because few data really demonstrate the existence of classical metapopulations. Harrison (1991) could cite only pool frogs *(Rana lessoniae)* in Sweden and waterflies *(Daphnia)* in rock pools as unequivocal cases (for some new examples, see Harrison and Taylor, this volume). The endangered Glanville fritillary butterfly *(Melitaea cinxia)* in Finland is another good example (Hanski *et al.*, 1995a,b) and may represent many other butterfly species (Hanski and Thomas, 1994; Hanski and Kuussaari, 1995; Thomas and Hanski, this volume). In such instances, an understanding of metapopulation dynamics is crucial to effective conservation. Research is required on local extinction and migration rates and how these are affected by patch size and isolation (C. D. Thomas *et al.*, 1993; Hanski *et al.*, 1995b).

Harrison (1994b) suggests that the species most convincingly conforming to the Levins model occupy habitats that inevitably change because of succession (see also Thomas and Hanski, this volume). In many such species, the extinction of local populations is deterministic rather than stochastic, but this fact does not fundamentally undercut the Levins conception of metapopulation dynamics. The endangered Furbish lousewort, for example, requires a riverside habitat that is neither too little nor too heavily disturbed (Menges, 1990). However, any local population is ultimately destroyed by ice scour and bank slumping, so the metapopulation requires a supply of temporarily suitable sites that are not too isolated to be colonized. A metapopulation analysis (Menges, 1990) including observations on local extinction and recolonization suggests that the species is in decline rather than at equilibrium and that tempering of the disturbance (flow) regime will likely exacerbate the situation. Further, in this species as in the Levins scenario in general (Lande, 1988a; Hanski, this volume; Nee *et al.*, this volume), restriction of conservation measures to occupied sites only would be fatal. A static, nonmetapopulation view of nature would not have led to the recognition of the importance of currently unoccupied habitat. Much of the history of refuge establishment consists simply of locating apparently healthy populations and preserving their sites (Simberloff, 1988).

Metapopulation models have been used to deduce the minimum viable metapopulation (MVM) size under certain assumptions (Hanski *et al.*, 1996b). This concept is analogous to the minimum viable population (MVP) size (Shaffer, 1981), but with the critical difference that MVM involves both the minimum viable number of populations and the availability of suitable habitat patches (Hanski *et al.*, 1996b). In practice, use of these concepts may degenerate into specious "magic numbers." A more constructive approach is to use metapopulation models to rank alternative scenarios of landscape change in terms of persistence of a focal species. One may ask, for instance, whether the entire removal of one large habitat patch is more detrimental to a metapopulation than reducing the areas of several patches (Hanski, 1994a,b; Hanski *et al.*, 1996c; Wahlberg *et al.*, 1996; note the connection to the SLOSS rule, Fig. 2b). The theory of island biogeography inspired the rules of refuge design discussed above (Fig. 2). The analogous contribution from metapopulation theory is predictions about the relative performance of particular species in particular fragmented landscapes based on relatively simple but spatially realistic models (Fig. 3, Hanski, 1996b). There are two reasons to expect the latter sorts of predictions to be more helpful than the island bioge-

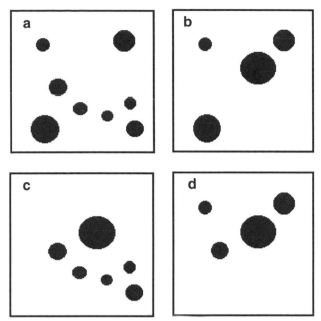

FIGURE 3   Four examples of the same landscape fragmented in different ways (scenarios a to d). Spatially realistic metapopulation models, such as the incidence function model (Hanski, 1994a), can be used to rank the alternative scenarios in terms of the persistence time of the focal species. Note the conceptual link to the SLOSS rule (Fig. 2b).

ographic rules of refuge design. First, the rules of refuge design (Fig. 2) are static, even those actually flowing from the theory. For example, the fundamental concept in rule a (Fig. 2a) is the species–area relationship, which in applications is seen as meaning a fixed number of species in a fixed area. In contrast, the metapopulation predictions explicitly address the dynamics of species survival. Second, the rules of refuge design contrast fixed general alternatives (such as in Fig. 2b), whereas the spatially realistic metapopulation models practically force one to compare specific fragmented landscapes (Fig. 3).

We now turn to potential misuses of the metapopulation concept in conservation. To start with, if a species is structured as a mainland–island (Levitt–Boorman) metapopulation, population turnover in the peripheral "island" populations may be irrelevant to the persistence of the metapopulation as a whole, though the dynamics are crucial to the persistence of the peripheral populations (Doak and Mills, 1994; Harrison, 1994b, Simberloff, 1994b). More generally, emphasis on metapopulation models can potentially harm conservation by drawing attention away from single populations on the grounds that no one of these is crucial to a species' persistence and it is the ensemble that matters (Harrison, 1994b).

Another example of the metapopulation concept used in misguided attempts to deemphasize single populations is the hype surrounding movement corridors (though we observe that, strictly speaking, metapopulation models tend to emphasize connectance among habitat patches, not corridors). One rule of refuge design associated with island biogeographic theory is that a set of refuges connected by corridors will contain more species than an otherwise identical set without corridors (Fig. 2f). In the original formulation of this rule, the focus was on a community-level statistic, species richness, and corridors were assumed to increase this statistic by increasing immigration rate. However, for most proposed corridor systems, there is scant evidence that the corridors would be used for movement or that they would actually forestall extinction (Hobbs, 1992; Simberloff *et al.,* 1992). Even more troubling, investment in corridors can be expensive and can detract from efforts to protect particular populations that require a specific refuge that is not part of a network (Simberloff *et al.,* 1992).

Some examples of classical metapopulations in human-fragmented landscapes may represent transient, nonequilibrium situations, in which a previously more continuous population becomes divided into smaller units, with consequent local extinctions, but no functional metapopulation was created, merely an assemblage of populations all slowly declining to extinction. It seems likely that almost any gradual extinction would appear, during some parts of the decline, as a nonequilibrium metapopulation situation (Simberloff, 1994b). Even in this case, understanding its current dynamics can aid in the maintenance of the species in a fragmented or otherwise changed landscape (Harrison, 1991). In SLOSS terminology (Fig. 2b), a single large population might have been better, but if all we have left is several small ones, their interactions may be crucial to their survival. For example, the metapopulation analysis by Beier (1993) of cougars *(Felis concolor)* in the Santa Ana Mountains of California showed that the species currently exists as a collection of small populations loosely linked by riparian corridors, and his radiotelemetry data on movement combined with a simulation model suggested how loss of particular populations and corridors could affect the entire metapopulation. Data on sources and sinks in source–sink metapopulation can also be key to maintaining a species. A particular worry about nonequilibrium metapopulations in increasingly fragmented landscapes is that we might not recognize them as such (Hanski, this volume), which would give us a misleadingly rosy picture of the ability of species to persist in present landscapes.

Attempts to model the minimum number of populations necessary to maintain a viable metapopulation are hampered by assumptions that are hard to verify and data that are difficult to gather. These are, of course, problems with all population models that aim at quantitative predictions, and the problems become even more severe with spatially realistic models that might guide specific management plans (Doak and Mills, 1994). The history of conservation biology is marked by many examples of misused minima (Simberloff, 1988; Crome, 1993): as soon as a minimum is set for any variable, forces opposed to conservation use it to see how much of nature they can get rid of. Thus, as tentative and general as the MVM model is, someone may attempt to manage for a specific minimum based on this model. As discussed above, a less controversial use of spatially realistic metapopulation models is simply to rank alternative management scenarios (Fig. 3) and to recognize that making long-term predictions about (meta)population persistence time in our rapidly changing world is practically hopeless.

The shift from island biogeographic to metapopulation thinking united ecologists and geneticists in focusing on populations. Genetic concerns have been prominent in the interest in metapopulations. Kimura and Crow (1963) first pointed out that occasional migration between local populations can maintain genetic variation better than would a single large population, essentially because drift is likely to fix different alleles in different populations. However, the situation becomes more complicated when we allow for local extinctions and recolonizations, and recently much effort has gone into modeling the way that such population turnover affects the maintenance of genetic diversity in metapopulations (Wade and McCauley, 1988; Hastings and Harrison, 1994; Barton and Whitlock, this volume). On theoretical grounds, one might expect species that naturally exist in metapopulations not to be prone to inbreeding depression because they lack genetic load (Harrison, 1994b), while local populations in a recently fragmented large population might be particularly susceptible to inbreeding depression because heterozygosity would quickly decline (Simberloff, 1988; Hedrick and Gilpin, this volume). Under the latter circumstances, maintaining movement among populations might seem particularly important, and indeed many management plans for declining populations call for measures to enhance population interaction, such as translocation and corridors, specifically to avoid inbreeding depression (e.g., U.S. Department of Agriculture, 1995). However, field evidence for inbreeding depression or other problems in recently fragmented populations is scarce (Harrison, 1994). Lande (1988b) argues more generally that the importance of genetic threats in conservation has been overblown. His view is that, in naturally small populations, the genes causing threatening inbreeding depression would have been selected out, while in recently reduced populations, ecological threats are more immediate. Despite this widely cited statement, genetic principles still underpin many viability analyses and management plans (Harrison, 1994b). Perhaps the very fact that genetic modeling is feasible ensures that it will be done, particularly if ecological modeling, even if potentially more useful, is more problematic. The latest round of papers (e.g., Lynch et al., 1995) appears to strengthen the genetic argument, but the most urgent need is for relevant field studies.

Thompson (1996) contends that metapopulations may be crucial to the conservation of various coevolutionary interactions, such as those between pathogens or parasites and their hosts. In models, locally unstable population dynamics can be stabilized by the addition of metapopulation structure (Hassell et al., 1991a; Nee et al., this volume). In other cases, the metapopulation structure stabilizes evolutionary dynamics of the interaction. For example, under certain circumstances, the coevolutionary dialog between a pathogen and its host can lead to the extinction of the host, if a new virulence gene in the pathogen spreads rapidly enough. A metapopulation structure can then prevent the gene from eliminating the entire species. Wild flax *(Linum marginale)* and flax rust *(Melampsora lini)* may be a natural example in which the host metapopulation structure serves this function (Burdon and Thompson, 1995). Frank (this volume) presents a thorough discussion of these issues.

The focus on metapopulations, combined with that on genetics, has led to the population and the species becoming the dominant levels of concern in conservation. It is striking that the recent explosion of interest in ecosystem management is quite antithetic to a primary interest in populations and to single-species management (Simberloff, 1996). In fact, a key motivation of ecosystem management is that research on species after species will be hopelessly expensive and inefficient, and so will management based on such research. Of course, both ecosystem management and metapopulation models share a concern with landscapes and regions, rather than highly local settings, and one could imagine a landscape with a distribution of habitat patches that would maintain many metapopulations simultaneously. Also, the emphasis in ecosystem management on maintaining processes rather than species (Simberloff, 1996) can accommodate concerns about the coevolutionary processes. Nevertheless, the research programs and primary goals of these two approaches differ fundamentally and they will surely compete for both research funding and influence in specific management plans in the future.

V. CONCLUSIONS

The changing pattern of citations of the key words "island biogeography" and "metapopulation" represents a fascinating example of a paradigm shift in population biology. This example is the more striking because the respective theories are so closely related that whatever evidence can be mustered for, or against, one theory is likely to serve the same function with respect to the other theory. We have discussed in this chapter how it is largely the wider context that has made the difference. One apparently important issue is the spatial scale. The dynamic theory of island biogeography was originally developed to explain patterns at large spatial scales, whereas the metapopulation concept is associated with fragmentation of our ordinary landscapes. Though the difference is in perception only, it matters.

Metapopulation models have contributed important insights to conservation, and they have inspired field studies focused on collecting key data on demography and movement. Nonetheless, the temptation to apply the metapopulation approach blindly to systems for which there is no supporting evidence can be counterproductive. Metapopulation maintenance may be crucial to a limited range of species, probably dominated by those characteristic of successional habitats. The role of metapopulation dynamics in forestalling genetic deterioration is particularly unverified.

2

Empirical Evidence for Metapopulation Dynamics

Susan Harrison   *Andrew D. Taylor*

I. INTRODUCTION

Underlying the many refinements and elaborations of metapopulation theory is the fundamental idea that the persistence of species depends on their existence as sets of local populations, largely independent yet interconnected by migration. Population structure at this large spatial scale is thought to alleviate the risks of widespread extinction that arise from unpredictable physical environments and from strong interactions among species. This concept has long attracted many ecologists, but more for its plausibility than because of any compelling empirical evidence. Support for metapopulation theory has mostly consisted of anecdotal accounts of local extinctions or asynchronous population fluctuations, combined with much theoretical evidence that metapopulation effects could occur. Recently, however, as interest in metapopulation dynamics and its conservation applications has grown, the number of more substantial studies has steadily increased.

Here we review the current body of empirical evidence and ask whether and how it supports metapopulation theory. We begin with a brief review of the theory and its origins, to lay the groundwork for specific criteria by which to evaluate the evidence. In this review, we highlight differences between metapopulation theory for single and multiple species, which will lead to somewhat different criteria in the two cases. We leave aside the genetic and evolutionary aspects of

Metapopulation Biology
27

metapopulation dynamics, which have received comparatively little empirical work (but see Olivieri et al., 1990; Olivieri and Gouyon, this volume; McCauley, 1991; Harrison and Hastings, 1996; Barton and Whitlock, this volume, for reviews).

Single-species metapopulation theory arose largely from early observations of species in patchy and unpredictable environments. The archetypal "shifting mosaic" or "blinking lights" species were insects whose populations were small or insular, fluctuated widely, and were prone to extinction either because of climatic events or because their habitat was transient (e.g., Andrewartha and Birch, 1954; Ehrlich and Birch, 1967; Ehrlich *et al.,* 1972). Recent examples in the same vein include herbs on riverbanks (Menges, 1990), amphibians in small ponds (Gill, 1978a; Sjögren, 1991; Sjögren Gulve, 1994), snails on rocky outcrops (Spight, 1974), insects on weedy plants (van der Meijden,1979a; van der Meijden and van der Veen-van Wijk, this volume), and many cases of butterflies vulnerable to bad weather or habitat change (Shapiro, 1979; Harrison *et al.,* 1988; Thomas and Harrison, 1992; Thomas and Jones, 1993; Hanski *et al.,* 1994; Hanski and Kuussaari, 1995; Hanski, this volume; Thomas and Hanski, this volume).

Single-species metapopulation models, beginning with Levins (1970), demonstrate that such sets of transient demes may persist through a balance between local extinction and recolonization (reviewed by Hastings and Harrison, 1994; Hanski, this volume). Here we denote as "classical" single-species metapopulations sets of local populations that are all subject to extinction and persist at the metapopulation level through recolonization (Fig. 1a). A very basic property of

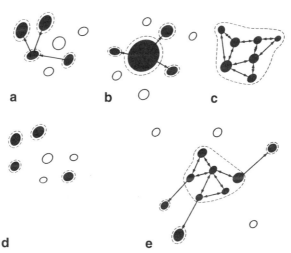

FIGURE 1   Different types of metapopulation. Filled circles, occupied habitat patches; empty circles, vacant habitat patches; dotted lines, boundaries of local populations; arrows, dispersal. (a) Classic (Levins); (b) mainland–island; (c) patchy population; (d) nonequilibrium metapopulation; (e) intermediate case combining features of (a), (b), (c), and (d).

classical metapopulations is that persistence requires an adequate rate of migration among patches. Probabilities of metapopulation persistence also increase with the number of habitat patches and local populations. Besides resonating with ecologists' interpretations of many natural systems, this verbal and mathematical model is increasingly seen as relevant to how species persist, or fail to do so, in landscapes recently fragmented by humans (Opdam, 1990; Fahrig and Merriam, 1994; Harrison, 1994b).

Theory on the metapopulation dynamics of multiple interacting species arose not so much from natural history as from mathematical and laboratory studies showing the instability of predatory or competitive interactions in simple environments (Nicholson and Bailey, 1935; Gause, 1935; Huffaker, 1958). What we may now all the "classical" multispecies model was first demonstrated in Huffaker's (1958) mite experiments, showing that a predator and its prey undergo oscillations and crashes within patches of habitat, yet coexist stably in a universe of interconnected patches. Mathematical models building on the same extinction-and-colonization format developed for single species, as well as models of other types, have explored many aspects of spatial subdivision as a stabilizing force in both predatory and competitive interactions (reviewed by Hastings and Harrison, 1994; Nee *et al.*, this volume).

Metapopulation theories for single and multiple species thus address related issues and are expressed in similar models, yet arose from different concerns and arrive at partly conflicting answers. In single-species theory, the problem is the patchiness of habitats and the harshness of the abiotic environment, and the solution is migration and recolonization. Multispecies theory addresses the problem of intrinsically unstable interactions and identifies spatial subdivision as a solution. Therefore, migration among patches always promotes persistence in models of classical single-species metapopulations (though less so in models that include the effects of emigration on local populations; Hanski and Zhang, 1993; Olivieri and Gouyon, this volume), but too much migration leads to instability in multispecies metapopulations.

We note that the above difference is not really due to the number of species involved, but rather to the assumed causes of local instability or extinction. If local dynamics are intrinsically unstable, high rates of migration may destabilize a single-species metapopulation (Allen *et al.*, 1993). Conversely, if multispecies systems are subject to frequent local extinctions from external causes, subdivision may possibly lose its stabilizing effect, an issue that deserves more investigation.

We may now identify criteria for judging the empirical evidence on metapopulations. For single-species studies, we ask first whether all local populations are prone to local extinction and, second, whether persistence at the metapopulation level requires recolonization. For multispecies systems, we ask, first, whether strong interactions between either competitors or predators and prey cause local extinctions or population oscillations and, second, whether subdivision increases the persistence or the temporal stability of the system as a whole.

Metapopulations may be defined more broadly than we have just done, to include any systems in which populations are subdivided but exchange some migrants. In fact, since we find few systems that conform well to the classical models, we return under Discussion to consider the implications of a broader view. However, like Hanski (this volume), we take the classical models as a point of departure, since their assumptions are implicit in most cases in which the term is used, including many of the more elaborate metapopulation models. Moreover, classical models yield the strongest predictions about metapopulation dynamics; in other types of metapopulation, as we will show, the causes of persistence or coexistence lie more at the local level.

Finally, we note some important practical reasons for giving careful scrutiny to metapopulation ideas. In conservation biology, the metapopulation model is sometimes used to support the need for numerous reserves, corridors, or a landscape-level approach, goals with which few conservationists would argue. However, in other cases this model is used more controversially, to justify strategies that would preserve only a handful of well-spaced fragments of a habitat that is presently continuous (see Harrison *et al.*, 1993; Harrison, 1994b; Doak and Mills, 1994; Noon and McKelvey, 1996; Gutierrez and Harrison, 1996).

II. SINGLE-SPECIES METAPOPULATIONS

In reviewing empirical studies, we examine each of the critical assumptions of classical metapopulation models in turn, first identifying studies that appear not to meet them and then proceeding toward studies that appear to exemplify most or all of them. Our purpose is not to criticize individual studies, nor to impose categories for their own sake, but to ask whether available evidence suggests any systematic pattern.

A. Are All Local Populations Subject to Frequent Extinction?

1. Definition and Causes of Local Extinction

Since populations show spatial structure at a hierarchy of scales (e.g., Amarasekare, 1994), the definitions of local populations and hence local extinction are usually partly arbitrary. However, as a minimal criterion, local extinction may be defined as the extirpation of any population segment sufficiently closed to immigration that, once extinct, typically remains so for several generations or more. This serves to exclude subpopulations so tightly connected to others that their "extinction" is caused more by the movement of organisms than by their mortality (to see why this is an important distinction, imagine birds in an orchard, undergoing "local extinction" every time they fly out of a tree).

In the much-used scheme devised by Shaffer (1981), local extinction has four general causes. One of these, demographic stochasticity, is expected to affect only populations below a threshold size, and another, loss of genetic variation, acts comparatively slowly. These factors may help to finish off a declining population or impede the establishment of new ones, but are unlikely to be ultimate causes of local extinctions. In contrast, environmental stochasticity (including "catastrophes") and deterministic threats (e.g., loss of habitat) may extirpate populations of a wide range of sizes and thus are the most likely ultimate causes of natural local extinctions, as a variety of empirical evidence confirms (Leigh, 1981; Karr, 1982; Pimm *et al.,* 1988; Schoener, 1983; C. D. Thomas *et al.,* 1992; Thomas and Hanski, this volume).

Environmental stochasticity often operates at a regional scale; for example, weather causes insect populations to fluctuate in synchrony over broad geographic areas (Hanski and Woiwod, 1993), and a single drought or freeze may eliminate multiple conspecific butterfly populations (Ehrlich *et al.,* 1972, 1980). Such "regional stochasticity" reduces the likelihood of persistence for classical metapopulations (Hanski, 1991). In turn, it increases the potential importance of other mechanisms that enable persistence, such as refuge habitats or resistant life stages. Price and Endo (1989) concluded that because Stephens' kangaroo rat undergoes extreme regionwide population fluctuations in response to weather, a single large reserve is much preferable to a proposed design of multiple small reserves connected by corridors.

Deterministic causes of local extinction include natural disturbance and succession and human destruction of natural habitats. These may lead to classical metapopulation dynamics, but do not always do so, for several reasons. Species adapted to successional habitats may be such good dispersers that their populations are not very subdivided. Species subjected to longer-term habitat change may shift their spatial distributions over time without ever approaching a dynamic balance between extinction and recolonization. Habitat fragmentation may produce patches that are too small to support populations or too isolated to interact with other patches.

2. Some Populations Are Highly Persistent

Most studies of natural local extinctions have taken place on small islands near the shore of a lake or ocean (e.g., Pokki, 1981; Peltonen and Hanski, 1991; reviews in Schoener, 1983; Diamond, 1984). Local extinctions affect the small (island) populations, but the system persists for essentially the same length of time as do its larger and more persistent (mainland) populations (Fig. 1b). Many metapopulations have an essentially similar, mainland and island structure, owing to high variation in the sizes of habitat patches or populations. Examples include metapopulations of spiders on Bahamanian islands (Schoener and Spiller, 1987a,b; Spiller and Schoener, 1990), checkerspot butterflies on patches of serpentine soil (Harrison *et al.,* 1988), and many others (reviewed by Schoener, 1991; Harrison, 1991). For a system to have mainland–island dynamics, there need not be a single mainland of extreme size. Substantial variance in patch or population size means that local extinctions will tend to strike the smallest populations, which are the ones with the least impact on metapopulation persistence.

Heterogeneity in habitat quality, rather than patch or population size, may produce a similar effect. A recently popular idea is that dispersal from "source" populations in high-quality habitat may permit "sink" populations to exist in inferior habitat (Pulliam, 1988; Pulliam and Danielson, 1991). Unlike island populations, which are merely small, sinks cannot support positive population growth because of their poor quality. This idea remains largely untested, but several insect studies provide suggestive examples, with the sinks ranging from marginal habitats that are occupied only during rare favorable years (e.g., Shapiro, 1979; Murphy and White, 1984), to areas in which populations flourish most of the time but cannot survive catastrophes (e.g., Strong *et al.,* 1990; Singer *et al.,* 1994).

3. Populations Are Not Subdivided Enough to Permit True Local Extinction

For local extinction to occur, populations on separate patches must be reasonably isolated from one another, with most recruitment coming from within the patch rather than from immigration. At the opposite extreme are systems in which progeny from all patches are completely mixed and reassorted among patches in each generation. Here the term "patchy population" is used for systems toward the latter end of the continuum (Fig. 1c). Sharp distinctions are difficult in practice, but if the average individual inhabits more than a single patch in its lifetime, the patches clearly do not support separate populations. Local "extinctions," presences followed by absences, may simply be the result of individuals' foraging behavior or responses to conspecifics (e.g., the birds in an orchard). Importantly, unlike a metapopulation, the persistence of a patchy population is not highly sensitive to the distances or rates of movement among patches.

Invertebrates that specialize on fallen fruit, rotting logs, dung, carrion, or water-filled tree holes are sometimes regarded as forming classical metapopulations, colonizing and becoming extinct on their transient resource patches. However, such species are typically highly mobile; each patch supports only one generation of the insect, and adults oviposit on numerous patches in their lifetimes (e.g., Kitching, 1971; Kaitala, 1987; Hanski, 1987). Although weedy host plants are a slightly more permanent habitat than dung or carrion, the specialist insects feeding on milkweed (Solbreck, 1991; Solbreck and Sillen-Tullberg, 1990) and ragwort (Harrison *et al.,* 1995) appeared to disperse so well that their populations were effectively unsubdivided across large arrays of patches (but see below and van der Meijden and van der Veen-van Wijk, this volume). Of course, at some larger scale (e.g., among different fields or forests) they may possibly show classical or nonclassical metapopulation structure.

Migration has long been considered an important adaptation to environments that vary in space and time (e.g., den Boer, 1968; Southwood, 1977). Sessile marine invertebrates with planktonic larvae show perhaps highest migration of any organisms, relative to the scale of the patches on which recruitment and growth occur, and they appear to persist longer in evolutionary time than comparable taxa with nonplanktonic larvae (Jablonski, 1991). Conversely, the evolution of flightlessness in insects is strongly linked to stable, continuous habitats (Wagner and Liebherr, 1992). Thus, it is perhaps to be expected that in many cases, species in patchy and risky environments will disperse too well to form classical metapopulations on patches of their habitat.

B. Does Recolonization Balance Local Extinction?

Migration and colonization in metapopulations have been reviewed by Ebenhard (1991), Hansson (1991), and Ims and Yoccoz (this volume). In classical models, there is a threshold rate of migration for the metapopulation to persist. Above this level, patch occupancy achieves a stable equilibrium, arising from the fact that every local extinction makes an empty habitat available for colonization, in strict analogy to a density-dependent birth rate. There are several reasons why this assumption may not always hold.

1. Nonequilibrium (Declining) Metapopulations

Rather than being part of a steady-state process, local extinctions may occur in the course of a species' decline to regional extinction, with recolonization occuring infrequently or not at all (Fig. 1d), typically as the species' habitat is undergoing reduction and fragmentation. A natural example is the series of extinctions of mammal populations caused by the isolation of mountaintop habitats during post-Pleistocene climate change (Brown, 1971). Many more examples can be found among species in habitats fragmented by humans, such as salamanders (Welsh, 1990) and woodpeckers (Stangel *et al.*, 1992) on remnant patches of old-growth forest. Hanski (this volume) discusses nonequilibrium dynamics in a butterfly metapopulation.

The conservation of species in fragmented habitats is an important area for the application of metapopulation concepts. In some cases, however, remnant populations are so isolated that there is little potential to manage them as an interconnected network (Harrison, 1994b), while in others, creating corridors or a dispersal-friendly matrix may be feasible (e.g., Noon and McKelvey, 1996).

2. Nonequilibrium (Habitat-Tracking) Metapopulations

Local extinctions are not always stochastic as most metapopulation theory assumes, but rather may occur when disturbance, succession, or long-term habitat change cause the loss of suitable habitats. In turn, colonization may occur only when new patches of habitat are created near existing populations. For example, the spatial distribution of many butterflies appears to be sensitive to vegetation age and height, which are governed by grazing and other transient disturbances (Thomas and Harrison, 1992; Thomas and Jones, 1993; Thomas and Hanski, this volume). Thomas (1994c) argues that deterministic extinction may be the rule and stochastic extinction the exception in real metapopulations.

The spatial dynamics created by disturbance and succession are interesting in their own right and are a subject of key importance for the conservation of many species. However, local extinctions caused by habitat loss violate an important premise of the classical model, namely that extinctions make habitats available for recolonization. The stable metapopulation-level equilibrium predicted by the classical model arises from the inverse relationship between patch occupancy and patch availability, which creates density-dependent regulation of patch occupancy. In contrast, when patches are created and destroyed by extrinsic forces, no such regulation occurs. The species' abundance and distribution will simply track the availability of habitat and will remain roughly constant only if the rates of habitat loss and renewal happen to be roughly equal (Thomas, 1994c).

C. Classical Metapopulations, Intermediate Cases, and Other Possibilities

The pool frog *Rana lessonae* in ponds along the Baltic coast of Sweden (Sjögren, 1991; Sjögren Gulve, 1994) and the butterfly *Melitaea cinxia* (Hanski *et al.*, 1994, and this volume) on granite outcrops in southwest Finland, form metapopulations in which all populations are susceptible to relatively frequent extinction, migration among populations is limited (i.e., most recruitment is local), and extinctions create vacant habitats which are recolonized. These two well-studied systems appear to conform reasonably closely to the classical concept of metapopulations in an extinction–colonization balance.

Many other systems resemble classical metapopulations in certain ways, for example having patchy distributions with no obvious "mainland" patches (Hanski and Kuussaari, 1995), or having local populations that do not appear to be self-sustaining (Stacey and Taper, 1992, this volume). However, based on the foregoing evidence, we would argue that only after much detailed study can any natural system be described as a classical metapopulation.

1. Mixed Structures

Many studies of species distributions in patchy habitats reveal that patches are more likely to be occupied the nearer they are to other occupied patches (Brown and Kodric-Brown, 1977; Fritz, 1979; Opdam, 1990; Laan and Verboom, 1990; Lawton and Woodroffe, 1991; C. D. Thomas *et al.*, 1992). This suggests the possibility that many real metapopulations combine features of all different kinds of metapopulation structure, along gradients from clustered central patches to isolated peripheral ones (Fig. 1e). Central patches are united by dispersal into a single population, slightly more isolated ones undergo extinction and recolonization, and still more isolated patches are usually vacant.

In other cases, the metapopulation structure of a species may vary among regions, because of differences in the configuration of habitat. Nine metapopulations of the silver-studded blue butterfly *(Plebejus argus)* in Wales show a continuum from nearly equal-sized patches to a mainland–island configuration (Thomas and Harrison, 1992). The butterfly *Euphydryas editha* forms a mainland–island metapopulation in coastal California, but shows a mixture of classical and patchy population features in montane California (Harrison *et al.*, 1988; Singer and Thomas, 1996). Insects on patches of ragwort show little population subdivision in British grasslands (Harrison *et al.*, 1995), but the effects of subdivision are significant in Dutch dunes (van der Meijden, 1979; van der Meijden and van der Veen-van Wijk, this volume).

The northern spotted owl *(Strix occidentalis caurina)* occupies a once-continuous but now coarsely fragmented forest habitat, while the Californian subspecies *(S. o. occidentalis)* lives (in part) in still-continuous but selectively logged forests, and the Mexican subspecies *(S. o. lucida)* inhabits insular mountaintops. This natural and unnatural variation in habitat structure, and presumed metapopulation structure, is a central feature of conservation strategies for the spotted owl (Noon and McKelvey, 1996; Gutierrez and Harrison, 1996).

2. Metapopulations with Little Turnover

If local populations fluctuate fairly independently of one another, but exchange low to moderate numbers of immigrants, metapopulation structure may have an important stabilizing effect at the regional level even without population turnover. We know of no good examples of this possibility, but it could be tested by comparing the magnitude of fluctuations in conspecific populations varying in their degree of isolation.

3. "Local" Spatial Dynamics

A growing number of empirical studies suggest that even in relatively continuous habitat, the dynamics and persistence of populations may be strongly affected by small-scale habitat heterogeneity, localized interactions, and limited migration (e.g., Weiss *et al.*, 1988; Harrison, 1994a; Amarasekare, 1994). This is an important class of phenomena, but one that lies outside the conceptual domain of metapopulation dynamics.

III. MULTISPECIES METAPOPULATIONS

We now review empirical studies in which it has been proposed that predators and prey or competitors coexist through multispecies metapopulation dynamics (see earlier reviews by Bengtsson, 1991, and Taylor, 1991). Here, the two "classical" conditions we examine are that the interspecific interaction leads to local extinction or instability and that both (or all) species have something like a classical metapopulation structure at similar spatial scales, leading to greater stability or persistence at the regional level than within each local patch. Once again, we begin by identifying ways in which these conditions may not be met in some natural systems.

A. Is the Interaction Locally Unstable?

1. All Local Populations Are Stable or Persistent

Several recent studies have tested the metapopulation explanation for coexistence by asking whether local populations of predators and prey are destabilized by being isolated. Murdoch *et al.* (1996) found that populations of the red scale *Aonidella aurantii* and its parasitoid *Aphytis melinus* did not fluctuate more on caged than on uncaged citrus trees. Similarly, C. J. Briggs (unpublished manuscript) did not find a significant increase in the temporal variability of populations of the midge *Rhopalomyia californica* or its parasitoids on caged versus uncaged coyote bushes *(Baccharis pilularis)*. However, the latter experiments lasted only 3–10 insect generations, possibly too short for the anticipated effects to become statistically significant.

The interaction between prickly-pear cactus (*Opuntia* spp.) and the moth *Cactoblastis cactorum*, which was successfully introduced to control the cactus in Australia, has been described as a classical case of coexistence through extinction and colonization dynamics (Dodd, 1959; A. J. Nicholson, as quoted in Monro, 1967). However, more recent observations suggest that both species persist locally and that the interaction is stable because some plants always survive attack by the moth (Monro, 1967, 1975; Caughley, 1976; Osmond and Monro, 1981; Myers *et al.*, 1981).

Local populations may survive strong predation or competition because of a cryptic life stage. For example, early-successional plants may coexist with superior competitors by "recolonizing" newly disturbed sites from seed banks, rather than by dispersal. Resting stages appeared to explain the recolonization by waterfleas (*Daphnia* spp.) of cattle tanks from which they had been eliminated by predatory bugs (*Notonecta* spp.) (Murdoch *et al.*, 1984). Alternatively, prey populations may survive simply because predators leave a patch before eliminating all prey; examples include olive scales and their parasitoids (Huffaker *et al.*, 1986; Taylor, 1991) and cottony cushion scale, vedalia beetles, and parasitoids (Quezada, 1969).

Local extinctions may occur primarily for reasons other than the interspecific interaction. Hanski and Ranta (1983) and Bengtsson (1989, 1993) hypothesized that three competing *Daphnia* species in rock pools in the Baltic Sea coexisted through extinction and recolonization. In studies with artificial pools, Bengtsson (1989, 1993) showed that extinction rates were higher in three-species pools than in two- or one-species pools. However, species pairs and even triplets could coexist for 4–7 years even in very small pools. Bengtsson concluded that some apparent extinctions were really pseudoextinctions caused by a cryptic resting stage, and that most natural extinctions were probably caused by predation, low resource levels, salinity, or desiccation. We note that if extinctions are caused by both the competitive interaction and the extrinsic forces, it creates the interesting possibility that subdivision has both positive and negative effects on stability.

The parasitoids *Hyposoter horticola* and *Cotesia melitaearum* parasitize up to 90% of the larvae of the butterfly *Melitaea cinxia*, and may contribute to the observed local extinctions of *M. cinxia* (Hanski *et al.*, 1994; Lei and Hanski, 1997), but their importance relative to drought and other factors is not yet clear. Other factors which may stabilize this host–parasitoid interaction include spatial density dependence in the mortality caused by a generalist hyperparasitoid (Lei and Hanski, 1997).

Recent work by Antonovics *et al.* (1994) has shown that the extinction and colonization of populations of white campion *(Silene alba)* affects both the incidence of its anther smut disease *(Ustilago violacea)* and the distribution of the plant's disease-resistance genotypes among populations. However, there is no evidence yet that the disease affects rates of local extinction in the plant.

2. Some Prey Populations Are Stable or Persistent ("Refuges")

In a close analogy to source–sink dynamics, a prey or inferior competitor may persist because it has a particular type of habitat in which it escapes its predator or superior competitor. The barnacle *Balanus balanoides* suffers heavy predation by the snail *Urosalpinx cinerea* in the subtidal zone, but persists and recruits in the intertidal zone where the snail is absent (Katz, 1985). European red mites have a refuge from predation in sprayed orchards where their main predators are scarce (Walde, 1991, 1994). However, experiments showed that refuges do not explain the stability of the interaction between red scale and the parasitoid *Aphytis melinus* (Murdoch *et al.,* 1996).

In a slight modification of the refuge pattern, interactions may be stable in some habitat types but not in others. For example, the winter moth *Opheroptera brumata* appears to coexist stably with its parasitoids in apple orchards, and to disperse from orchards into forests where local extinctions are frequent (Murdoch *et al.,* 1985; MacPhee *et al.,* 1988).

Finally, patch size may create effective refuges. Lizards *(Anolis* spp.) contribute to local extinctions of spider populations, but sufficiently large islands support stable populations of both spiders and lizards (Schoener and Spiller, 1987a,b; Schoener, 1991). Conversely, pool frogs *(Rana lessonae)* persist better in small than in large ponds, since large ponds support pike *(Esox lucius)*, which are a major cause of local extinctions of the frog (Sjögren, 1991; Sjögren Gulve, 1994).

3. Predator Populations Are Stable (Generalists)

Even if a prey species exists as a metapopulation, and its predator causes local extinctions or instability, the predator may persist stably if it has alternate prey. The predatory mite *Typhlodromus pyri* frequently eliminates the European red mite *(Panonychus ulmi)* from individual apple trees, and migration among trees enhances the persistence of *P. ulmi*. Nonetheless, *T. pyri* is consistently abundant, since it can feed on pollen and the apple rust mite *(Aculus schlectendali)* as well as *P. ulmi* (Walde, 1991, 1994; Walde *et al.,* 1992). Similar examples include *Daphnia* and *Notonecta* (Murdoch *et al.,* 1984), the oak gall wasp *Xanthoteras politum* and its parasitoids (Washburn and Cornell, 1981), and spiders and lizards (Schoener and Spiller, 1987a,b; Spiller and Schoener, 1990). It is possible, though by no means proven, that these systems function as single-species metapopulations for the prey.

B. Do Both (All) Species Show Metapopulation Structures at Similar Spatial Scales?

1. Prey Is Subdivided but Predator Is Not

Interacting species often differ in their mobility, with predators or parasitoids usually being better dispersers than their hosts or prey. This is clearly the case for goldenrod aphids *(Uroleucon nigrotuberculatum)* and their ladybird beetle predator *(Coccinella septempunctata)* (Kareiva, 1984, 1987) and may also be true for the red scale, olive scale, cottony cushion scale, gall midges, and their respective parasitoids, discussed above. In subdivided experimental populations of the intertidal snail *Nucella* (= *Thais*) *emarginata*, local extinctions were caused by predators (gulls, geese, and crabs) that are both mobile and generalists (Quinn *et al.,* 1989). If a predator is either very mobile or a generalist, it will be present on most prey patches most of the time, and subdivision would not appear to explain coexistence. Again, it is possible for such a system to act as a single-species metapopulation for the prey.

2. None of the Species Is Subdivided

Mosquitos have been described as coexisting with their aquatic predators through extinction and recolonization, but all individuals disperse from their natal patch and oviposit on many patches (Murdoch *et al.,* 1985). A metapopulation explanation for coexistence has been proposed for kangaroo rats and smaller seed-eating rodents by Valone and Brown (1995), but these authors note that "the responses observed largely represent habitat selection by individual rodents" and that typical persistence times are on the order of a few months, indicating that these dynamics take place within rather than among local populations.

Some evidence suggests that cinnabar moths *(Tyria jacobaeae)* may coexist with their patchily distributed host plant (ragwort, *Senecio jacobaea*), and with certain of their parasitoids and competitors, through classical multi-species metapopulation dynamics (van der Meijden, 1979a; van der Meijden and van der Veen-van Wijk, this volume; Crawley and Pattrasudhi, 1988; McEvoy *et al.,* 1993). However, one test of this hypothesis in British grasslands found that the moth disperses so well, relative to the distances between ragwort patches, as to preclude a metapopulation explanation for coexistence (Harrison *et al.,* 1995).

C. Classical Multispecies Metapopulation Dynamics

Interactions between herbivorous and predatory mites in greenhouses present an interesting mixture of patchy-population and metapopulation features (e.g. Nachman, 1988, 1991; Sabelis and Laane, 1986; Sabelis *et al.,* 1991; van de Klashorst *et al.,* 1992). Metapopulation dynamics are suggested by the facts that mites have very limited mobility between adjacent plants, and suitable plants are frequently unoccupied. However, movement is more frequent and more directed than most metapopulation models assume: individuals may occupy many plants in their lifetimes, emigration by both predators and prey is density-dependent, and predator dispersal may respond to chemical signals by the prey. To date there has been no direct, conclusive test of the importance of metapopulation structure in stabilizing mite and plant systems, e.g., by comparing the persistence of predators and prey on closely versus widely spaced arrays of plants.

Conclusive experimental tests of classical multispecies metapopulation dynamics are exceptionally difficult, and few have been done. (However, see Holyoak and Lawler, 1996, for an excellent recent example.) In certain studies mentioned above, such as that of *Baccharis*-feeding insects, there is a tantalizing suggestion of metapopulation effects, but strong tests are precluded by too few patches and/or generations. In others, such as that of *Melitaea cinxia* and its parasitoids, suggestive patterns are beginning to emerge. Finally, in such studies as that of rock pool *Daphnia,* some of the conditions for metapopulation coexistence are met, but it is difficult to assess their importance relative to other factors, such as dormant life stages and abiotic causes of extinction.

IV. DISCUSSION

Large-scale spatial population structure—metapopulation structure in the broad sense—is clearly important in a large number and variety of natural systems. Real populations often behave very differently than they would if they were unsubdivided, and both the natural and the human-caused discontinuity of habitats have large effects on how populations function. Thus, a metapopulation approach is essential to understanding and managing many natural phenomena. Our aim here is not to discount such an approach, but rather to use empirical evidence to refine and clarify our notions of how real metapopulations work. The results cast some doubt on the classical models of both single- and multispecies metapopulations.

We find that natural metapopulations have a variety of structures (Fig. 1), with implications for persistence and coexistence that are correspondingly varied. These different structures are of course not discrete entities, but rather lie along continuua in terms of patch structure and migration rates (Fig. 2). Our review illustrates that when natural systems deviate substantially from the classical, extinction-and-colonization structure, their essential behavior changes considerably as well; in all cases, persistence and/or coexistence become more dependent on local (within-population) processes and less so on metapopulation ones. It is therefore crucial to avoid labeling a system as a metapopulation under a broad definition—e.g., because habitat is patchy, some local extinctions occur, or populations in different areas fluctuate out of synchrony—and then applying to it conclusions that follow from a narrower (classical) definition.

We also conclude that classical metapopulations form a minority, even among the modest number of systems that have been well studied in metapopu-

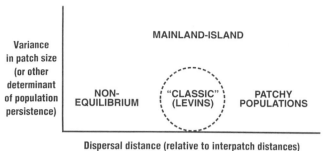

FIGURE 2   Relationships among different types of metapopulation.

lation terms. We find only a few examples of single species existing in a balance between the extinction and colonization of populations and almost none of systems in which multiple species coexist through tightly coupled metapopulation dynamics at comparable spatial scales. Is this apparent scarcity real? Clearly, the great difficulty of studying among-population processes in nature is a major obstacle to reaching firm conclusions. However, it may also be that systems in which migration among patches is too high or too low, or in which some populations are highly persistent (Fig. 2), are truly much more common than ones approximating the classical structure.

If we broaden our view of metapopulations sufficiently to include the nonclassical kinds, how does this affect our perception of the prevalence and importance of metapopulations? This is clearly an area of active work, as this volume illustrates, but we will advance some of our own speculations. Regional persistence may be an overemphasized concern, arising from models that were based on a simple birth-and-death analogy and that overlooked variation in patch size, detailed local population dynamics, explicit spatial relations among patches, and spatio-temporal correlation in the environment. Adding such real-world refinements to models may have the general effect of reducing the relative importance of migration and recolonization, and increasing that of local population processes, for regional persistence. If this is true, then what other types of "metapopulation effect" may we seek in nature?

When levels of migration among patches are moderate, and patches vary in their degree of spatial isolation, a system may be demographically unified in central patches and exhibit rescue effects or extinction and recolonization on increasingly marginal ones. The defining feature of such a metapopulation is not the dependence of regional persistence upon local extinction and recolonization, but the strong effect of patch structure and dispersal on *local* population persistence and/or regional *distribution*. This type of metapopulation structure seems to us a highly plausible one.

Consideration of variation in patch size or quality leads into the realm of mainland–island and source–sink dynamics, where again the appropriate questions are not about regional persistence, but about regional distribution, and about local persistence in habitats too small (islands) or poor in quality (sinks) to support long-lived populations. Mainland–island dynamics are well documented, but source–sink dynamics are virtually untested; how frequently species are found in habitats where they are unable to replace themselves without immigration, and how much this affects their overall demography, remains a very open area for empirical research. The related refuge models of predation and competition, in which coexistence is made possible by the net dispersal of the victim species from habitats of low to those of high predation or competition (e.g., Hochberg and Holt, 1995), also deserve more empirical study.

For multispecies systems, another important area for exploration is the effect of trophic complexity. Both theory and empirical work have emphasized two or three tightly coupled species, but real food webs are nearly always more complex than this. The *M. cinxia* system comprises four important species on three trophic levels, plus some eight minor species of which some are generalists (G. Lei and Hanski, unpublished manuscript). Four spider and three lizard species and their shared prey interact on Bahamanian islands (Schoener and Spiller, 1987a,b). Even relatively simple biocontrol systems typically include several important natural enemies (e.g., Quezada, 1969; Murdoch *et al.*, 1996). Metapopulation models have yet to address systems of many species at multiple trophic levels, each with different population dynamics and dispersal characteristics, each coupled to other species to varying degrees (but see Holt, this volume). Whether spatial subdivision retains its potentially stabilizing effect in these situations, or has other important consequences, is an open area for theoretical and empirical work.

In all of these extensions of metapopulation dynamics, an approach that combines empirical work and modeling will be very helpful. For example, no general guidelines are available for empiricists to decide how much migration is enough, too little or too much for metapopulation effects to occur. Merely observing some asynchrony in local population fluctuations is not adequate, since this will be shaped not only by migration, but by patterns of environmental variability and by the sampling regime. Moreover, the critical level of migration will depend greatly on the exact hypothesis, or type of metapopulation behavior, that is of interest. Thus, combining field work with moderately detailed system-specific models will be valuable in many cases.

In conclusion, this review of empirical studies makes it clear that a great variety of spatial population structures exists in nature, and recognition of this diversity suggests changes in how both empirical and theoretical metapopulation research are approached. Whether experimental or observational, single- or multispecies, empirical studies need to take fully into account the different types of metapopulation structure that are possible, perhaps treating them as alternative hypotheses to test. While empiricists attempt to better characterize metapopulations in nature, an important task for theorists is to continue exploring the ways in which metapopulation behavior changes as patch configuration, dispersal, and local population dynamics are altered in realistic ways. Through this

combined effort, we will be much better able to identify the ways that within-
and between-population processes interact to determine the behavior of natural
systems.

ACKNOWLEDGMENTS

We thank Ilkka Hanski, Daniel Simberloff, Chris Thomas, and an anonymous reviewer for
helpful comments on an earlier draft.

Metapopulation Dynamics and Landscape Ecology

John A. Wiens

I. INTRODUCTION

> *The fusion of metapopulation studies and landscape ecology should make for an exciting scientific synthesis* (Hanski and Gilpin, 1991)

The synthesis of metapopulation studies and landscape ecology anticipated by Hanski and Gilpin has barely yet begun. There are at least two reasons for this (Wiens, 1995a). First, as many of the chapters in this volume illustrate, metapopulation theory continues to be tied to a view of spatial patterning of environments in which patches are embedded in a featureless background matrix. Second, landscape ecology seems still to be in the process of defining what it is about and describing complex spatial patterns, but it has not developed much theory to deal with spatial patterning. By focusing on some shared areas of interest, perhaps the synthesis of these disciplines can be accelerated.

In this chapter, I consider the relationship between the emerging (but yet immature) discipline of landscape ecology and the emerged (but perhaps adolescent) discipline of metapopulation dynamics. I will argue that considerations of metapopulation structure may often be incomplete unless they are framed in the context of the underlying landscape mosaic.

II. APPROACHES TO PATCHINESS

Ecologists have always known that nature is patchy and heterogeneous, even if much of their theory has not treated it so. Habitats in areas used by humans occur as sharply defined blocks or fragments, and the patchwork nature of the landscape mosaic is especially evident in such environments. Even in more natural settings, however, habitats are heterogeneous at virtually any scale of resolution. Although patch boundaries in such situations may sometimes be indistinct gradients rather than sharp discontinuities (Wiens, 1992b), the spatially variable character of environments still remains. In this chapter I will follow the convention that has become widespread in ecology of considering such variation under the rubric of "patchiness," even though "patches" are not always evident in nature.

Dealing with such spatial heterogeneity has been a major challenge in both empirical and theoretical ecology. Faced with the daunting complexity of spatial patterns in the real world, field ecologists historically tended to focus on patterns and dynamics of ecological systems within relatively homogeneous habitat types (e.g., watersheds, woodlots) or aggregated spatial variation into dimensionless indices of heterogeneity or dispersion. More recently, it has become fashionable to map spatial patterns at broad scales using geographic information systems and spatial statistics, but the link between such technologies and ecologically important questions is not always apparent.

Spatial variance also strains the capacities of analytical models and theory if it is viewed explicitly (i.e., by location) rather than averaged as "noise." As a consequence, many theoreticians concerned with heterogeneity have contented themselves with simple models in which spatial patterning is collapsed into patches and an ecologically neutral "matrix" (Kareiva, 1990b; Wiens, 1995a). Such patch–matrix theory is usually spatially implicit (Hanski, 1994c), in that the locations of patches in the matrix are not specified (Wiens, 1996a). The interesting dynamics occur in the patches, which are usually considered to be internally homogeneous; the matrix is viewed as inhibiting interactions among patches (e.g., migration, colonization, gene flow, prey discovery by predators).

Traditional metapopulation theory is an elaboration on this patch–matrix theme. Levins' metapopulation model (1970; Hanski, this volume) considered the habitat of a population to be subdivided into an infinite number of similar patches occupying undefined locations in a background matrix. As metapopulation modeling has progressed, however, details about patch sizes, patch clumping, individual movement capacities, local patch dynamics, and explicit patch locations have been incorporated (Hanski, 1994a,c; see Hanski, this volume; Gyllenberg *et al.*, this volume).

Most patch theory deals with the dynamics of populations occupying a patchy environment (Wiens, 1976; Levin, 1976; Kareiva, 1990b; Shorrocks and Swingland, 1990). Another approach to heterogeneity has focused on the dynamics of the patches themselves. Although the spatial pattern of some patches, such as the islands considered in island biogeography theory, may be relatively static in ecological time, the patch structure of most environments is not. Patches are destroyed and generated by disturbances at multiple scales. They undergo change through successional development. These "patch dynamics" (Pickett and White, 1985) produce changes in the spatial patterns and relationships of patches in a matrix. Attempts to model these dynamics have generally followed analytical approaches (patch demography; Levin and Paine, 1974; Hastings, 1991) or have simulated the spatial and temporal dynamics of patchy environments (e.g., Fahrig, 1990). Most of this work has followed the patch–matrix conceptualization of spatial patterns.

The recent emergence of landscape ecology as a discipline (Risser *et al.*, 1984; Forman and Godron, 1986; Merriam, 1988; Turner, 1989; Wiens, 1992a; Wiens *et al.*, 1993; Hobbs, 1995) offers the prospect for going beyond a simple patch–matrix approach to adopt a more realistic, spatially textured view of heterogeneity. In landscape ecology, the "matrix" is itself spatially structured, and spatial relationships play an active role in determining the dynamics within the "patches" of interest. Patches are viewed as components in a landscape mosaic, and what happens within and among the patches in a landscape may be contingent on the composition and dynamics of other elements of the landscape mosaic (Wiens *et al.*, 1993; Andrén, 1994; Wiens, 1995a, 1996a).

III. WHAT IS LANDSCAPE ECOLOGY?

One of the first tasks of an emerging discipline is to define its topic and itself. "Landscape" has been defined as "a heterogeneous land area composed of a cluster of interacting ecosystems" (Forman and Godron, 1986), "a mosaic of heterogeneous land forms, vegetation types, and land uses" (Urban *et al.*, 1987), or "a spatially heterogeneous area" (Turner, 1989). Accordingly, "landscape ecology" is "a study of the structure, function, and change in a heterogeneous land area composed of interacting ecosystems" (Forman and Godron, 1986) or "the investigation of ecosystem structure and function at the landscape scale" (Pojar *et al.*, 1994). It emphasizes "broad spatial scales and the ecological effects of the spatial patterning of ecosystems" (Turner, 1989) and "offers a way to consider environmental heterogeneity or patchiness in spatially explicit terms" (Wiens *et al.*, 1993).

If these definitions are a bit nebulous, it may reflect the multifarious historical development of landscape ecology and continuing uncertainty or disagreement over what it is really about. Landscape ecology began in northern Europe during the 1960s as a merging of holistic ecology with human geography, with infusions from land-use planning, landscape architecture, sociology, and other disciplines (Turner, 1989; Wiens *et al.*, 1993) (Fig. 1, top). From the outset, the emphasis was practical and applied: the focus was on the interaction of humans with their environment at a broad (landscape) spatial scale. In the early 1980s, the discipline colonized North America (and other continents, most notably Australia). The

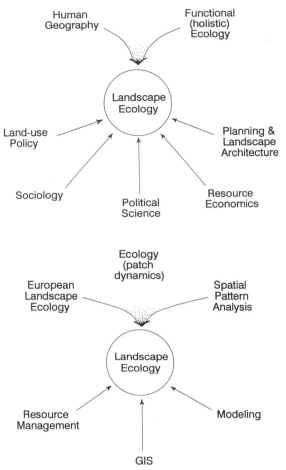

FIGURE 1   Contributors to the historical development of landscape ecology in Europe (top) and North America (bottom).

beachheads in North America were small and initially somewhat isolated. Perhaps through founder effects or mutations, the development of landscape ecology there followed a somewhat different trajectory (Fig. 1, bottom). The linkage with traditional ecology was much stronger than in Europe, and as a consequence the questions asked and approaches used differed considerably. There was a more self-conscious emphasis on concepts (Wiens, 1995a), a greater reliance on quantitative procedures (Turner and Gardner, 1991), and an application of the landscape perspective to a broad range of basic as well as applied problems.

These pathways of historical development have led to three rather different views of the primary focus of landscape ecology. Continuing in the European tradition, one view portrays landscape ecology as "a new holistic, problem-solving approach to resource management" (Barrett and Bohlen, 1991). It is a synthetic, holistic, human ecology. The second view, which has become most prevalent among ecologists, treats "landscape" as a level of organization (e.g., O'Neill *et al.*, 1986; Gosz, 1993) or as a scale of investigation (i.e., tens to thousands of ha; Forman and Godron, 1986; Hansen *et al.*, 1993; Hobbs, 1994; Pojar *et al.*, 1994). In the latter case, the questions are often no different from those that ecologists have always asked; they are just asked at a much broader scale. The third view more explicitly emphasizes the structure and dynamics of landscape mosaics and their effects on ecological phenomena (Turner, 1989; Wiens *et al.*, 1993; Wiens, 1995a). Rather than restricting the focus to broad scales, the scale of investigation is dictated by the organisms studied and the questions asked (Wiens, 1989a; Haila, 1991; Pearson *et al.*, 1996). In this view, landscape ecology is more than just spatially explicit ecology, because the patterns and interactions of entire mosaics are the focus of investigations.

This diversity of views suggests that landscape ecology is "a science in search of itself" (Hobbs, 1994). In addition to being a young discipline, it is also intellectually immature, in that it lacks conceptual unity (cf. Lochle, 1987; Hagen, 1989). It has no well-defined theoretical framework (Turner, 1989; Wiens, 1995a) and tends to be more qualitative than quantitative (Wiens, 1992a). Despite all of this, several prevailing themes of landscape ecology have emerged:

• Elements in a landscape mosaic (patches) vary in quality in both space and time. In a landscape, patch quality is a continuous rather than a categorical (i.e., suitable vs unsuitable, or patch–matrix) variable. Patch quality can be viewed as a spatially dependent cost–benefit function (Wiens *et al.*, 1993; Wiens, 1996a).

• Patch edges or boundaries may play critical roles in controlling or filtering flows of organisms, nutrients, or materials over space (Wiens *et al.*, 1985; Holland *et al.*, 1991; Hansen and di Castri, 1992). What happens at boundaries may have important effects on both within-patch and between-patch dynamics.

• The degree of connectivity among elements in a landscape mosaic has major consequences on patch interactions and landscape dynamics (Lefkovitch and Fahrig, 1985; Taylor *et al.*, 1993). How disturbances propagate over a landscape, for example, may be dictated by landscape connectivity as well as boundary effects (Turner *et al.*, 1989). Connectivity involves much more than corridors.

• Patch context matters. What happens within a patch is contingent on its location, relative to the structure of the surrounding mosaic. A patch of the same habitat may be of quite different quality, depending on the features of adjacent or nearby elements of the landscape (Wiens *et al.*, 1993). Contrary to island biogeography theory (or, implicitly, patch-matrix theory), no patch is an island (cf. Janzen, 1983). It is this contextual dependency that requires landscape ecology to be spatially explicit.

IV. HOW IS LANDSCAPE ECOLOGY RELEVANT TO METAPOPULATION DYNAMICS?

To see how these themes may relate to metapopulation dynamics, we must review briefly the essential features of metapopulation theory (see Hanski and Simberloff, this volume). "Metapopulations" have been defined in various ways, but generally a metapopulation is spatially subdivided into a series of local (patch) populations. The classical view emphasizes a balance between extinctions and recolonizations of local populations that facilitates long-term persistence of the metapopulation (Levins, 1970; Hanski and Thomas, 1994; Hanski, this volume). The dynamics of local populations are density-dependent within patches but asynchronous among patches, and migration (dispersal[1]) among patches links them together. Interpatch movement is the key. If migration is large relative to interpatch distances (and other spatially uncorrelated sources of population variability are not important), the dynamics of local populations will be mixed together and they will act as a single large population. On the other hand, if movement among patches is infrequent it may not be adequate to ensure recolonization of habitat patches in which local populations have suffered extinction, dooming the entire metapopulation to global extinction.

The contrast between this classical view of metapopulations and a landscape-based view is perhaps most apparent graphically. In a traditional (theoretical) metapopulation, local populations occur in habitat patches in a featureless matrix (Fig. 2A). Not all patches are occupied at a given time, and local populations wink into and out of existence as extinction and recolonization occur. Patches may vary in size or shape, but the primary determinants of patch-colonization probability are movement rates and interpatch distances. Making a metapopulation model spatially explicit is therefore necessary, but not sufficient, to cast it in a landscape context. In reality, the local populations of a metapopulation occur in habitat patches that are immersed in a complex mosaic of other habitat patches, corridors, boundaries, and the like (Fig. 2B). The most obvious effects of this landscape structure are on individual movement patterns among patches and, consequently, on patch-recolonization probabilities. In a landscape mosaic, interpatch distances are not Euclidean (e.g., Fig. 2A), but are a complex function of boundary permeabilities and relative patch viscosities to moving organisms (e.g., Fig. 2B; Wiens *et al.*, 1993). Other aspects of metapopulation structure, such as the dynamics of the patches themselves (and, consequently, patch-extinction probabilities), may also be influenced by landscape structure.

Because very little empirical work that directly links landscape ecology to

---

[1] To be consistent with usage elsewhere in this volume, I use "migration" rather than "dispersal" to refer to one-way movements of individuals beyond their home ranges. Although "migration" is customarily used in this sense by geneticists and entomologists, to an ornithologist like myself the term has a specific meaning that is different from "dispersal." In examples dealing with birds, I will therefore use "dispersal" rather than "migration." Stenseth and Lidicker (1992b) discuss these terminological issues.

A

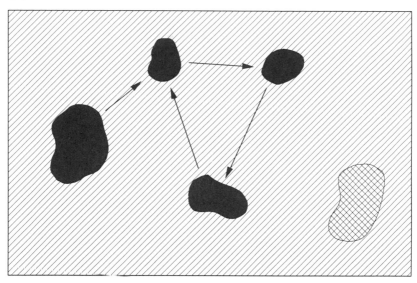

B

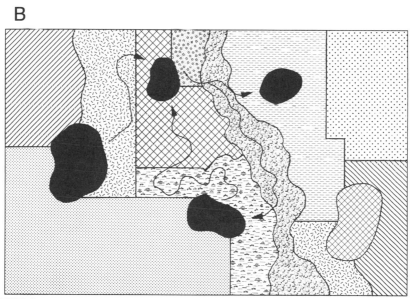

FIGURE 2 (A) Metapopulations in theory. The solid patches are occupied and are linked by inter-mittent migration, whereas the hatched patch is suitable habitat that is presently unoccupied. The background matrix has no effect on interpatch movements, although the distance between patches and their arrangement may. (B) Metapopulations in reality. The patches are the same, but the "matrix" is a landscape mosaic of various patches and corridors. Movement pathways among suitable patches, and the probability that migrating individuals will reach the patches, are affected by the explicit spatial configuration of the landscape.

metapopulation dynamics has been done, a discussion of how the major themes of landscape ecology—spatial and temporal variations in patch quality, boundary effects, landscape connectivity, and patch context—affect the three components of metapopulation dynamics (local extinction, interpatch movement, and recolonization) must necessarily be somewhat abstract and conceptual. It may be useful, therefore, to preface this discussion with a few examples of the effects of landscape structure in the real world. Additional examples are provided by Angelstam (1992), Fahrig and Freemark (1993), and Hobbs (1995).

A. Some Examples of Landscape Effects

Some of the effects of landscape structure are related to patch characteristics such as patch size or spacing. For example, the size of habitat patches has been related to the persistence of local populations of forest birds (Verboom *et al.,* 1991a; Villard *et al.,* 1992), and the degree of spacing of habitat patches has been shown to affect the likelihood of recolonization of vacant patches by the Glanville fritillary *(Melitaea cinxia)* in Finland (Hanski *et al.,* 1995a). Both patch size and spacing influenced the use by brown kiwis *(Apteryx australis)* of remnant forest fragments in an agricultural matrix in New Zealand (Potter, 1990). Kiwis are flightless, so they must walk between isolated remnants. All fragments less than 80 m from other forest remnants were used by the birds, regardless of their size. Movements of more than a kilometer from the reserve, however, were accomplished by using small fragments as "stepping stones." In this situation, the spatial interspersion of habitat patches was a critical factor in determining the effects of patch isolation and, consequently, the potential for metapopulation dynamics.

Patch edges and their configuration may also be important. The emigration of Glanville fritillaries from patches of suitable habitat, for example, increases with the proportion of the patch boundary that is bordered by open fields (Kuussaari *et al.,* 1996). Gates and Gysel (1978) found that the abundance of passerine birds increased at the boundary between fields and forests, and they suggested that individuals might be drawn to the edge as nesting habitat because of greater food availability there. Numerous studies (e.g., Angelstam, 1992; Andrén, 1992, 1995), however, have documented that predation rates may be greater at such ecotones, presumably due to predators living in adjacent areas. For some species, edges may function as an "ecological trap" by attracting individuals to areas in which predation losses are great (Gates and Gysel, 1978). Predation risks at habitat edges vary as a function of the surroundings (Wilcove, 1985; Angelstam, 1992; Wiens, 1995b), so the landscape context of patches is also important. Pearson's (1993) work on habitat occupancy by birds in the Georgia Piedmont also illustrates the effects of landscape context. There, the composition of the surrounding matrix explained as much as 74% of the variance in habitat occupancy by some species but was unimportant for other species. The demographic consequences of such edge- and context-related effects have received very little attention, but they may have important effects on metapopulation dynamics, especially where populations are subdivided among many small habitat patches and predation risk is significant.

The effects of corridors linking elements in a landscape mosaic have also been documented by field studies (although not to the degree that the widespread adoption of corridors as a management option would lead one to believe; Bennett, 1990, 1991; Hobbs, 1992). In Western Australia, for example, Carnaby's cockatoos *(Calyptorhynchus funereus)* use roadside vegetation as a pathway for foraging movements among woodland patches in their large home ranges (Saunders, 1990). Where woodland patches are not linked or are not visually apparent to the cockatoos, they are not used, even though food may be available there. On the other hand, singing honeyeaters *(Lichenostomus virescens),* which are habitat generalists, readily fly across farmland with little vegetation (Merriam and Saunders, 1993) and apparently make little use of corridors. Osborne (1984) found that hedgerow area was the best predictor of bird species richness in an area of Great Britain, and the presence of red squirrels *(Sciuris vulgaris)* in wooded fragments in The Netherlands was positively related to the amount of hedgerow surrounding the fragments (Verboom and van Apeldoorn, 1990). In Australia, the occupancy of corridors by arboreal marsupials could not be predicted by habitat features within the corridor but required additional information on the composition of the surrounding landscape (Lindenmayer and Nix, 1993).

B. Movement and Migration

Individual movement is the most important unifying element in both metapopulation dynamics and landscape ecology (Saunders *et al.,* 1991; Wiens, 1992b, 1995a; Wiens *et al.,* 1993; Ims, 1995). Moreover, how fast and how far organisms move imposes a scale on the environment: highly vagile animals integrate heterogeneity over broader scales than do sessile individuals and therefore perceive the environment with a coarser filter or "grain" (Wiens, 1985; Fahrig and Paloheimo, 1988; Kotliar and Wiens, 1990; De Roos *et al.,* 1991; With, 1994). At the outset of any field study or modeling exercise, then, the mean and shape of a species' migration function determine the scale(s) at which population responses to environmental patchiness must be investigated.

In the tradition of island biogeography theory, most metapopulation models use interpatch distance and migration rates as the major determinants of patch-colonization probabilities (e.g., Hanski, 1994a). The Fahrig and Paloheimo (1988) simulation studies of the effects of the spatial configuration of patches on population abundances in a metapopulation, for example, indicated that migration distance, rather than migration rates alone (or demographic features such as birth rate), was critically important, especially when interpatch distances were great. In contrast, when Liu *et al.,* (1995) modeled Bachman's sparrow *(Aimophila aestivalis)* population dynamics, they found that demographic parameters were more important than mortality during dispersal (although not necessarily dispersal rate or distance). These differences may stem from differences in model structure, but they may also reflect basic differences in the life histories of the organisms modeled.

Traditional metapopulation models usually do not consider the details of movement in even an abstract sense. Movement is modeled as transition probabilities among cells in a grid (Liu *et al.*, 1995) or movement rates and distances are simply specified or are drawn from frequency distributions. Whether movement through the matrix between patches is directional (e.g., Fig. 2A) or follows a diffusion, correlated random walk, or some other algorithm (e.g., Okubo, 1980; Turchin, 1989; Johnson *et al.*, 1992a) is not considered, even though the differences among these movement patterns can produce substantial differences in the probability of encountering a patch in the matrix. This is especially true if migration rates are low or if the number of individuals available to migrate is quite limited (as may occur when local populations are small).

Movement patterns such as diffusion or random walks are handy modeling devices that may have some relevance to how real organisms move through a featureless matrix, but they are of limited value (other than as neutral models) in specifying how individuals might respond to a complex landscape mosaic (e.g., Fig. 2B). Conceptually, the movements of individuals through a landscape may be viewed as a consequence of their movements within individual patches and their movements between patches (Fig. 3A; Wiens *et al.*, 1993). Within-patch movement patterns vary among different patch types. The probability that an individual will encounter a patch boundary during a specified time interval is a function of these patch-specific movements and of patch size and shape (perimeter : area ratio). Whether or not an individual will cross a patch boundary upon encountering it is a function both of features of the boundary itself (boundary "permeability"; Stamps *et al.*, 1987; Wiens, 1992b) and of the characteristics of the adjoining patch (patch context). [This is where another behavior, patch or habitat choice, becomes important.] Both costs (e.g., predation risk, physiological stress) and benefits (e.g., shelter, food availability, mating opportunities) may differ among elements in a mosaic, and movement patterns within and between patches may reflect these relative costs and benefits (i.e., patch quality), at least in part (Wiens *et al.*, 1993; Wiens, 1996a). Some simulation models of metapopulation migration in patchy environments (e.g., Pulliam *et al.*, 1992; Adler and Nuernberger, 1994) vary migration costs as a function of distance or incorporate differences in patch quality.

To make such an individual-based conceptualization of mosaic movements relevant to metapopulation dynamics, it must be extended to the scale of population rather than individual patches (Fig. 3B). In simple terms, this is a matter of shifting the scale from that of movements and patches defined by individual home ranges to the broader-scale movements of populations (i.e., migration) and the scale of patchiness represented by interactions within a local population (i.e., nodes in a metapopulation). Exactly how the translation from individual movements to population distribution and interactions should be accomplished is one

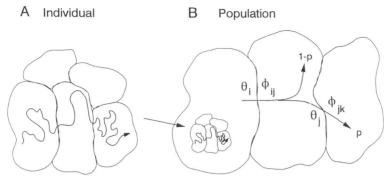

FIGURE 3   (A) Patterns of movement of an individual among elements of a landscape in its home range. The movement pathway consists of within-patch and between-patch components; both may be affected by the characteristics of patches and by their spatial configuration (patch context). (B) Extension of individual movements to the population level. A local population may occupy patch i, within which individuals move according to the local habitat heterogeneity within that patch. These movements (characterized by a function, θ_i) determine the probability that individuals will encounter the boundary between patch i and patch j during a given time interval. The probability that individuals encountering the boundary will cross into patch j, ϕ_{ij}, is a function of the permeability of the boundary and the behavior of the organisms (e.g., patch choice). Within patch j, a proportion of the dispersing individuals $(1-p)$ may die or establish residency in the patch. Movements within patch j (θ_j) determine the probability that the boundary between patch j and another element in the landscape (patch k) will be encountered; ϕ_{jk} determines p, the proportion of dispersers from patch i that move into patch k. Values of θ and ϕ are patch specific (as is patch density, which may have density-dependent effects on movement and migration). Developed from Wiens *et al.* (1993).

of the most vexing problems confronting a metapopulation-landscape synthesis. It is part of the more general problem of translating across scales in ecology (Wiens, 1989a; King, 1991; Rastetter *et al.,* 1992).

My colleagues and I have used systems of small animals (insects) moving through grassland "microlandscapes" as experimental model systems (Ims *et al.,* 1993; Ims, 1995) to investigate how movements are affected by mosaic structure, following the framework of the model of Wiens *et al.* (1993). Initial studies of tenebrionid beetles (*Eleodes* spp.) indicated that individuals moved differently in microlandscapes of a few square meters that differed in internal heterogeneity, as measured by the fractal dimension of the landscape pattern (Wiens and Milne, 1989). Movement alternated between matching the predictions of an ordinary diffusion model and those of anomalous diffusion depending on movement "rules," landscape pattern, and spatial and temporal scales (Johnson *et al.,* 1992a). In particular, diffusion exponents changed significantly at spatial scales corresponding to the size of vegetation patches (a radius of ≈ 42 cm), suggesting that the effects of spatial heterogeneity on beetle movements at finer scales differed fundamentally from those at broader scales. Other work (Crist *et al.,* 1992) demonstrated that variations in vegetation structure within 25 m² areas had significant effects on beetle movements and that these effects differed among *Eleodes* species. The net displacement of individuals per unit time, for example, was greater in areas dominated by bare ground and by continuous low grass cover than in more heterogeneous areas that contained cacti or shrubs, and larger beetle species exhibited greater displacements in a given habitat type than did smaller beetles. The relative complexity (fractal dimension) of the movement pathways, however, was insensitive to variation among species or habitat types, at least at the 25 m^2 scale of resolution. On the other hand, broader comparisons among beetles, harvester ants, and grasshoppers in the same landscape mosaics revealed significant differences in fractal dimensions of pathways (Wiens *et al.*, 1995), indicating fundamental differences in the ways these taxa respond to landscape heterogeneity at this scale.

These studies were conducted at relatively fine, "within-patch" scales and recorded how individual animals responded to landscape patterns. To determine how such movements might translate into patterns of population distribution at broader spatial scales, With and Crist (1995) used a cell-based simulation model to project the dispersion patterns of populations of grasshoppers over a broader mosaic. Individuals moved within a cell of a given habitat type according to the empirically observed movement parameters for that habitat. Movement characteristics changed when individuals entered cells of a different habitat type, according to a specified transition probability (this corresponds to the between-patch component, ϕ, of Fig. 3B). The landscape mosaic was dominated (65% coverage) by a single habitat type. Under certain specifications of transition probabilities, a large species, *Xanthippus corallipes,* moved rapidly through this cover type. As a consequence, it had increased patch-residence time (and an aggregated distribution) in the remaining 35% of the landscape. A smaller species, *Psoloessa delicatula,* was much more sedentary and preferred a habitat comprising only 8% of the landscape. Given its low vagility, there was a low likelihood of individuals of this species locating and aggregating within cells of the relatively rare, preferred habitat. The model simulations suggested that the distribution of this species would not diverge from the random distribution used to initiate the simulations. In fact, in the field both species exhibited the general dispersion patterns predicted by the model.

How do these observations and model analyses of patch-specific movements relate to the four components of landscape ecology (patch quality, boundary effects, patch context, and connectivity)? The differences in within-patch movement patterns may indicate differences in patch quality, but the sensitivity of model predictions to the value of transition probabilities between patch types indicates that knowledge of within-patch movement patterns by itself is not adequate to predict broad-scale population distributions. Something else is needed. The most likely factors affecting the translation from individual, within-patch movements to population distribution over a landscape are patch boundary effects and the influences of patch context. If individual beetles react behaviorally to the patch boundary itself, the likelihood of moving from one patch to another will be altered. If patch context is important, then the particular characteristics of what is beyond a given patch boundary will further modify transition probabilities.

Landscape controls over movement patterns have yet to receive detailed attention in either models or field studies. Moreover, all of these approaches consider the structure of the landscape mosaic to be fixed; patch dynamics in time would add another level of realism (and further computational complications) to the research program.

One aspect of landscape structure that is implicit in the spatial arrangement of mosaic elements and the transition probabilities among them is connectivity. Landscape connectivity refers to the degree to which the landscape facilitates or impedes movement among patches (Taylor *et al.*, 1993). Corridors of similar habitat linked together are thought to enhance connectivity (Bennett, 1990; Hobbs, 1992), but dissimilar habitat patches among which transition probabilities are high may also result in high connectivity. Through the patterns of connectivity that characterize a landscape, movement pathways are directed in spatially non-random manners (Fig. 2B), which can either increase or decrease the likelihood that movement among specific patches in the landscape (e.g., subpopulations in a metapopulation) will occur.

Connectivity is related to the coverage of a given habitat type in the landscape, but the relationship is strongly nonlinear. If a continuous habitat is broken into fragments by habitat conversion, the initial effects are due primarily to the loss of habitat coverage alone. As coverage drops below some threshold value, however, the effects of patch isolation begin to be more important. In landscapes with a low proportion of suitable habitat, further decreases in coverage result in a rapidly increasing distance between habitat patches and even greater isolation effects (Fig. 4). For example, Andrén (1994) found that habitat loss was a good predictor of fragmentation effects on birds and mammals in landscapes with >30% coverage of suitable habitats, but in more highly fragmented landscapes the effects of patch isolation and size also became important.

Such threshold effects also emerge in simulation studies based on percolation theory. In simple percolation models, a landscape mosaic is divided into suitable and unsuitable habitat patches (cells) that are distributed over the landscape at random, with a specified coverage or proportion, p, of the suitable patches (Gardner *et al.*, 1987, 1989). Above some critical threshold, p_{crit}, cells of the suitable habitat are likely to form a continuous cluster that spans the landscape. An organism in this "percolating cluster" will be able to move or "percolate" across the landscape; connectivity is high (O'Neill *et al.*, 1988). For a random landscape in which organisms move only to adjacent (but not diagonal) cells, p_{crit} has a value of 0.5928. If the landscape pattern is generated using a nonrandom algorithm (e.g., fractal curdling; Lavorel *et al.*, 1993; With *et al.*, in press), the value of p_{crit} is lower (0.29–0.50). Similar reductions in p_{crit} occur with changes in the movement patterns to allow individuals to move to any adjacent cell or to cross

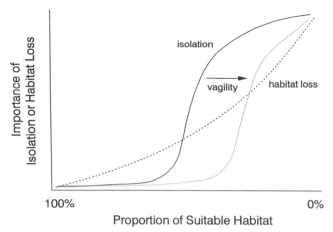

FIGURE 4   A hypothesized relationship between the proportion of suitable habitat in a landscape and the relative importance of habitat loss and patch isolation to individual movement or population dynamics. As the availability of suitable habitat decreases, the importance of habitat loss increases monotonically. The effects of patch isolation (the inverse of landscape connectivity) are relatively slight when coverage of suitable habitat is high, but increase sharply when a connectivity threshold is passed (p_{crit} in percolation theory parlance). Increases in individual vagility will move this threshold to lower coverage values of the suitable habitat.

gaps where suitable cells are not immediately adjacent (Dale *et al.*, 1994; Pearson *et al.*, 1996). Field experiments with *Eleodes* beetles moving through random landscapes (Wiens *et al.*, in press) indicated a threshold change in movement patterns when coverage of grass in a bare-ground matrix increased from 0 to 20%.

Changes in either the spatial pattern of the landscape or the scale over which individual organisms "perceive" landscape patterns (as judged by their movements) can therefore produce high connectivity in a mosaic even when the favored habitat type occupies a relatively small proportion of the landscape. Differences in vagility among organisms (e.g., the grasshoppers studied by With and Crist, 1995) may also affect the location of a percolation threshold (Fig. 4), as Fahrig and Paloheimo (1988) also suggested in a somewhat different context. Details of the spatial arrangement of habitat patches in the mosaic, such as those modeled by Lefkovitch and Fahrig (1985) or Adler and Nuernberger (1994), are likely to become important only around or below this threshold.

Most models that link animal movements to landscape structure assume that movement parameters are fixed species traits and that migration can adequately be represented using average values. Individuals do vary in movement characteristics, of course, and the effects of this variation may be profound. For example, Lens and Dhondt (1994) found that crested tit *(Parus cristatus)* young dispersed 1 week later from small, isolated pine stands than did those in large pine forests.

Chicks from second broods were also more likely to disperse into less suitable habitat fragments than were young from first broods. Collectively, these movement characteristics reduced the probability that second-brood young would be integrated into winter flocks, which would affect their overwinter survival probabilities. In another vein, the simulation studies of Goldwasser *et al.* (1994) suggested that variability among individuals could markedly increase the rate of spread of a population, even if only a few individuals in the population migrated rapidly. The prospect that individual movement behavior may be facultatively adjusted to landscape patterns such as the interspersion or isolation of suitable habitat patches (Matthysen *et al.,* 1995; Fahrig and Merriam, 1994) may further complicate attempts to model migration dynamics in heterogeneous landscapes. Nonetheless, it is apparent that the complex interplay between fine-scale movement patterns, broad-scale migration dynamics, and the nonlinear effects of landscape-mosaic structure may have fundamentally important effects on the interpatch movements that lie at the heart of metapopulation dynamics.

C. Local Extinction and Recolonization

In addition to interpatch movement, the extinction of local populations in habitat patches and the subsequent recolonization of those patches are what drive metapopulation dynamics. Local population extinctions are often associated with the stochastic dynamics that characterize small populations. Deterministic local habitat changes, however, can produce patch dynamics in the landscape that also result in the extinction of local populations (Thomas, 1994c). If this is the case, the local patch environment may remain unsuitable for some time after extinction occurs. Under these conditions, the persistence of the metapopulation depends on how well the organisms can track the shifting spatial locations of suitable habitat patches. Because the location of suitable patches may be unpredictable in time as well as in space, how organisms move through the landscape mosaic and the scales on which they perceive environmental patchiness become all the more important.

The pattern of interspersion of suitable habitat patches through a landscape mosaic also influences extinction and colonization probabilities. The degree to which a patch is connected to other suitable areas or is isolated may have little direct effect on extinction, although it may influence the immigration flow and therefore determine the magnitude of the "rescue effect" (Brown and Kodric-Brown, 1977). Colonization, on the other hand, is clearly related to the interplay between individual migration abilities and *both* the distribution (i.e., isolation) and the connectivity of habitats in the landscape. If fragmentation alters the landscape so that the interspersion of habitat patches no longer coincides with the migration patterns of a species, metapopulation dynamics may be disrupted. To some degree, this situation characterizes the Glanville fritillary in Finland (Hanski *et al.,* 1995a).

D. When Is a Landscape Approach Necessary?

In all but a few situations, landscapes, rather than patches in a featureless matrix, are reality. Given this, one might conclude that any attempt to model or understand metapopulation dynamics that does not explicitly include landscape structure would be futile. The essence of theory, however, is simplification of reality. Good theory simplifies in a way that does not violate reality too much, while incorporating its essential features. In this sense, patch–matrix theory represents a significant improvement over theories based on spatial homogeneity (Wiens, 1995a). When can the details of landscape structure reasonably be ignored or simplified?

Green (1994) and Fahrig (Fahrig and Paloheimo, 1988, personal communication) have addressed this question using simulation models. Green considered the effects of habitat connectivity in relation to population and community persistence and concluded that in highly connected landscapes one could treat the entire landscape as a single element (in which case metapopulation theory is no longer very relevant). If the landscape is strongly disconnected, on the other hand, it may be possible to treat each element as a separate unit and ignore all but the most basic descriptors of patch structure (e.g., patch size and separation). Closer to the percolation threshold (Fig. 4), on the other hand, the explicit spatial arrangement of patches in the landscape and the details of individual movements and patch transition probabilities may become much more important. Fahrig's simulation analyses suggested that a landscape approach may not be required when suitable habitat is abundant and widespread, when individual movement distances are large relative to interpatch distances (i.e., the "grain" of the environment is finer than that of the organisms), when movement patterns do not differ greatly among different elements of the landscape (i.e., transition probabilities are roughly equal and high), or when the habitat pattern is ephemeral. In most of these situations, either the environment approaches homogeneity or the organisms treat it as such. If this occurs at a broad, population scale, then it is unlikely that metapopulation dynamics will develop. The kind of interplay between local patch structure, individual movements, and local extinction and recolonization that is the essence of metapopulation dynamics would seem to require a certain form of patchiness, one that is in the vicinity of the connectivity threshold and does not meet the conditions specified in Fahrig's analysis. Under these conditions, attention must be given to the details of landscape structure.

V. METAPOPULATIONS, LANDSCAPES, AND CONSERVATION

The relevance of metapopulation dynamics to conservation issues is treated in detail in many other chapters in this volume, so I will not dwell on it here. If metapopulations are to be viewed in a landscape context, however, some implications for conservation practice cannot be ignored.

The traditional focus of conservation has been on reserves, and much of the debate about reserve design has dealt with the size, shape, and number of reserves. Reserves have usually been viewed as habitat islands (patches) in a background matrix. Metapopulation theory has become important in conservation biology because it fits neatly into this patch–matrix tradition and because the widespread occurrence of habitat fragmentation has subdivided populations (Wiens, 1995b, 1996b), creating spatial patterns that appear to match those of metapopulations. Metapopulation theory also predicts stability solutions, offering the hope of population persistence in the face of local extinctions.

Habitat fragmentation, however, involves much more than changes in the size and isolation of habitat patches. When a landscape is fragmented, habitats are replaced by other habitats, patch boundaries are often sharpened and patch context changed, connectivity patterns are altered, and the cost–benefit contours of the landscape shift. Simple island biogeography theory does not deal with such complexity of spatial patterns, and this is one reason why its value in conservation efforts is quite limited (Simberloff and Abele, 1982, 1984; Sobcrón, 1992; Haila *et al.,* 1993; and Wiens, 1995b, give other reasons). Island theory, for example, predicts a loss of species with a reduction in island (patch) area—the well-known species–area (S–A) relationship. A scatter of points above the S–A curve has been interpreted as evidence of community "supersaturation," which will inevitably lead to a loss of species ("faunal relaxation"), whereas points lying much below the curve have been explained as results of island disturbance (e.g., volcanic eruptions) or extreme isolation (see Wiens, 1989b). Because terrestrial habitat patches are immersed in a landscape mosaic, it seems more likely that such scatter represents (at least in part) the effects of connectivity, patch context, or edge conditions (Fig. 5). The specific ways in which landscape configuration might affect species–area relationships have not been explored.

These and other considerations have led to challenges to the "reserve mentality" (Brussard *et al.,* 1992), the belief that conservation problems are solved by establishing reserves and ignoring the surroundings. Reserves are necessary, to be sure, but areas outside of reserves may also play important roles (Noss and Harris, 1986; Saunders *et al.,* 1991; Woinarski *et al.,* 1992; Barrett *et al.,* 1994; Hanski and Thomas, 1994; Turner *et al.,* 1995; Wiens, 1995b, 1996b). For habitat generalists or species that move widely, management of landscape mosaics over large areas may be essential. In Australia, for example, the endangered Gouldian finch *(Erythura gouldiae)* has a limited and patchy distribution (Woinarski *et al.,* 1992). Large breeding populations still exist in several areas, and these can be protected by reserves. However, the population leaves these areas in postbreeding movements, with transient groups appearing in widely spaced (and unpredictable) locations over the landscape. Management by a series of static reserves will not work during this phase, when considerable mortality occurs.

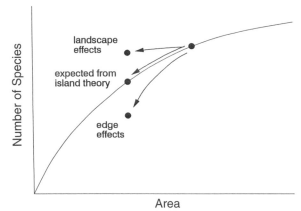

FIGURE 5   The species–area relationship. If the area of a habitat in the landscape is reduced (e.g., by fragmentation), island biogeography theory predicts that a new equilibrium number of species that is appropriate to the new habitat area will be reached. Landscape effects (e.g., connectivity, patch context), however, can reduce the species loss by providing habitat refugia or increasing the likelihood that local habitat patches will be rapidly recolonized. On the other hand, edge effects (e.g., low boundary permeability, increased predation mortality in habitat edges) may reduce species number below that expected from equilibrium island theory. Some of the scatter of points about reported species–area relationships may reflect the effects of such mosaic features.

The solution to such problems may be to shift from reserve management to "mosaic management," in which reserves are combined with areas that receive varied (and perhaps intense) human use. If one wishes to enhance a metapopulation structure in an area, for example, it may be necessary to manage not only the habitat patches that contain (or could contain) local populations but the landscape features that facilitate or impede interpatch movement as well. Too often, such considerations are cast in terms of corridors of like habitat linking patches together (e.g., the management plan for northern spotted owls *(Strix occidentalis caurina);* J. W. Thomas *et al.,* 1990), rather than evaluating overall landscape connectivity. Proper mosaic management requires that attention be given to *all* of the features of a landscape and how they interact to determine the fate of local populations in habitat patches. I maintain that the key to accomplishing this objective lies in understanding how landscape structure affects movement patterns within and among patches (Wiens, 1996b).

VI. CONCLUSIONS

The main message of this chapter is that landscape structure may often be an important component of metapopulation dynamics. Variations in patch quality in space and time, the form and permeability of patch boundaries, the composition and characteristics of surrounding mosaic elements, and the connectivity among landscape components may all influence the dynamics of local populations and, especially, the ways in which populations are linked by movements of organisms. The synthesis of landscape ecology with metapopulation dynamics *is* important.

Although I have emphasized the contributions that landscape ecology can make in developing an understanding of metapopulation dynamics, the relationship between these disciplines should not be one-sided. Metapopulation dynamics may also contribute to the development of landscape ecology, in two ways. One is by emphasizing the *dynamics* that occur in a landscape. The spatiotemporal patterns of local extinctions and patch recolonizations create a shifting distribution of populations among patches. Understanding what controls these dynamics addresses issues of spatial relationships and mosaic composition that are at the heart of landscape ecology. Moreover, an emphasis on these dynamics can draw attention away from the map-based descriptions that characterize some approaches to landscape ecology.

The second way in which metapopulation dynamics can contribute to landscape ecology is in the area of theory. In contrast to many other areas of ecology, landscape ecology has developed rather little theory. The lack of theory may stem in part from the diverse historical roots of the discipline (Fig. 1), but it may also reflect the complexity of landscapes and their linkages. The variety of landscape patterns is virtually unlimited, and thus there is no single mosaic pattern (or small set of patterns) about which theory can be generated (Wiens, 1995a). In contrast, patch theory has developed at least in part because "patchiness" can be collapsed into simple patterns of patches and matrix (or so we believe). Further development of landscape ecology as a predictive rather than a descriptive science requires concepts or theories that link landscape patterns to their consequences. As metapopulation theorists continue to add complexity and realism to simple patch–matrix models, they come closer and closer to developing true mosaic models. Quite beyond the value of such models in enhancing our understanding of metapopulation dynamics, they may provide a wedge that landscape ecologists can use to develop models of landscape interactions. A linkage of metapopulation theory with percolation theory might be especially fruitful (see With, in press).

Throughout this chapter I have emphasized the importance of understanding movement. Whether or not a spatially subdivided population functions as a metapopulation depends on how individuals move among patches. How individuals migrate is, in turn, affected in a myriad of ways by landscape structure. Understanding these effects on movements is of fundamental importance, yet we know very little about movement in an ecological context (May and Southwood, 1990; Opdam, 1991; Dunning *et al.*, 1995). Existing theory will not provide much help here. Instead, we must focus our attention on well-designed empirical studies of how individual movements are affected by the explicit spatial patterning of environments. Such investigations can provide the information and insights necessary to bring metapopulation dynamics and landscape ecology together.

ACKNOWLEDGMENTS

Diane Debinski, Mike Gilpin, Andy Hansen, Ilkka Hanski, Nancy McIntyre, and an anonymous reviewer offered a wide variety of helpful comments on an initial draft of the manuscript, and conversations with Ilkka were particularly useful in focusing my thinking about metapopulations and landscapes. My research on landscapes and spatial heterogeneity has been supported by the U.S. National Science Foundation, most recently through Grant DEB-9207010.

METAPOPULATION THEORY

The six chapters in this section review much of the existing theory in metapopulation biology, covering both ecology and genetics and single-species and multispecies theory. Theory is also developed and discussed in Part III, but with a focus on particular processes rather than on metapopulation dynamics in general. The final chapter by Giles and Goudet in Part IV adds a useful discussion on the theory underlying genetic population differentiation in metapopulations.

The first four chapters ascend from simple (Hanski) and more complex ecological models (Gyllenberg, Hanski, and Hastings) of single species to models of competition and pre-dation (Nee, May, and Hassell) to models of communities (Holt). The first two chapters are entirely rewritten versions of respective chapters in the previous volume (*Metapopulation Dynamics,* 1991), giving the interested reader a very concrete opportunity to judge the kind of change and even progress that has occurred in the past 5 years.

Hanski's chapter covers some of the basic issues about the ecological dynamics of single-species metapopulations. To start with, how commonly do species persist in fragmented landscapes as classical metapopulations? As also highlighted by Harrison and Taylor in their chapter, the answer is not well known because of scarcity of appropriate field studies on large enough spatial scale for long enough time. What is clear by now, however, is that some very good examples of species persisting as classical metapopulations do exist (see also Part IV). What is the minimum amount of suitable habitat necessary for metapopulation survival, and what is the minimum viable metapopulation size? These are likely to be controversial questions, like questions about the minimum viable population size in unbroken habitats. These are, however, the kinds of quantitative questions that ecologists will be asked, and it is our duty to clarify not only the answers to these questions but also the various kinds of uncertainties that are associated with particular answers and the risks associated with practical applications. More generally, ecologists will be asked to predict, in quantitative terms, the dynamics of particular species in particular fragmented landscapes, including the expected time to metapopulation extinction. Hanski reviews some of the modeling approaches that have been developed for this purpose. Among other things, these models clearly demonstrate that it may be very misleading to assume that a metapopulation occurs at a stochastic steady state in a rapidly changing landscape, a conclusion that has weighty implications for conservation.

Patch models of metapopulation dynamics, such as the well-known Levins model, are often criticized for excessive simplicity. The chapter by Gyllenberg, Hanski, and Hastings extends the deterministic single-species theory to structured models, where the quantity of interest is not just the fraction of occupied patches, like in the Levins model, but rather the distribution of local population sizes. The structured models include the effects of birth, death, immigration and emigration on metapopulation dynamics, though still retaining the abstraction of infinitely many patches and equal connectance among the patches. Even with these simplifications, the mathematics become very complicated. It is encouraging that one key prediction, the possibility of multiple equilibria, stemming from the theory of structured populations, has been recently supported by a large-scale field study (Gyllenberg, Hanski, and Hastings).

Nee, May, and Hassell extend the single-species models to pairs of competitors and mutualists and to predator–prey interactions (broadly interpreted). Theory makes it clear that the spatial structure of populations often matters and often makes it easier for species to coexist, which has been the main incentive for developing much of this theory in the first place (see also the chapter by Harrison and Taylor in Part I). Metapopulation-level coexistence may take striking forms, such as the emergence of spatially chaotic patterns of local abundance in spatially explicit predator–prey models with restricted movements. A great challenge here remains to relate the theory to the dynamics of real metapopulations. Another central theme addressed by Nee, May, and Hassell is the consequences of habitat destruction on persistence of single-species, competitive and predator–prey metapopulations. Observing that an analogous issue has for a long time been in the center of epidemiological theory, Nee, May, and Hassell discuss under which circumstances the "eradication threshold" of a metapopulation, essentially the minimum amount of suitable habitat as discussed by Hanski, can be estimated simply by measuring the amount of unused habitat at equilibrium, the limiting resource for metapopulation growth. This is clearly a theme of great importance to conservation biologists, but also an area where extra caution is needed in translating the theoretical results to practical recommendations (Hanski).

Holt extends the predator–prey metapopulation models to chains of three species, and to landscapes with two kinds of habitat patches, with a possibility of habitat specialization. His analysis confirms that it is difficult to survive in sparse habitats, and the species doing so are either extreme specialists (low extinction rate, high colonization rate) or, on the contrary, habitat generalists. Species at higher trophic levels are even more constrained, as the suitable patches for specific predators are always subsets of patches available for the prey (prey is generally absent in some patches). In a spatial mosaic of several habitat types, surprising patterns are possible, such as a generalist predator excluding a specialist prey from particular habitat type and itself surviving on the alternative prey in some other patch types. This outcome would be difficult to observe, as both the prey and the predator are now absent from the focal habitat type! Including both complex landscapes and complex communities in the same models, Holt's analyses complement the results of Nee, May, and Hassell and take a step toward a better understanding of metapopulation and metacommunity dynamics in the kind of mosaic landscapes that Wiens painted in his chapter in Part I.

One of the consequences of increased habitat fragmentation is often thought to be reduced potential for maintaining genetic variation in local populations and across the entire metapopulation. The equilibrium level and rate of change in genetic variation, measured for instance by heterozygosity levels, are generally functions of the effective size of the population; hence one important way habitat fragmentation may affect genetic variation is by changing the effective population sizes. In metapopulations, one may distinguish between effective population sizes at the local and metapopulation levels, respectively. Hedrick and Gilpin explore with numerical simulations the effective metapopulation size, taking as their starting point the Levins model with a finite number of habitat patches. They examine how the various model parameters, such as the number of patches, population turnover rate, patch carrying capacity and gene flow affect the effective sizes of local populations and the entire metapopulation. Consistent with theory (Barton and Whitlock), they find that, under the assumptions of their model, the effective metapopulation size is greatly reduced by high extinction rate and a small number of founders originating from just one or a few existing populations. Thus, metapopulation dynamics per se and its key parameters, such as propagule size, have significant genetic consequences. This theme is explored further in the context of an empirical case study by Gilet and Goudet in Part IV. Hedrick and Gilpin infer from the generally high levels of heterozygosity observed in nature for allozyme markers that metapopulation dynamics in the form explored in their model have not been of overriding importance in many species; otherwise heterozygosity levels should be much lower. However, they caution that increased habitat fragmentation may have recently forced species to conform to a metapopulation structure, possibly triggering a course of rapidly declining genetic variation. This is an argument analogous to that advanced by Hanski and by Nee, May, and Hassell in their chapters about nonequilibrium metapopulations on their way to extinction; past habitat destruction may already have reduced the amount of suitable habitat below a critical treshold, and it is only a matter of time before the actual extinctions will occur. These conclusions reflect the relatively slow time scale of metapopulation dynamics.

Barton and Whitlock present a comprehensive review of the consequences of spatial population structure on the genetic composition of metapopulations. The consequences of spatial structuring of populations on adaptation and speciation have been a controversial issue ever since Fisher and Sewall Wright established the fundamental results. In the metapopulation context, Wright's shifting balance between the processes of random drift, selection, and migration is particularly intriguing. Barton and Whitlock conclude that though the shifting balance process is possible, there are several factors which make it unlikely. Migration rate should not be too great to prevent populations from drifting to the domain of new adaptive peaks; but migration rate must be sufficiently high to allow the new peaks to spread in the metapopulation. Small population size is generally beneficial for a peak shift, but small populations are prone to local extinction, and generally send out fewer emigrants, than large populations, which makes spreading of the new peak into the metapopulation more difficult. No grand conclusion on the shifting balance process is yet possible. The message that Barton and Whitlock put forward is that the standard simple measures of genetic population structure, such as effective size or F_{st}, are not sufficient, but empirical studies should strive toward a much more comprehensive picture of the distribution of genotypes across populations in a metapopulation and of the ecological and selective forces that are responsible of these distributions. Studies of population differentiation based on neutral markers have only a limited value.

Metapopulation Dynamics

From Concepts and Observations to Predictive Models

Ilkka Hanski

I. INTRODUCTION

The concepts of metapopulation dynamics and metapopulation persistence in fragmented landscapes have become well established in ecology during the past 5 years (Hastings and Harrison, 1994; May, 1994; Harrison, 1994b; Hanski, 1994b; Kareiva and Wennergren, 1995). The accelerating loss and fragmentation of natural habitats (Morris, 1995), of which most of us are personally and painfully aware, makes it tempting to suggest that, in an increasing number of species, the spatial structure of populations is somehow consequential to their dynamics. Many studies have demonstrated that small populations in small habitat fragments have a high risk of extinction (Schoener and Spiller, 1987b; Kindvall and Ahlén, 1992; Hanski, 1994b); hence if only small fragments remain, long-term persistence becomes necessarily a regional issue. We have now an extensive theoretical literature on metapopulation dynamics (Hanski, 1985, 1994a,b; Gilpin and Hanski, 1991; Hastings, 1991; Gyllenberg and Hanski, 1992; Hanski and Gyllenberg, 1993; Hastings and Higgins, 1994; Tilman *et al.*, 1994; Hassell *et al.*, 1994; Durrett and Levin, 1994; Hastings and Harrison, 1994) and a large number of useful empirical studies (Harrison *et al.*, 1988; McCauley, 1989; Nachman, 1991; Harrison, 1991; Sjögren, 1991; Sjögren Gulve, 1994; Whitlock, 1992b; Thomas and Harrison, 1992; Bengtsson, 1993; Hanski *et al.*, 1994, 1995a; many chapters in this volume). Nonetheless, it is fair to say that our understanding of metapopulation dynamics in real fragmented landscapes is still restricted (Harrison and Taylor, this volume), largely because of the practical problems of conducting sound empirical research at a sufficiently large spatial scale.

Since the publication of the previous volume on metapopulation dynamics (Gilpin and Hanski, 1991), it has become evident that a broadly defined concept of metapopulations is needed to embrace the range of existing spatial population structures (Harrison, 1994b; Harrison and Taylor, this volume; Hanski and Simberloff, this volume). The classical metapopulation concept of Levins (1969a, 1970), which assumes a large number of small and hence extinction-prone local populations connected by not-too-much migration, is now seen as a special case, possibly an uncommon special case (Harrison, 1991, 1994b). This chapter is nonetheless focused on metapopulations essentially agreeing with the classical concept. This is for two reasons: First, it is too early to conclude that Levins-type metapopulations are exceptional; a large fraction of rare and specialized species in many lanscapes may fall into this category (Hanski, 1994c; Hanski and Hammond, 1995). Second, a better understanding of the classical case should enhance our understanding of metapopulation dynamics more generally.

In this chapter, I pose four broad questions:

1. How commonly do species persist in fragmented landscapes as classical metapopulations? This is the fundamental empirical question which I cannot answer here, but I give one well-researched example which highlights some of the reasons why the answer is not better known.

2. What is the minimum amount of suitable habitat necessary for metapopulation survival, and what is the minimum viable metapopulation size?

3. Can we make quantitative predictions about the dynamics of particular metapopulations in particular fragmented landscapes?

4. How common are nonequilibrium metapopulations, in which the rates of local extinction and recolonization are not in balance?

Recognizing the wide interest that these issues have aroused in conservation biology (Western and Pearl, 1989; Falk and Holsinger, 1991; Fiedler and Jain, 1992; Harrison, 1994b), I summarize, toward the end of this chapter, four messages for conservation stemming from the answers to these questions. The final remarks are concerned with the perennial question about density dependence in population dynamics.

II. AN EXAMPLE OF CLASSICAL METAPOPULATION DYNAMICS WITH RAMPANT POPULATION TURNOVER

One could argue that it is futile to search for criteria by which metapopulations of various kinds (Hanski and Simberloff, this volume; Harrison and Taylor, this volume) could be identified, to answer whether a particular system is a "metapopulation" or not. This is futile because populations in nature exhibit continuous variation in their spatial structures and also because the real issue is not so much to classify populations living in fragmented landscapes but to find ways of understanding and predicting their dynamics. This being said, it is also clear that different approaches to population dynamics are likely to be most effective in different kinds of systems. In this spirit, I suggest that if a system satisfies the following four "conditions" then a metapopulation approach based on Levins's (1969a) original concept is likely to be helpful. I apply these conditions to an example from the work of my research group on a species of butterfly, the Glanville fritillary *Melitaea cinxia,* which we have studied on the Åland islands in southwestern Finland.

Condition 1

The suitable habitat occurs in discrete patches which may be occupied by local breeding populations. The habitat type suitable for *M. cinxia* is dry meadows, which occur as discrete and small patches on Åland islands (Fig. 1), with the mean, median, and maximum areas of 0.13, 0.03, and 6.80 ha, respectively ($n = 1502$; Hanski *et al.*, 1995a). An estimated 60–80% of butterflies spend their entire lifetime in the natal patch (Hanski *et al.*, 1994; Kuussaari *et al.*, 1996); hence meadows have local populations, not just ephemeral aggregations of individuals.

Condition 2

Even the largest local populations have a substantial risk of extinction. If not, then the metapopulation would persist simply because of the persistence of the largest population(s), and we would have an example of mainland–island metapopulations (which are common in nature; Harrison, 1991, 1994b). In *M. cinxia,* the largest local population of 377 extant populations in 1994 had ca 500 butterflies. In this and related butterflies, populations with several hundred individuals have been observed to go extinct in only a few years (Harrison *et al.,* 1991; Foley, 1994; Hanski *et al.,* 1995a), hence the large metapopulation in Fig. 1 has no "mainland" populations.

Condition 3

Habitat patches must not be too isolated to prevent recolonization. If they were, we would have a nonequilibrium metapopulation heading toward global extinction. Such metapopulations are common; Hanski and Kuussaari (1995) conclude that 10 of the 94 resident butterfly species in Finland represent the nonequilibrium case due to recent loss of habitat. However, *M. cinxia* on Åland islands is not one of them, as the mean nearest-neighbor distance between suitable habitat patches is only 240 m (median 128 m, maximum 3870 m; Fig. 1), and the

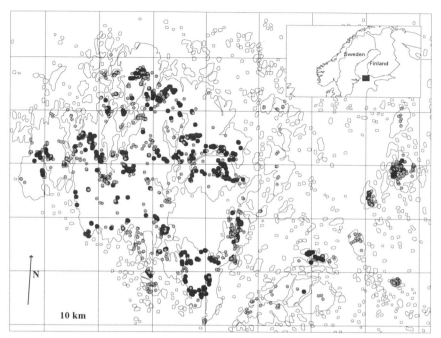

FIGURE 1   Map of Åland islands in southwestern Finland, showing the locations of the habitat patches (dry meadows) suitable for the Glanville fritillary *Melitaea cinxia* (dots). Patches that were occupied in late summer 1993 are shown by black dots. The size of the grid is 100 km² (modified from Hanski *et al.*, 1995a).

mean, median, and maximum distances moved by migrating butterflies among habitat patches in one 50-patch network were 590, 330, and 3050 m respectively (Hanski *et al.*, 1994).

Condition 4

Local populations do not have completely synchronous dynamics. If they have, the metapopulation would not persist for much longer than the local population with the smallest risk of extinction. In *M. cinxia*, we have demonstrated substantial asynchrony in the dynamics of populations within an area of 5 by 5 km² (Hanski *et al.*, 1995a). The most recent results suggest dynamics that may be somewhat correlated across areas up to some tens of square kilometers, but at the scale of the entire metapopulation changes in population size occur in opposite directions (Fig. 2). The question about spatial synchrony and its causes is a complex one (Thomas and Hanski, this volume), but the point which I wish to make here is that in our butterfly metapopulation there is certainly enough asynchrony to make simultaneous extinction of all local populations a very unlikely event under the prevailing environmental conditions.

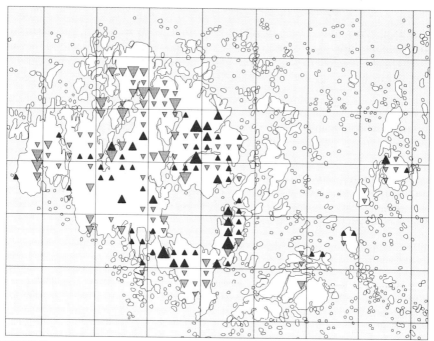

FIGURE 2   Observed changes in the population sizes of *M. cinxia* on Åland islands from 1993 till 1994. The study area was divided into 2 by 2 km² squares for the purpose of this analysis. The symbol indicates the sign and the magnitude of the change in the number of larval groups per 4 km² square between the 2 years (stippled triangles, decrease; black triangles, increase; logarithmic scale) (data from I. Hanski, J. Pögry, and T. Pakkala, unpublished).

III. CLASSICAL METAPOPULATION DYNAMICS: THE LEVINS MODEL

The purpose and "validity" of simple models in population ecology is often misunderstood. Their purpose is not to replicate in the model as many details of real populations as possible. Models which do that are not simple and their purpose is different (Section V). The purpose of simple models is to isolate, for a theoretical study, some feature of real populations that happens to be of interest. A simple model is not invalid just because all known real examples deviate in some respect from model assumptions; these differences may be immaterial for the purpose that the model was constructed. A simple model is defective if it fails to incorporate the critical variables and processes affecting the phenomenon under scrutiny or if it makes some critically unrealistic assumptions.

In this spirit, I suggest that the well-known Levins model (1969a, 1970), the mother of all metapopulation models with population turnover, provides a valuable theoretical framework for studying systems such as shown in Fig. 1 and satisfying the four conditions detailed in the previous section. The Levins model assumes a large number of discrete habitat patches, ideally of the same size, and all connected to each other via migration. In reality, not all populations are directly connected to each other, because migration distances are restricted, but this makes no important difference to the steady-state behavior of the model unless the network of habitat patches is strongly spatially heterogeneous. In the Levins model, habitat patches are scored only as occupied or not, as shown in Fig. 1, and the actual sizes of the local populations are ignored. The model therefore applies best to situations in which local dynamics occur at a fast time scale compared with metapopulation dynamics, either because the habitat patches are relatively small and hence local populations quickly reach the local "carrying capacity" or because colonization rate is low. All extant populations are assumed to have a constant risk of extinction. The rate of colonization is assumed to be proportional to the fraction of currently occupied patches (sources of colonists), denoted by P, and to the fraction of currently empty patches (targets of colonization), $1 - P$. With these assumptions, the rate of change in P in continuous time is given by

$$\frac{dP}{dt} = cP(1 - P) - eP, \tag{1}$$

where c and e are the colonization and extinction parameters, respectively. The equilibrium value of P is given by

$$\hat{P} = 1 - \frac{e}{c}. \tag{2}$$

The Levins model thus predicts that the fraction of occupied habitat at equilibrium increases with decreasing value of the ratio e/c, and the metapopulation is predicted to persist (P is positive) as long as $e/c < 1$. In spite of its simplicity, the Levins model is most useful in highlighting a key feature of metapopulation dynamics: for the metapopulation to persist, recolonization must occur at a sufficiently high rate to compensate for extinctions and to allow an increase from small metapopulation size. More specifically, condition $e/c < 1$, or $1 < c/e$, implies that a local population surrounded by empty patches must cause the establishment of at least one new population during its lifetime ($1/e$) for the metapopulation to persist.

Equation (2) leads to some straightforward but important predictions when we recognize that, very generally and not surprisingly, the risk of population extinction decreases with increasing patch area, and the probability of colonization decreases with increasing distance from the extant populations (Hanski, 1991, 1994b). The Levins model predicts that the fraction of occupied habitat at equilibrium (P) decreases with decreasing average size and decreasing density of habitat patches in a patch network. The results in Fig. 3 support these predictions for *M. cinxia.* This species went extinct on the Finnish mainland in the late 1970s and from many other regions in northern Europe during the past decades (Hanski and Kuussaari, 1995). The most probable reason for these metapopulation-level

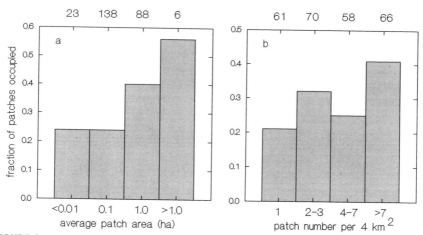

FIGURE 3 Effects of average patch size and regional density on the fraction of occupied patches in *M. cinxia* on Åland islands (Fig. 1). The study area was divided into 2 by 2 km² patches (as in Fig. 2). (a) Squares are divided into four classes based on the average patch area in the square; (b) squares are divided into four classes based on the number of patches per square (patch density). Note that the fraction of occupied patches is high in the squares where patches are large and where patch density is high (both effects are highly significant; statistical analysis in Hanski *et al.*, 1995a).

extinctions is decreased density of suitable habitat patches, forcing the equilibrium metapopulation size to zero ($e/c > 1$).

The large *M. cinxia* metapopulation shown in Fig. 1 is a good example of Levins-type metapopulations. However, how representative is this example? Hanski and Kuussaari (1995) attempted to answer this question for Finnish butterflies. By our count, 57 of the 94 resident Finnish species may belong to this category. This figure includes much uncertainty, though, because the spatial population structures of all the 93 species apart from *M. cinxia* are not well known. Collecting the kind of large-scale information we have collected for *M. cinxia* (Fig. 1) is expensive, obtaining funding for this kind of work is difficult, and the life cycles and larval biologies of most species make them much harder to study than *M. cinxia* on a large spatial scale. These are some of the reasons why we do not know how common Levins-type metapopulations are in nature.

Promising candidates for species with Levins-type metapopulations can be found in forest insects living in small and patchy microhabitats such as dead tree trunks. One such example involving beetles specializing on dead aspen trees in boreal forests is described in detail by Siitonen and Martikainen (1994; see also Hanski and Hammond, 1995). As most insects live in forests, and as most forest-living insects are more or less specialized on discrete microhabitats, Levins-type metapopulation structures may be common in insects (see also van der Meijden and van der Veen-van Wijk, this volume). Other examples include *Daphnia* water fleas in rock pools (Hanski and Ranta, 1983), frogs in ponds (Sjögren, 1991;

Sjögren Gulve, 1994), passerine birds in small woodlots (nuthatch; Verboom *et al.*, 1991b), and small mammals in small patches of suitable habitat (pika; Smith, 1980; Smith and Gilpin, this volume).

IV. MINIMUM VIABLE METAPOPULATION SIZE

The minimum viable population (MVP) size has become a well-established concept in population and conservation biology. MVP is intended to be an estimate of the minimum number of individuals in a population which has a good chance of surviving for some relatively long period of time, for instance, 95% chance of surviving for at least 100 years (Soulé, 1980). Though MVP is difficult to apply in practice (Soulé, 1987; Lande,1988b), it is a useful concept in highlighting the need for a quantitative analysis of the risk of population extinction.

In the case of Levins-type metapopulations, consisting of extinction-prone local populations, an analogous concept of minimum viable metapopulation (MVM) size may be defined as the minimum number of interacting local populations necessary for long-term persistence (Hanski *et al.*, 1996b). Apart from MVM, one also has to consider the minimum amount of suitable habitat (MASH) necessary for metapopulation persistence, because not all suitable habitat may be simultaneously occupied by a metapopulation persisting in a balance between local extinctions and recolonizations (that is, P is generally less than 1).

The original Levins model cannot be used to answer questions about MVM, because Eq. (1) is a deterministic model and only applicable to large networks of habitat patches in which the stochasticity involved in local extinctions and recolonizations becomes drowned by large numbers. In reality, many metapopulations live in small patch networks. Such metapopulations may go extinct when all local populations happen to go extinct at the same time, even if the expected colonization and extinction rates would allow long-term persistence by Eq. (1) or by some other deterministic model.

Gurney and Nisbet (1978; summarized in Nisbet and Gurney, 1982) have analyzed a stochastic version of the Levins model. Their analysis yielded the following approximation for the expected time to metapopulation extinction, T_M,

$$T_M = T_L e^{(H\hat{P}^2)/(2(1-\hat{P}))}, \qquad (3)$$

where T_L is the expected time to local extinction, H is the number of suitable habitat patches, and P is the fraction of occupied patches at a stochastic steady state. If one defines long-term metapopulation persistence as $T_M > 100\, T_L$, Eq. (3) leads to the following condition for reasonably large H (Gurney and Nisbet, 1978):

$$\hat{P}\sqrt{H} \geq 3. \qquad (4)$$

For example, if there are 50 habitat patches, Eq. (4) says that the colonization and extinction rates must be such that $P > 0.42$ for the metapopulation to persist for longer than roughly 100 T_L. Assuming a good colonizer, for which P is large, the critical minimum patch number is of the order of 10 (however, the approximation becomes less satisfactory for small H). Empirical results for *M. cinxia* and for other butterfly species (Thomas and Hanski, this volume) are in broad agreement with these predictions, suggesting that a minimum of 10–20 small and well-connected habitat patches are needed for long-term persistence.

Hanski *et al.* (1996b) have studied the stochastic Levins model numerically, incorporating such realistic features as variation in patch areas and the rescue effect (decreased risk of extinction due to immigration; Brown and Kodric-Brown, 1977; Hanski, 1991). Figure 4 gives the predicted time to metapopulation extinction in the model parameterized with data on *M. cinxia* (Hanski *et al.*, 1996b). These results are in good agreement with the analytical results of Gurney and Nisbet (1978) and strengthen the conclusions about the minimum numbers of habitat patches and local populations necessary for long-term metapopulation persistence. Notice that the condition about metapopulation lifetime combines characteristics of the species, as reflected in the value of P, with the properties of the landscape (patch number H). Hence the concepts of MVM and MASH cannot be applied independently. Equation (3) is not very sensitive to varying assumptions about metapopulation dynamics, because the effects of these assumptions are reflected in the value of P, itself a part of the condition. For instance, making migration more restricted in space will lower the colonization rate

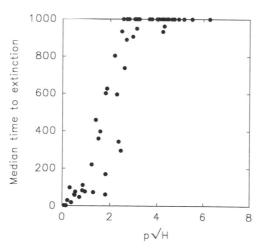

FIGURE 4   The relationship between the expected time to metapopulation extinction T_M and the product $P\sqrt{H}$ in simulations of an incidence function model parameterized with data from real metapopulations of the butterfly *M. cinxia* (the median time to local extinction $T_L = 3.3$ in these examples; details in Hanski *et al.*, 1996b).

and hence P, making metapopulation survival less likely. Even a very large number of habitat patches is not sufficient for metapopulation persistence if these patches are spread thinly across a large area! The above models include colonization–extinction stochasticity (Hanski, 1991) but they assume no environmental stochasticity and no regional stochasticity (spatially correlated environmental stochasticity). Regional stochasticity increases fluctuations in metapopulation size and decreases metapopulation lifetime (Hanski *et al.*, 1996b). However, for regional stochasticity to have a really significant effect the mean and the variance of the extinction rate must be high (Hanski, 1989; Harrison and Quinn, 1990).

V. PREDICTIVE MODELS OF METAPOPULATION DYNAMICS

The most obvious question that a metapopulation biologist may expect to be asked is whether some species X is likely to persist, as a metapopulation, in some particular set of habitat patches Y. In the context of conservation biology, the set of patches Y is often a subset of some larger number of larger patches, and the ecologist is asked to predict whether species X, present in the current patch network, would still persist if some patches were removed or their areas were reduced.

Analytical models of metapopulation dynamics, whether simple or more complex (Hanski, 1985, 1991; Hastings and Wolin, 1989; Hastings, 1991; Gyllenberg *et al.*, this volume; Verboom *et al.*, 1991a; Gyllenberg and Hanski, 1992; Hanski and Gyllenberg, 1993), are not helpful in answering such questions, because these models, intended for examining the balance between colonizations and extinctions more generally, do not incorporate specific information about patch qualities and locations and hence cannot be used to generate predictions for particular metapopulations. What are needed for the purpose of making specific predictions are spatially realistic metapopulation models. There are currently three main types of such models [I omit here a discussion of spatially explicit but not realistic approaches (Hanski, 1994c), such as cellular automata; see Caswell and Etter, 1993; Durrett and Levin, 1994; Nee *et al.*, this volume].

A. Spatially Realistic Simulation Models

Spatially realistic simulation models generalize models of local dynamics to several local populations connected by migration. Dynamics in each local population are modeled separately, complemented with specific assumptions about migration. The model can be linked with GIS-based information about particular landscapes (Akçakaya, 1994). Spatially realistic simulation models have been constructed to study the dynamics of, e.g., the spotted owl in California (McKelvey *et al.*, 1993; Lahaye *et al.*, 1994; Lamberson *et al.*, 1994; but see Harrison *et al.*, 1993) and the dynamics of butterfly metapopulations (Hanski *et al.*, 1994; Hanski and Thomas, 1994; Thomas and Hanski, this volume). There is no limit to the amount of "realism" incorporated in these models, but the realism comes with a cost, a large number of assumptions and parameters, which may be hard to verify and estimate. Nonetheless, spatially realistic simulation models may provide the most effective modeling framework especially for vertebrates with much information to parameterize the model and often with only few habitat patches and local populations. Thomas and Hanski (this volume) discuss the application of spatially realistic models to butterfly metapopulations (see also Hanski *et al.*, 1994; Hanski and Thomas, 1994).

B. State Transition Models

The two other approaches to spatially realistic metapopulation modeling, state transition and incidence function models, are patch occupancy models like the Levins model; hence only the presence or absence of a species in habitat patches is considered. Both state transition and incidence function models are discrete time stochastic models. From the practical point of view, the fundamental difference between the two is in the kind of data that are needed for model parameterization. State transition models are parameterized with data on observed *rates* of extinction and colonization, whereas incidence function models can be parameterized with data on *patterns* of patch occupancy. Pattern data are generally much easier to obtain than adequate data on colonization and extinction rates; hence the incidence function models can probably be used more widely. For this reason, and because I have a personal interest in the incidence function models, I describe their structure and application in greater detail (below).

Sjögren Gulve and Ray (1996) construct a state transition model using logistic regression to estimate the dependences of extinction and colonization probabilities on population size, isolation, and patch attributes. Having estimated the regression parameters, metapopulation dynamics may be iterated from an arbitrary starting configuration of patch occupancies by generating patch-specific extinction and colonization probabilities in each generation. An advantage of this approach is that it is straightforward to incorporate any empirically observed effects of habitat quality on extinctions and colonizations. The greatest disadvantage is that the model is parameterized with data on observed extinctions and colonizations; hence practical applications are restricted to large metapopulations with high turnover rate. The model parameters can be estimated from nonequilibrium metapopulations (which is an advantage), but the estimated extinction and colonization probabilities are sensitive to any temporal variation in these probabilities (regional stochasticity). For instance, if the extinction probabilities are estimated over a time interval during which exceptionally many populations happened to go extinct, the model prediction would extend the exceptionally high extinction rate to the future. Sjögren Gulve and Ray (1996), Thomas and Jones (1993), and Kindvall (1996a) apply a state transition model to metapopulations of a frog, a butterfly, and a bush cricket, respectively.

C. Incidence Function Models

Incidence function (IF) models (Hanski, 1994a,b) are based on a linear first-order Markov chain in which each habitat patch has constant transition probabilities between the states of being empty or occupied. Thus if patch i is presently empty it becomes recolonized with a patch-specific probability C_i in unit time (typically 1 year in practical applications). If patch i is presently occupied, the population goes extinct with a patch-specific probability E_i in unit time. With these assumptions, the stationary probability of patch i being occupied, called the incidence of the species in patch i, is given by

$$J_i = \frac{C_i}{C_i + E_i}. \tag{5}$$

From here we proceed in three steps (details in Hanski, 1994a,b):

(1) Specific assumptions are made about the effects of landscape structure on the colonization and extinction probabilities. Often it is realistic to assume that the extinction probability depends on patch area (because the extinction probability depends on population size which depends on patch area) but not on isolation. A convenient functional form is:

$$E_i = \min\left[\frac{\mu}{A^x}, 1\right], \tag{6}$$

where A_i is the area of patch i and μ and x are two parameters. In this formulation, there is a minimum patch area A_0 such that the extinction probability equals 1 for patches smaller or equal to A_0. The extinction probability is related to patch area for convenience, because data on patch areas are easy to obtain. The variable of fundamental interest is local population size, but it is often reasonable to assume that there exists a linear (Kindvall and Ahlén, 1992) or some other simple relationship (Hanski *et al.*, 1996c) between patch area and local population size; hence patch area can be used instead.

The model parameters can be interpreted in terms of an extinction model. Assuming realistically that extinctions are due to environmental stochasticity, and that the population has a positive growth rate at low density, the value of parameter x in Eq. (6) is related to the mean population growth rate $\bar{r}$ and the variance in growth rate V_e as $x = 2\bar{r}/V_e - 1$ (Lande, 1993; see also Foley, this volume). The value of x thus reflects the effective strength of environmental stochasticity ($\bar{r}/V_e$), large values of x indicating weak stochasticity.

The colonization probability C_i is an increasing function of the numbers of immigrants M_i arriving at patch i in unit time. In the case of mainland–island metapopulations (Hanski and Gyllenberg, 1993; Hanski and Simberloff, this volume), with a permanent "mainland" population as the sole or main source of colonists, a reasonable simple functional form is

$$C_i = \beta e^{-\alpha d_i}, \tag{7}$$

where d_i is the distance of patch (island) i from the mainland, and α and β are two parameters. For common species, which recolonize a little isolated patch (d_i close to zero) without delay, Eq. (7) may be simplified by setting $\beta = 1$.

In the case of metapopulations without a mainland, M_i is the sum of individuals originating from the surrounding extant populations. Taking into account the sizes and distances of these populations, we may assume that

$$M_i = \beta S_i = \beta \sum_{j=1}^{n} p_j\, e^{-\alpha d_{ij}} A_j, \qquad j \neq i, \tag{8}$$

where p_j equals 1 for occupied and 0 for empty patches, d_{ij} is the distance between patches i and j, and α and β are two parameters as in Eq. (7). The sum in Eq. (8) is denoted by S_i for convenience. If there are no interactions among the immigrants in the establishment of a new population, C_i would increase exponentially with M_i. Often, though, the probability of successful establishment of a new population depends on propagule size in a nonlinear manner (Schoener and Schoener, 1983; Ebenhard, 1991), and an s-shaped increase in C_i with increasing M_i is better justified,

$$C_i = \frac{M_i^2}{M_i^2 + y^2}, \tag{9}$$

where y is an extra parameter (notice that when Eq. (8) is substituted into Eq. (9), only the parameter combination $y' = y/\beta$ can be estimated). The colonization probabilities do not remain constant when the pattern of patch occupancy (the p_j values) changes, but this violation of the assumption of Eq. (5) is generally of little importance when the metapopulation is at a steady state (Hanski, 1994a).

One could make some other assumptions about the functional forms of C_i and E_i. For instance, it is possible to include in the model the effects of other patch attributes apart from area (Moilanen and Hanski, 1997). The essential point is that with such assumptions Eq. (5) is transformed into a parameterized model which can be fitted to empirical patch occupancy data. Assuming that patches may be rescued from extinction by immigration, Hanski (1994a) arrived at

$$J_i = \frac{C_i}{C_i + E_i - C_i E_i} = \frac{1}{1 + \mu'/(S_i^2 A_i^x)}, \tag{10}$$

where $\mu' = \mu y'$ for patches greater than A_0.

(2) The second step is to estimate the model parameters, α, x, μ, and y', using nonlinear maximum-likelihood regression or some other technique. In parameter estimation, the observed occupancies p_i are regressed against the incidences J_i (Hanski, 1994a). Minimally, one needs the following data from one metapopulation at a stochastic steady state: patch areas A_i, their spatial coordinates (to calculate the pair-wise distances d_{ij}), and the state of the patches at one point (year) in time (the p_j values). If more information is available, it can be used to

obtain more robust parameter estimates. For instance, Hanski (1994a) used mark–recapture data to estimate α independently, leaving only three parameters to estimate from occupancy data. The critical assumption at this stage is that the metapopulation from which the parameter values are estimated is at a stochastic steady state, that is, that there is no long-term increasing or decreasing trend in metapopulation size.

The values of the model parameters summarize essential information about metapopulation processes. Thus, the value of μ gives the probability of extinction per unit time in a patch of unit size, x gives the rate of change in extinction probability (and its inverse, expected time to extinction) with increasing patch area, α describes the effect of distance on migration rate, y gives the colonization efficiency, and β is a compound parameter, including emigration rate and population density (but note that, with occupancy data, one cannot estimate y and β independently; see Eq. (10)).

If one allows for the rescue effect, as was done in Eq. (10), the values of μ and y' cannot be estimated independently (Hanski, 1994a). To tease apart their values one may use either information on population turnover between 2 or more years, as explained in Hanski (1994a); or one may estimate (or guess) the minimum patch area A_0 (then $\mu = A_0^x$ and $y' = \sqrt{(\mu'/A_0^x)}$). The latter assumption will affect the predicted rates of extinction and colonization, but not the J_i values nor metapopulation size at steady state.

(3) Having estimated the model parameters, one may proceed to numerically iterate metapopulation dynamics in the same or in some other patch network to generate quantitative predictions about nonequilibrium (transient) dynamics and the stochastic steady state (Hanski, 1994a,b). This is the step of the greatest interest with many possible applications.

D. Tests and Applications of Incidence Function Models

Perhaps the most direct test of the model involves a comparison between the predicted and observed rates of extinction and colonization. I was able to do that in a long-term study of shrew populations on small islands in lakes in Finland (Hanski, 1992a). Incidence functions for three shrew species were parameterized with occupancy data from 68 islands. Using the estimated parameter values, I then predicted the per-year colonization and extinction probabilities in another set of 17 islands, which were censused for 5 years. The observed rates matched remarkably well the predicted rates in all three species, which represent practically independent replicates for the purpose of this test (Table I; interspecific competition affects only little if at all the extinction and colonization rates; Peltonen and Hanski, 1991). Using data on population densities and the estimated x values, I further inferred the relationship between the expected time to population extinction (the inverse of the extinction probability) and the expected population size conditional on no extinction (Fig. 5). Notice the dramatic differences among the species. This kind of information should be useful when assessing the relative

TABLE I   Parameter Estimates of a Mainland–Island Incidence Function Model, and the Predicted and Observed per-Year Extinction and Colonization Rates, in Three Species of *Sorex* Shrews on Small Islands in Lakes (Peltonen and Hanski, 1991; Hanski, 1992a)[a]

Species	Model parameters				Predicted		Observed	
	x	SE	μ/C	SE	Col	Ext	Col	Ext
araneus	2.30	0.68	0.79	0.22	0.26	0.04	0.20	0.04
caecutiens	0.91	0.24	17.67	11.36	0.03	0.28	0.05	0.33
minutus	0.46	0.16	4.09	1.51	0.18	0.53	0.13	0.46

[a] Isolation varied relatively little among the islands; hence the colonization probability C_i was assumed to be constant for all i. Parameters were estimated from a single survey of 68 islands (Hanski, 1992a). To tease apart the values of μ and C, I assumed that the minimum island area for occupancy, A_0, is 0.5 ha. The predicted extinction probability was calculated for an island of 1.6 ha, the average size of the 17 islands from which the observed colonization and extinction rates were measured in a 5-year study (Peltonen and Hanski, 1991).

importance of small and large reserves for the conservation of different kinds of species.

Rapidly increasing time to extinction with expected population size in Fig. 5 is associated with large x values. Recalling that the value of x is related to the strength of effective environmental stochasticity (above), the results in Fig. 5 illustrate the point that different forms of stochasticity lead to different relationships between time to extinction and expected population size (Goodman, 1987; Lande, 1993). In shrews (Fig. 5), as well as in land birds on oceanic islands (Fig. 6), there is a positive relationship between the x value and body size. Following the above line of reasoning, this suggests that small vertebrates are more sensitive to environmental stochasticity than large ones (Pimm, 1991), probably because small individuals have small body reserves and are hence more vulnerable to starvation than large ones (Hanski, 1992a). In invertebrates, we would not expect such a simple relationship between starvation time and body size; hence it is not surprising that Nieminen (1996) found no relationship between body size and the x value in herbivorous moths. The message from here is that the incidence function models can be used to draw interesting inferences about the rate and causes of population extinction from knowledge of the pattern of patch (or island) occupancy.

The examples in Figs. 5 and 6 and in Table I come from mainland–island metapopulations, where the colonization probability is a function of the distance to the mainland (Eq. (7)). In metapopulations without permanent mainland populations, the colonization probability has to be modeled with a more complex expression like the one given by Eq. (8), but the principle remains the same. Hanski *et al.* (1996c) used a small subset of the data shown in Fig. 1 to parameterize an incidence function model for the Glanville fritillary butterfly. Using

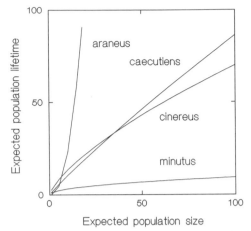

Expected population size

FIGURE 5 The relationship between the expected time to population extinction (inverse of the per-year extinction probability) and the expected population size (conditional on no extinction) in four species of *Sorex* shrews, the three European species in Table I, and *S. cinereus,* a North American species similar to *S. caecutiens* (from Hanski, 1993). The results are based on the parameter values of an incidence function model estimated from a snapshot pattern of island occupancy.

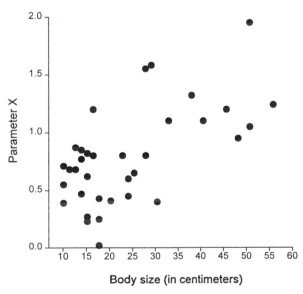

Body size (in centimeters)

FIGURE 6 The relationship between the value of parameter *x* in the incidence function model and body size in birds on oceanic islands (reprinted with permission of University of Chicago Press from Cook and Hanski, 1995).

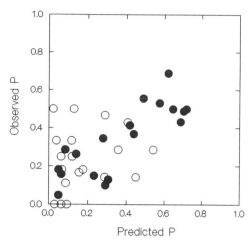

FIGURE 7   Comparison between the predicted and observed fraction of occupied patches (*P*) in *M. cinxia* metapopulations on western Åland islands. The *P* values were calculated for 4 by 4 km² squares. Open dots are for squares with < 15 habitat patches; black dots are for squares with ≥ 15 patches (from Hanski *et al.*, 1996c).

these parameter values, we then predicted the fraction of occupied habitat in the rest of the study area. For a part of the Åland islands, the prediction failed, perhaps because of some environmental differences (Moilanen and Hanski, 1997; Hanski *et al.*, 1996c), but in most of the study area the observed fraction of occupied patches matched well the predicted one (Fig. 7). Many conservation applications do not require quantitatively correct predictions, as the practical task is often to rank alternative management options in terms of their likely population dynamic consequences. Incidence function models should be very helpful in such applications.

VI. NONEQUILIBRIUM METAPOPULATIONS

Traditionally, the focus of population dynamic modeling has been in the equilibrium behavior of a hypothetical or some real population. Metapopulation modeling is no exception. Unfortunately, especially in the case of metapopulations, it takes a long time to reach the equilibrium following any major perturbation (for an extreme example, see Hastings and Higgins, 1994). Many of the metapopulations which we care about may not have had time to reach an equilibrium in a rapidly changing landscape. In a declining patch network the discrepancy between the prevailing state of the metapopulation and the equilibrium state imposes a "debt of extinctions" (Hanski, 1994c; Tilman *et al.*, 1994), extinctions which are expected to occur in the course of time even if the environment would not change any further. It goes without saying that this is a serious problem for conservationists, who are typically forced to operate within a time frame too short to address any long-term consequences, however likely they may be.

Perhaps the only general statement that can be made about nonequilibrium dynamics is that the discrepancy between the equilibrium and the existing state of a metapopulation is likely to be greatest in networks with relatively large and isolated patches, because then the turnover rate and hence the rate of approach to equilibrium are low. Extreme examples are the gradual decline of species number on land-bridge islands (Diamond, 1984) and on mountaintop habitats following postglacial climate change (Brown, 1971). Of greater concern, though, is the possibility that many metapopulations on much smaller spatial scales may not be at equilibrium.

I illustrate such nonequilibrium dynamics with another example on the butterfly *M. cinxia*. Figure 8 shows the loss and increasing fragmentation of suitable habitat for this species within an area of ca 25 km² during the past 15–20 years. During this period, the total area of suitable habitat declined to one-third of its original extent, and the number of distinct patches declined from 55 to 42, largely due to decreased grazing pressure on the meadows.

Figure 9 shows the predicted change in the fraction of occupied patches during the past 20 years. These results suggest that, so far, the butterfly has tracked rather closely the amount of suitable habitat, apparently because the amount of habitat and the total expected metapopulation size have remained large. However, one should not draw the conclusion that the same result would necessarily apply

1:30000

FIGURE 8   A map of the habitat patches within a 25 km² area in the northern part of the Åland islands (Fig. 1), showing the presumed extent of the suitable habitat for the butterfly *Melitaea cinxia* ca 20 years ago and today (shaded) (data from Frank Hering, personal communication).

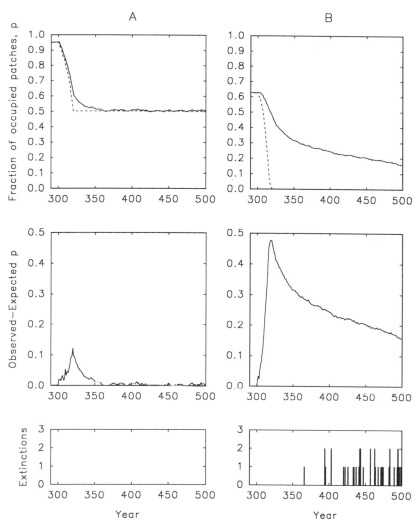

FIGURE 9   (A) Metapopulation size of the butterfly *Melitaea cinxia* as measured by the fraction of occupied patches *P* in the landscape shown in Fig. 8. The results were obtained by iterating the incidence function model parameterized with field data. The model iteration was started by assuming the patch network 20 years ago (Fig. 8). During a period of 20 years (from year 300 to 320 in the figure), this network was reduced to its present size (Fig. 8) as described in detail by Hanski *et al.* (1996b). The broken line gives the expected (equilibrium) metapopulation size, whereas the continuous line gives the actual metapopulation size in the declining network (the lines give the average *P* value in 200 replicate simulations). (Middle) Difference between the actual and equilibrium metapopulation sizes; (bottom) numbers of metapopulation extinctions in the 200 simulations (no extinctions in this case). (B) As for A, but now starting with the current patch network (Fig. 8) and halving the area of each patch in 20 years. Note that the equilibrium metapopulation size drops to zero, but it takes decades for most metapopulations to reach the equilibrium (extinction). (Top) The *P* value in the beginning of simulation is higher than the final value in A because the number of habitat patches is now smaller (reprinted with permission of University of Chicago Press from Hanski *et al.*, 1996b).

to all scenarios of habitat loss even in this species. The following example makes the point. Let us assume that each of the present patches (Fig. 8) would loose another 50% of its area in another 20 years. Figure 9 shows that such further loss of habitat would soon lead to a patch network in which the equilibrium state is metapopulation extinction. However, now the actual extinction is predicted to take tens or even hundreds of years, because the last local populations to go extinct are typically the largest ones with the smallest risk of extinction. The inevitable decline to extinction may become temporarily halted for long periods of time, with the number of occupied patches fluctuating without any obvious trend (Hanski *et al.*, 1996b).

VII. FOUR CONSERVATION MESSAGES

A. Metapopulation Survival in the Current Landscape May Be Deceptive

The previous section described one important message for conservation: many lanscapes may have changed so fast in the recent past that the respective metapopulations are far from equilibrium. In the worst case, the current patch network is already too fragmented to support a viable metapopulation, which is therefore committed to extinction unless the loss and fragmentation of habitat is reversed.

Hanski and Kuussaari (1995) estimated that 10 of the 94 resident butterfly species in Finland are presently represented by a nonequilibrium metapopulation on its way to extinction. Generally, it is not known how many metapopulations have already reached the state of "living dead," though we have little doubt that many have. Conservationists should dismiss the false belief that protecting the landscape in which a species now occurs is necessarily sufficient for long-term survival of the species.

B. More Than 10 Habitat Fragments Are Needed

Assuming that a network of small habitat fragments is established for the protection of some species, a natural question to ask is how many fragments should be created/retained. The blunt message that one is not enough is brought home by the fate of British butterflies on protected small reserves: tens of isolated populations of rare and endangared butterflies went extinct in 20 years, including all populations of three species (Warren, 1992; see also Thomas *et al.*, 1992).

Mathematical models reviewed in Section IV and limited data on butterflies (Thomas and Hanski, this volume) suggest that an adequate successful network of small habitat fragments should have a minimum of 10–15 well-connected fragments. Even this number may be insufficient if regional stochasticity is strong and local dynamics are strongly correlated. It is necessary to emphasize, though, that as long as even one population survives there is hope. Metapopulation decline may advance so slowly that there is time to act if there is wish to act. In the case of metapopulations on the brink of extinction, intervention in the form of managed recolonizations is likely to become an increasingly necessary, and accepted, form of management.

C. Ideal Spacing of Habitat Fragments Is a Compromise

Even a large number of small habitat fragments is no guarantee of metapopulation survival if the patches are located so far from each other that recolonization and population rescue from extinction by immigration are unlikely. A tentative practical answer to the question of minimum density of suitable habitat patches necessary for long-term survival has been sought from the Levins model, Eq. (1). To model habitat loss, assume that fraction $1 - h$ of the patches is permanently destroyed. The colonization rate becomes lowered because the density of empty but suitable patches is decreased from $1 - P$ to $h - P$, and the model becomes (May, 1991; Nee, 1994; Nee and May, 1992; Lawton *et al.*, 1994; Moilanen and Hanski, 1995)

$$\frac{dP}{dt} = cP(h - P) - eP. \tag{11}$$

At equilibrium, the fraction of empty patches (out of all patches, including the destroyed ones) is given by

$$h - P* = \frac{e}{c}. \tag{12}$$

Thus the fraction of empty patches out of all patches remains constant as long as the metapopulation does not go extinct, which happens when $h < e/c$. This is a seemingly very useful result, because it gives an estimate of the critical minimum patch density from the very limited information of the number of empty patches in a landscape in which the metapopulation still survives; no detailed knowledge of metapopulation dynamics is required (Nee, 1994). In practice, though, this rule of thumb is liable to yield an underestimate, and possibly a severe underestimate, of the critical patch density, because of three reasons (Hanski *et al.*, 1996b): the rescue effect, colonization–extinction stochasticity in small patch networks (Section IV), and nonequilibrium metapopulation dynamics, when a metapopulation is approaching the equilibrium from above (Section VI).

Increased patch density facilitates colonization and is hence helpful, but if habitat fragments are located close to each other the degree of spatial synchrony in local dynamics may become elevated (Fig. 2), which has a negative effect on long-term survival. In theory, a row of well-connected habitat fragments might often provide a better chance of long-term survival than a tight cluster, but such considerations are seldom practical. The main recommendation is simply to provide sufficient connections among habitat fragments by maintaining their density at such a level that recolonization occurs within a few generations. If recolonization rate appears to be worryingly low, one may have to consider the merits of managed recolonizations.

D. Substantial Variance in Habitat Quality Is Beneficial

A major cause of spatial synchrony in population dynamics is spatially correlated weather effects (Thomas and Hanski, this volume). This is not the entire story, though, because the effect of weather often interacts with attributes of habitat patches. For instance, in the butterfly *M. cinxia* dry meadows with low vegetation are generally favorable for larval growth and survival, but in very dry summers the host plants may wither on the dryest meadows and larval mortality is greatly increased. Most likely, an important reason why populations in large habitat fragments have a low risk of extinction, apart from the large expected population size, is the greater heterogeneity of habitat quality in large than small patches. Figure 10 gives an empirical example which suggests that the risk of local extinction decreases with increasing within-patch heterogeneity.

It is not often possible to substantially change within-patch heterogeneity, but when multiple reserves are selected there may be the option of including more

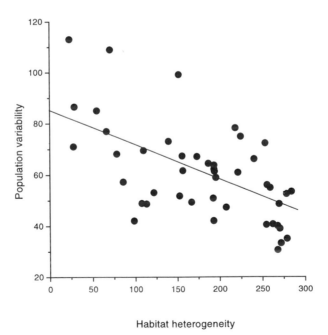

FIGURE 10   The relationship between temporal variation in population size (CV) and a measure of habitat heterogeneity in the bush cricket *Metrioptera bicolor*. Each symbol refers to one population (from Kindvall, 1996b).

or less variation in habitat quality among the selected patches. Though it may be tempting to aim at maximizing the "quality" of the preserved areas, there are good reasons to preserve a range of habitat qualities, to buffer the metapopulation against the adverse effects of environmental and regional stochasticities (Thomas and Hanski, this volume), and possibly also to maintain greater genetic diversity (Hoffman and Parsons, 1991).

VIII. CONCLUDING REMARKS

A better understanding of population dynamics is of fundamental intrinsic interest (Hassell and May, 1990) as well as necessary for improved conservation and management of natural populations (Caughley and Sinclair, 1994). Population ecologists have made great progress in the past decades using experimental, observational, and theoretical approaches (Price and Cappuccino, 1995), but fundamental questions about population regulation have remained controversial. Some field ecologists continue to resent the conclusion that density-dependent population regulation is necessary for long-term persistence of populations (den Boer, 1987, 1991; Wolda and Dennis, 1993), and others have doubted how generally density-dependent regulation occurs in natural populations (Strong, 1983; Stiling, 1987).

Den Boer (1968, and later papers) has championed the view that species may persist at a spatial scale larger than the local population thanks to the "spreading of the risk" process, involving movements among asynchronously fluctuating local populations. "The consequences of this spreading of the risk in space will be a relative reduction in the amplitude of fluctuations of animal numbers in the entire population" (den Boer, 1968). However, it is simply not possible to have long-term persistence even in a metapopulation without some density dependence in local dynamics, given that local population sizes are restricted, as they always are, below some maximum value (Hanski *et al.*, 1996a). In this respect, there is no difference between the dynamics of a single population and the dynamics of a metapopulation, regardless of the spreading of the risk. However, den Boer is correct to the extent that the incidence of density dependence may be low in some persisting metapopulations, in comparison with the incidence of density dependence necessary for long-term persistence of isolated local populations. In metapopulations, the combination of long persistence time with little density dependence is associated with high turnover rate and frequent local extinctions and colonizations (Hanski *et al.*, 1996a). Metapopulation persistence of assemblages of unstable local populations may explain some failures to detect statistically significant density dependence in natural populations (den Boer, 1987; Stiling, 1987; Gaston and Lawton, 1987), though a much more important reason for such failures is simply short runs of data that have been analyzed (Hassell *et al.*, 1989; Woiwod and Hanski, 1992).

5 Structured Metapopulation Models

Mats Gyllenberg   *Ilkka Hanski*

Alan Hastings

I. WHY USE STRUCTURED MODELS?

The simplest mathematical model of classical metapopulation dynamics with local population turnover is the one originally formulated by Levins (1969a, 1970; see Hanski, this volume),

$$\frac{dP}{dt} = \beta P(1 - P) - \mu P, \qquad (1.1)$$

where P denotes the fraction of occupied habitat patches, μ is the extinction rate per extant local population, and β is the colonization rate per empty patch and extant local population (to conform with the established notation of structured population models we have used here β and μ instead of c (or m) and e, respectively, which are the usual symbols for the colonization and extinction rates in the ecological literature and which are also used elsewhere in this volume). This simple model nicely captures the key idea of a metapopulation of extinction-prone local populations persisting in a balance between local extinctions and recolonizations of empty habitat patches (Hanski, this volume). The model predicts a threshold patch density necessary for long-term metapopulation persistence, a conclusion that is of fundamental significance for conservation (Lande,

1987; Nee and May, 1992; Hanski, this volume). Formally, Eq. (1.1) is identical with the susceptible–infected–susceptible *(SIS)* model of mathematical epidemiology (see, e.g., Anderson and May, 1991), with empty patches in a patch network playing the role of susceptible individuals in a population, and occupied patches corresponding to infected individuals. The agreement between the Levins model and the basic *SIS* model is more than coincidental, as the phenomena studied in metapopulation dynamics and in epidemiology share the same basic processes (Cohen, 1970; Levin, 1974; May, 1991; Lawton *et al.,* 1994).

The elementary *SIS* model assumes homogeneous mixing of a large number of individuals, with all infected individuals being equally infectious. The Levins model Eq. (1.1) is based on similar simplifying assumptions (Hanski, this volume). In particular, the Levins model is *unstructured* in that it assumes that all habitat patches and local populations are identical in all respects. This assumption involves several ecologically significant elements. First, the spatial arrangement of patches is ignored; every local population exerts the same colonization pressure on each empty patch regardless of its spatial location. This sort of assumption may be a reasonable approximation in models of disease spread in a population of freely moving individuals, but it is less likely to be satisfactory for habitat patches and local populations with fixed spatial locations. For the purpose of predicting the equilibrium metapopulation size this equal-connectance ("mean field") assumption is not badly misleading, especially if migrants move relatively long distances (Durrett and Levin, 1994; Caswell and Etter, 1993). Not surprisingly, if one is instead interested in the origin and maintenance of spatial patterns, the spatial arrangement of patches and populations becomes critical (Hassell *et al.,* 1991; Durrett and Levin, 1994).

Second, the Levins model assumes that all patches are of the same size and quality, whereas in nature this is hardly ever true (Harrison, 1991). Once again, though, this is not a sufficient reason to move to more complex models. In the first place, there are interesting natural systems in which variation in patch size is not very great, for instance dead tree trunks that are habitat patches for thousands of specialist insect species (Hanski and Hammond, 1995). Even if there is substantial variation in patch size and quality, the qualitative lessons from the Levins model still apply as long as the largest patches are not so large that the respective local populations are effectively immune to extinction (mainland–island metapopulations; Hanski and Simberloff, this volume).

Third, since all local populations are considered to be equal, the Levins model ignores local dynamics, and it assumes that emigration and immigration have no effect upon local dynamics. This assumption conflicts with a wide range of observations from natural populations (Brown and Kodric-Brown, 1977; Hanski, 1991; Hanski *et al.,* 1996b). In particular, this simplifying assumption means that the Levins model is really appropriate only for metapopulations with the migration rate within a relatively narrow range: enough migration to allow recolonizations, but not too much migration to have an effect on local dynamics (Harrison, 1994b). As the patch networks in nature come in all shapes and sizes, it is clearly desirable to be able to relax this assumption. Finally, being a deterministic patch model (Gilpin and Hanski, this volume), the Levins model tacitly assumes a very large (effectively infinite) number of patches.

There is a clear analogy between the simplifying assumptions on which the Levins model is based and the corresponding assumptions of classical models in population ecology, such as the logistic equation. Classical population models are concerned with the total number (or density) of individuals in a population but neglect any differences among individuals (age, size, sex, etc.). To take these differences into account one has to turn to *structured* population models, which allow one to use information about individual behavior to draw conclusions about the dynamics of a population. The book by Metz and Diekmann (1986) presents a comprehensive introduction to the philosophy of using structured population models as well as a wealth of examples. More recently, Diekmann *et al.* (1993a,b, 1995a,b) have developed a slightly different approach to structured population models, which we apply in this chapter.

Our concept of a metapopulation is a population of populations (for alternative approaches, see Hanski and Simberloff, this volume; Harrison and Taylor, this volume; Hanski, 1996c; Hastings and Harrison, 1994). As pointed out by Diekmann *et al.* (1988, 1989), the theory of structured populations can be applied to metapopulations in a relatively straightforward manner if one makes the analogy between local populations and individuals and between local populations and metapopulations, respectively. In more detailed metapopulation models, where for instance dynamical changes in patch quality are included, one has to replace, in this analogy, local populations by some other kind of local entities.

One of the first structured metapopulation models was presented by Levin and Paine (1974, 1975). Their model was structured by the age and the size of a patch. Extinction was assumed to be age-dependent and size-dependent, but colonization (establishment of new populations) was not modeled explicitly and the effect of migration on local dynamics was not considered. Hastings and Wolin (1989) used a McKendrick-type model in which local populations are structured by age (time since colonization). They assumed that the size of a local population is a function of its age, and they could thus predict the size distribution of local populations. In this framework, it was not convenient to model the effect of migration on local dynamics, since migration does not affect the age of a population. Gyllenberg and Hanski (1992) chose local population size as the structuring variable and could thereby model explicitly the within-patch consequences of migration. Later they extended their model to account for variation in patch quality (Hanski and Gyllenberg, 1993). More recently, Val *et al.* (1995) have made a detailed analysis of the effect of migration upon local dynamics using similar structured models.

In this chapter we present a unified treatment of a large class of deterministic structured metapopulation models and illustrate the mathematical framework with several examples. Being deterministic, the models continue to assume an infinite number of patches and local populations, and the results are hence applicable to large metapopulations. Deterministic metapopulation models with a finite number of patches have been investigated by Levin (1974), Holt (1985), Davis and Howe (1992), Hastings (1993), Gyllenberg *et al.* (1993, 1996), Doebeli (1995), and many others. All these papers are concerned with the effect of migration on local dynamics, with a special focus on how migration may synchronize and stabilize local dynamics. On the other hand, these models ignore local extinctions and recolonizations (for stochastic models of finite metapopulations, see Gyllenberg and Silvestrov, 1994; Hanski, 1994a). It is practically impossible to incorporate the spatial arrangement of patches into deterministic structured models of the type treated in this chapter. Perhaps the most that can be done is to analyze the qualitative effects of spatial aggregation of habitat patches (Adler and Nürnberger, 1994).

This chapter has two parts. The first part (Sections I–V) gives a nonmathematical description of the basic principles of modelling structured populations and shows by examples the kind of results that can be obtained by such models. An empirical example illustrates the relevance of structured models. The mathematical formalism is outlined in Section VI. This approach was first used by Diekmann *et al.* (1993b, 1995b; see also Diekmann *et al.*, 1993a,c, 1995a) for "ordinary" structured populations.

II. MODELING STRUCTURED METAPOPULATIONS

We consider structured metapopulation dynamics as the study of the interrelation between processes at the local level and on the metapopulation level under the influence of the environment. We shall interpret "environment" in a wide sense and it will often be convenient to include for instance the fraction of empty patches among the environmental variables. Recall that in Levins's model the rate at which a local population gives rise to new local populations depends on the fraction of empty patches. The crucial point is that all (nonlinear) feedback takes place through the environment. By a "virgin" environment we understand an environment with no local populations and where the patch quality distribution has settled down to an equilibrium.

The most essential features of classical metapopulation dynamics are recurrent local extinctions and colonizations of empty patches. The Levins model is concerned with only these two processes and treats them directly at the level of the metapopulation, thus entirely ignoring local dynamics.

Modeling structured metapopulations and analyzing the models take place in three steps. First one has to model mechanisms at the local level, that is, at the level of patches and local populations. In the second step one lifts the model to the level of the metapopulation by simple book-keeping, and finally one studies population dynamical phenomena at this level.

Local dynamics may include both the dynamics of local populations and the dynamics of patches. Local populations grow or decline as a consequence of reproduction, death, emigration, and immigration. Patches may change in size and quality, be destroyed, and new patches may be created. In order to model these local processes one has to start by specifying the *basic local entity* corresponding to an individual in ordinary population dynamics and by choosing variables that adequately describe the local states. Here "adequate," of course, refers to quantities that affect processes like growth, migration, extinction, and colonization. If we, for instance, consider a metapopulation in a set of patches of dynamically changing quality a relevant choice of basic local entity would be a patch and its state would be described by the vector (x_1, x_2) where x_1 denotes the quality (e.g., resource density) of the patch and x_2 denotes the size of the local population inhabiting it. We call the set of all conceivable local states the *local state space* and denote it by Ω. In the example above Ω is (a subset of) the positive quadrant $\mathbf{R}_+^2$.

The following processes have to be modeled:

(i) patch quality dynamics, for instance how do resource density and patch area change with time;

(ii) local population growth;

(iii) extinction, patch destruction;

(iv) colonization, patch formation, and production of dispersers, that is, how many new basic local entities and with what local state at "birth" a given local entity will give rise to.

When modeling these four processes one has to describe how they depend on the local state and on the environmental state. Some of the processes, for instance the formation of new patches, may be the result of processes independent of the metapopulation. If that is the case, time enters the description of patch formation as an independent variable.

As the metapopulation affects its environment one has to close the loop by modeling the

(v) feedback mechanism.

Next we describe in some detail how the processes (i–v) should be modeled.

A. Colonization

Without recolonization of empty patches a metapopulation consisting of extinction-prone local populations certainly goes extinct. The foremost modeling task is therefore to prescribe the colonization process. To answer some simple but important questions like "will the metapopulation persist or go extinct?" a precise characterization of this process is sufficent.

The basic idea in the present approach is to model colonization by describing mathematically the expected cumulative number and structure of new local entities produced in the future by a given local entity whose present state is known and when the course of the environment is known. All this information is condensed into a mathematical object called the *colonization kernel* and rigorously defined in Section VI. To give an example, suppose that colonization is modeled as a two-step process: local populations produce dispersers which may colonize empty patches. Then the colonization kernel should contain at least the following information: Given a local population with a given state (e.g., size) at present time, it will on average produce so and so many dispersers in the future. Given a disperser having a given state at present it will on average colonize so and so many empty patches in the future. The "so and so" depend on the future development of the environmental state, which for the time being is considered to be known.

The populations on the patches colonized by a given local population will in turn give rise to new local populations that can be regarded as "second generation offspring" of the given ancestral population. By applying the colonization kernel to itself in a way to be made precise in Section VI, we obtain a mathematical description of this second generation or "grandchildren" to use an analogy with ordinary populations. This procedure can be repeated *ad infinitum* to obtain "great grandchildren," etc. Summing up over all generations we obtain what we shall call the *clan kernel*. The clan kernel Λ_E^c contains information about all "descendants" of a given ancestor population with respect to the time of colonization and state at colonization.

B. Patch Dynamics, Population Growth, and Extinction

The state of the basic local entity (e.g., local population, patch) changes during its lifetime because of population growth, changes in patch quality, and so on. We refer to all these processes by the common term *local state development*.

Our framework allows for stochastic local state development. As our second model ingredient we therefore choose the probability distribution of future local states of a local entity whose present state is known. Here we assume again that the future course of the environment is known. From this model ingredient many important quantities can be computed, for instance the *survival probability*, that is, the probability that a given local entity still exists at a given instant of time in the future.

The state of a local entity can move from its present state to a future state along different routes in the local state space. Assume that the local entity has a certain state before it reaches the final state. Using the aforementioned probability distribution one can compute the probability distribution of the final state, given that it had this intermediate state at a given time. Summing up over all intermediate states one should get the probability distribution of the final state, given the initial state. This leads to a consistency relation that the second model ingredient has to satisfy. The same argument also applies to colonization; hence we get a consistency relation combining the colonization kernel and the transition probabilities.

We emphasize that deterministic local development described by differential equations is included in the formalism as a special case. The consistency relation for the transition probability then reduces to the statement that the system of differential equations describing local development defines a dynamical system.

C. Combining Local Dynamics and Colonization

The colonization kernel introduced in Section II.A describes the cumulative number and structure *at time of colonization* of "descendants" of a given "ancestral" local entity. The new local entities will develop in time and ultimately we are interested in the structure of the whole metapopulation. As a first step in this direction we obtain a formula for the structure of all the "descendants" of a given "ancestor" by combining the colonization kernel with the transition probabilities of local development. Given a local entity with a given local state at a given time and given the course of the environmental state, this *derived* quantity yields the expected number of "descendants" with local state in a given set at a later time.

D. The Metapopulation Level

The state of the metapopulation is by definition the distribution of local states. In Section II.C we explained how one can derive an expression for the distribution of local states of all local entities descending from a given initial local entity. If the initial state of the metapopulation is known, that is, if the initial number of local entities and their local states are known, one obtains the state of the metapopulation at any future time simply by summing up the state distributions of all the clans stemming from the local entities present in the initial metapopulation. Thus lifting the model from the local level to the metapopulation level is a matter of straightforward book-keeping.

E. Feedback to the Environment

The model has been described so far under the assumption that the environmental state is a given function of time. In cases where the environment can be controlled for instance by the experimenter such a model fully decribes the time-evolution of the metapopulation. However, in natural metapopulations the local populations affect the environment and to obtain a complete model we have to specify the feedback mechanism. We do this by introducing a third ingredient giving the contribution to the environmental state of a basic local entity with a given state when the environment has a given state.

The solution to the full model including feedback to the environment is obtained in the following way. One starts by choosing an arbitrary (continuous) function of time to describe the development of the environmental state. One then keeps this function fixed while one constructs the operators giving the time-evolution of the structure of the metapopulation. Using the third model ingredient one calculates the course of the environment that this metapopulation gives rise to. The function so obtained is then used to construct the structure of the meta-population and the procedure is repeated *ad infinitum*. At each iteration one obtains a better approximation of the metapopulation structure.

III. STEADY STATES AND METAPOPULATION EXTINCTION

The most important questions one wants to answer using mathematical meta-population models are concerned with the long-term behavior of the metapopulation: Will the metapoulation persist or go extinct? Does the metapopulation structure tend to a steady state as time grows? What are the stability properties of the steady states? Can there be several alternative steady states, and if so, to which steady state will the metapopulation structure actually converge? The answers to these questions depend on the parameters of the model, which should reflect biologically relevant properties of the system under consideration. Pure mathematical analysis of the model can thus lead to major biological insights.

The traditional approach to modeling population dynamics is based on differential equations and the steady states are obtained by putting all time derivatives equal to zero and solving for the population state. The resulting system of equation can be very complicated and solving it is in many cases, if not impossible, at least a formidable task that does not give any insight whatsoever. One of the main advantages of the framework presented in this paper is that it allows us to derive important biological quantities more or less on the basis of their interpretation and not as a result of tedious formula manipulation. A clear example is the *basic reproduction number* which is of fundamental importance in connection with existence of steady states and questions concerning extinction and persistence.

The basic reproduction number $R(\bar{E})$ is the expected number of new local entities produced by one typical local entity during its lifetime, when the environmental state is kept at the constant value $\bar{E}$. It is intuitively clear that at equilibrium each local entity should exactly replace itself. We thus obtain the following necessary condition for a nontrivial steady state:

$$R(\bar{E}) = 1.$$

(A trivial steady state is a state with no local populations or local entities.)

By a *virgin* environment we understand the steady environment in the absence of local entities. Let E_0 denote the virgin environment and define

$$R_0 = R(E_0).$$

The basic reproduction number R_0 can be any nonnegative number. R_0 is the expected number of new local entities produced in a virgin environment by a typical local entity during its lifetime. If this number is greater than one, then the metapopulation will grow exponentially as long as it remains small. If, on the other hand, $R_0 < 1$, then the trivial steady state is locally stable and sufficiently small metapopulations will go extinct. However, extinction is not certain since there may exist a nontrivial attractor in addition to the trivial one.

The basic reproduction number as defined here is completely analogous to the basic reproduction ratio (sometimes called net reproductive rate) used in the context of models for infectious diseases (Diekmann *et al.*, 1990; Heesterbeek, 1992). For a definition and discussion of the basic reproduction number within the traditional approach to structured (meta)populations we refer to Diekmann (1993).

IV. EXAMPLES

In this section we give some examples which illustrate the use of structured metapopulation models. A mathematically detailed analysis of these examples is given in Section VI.

A. The Levins Model

The unstructured Levins model (1.1) is so simple that an explicit solution can immediately be written down using elementary methods. It is, however, instructive to consider it as a special case of a structured model and use it to illustrate the concepts introduced in the preceding sections.

In the Levins model (1.1) extinction is modeled by prescribing a constant hazard rate μ, which means that the lifetime of a local population is exponentially distributed with parameter μ. The expected lifetime of a local population is thus $1/\mu$. In a virgin environment all patches are empty and the colonization rate is β. It follows from the interpretation of the parameters that in a virgin environment the expected number of new local populations produced by one local poulation during its lifetime equals

$$R_0 = \frac{\beta}{\mu}.$$

We thus arrive at the well-known threshold condition

$$R_0 = \frac{\beta}{\mu} > 1$$

for metapopulation persistence.

B. A Simple Structured Model

The model presented in this section is essentially due to Hanski (1985) but we have reformulated it in terms of the general framework described in this chapter. We consider a fixed (but large) number of patches of the same quality. The basic local entity is the local population, which can be in any of two states x_1 and x_2, where x_1 corresponds to a small population and x_2 to a large population.

A state transition from x_1 to x_2 can be due either to population growth or to immigration of individuals from other populations. We thus explicitly assume that immigration affects local dynamics. The state transition from x_2 to x_1 is due to the effect of environmental stochasticity. Both small and large populations can go extinct as a consequence of a local disaster. The metapopulation state can be represented by a vector (m_1, m_2) with m_i denoting the fraction of patches with local population of size x_i, $(i = 1, 2)$.

We assume that large populations usually produce more dispersers and thus exert a higher colonization and immigration pressure than small ones. To model this we assume that the transition rate from x_1 to x_2 equals $\gamma_{12} + \alpha_1 m_1 + \alpha_2 m_2$, where γ_{12} describes intrinsic growth of local populations and $\alpha_1 m_1 + \alpha_2 m_2$ the effect of immigration. Here α_1 and α_2 are nonnegative weights.

The hazard rate of local extinction is μ_1 in x_1 and μ_2 in x_2 and the rate of partial disaster bringing a population from x_2 to x_1 is γ_{21}. A population in state x_i is expected to colonize empty patches at a rate $\beta_i(1 - (m_1 + m_2))$. Note that $(1 - (m_1 + m_2))$ is the fraction of empty patches. New populations are small; they have state x_1.

From the interpretation of the model parameters it is clear that in a virgin environment (all patches empty) the expected number of local populations produced by a local population during its sojourn in x_1 is

$$\frac{\beta_1}{\mu_1 + \gamma_{12}} \tag{4.1}$$

and during its sojourn in x_2 it is

$$\frac{\beta_2}{\mu_2 + \gamma_{21}}. \tag{4.2}$$

The probability that the transition from x_1 to x_2 occurs before extinction is

$$\frac{\gamma_{12}}{\mu_1 + \gamma_{12}} \tag{4.3}$$

and the corresponding probability for the transition from x_2 to x_1 is

$$\frac{\gamma_{21}}{\mu_2 + \gamma_{21}}. \tag{4.4}$$

Elementary probability considerations lead to the following expression for the expected number of new local populations produced by one local population placed into a virgin environment:

$$R_0 = \frac{1}{1 - (\gamma_{12}/(\mu_1 + \gamma_{12}))(\gamma_{21}/(\mu_2 + \gamma_{21}))} \frac{\beta_1}{\mu_1 + \gamma_{12}}$$

$$+ \frac{(\gamma_{12}/(\mu_1 + \gamma_{12}))}{1 - (\gamma_{12}/(\mu_1 + \gamma_{12}))(\gamma_{21}/(\mu_2 + \gamma_{21}))} \frac{\beta_2}{\mu_2 + \gamma_{21}}$$

$$= \frac{(\mu_2 + \gamma_{21})\,\beta_1 + \gamma_{12}\beta_2}{(\mu_1 + \gamma_{12})(\mu_2 + \gamma_{21}) - \gamma_{12}\gamma_{21}}$$

$$= \frac{(1 + \gamma_{21}/\mu_2)(\beta_1/\mu_1) + (\gamma_{12}/\mu_1)(\beta_2/\mu_2)}{1 + \gamma_{21}/\mu_2 + \gamma_{12}/\mu_1}.$$

The value of R_0 thus depends essentially on four parameter combinations, namely β_1/μ_1, β_2/μ_2, γ_{12}/μ_1, and γ_{21}/μ_2. R_0 is obviously increasing with β_1/μ_1 and β_2/μ_2. For most natural metapopulations one would assume that $\beta_2/\mu_2 > \beta_1/\mu_1$; that is, larger populations are less vulnerable to extinction and exert a greater colonization pressure on empty patches than small populations, and then R_0 is also increasing in γ_{12} but decreasing in γ_{21}. However, if $\beta_2/\mu_2 < \beta_1\mu_1$, then R_0 is decreasing in γ_{12} and increasing in γ_{21}. This result agrees with biological intuition. We emphasize that the approach employed in this paper allowed us to find the value of R_0 and in particular the threshold criterion $R_0 > 1$ for metapopulation persistence directly from the biological interpretation of the parameters. To arrive at the same result using the classical approach based on differential equations (Hanski, 1985) one would have to calculate all the eigenvalues of a matrix and determine the largest of them. The model treated by Hanski (1985) is only two-dimensional and the eigenvalues can be readily calculated but for higher dimensional systems the task is extremely tedious and sometimes even impossible. We point out that if $\beta_2/\mu_2 = \beta_1/\mu_1$, that is, if there is no difference in colonization capacity between the two size classes, then the threshold criterion $R_0 > 1$ reduces to the usual condition $\beta/\mu > 1$ for the Levins model.

The nontrivial steady states can also be found by this approach. The details are given in Section VI.C.2. Hanski (1985; see also Hastings, 1991) found that for certain parameter values there are two nontrivial steady states. This simple structured model thus predicts a fundamentally different qualitative behavior than Levins's model.

C. Models with Continuous Local State

In the previous section we analyzed a model in which a local population could be in two different states: "small" or "large." In many cases it is biologically more realistic to assume the local state to be a continuous variable representing for instance the size or age of a local population. Such models have been con-

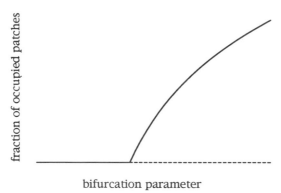

FIGURE 1   Bifurcation diagram in the case where the impact of migration on local dynamics is small. Stable equilibria are shown by continuous line, unstable by broken line.

structed and analyzed by Hastings and Wolin (1989), Gyllenberg and Hanski (1992), Hanski and Gyllenberg (1993), and Val *et al.* (1995).

In Section VI.C.3 we give a detailed description of a structured metapopulation model with continuous local state variable. Here we shall only briefly indicate what kind of behavior such a model can exhibit.

From the parameters of the structured model one can derive a quantity that measures the impact of migration upon local dynamics. If this impact is small, the model gives qualitatively the same prediction as the Levins model (where the impact is zero), but if it is sufficiently large, then there are multiple nontrivial steady states for certain values of the colonization parameter. Figures 1–3 show a sample of bifurcation diagrams obtained from structured metapopulation models with continuous local state. In these diagrams, the dependent variable is the frac-

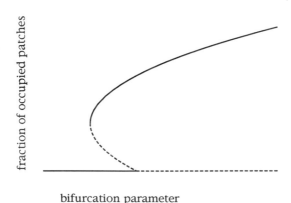

FIGURE 2   Bifurcation diagram in the case where the impact of migration on local dynamics is large. Stable equilibria are shown by continuous line, unstable by broken line.

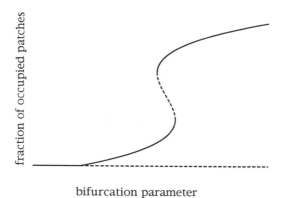

FIGURE 3 Bifurcation diagram with large differences in patch quality and large impact of migration on local dynamics (Hanski and Gyllenberg, 1993). Stable equilibria are shown by continuous line, unstable by broken line.

tion of occupied habitat patches (the size of the metapopulation), which is shown as a function of the colonization parameter. The lines in the figure give the model steady states. Note that for some values of the colonization parameter there is only one steady state, which is then necessarily stable. For other values, there are two stable states separated by an unstable state. In these cases the model has multiple steady states.

D. An Empirical Example

An example strongly suggesting the kind of bifurcation pattern shown in Figs. 2 and 3 and hence alternative stable equilibria has been recently described for the butterfly *Melitaea cinxia* (Hanski *et al.*, 1995b). In Finland, this butterfly occurs on the Åland island in the Baltic, in a very large network of more than 1500 habitat patches (dry meadows) within an area of 3500 km² (see Fig. 1 in Hanski, this volume; for further ecological details see also Thomas and Hanski, this volume; Hanski *et al.*, 1995a).

Hanski *et al.* (1995b) tested the predicted bifurcation pattern by dividing the material consisting of 524 local populations in 1530 habitat patches into 65 semi-independent patch networks with weak interaction among the networks (for details, see Hanski *et al.*, 1995b, 1996c). These networks vary in the number, sizes, and density of habitat patches. They used the occupied fraction of the pooled patch area in a network, denoted by P_A, as the dependent variable, and the potential colonization rate β as the bifurcation parameter. The latter was measured as

$$\left(\frac{1}{n} \sum_{j=1}^{n} \sum_{i=1}^{n} e^{-d_{ij}} \sqrt{A_j} \right)^{1/2},$$

where d_{ij} is the distance between patches i and j in km, A_j is the area of patch j in ha (square root transformation was used to scale patch area to population size; Hanski *et al.* 1996c), and n is the number of patches. β thus defined is the average value of the expected immigration rate to the patches, calculated on the assumption that all patches are occupied (because β has to measure the potential, not actual, rate of colonization). The expected immigration rate depends on the sizes and distances of local populations from the focal habitat patch (Hanski, 1994a). The square root transformation in calculating β is used to spread the data points more evenly along the *x*-axis in Fig. 4. This definition of β simply says that with larger and less isolated habitat patches within a region the potential colonization rate is higher, which accords with direct mark–recapture results on butterfly movements (Hanski *et al.*, 1994).

The observed relationship between β and P_A (Fig. 4) resembles greatly the predicted bifurcation pattern (Fig. 2). Note that all networks with large β were practically fully occupied, and that P_A increased with β in the upper "branch" in Fig. 4, as predicted by the model. The distribution of the P_A values is strikingly bimodal (Fig. 4, Hanski *et al.*, 1995b), supporting the hypothesis of alternative stable equilibria. The existing metapopulations with less than ca 70% of the habitat occupied are predicted to be in transit toward one of the two stable equilibria (Fig. 4). Infrequent long-distance migration prevents permanent metapopulation

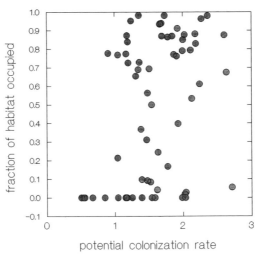

FIGURE 4   Empirical result for 65 semi-independent patch networks with at least 5 patches, all isolated by at least 1 km from other patches. These networks vary in the number (5 . . . 95), sizes (average area 0.02 . . . 0.56 ha) and density of patches (total area covered by the network 1.9 . . . 61 km^2). Each point on the graph refers to one network. The *y*-axis gives the fraction of pooled area in the network that was occupied, and the *x*-axis has a measure of potential colonization rate (see text; more details in Hanski *et al.*, 1995b).

extinction in these semi-isolated patch networks (see Fig. 1 in Hanski, this volume).

V. DISCUSSION

Intuitively, understanding the role of local population dynamics (other than just local extinction) in metapopulation dynamics requires the use of a structured metapopulation model. For example, how does the rate of growth of local populations (relative to the time scale of migration) affect metapopulation persistence? Virtually all questions related to genetic structure require the use of structured metapopulation models (Hastings and Harrison, 1994). Ultimately, the utility of structured metapopulation models will be determined by whether they provide "better" answers to these and other biological questions than the unstructured models.

In this paper we have presented a framework, originally brought forth by Diekmann *et al.* (1993b, 1995b; see also Diekmann *et al.,* 1993ac, 1995a), that allows us to treat a large variety of structured metapopulation models in a unified fashion. Within this framework, we have shown how to find steady states and how to answer fundamental biological questions concerning persistence and extinction of metapopulations. Here the notion of basic reproduction number plays a crucial role. It was defined in a purely mathematical way, but its clear and intuitive biological interpretation as the expected number of new local entities produced by a typical local entity during its lifetime made it possible to find steady states and criteria for persistence directly on the basis of the biological interpretation of the model parameters. This is a clear advantage compared with the traditional approach to structured metapopulation models as reviewed for instance by Hastings and Harrison (1994).

The most fundamental difference between the unstructured and structured metapopulation models is that with the structured models there can be multiple equilibria. The possible presence of multiple equilibria suggests that caution is required when using metapopulation models to make inferences for conservation. Simple metapopulation models have been used to make arguments about the effect of habitat destruction on the persistence of species (Tilman *et al.,* 1994). Yet, these arguments are usually based on models with a single stable equilibrium. As we have shown, structured metapopulation models can have a much more complex bifurcation structure with multiple equilibria (Figs. 2 and 3). What this means in practice is that unlike the simpler models where equilibria change smoothly as the environment changes, the structured metapopulation models can exhibit more drastic threshold effects: small changes in the environment can cause potentially large changes in the equilibrium levels of the species. The butterfly example which we described in Section IV.D strongly suggests that this is not idle speculation; multiple equilibria are likely to occur in real metapopulations with substantial migration between populations.

We have shown how the simple Levins model can be expressed within the framework we have developed here. Thus, it is not the framework, but different biological assumptions which lead to the presence of multiple equilibria and thresholds. With this potentially large difference in the behavior of the models, can we make any generalizations concerning the circumstances that lead to thresholds? Using the results developed here and elsewhere (Hanski, 1985; Gyllenberg and Hanski, 1992; Hanski and Zhang, 1993; Val *et al.*, 1995), it appears that the key ingredient for the presence of multiple equilibria is that immigrants affect the local dynamics and extinction of existing populations. From the perspective of individual fitness, multiple equilibria at the metapopulation level are generated by immigrants to empty patches or very small populations having smaller long-term fitness than immigrants arriving at somewhat larger populations (Gyllenberg and Hanski, 1992). At very high local densities, negative density dependence may again reduce fitness.

We conclude by emphasizing that conclusions about the role of habitat fragmentation and destruction on ecological communities (e.g., Tilman *et al.*, 1994) should be based on the framework of structured metapopulation models. As the amount of suitable habitat is reduced, the system may undergo a bifurcation removing the equilibrium which allows the persistence of the species. The development of these more complex models does lead to one major problem when one wants to apply the models to answer practical questions in conservation—there is a proliferation of parameters. This is a problem for which we do not have a general answer.

VI. THE MATHEMATICAL FORMALISM

A. Modeling

1. Colonization

Consider a basic local entity which at time t has state $x \in \Omega$. Suppose that s time units later (that is, at real time $t + s$) it gives rise to a new local entity of state $y \in \Omega$. We then call $(s, y) \in \mathbf{R}_+ \times \Omega$ the *colonization coordinates* of this local entity. We now define the *colonization kernel* Λ as follows: For each $(t, x) \in \mathbf{R}_+ \times \Omega$ and each subset A of $\mathbf{R} \times \Omega$.

$\Lambda_E(t, x)(A) =$ expected number of new local entities with colonization coordinates in A produced by a local entity which at time t had state x, given the course of the environment E.

This definition is perhaps most easily understood when we take $A = [0, s) \times \omega$ with $\omega \subset \Omega$. $\Lambda_E(t, x)([0, s) \times \omega)$ gives the expected *cumulative* number of new local entities with state at colonization in ω produced by a local entity with state x during s units of time since the reference time t. Notice that we do not condition on survival of the "ancestor" entity and thus Λ_E implicitly takes extinction into account. Λ_E is the basic ingredient of our general structured meta-population model.

The populations on the patches colonized by a given local population will in turn give rise to new local populations that can be regarded as "second generation offspring" of the given ancestor. It therefore makes sense to define $\Lambda_E^k(t, x)(A)$ exactly as $\Lambda_E(t, x)(A)$ but for the kth generation rather than for direct (first generation) "offspring." Since the $(k + 1)$th generation offspring are direct offspring of the kth generation, consistency requires that

$$\Lambda_E^{k+1}(t, x)(A) = \int_{\mathbf{R}_+ \times \Omega} \Lambda(t + \tau, \xi)(A_{-\tau}) \Lambda_E^k(t, x)(d\tau \times d\xi), \quad (6.1)$$

where

$$A_{-\tau} = \{(\sigma, \xi) \in \mathbf{R}_+ \times \Omega : (\sigma + \tau, \xi) \in A\}.$$

Since the time coordinate in A measures time relative to time t when the ancestor had state x, this translation in the time direction is necessary.

Formula (6.1) shows that one obtains Λ_E^k by iteration from Λ_E. Regarding the right-hand side of (6.1) as the definition of a "product" of the kernels Λ_E and Λ_E^k, Λ_E^k is indeed the kth power of Λ_E. Summing over all "generations," we define the *clan kernel*

$$\Lambda_E^c = \sum_{k=1}^{\infty} \Lambda_E^k. \quad (6.2)$$

2. Local State Development

We model local state development and survival of a basic local entity as a Markov process with extinction and/or patch destruction as transition to an absorbing state. As the second model ingredient we therefore choose the transition probabilities $u_E(t, x; s)(\omega)$ of this process. Specifically,

$u_E(t, x; s)(\omega) =$ probability that a basic local entity which has state x at
time t still exists s time units later and then has a state in
$\omega \subset \Omega$, given the course of the environment E.

The case of deterministic local development is included as a special case. In this case $u_E(t, x; s)$ is a measure concentrated at the point $X(t, x; s)$ satisfying an ordinary differential equation with initial condition

$$\frac{\partial}{\partial s} X(t, x; s) = g(X(t, x; s), E(s)),$$

$$X(t, x; t) = x.$$

Here $g(x, e)$ is the rate at which the local state changes when the local state is x and the environmental state has the value e.

The total probability mass $\mathcal{F}(t, x; s) := u_E(t, x; s)(\Omega)$ is the *survival prob-*

ability. It satisfies the initial value problem

$$\frac{\partial}{\partial s} \mathscr{F}(t, x; s) = -\mu(X(t, x; s), E(s)) \mathscr{F}(t, x; s),$$

$$\mathscr{F}(t, x; t) = 1,$$

where $\mu(x, e)$ is the extinction rate.

Being transition probabilities of a Markov process the measures $u_E(t, x; s)$ have to satisfy the Chapman–Kolmogorov equation

$$u_E(t, x; s)(\omega) = \int_\Omega u_E(t + \sigma, \xi; s - \sigma)(\omega) \, u_E(t, x; \sigma)(d\xi) \tag{6.3}$$

for all t, all $0 \leq \sigma \leq s$, and all $\omega \subset \Omega$. Equation (6.3) is a consistency relation. It has the following interpretation: Consider a basic local entity with state x and assume it has state ξ s time units later. The probability that its state belongs to ω at time $t + s$ is $u_E(t + \sigma, \xi; s - \sigma)(\omega)$. Summing up over all possible intermediate states ξ one has to get the probability of transition from x to ω as time elapses from t to $t + s$. Equation (6.3) expresses exactly this. The same argument holds for colonization, too, and therefore we get another consistency relation combining u and Λ:

$$\Lambda_E(t, x)(A) = \Lambda(t, x)(A \cap ([0, s) \times \Omega))$$

$$+ \int_\Omega \Lambda_E(t + s, \xi)(A_{-s})u_E(t, x; s)(d\xi). \tag{6.4}$$

3. Combining Local Dynamics and Colonization

Consider a basic local entity with state x at time t. We describe by $u_E^c(t, x; s)$ the expected size and structure of the whole clan of basic local entities descending from the given ancestor and including the ancestor in the clan. We thus *define*

$$u_E^c(t, x; s)(\omega) = u_E(t, x; s)(\omega)$$

$$+ \int_{[0, s) \times \Omega} u_E(t + \sigma, \xi, s - \xi)(\omega) \Lambda_E^c(t, x)(d\sigma \times d\xi). \tag{6.5}$$

The first term on the right-hand side of (6.5) gives the probability that the ancestor's state is in ω at time $t + s$ and the latter term describes the expected structure of descendants at time $t + s$. We emphasize that Λ_E^c was *constructed* from Λ_E in Eq. (6.2) so u_E^c defined by (6.5) can be regarded as constructed from our basic ingredients Λ_E and u_E.

It is clear from the interpretation that u_E^c should satisfy the Chapman–Kolmogorov equation and that Λ_E^c and u_E^c have to satisfy a corresponding consistency relation. And indeed, Diekmann *et al.* (1995b) proved that Eqs. (6.3) and (6.4) hold with u_E and Λ_E replaced by u_E^c and Λ_E^c, respectively.

4. The Metapopulation Level

The state of the metapopulation is by definition the distribution of local states. More precisely, the metapopulation state is a measure m on the local state space Ω such that for every $\omega \subset \Omega$, $m(\omega)$ gives the (expected) number of basic local entities with state in the set ω. If for instance the local entity is a patch and the local state is a vector (x_1, x_2), where x_1 denotes patch quality and x_2 denotes the size of the local population inhabiting the patch, then $m([\xi_1, \eta_1] \times [\xi_2, \eta_2])$ is the number of all patches with quality in the range $\xi_1 \leq x_1 \leq \eta_1$ supporting local populations with size in the range $\xi_2 \leq x_2 \leq \eta_2$.

We now define the *colonization operators* V_E and V_E^c and the *next state operators* U_E and U_E^c as follows:

$$(V_E(t + \tau, \tau)m)(\omega) = \int_\Omega \Lambda_E(\tau, x)([0, t) \times \omega)\, m(dx), \qquad (6.6)$$

$$(V_E^c(t + \tau, \tau)m)(\omega) = \int_\Omega \Lambda_E^c(\tau, x)([0, t) \times \omega)\, m(dx), \qquad (6.7)$$

$$(U_E(t + \tau, \tau)m)(\omega) = \int_\Omega u_E(\tau, \xi; t)(\omega)\, m(d\xi), \qquad (6.8)$$

$$(U_E^c(t + \tau, \tau)m)(\omega) = \int_\Omega u_E^c(\tau, \xi; t)(\omega)\, m(d\xi). \qquad (6.9)$$

In these formulae we assume that the state m of the metapopulation is given at time τ. The measure $V_E(t + \tau, \tau)m$ gives the expected *cumulative* amount and distribution with respect to state at colonization of new local entities produced by this initial metapopulation in the time interval $[\tau, \tau + t)$. $V_E^c(t + \tau, \tau)m$ has the same interpretation but it takes the whole clan into account. $U_E(t + \tau, \tau)$ tells us where the local entities of the initial metapopulation "have moved in Ω" during t time units since the initial time τ. $U_E^c(t + \tau, \tau)m$ is the solution of the metapopulation model. It represents the expected state of the metapopulation at time $t + \tau$ given the state m at time τ.

If the environment E is a constant function $\bar{E}$ of time, then both $\Lambda_{\bar{E}}(t, x)$ and $u_{\bar{E}}(t, x; s)$ are independent of t. It follows that the same is true of $\Lambda_{\bar{E}}^c(t, x)$ and $u_{\bar{E}}^c(t, x; s)$ and hence by (6.6)–(6.9) also of $V_{\bar{E}}(t + s, t)$, $V_{\bar{E}}^c(t + s, t)$, $U_{\bar{E}}(t + s, t)$, and $U_{\bar{E}}^c(t + s, t)$. In particular, the one parameter families

$$T_{\bar{E}}(s) = U_{\bar{E}}(t + s, t), \qquad (6.10)$$

$$T_{\bar{E}}^c(s) = U_{\bar{E}}^c(t + s, t), \qquad (6.11)$$

$$W_{\bar{E}}(s) = V_{\bar{E}}(t + s, t), \qquad (6.12)$$

$$W_{\bar{E}}^c(s) = V_{\bar{E}}^c(t + s, t), \qquad (6.13)$$

are well-defined and give a complete description of the time evolution of the metapopulation in the special case of a constant environment.

5. Feedback and the Full Model

We denote by $\gamma(x, e)$ the contribution to the environmental state of a basic local entity with state x when the environment has state e. Consider a metapopulation with state m at time 0. As explained in Section VI.A.4, it will have state

$$\int_\Omega u_E^c(0, \xi; t)(\cdot)m(d\xi)$$

at time t. Adding all the contributions of all the local entities we obtain the total contribution

$$\int_\Omega \gamma(x, E(t)) \int_\Omega u_E^c(0, \xi; t)(dx)m(d\xi),$$

which is fed to the environment through a possibly nonlinear function F. We thus arrive at the following equation:

$$E(t) = F\left(\int_\Omega \gamma(x, E(t)) \int_\Omega u_E^c(0, \xi; t)(dx)m(d\xi) \right). \tag{6.14}$$

Let us recapitulate the situation. The *ingredients* of the model are Λ_E, u_E, and γ, where Λ_E and u_E satisfy the consistency relations (6.3) and (6.4). The *model* consists of Eqs. (6.2), (6.5) and (6.14), which we repeat here:

$$\Lambda_E^c = \sum_{k=1}^{\infty} \Lambda_E^k, \tag{6.15}$$

$$u_E^c(t, x; s)(\omega) = u_E(t, x; s)(\omega)$$

$$+ \int_{[0, s)\times\Omega} u_E(t + \sigma, \xi, s - \xi)(\omega) \Lambda_E^c(t, x)(d\sigma \times d\xi), \tag{6.16}$$

$$E(t) = F\left(\int_\Omega \gamma(x, E(t)) \int_\Omega u_E^c(0, \xi; t)(dx)m(d\xi) \right). \tag{6.17}$$

In the model (6.15)–(6.17) m is the given *initial state* (at time 0) of the metapopulation. By a *solution* of the initial value problem (6.15)–(6.17) we understand

$$S(t)m := U_E^c(t, 0)m = \int_\Omega u_E^c(0, \xi; t)(\cdot)m(d\xi), \tag{6.18}$$

where E solves (6.15)–(6.17). In practice the solution is found by the method of *successive approximations:* One starts by guessing the course E of the environment. Using this initial guess one constructs Λ_E^c according to (6.15) and substitutes it into the right-hand side of (6.16). The u_E^c obtained in this way is substituted into (6.17) and one considers this relation as the definition of a new approximation of the environment E. Repeating this procedure over and over again one gets a sequence of environmental functions which under suitable conditions converges to some function E. The solution of the full model is then given by (6.18) with the limit function E on the right-hand side.

B. Steady States

A *steady state* is by definition a solution that does not change with time. So a steady state consists of a *constant* function $\bar{E}$ satisfying (6.15)–(6.17) and a metapopulation state m such that

$$S(t)m = m \text{ for all } t \geq 0.$$

where S is defined by (6.18).

At first sight it seems like determining the steady states would be a difficult task in the present approach as compared with the traditional approach based on differential equations, where the steady-state condition is at least formally the simple $Am = 0$ for some nonlinear operator A. Nonetheless, we show that appearance is deceptive, and that the present approach is close to biology and allows for intuitive and helpful interpretations.

First we notice that a solution with $m = 0$ for all $t \geq 0$ is always a steady state. To see this simply define $\bar{E} = F(0)$. Then obviously $\bar{E}$ satisfies (6.15)–(6.17) and $S(t)0 = 0$. This steady state is called trivial, and it corresponds to metapopulation extinction. Next we consider nontrivial steady states. We assume that the environment is kept constant at $\bar{E}$, that is $E(t) = \bar{E}$ for all $t \geq 0$. The operator $W_{\bar{E}}(\infty)$, which according to (6.12) is given by

$$(W_{\bar{E}}(\infty)m)(\omega) = \int_{\Omega} \Lambda_{\bar{E}}(x)([0, \infty) \times \omega)m(dx), \tag{6.19}$$

contains all information about the state at colonization of the entire next "generation" produced by the initial metapopulation m. Intuitively it is clear that at a nontrivial equilibrium each local entity should on average exactly replace itself. Following Diekmann *et al.* (1995b), we make this idea precise by defining the *basic reproduction number* $R(\bar{E})$ as the spectral radius of the operator $W_{\bar{E}}(\infty)$. Positivity arguments that are usually biologically self-evident ensure that $R(\bar{E})$ is an eigenvalue of $W_{\bar{E}}(\infty)$ which is simple.

The eigenvector corresponding to the eigenvalue $R(\bar{E})$ is denoted by $b_{\bar{E}}$ and it gives the stable distribution of states at colonization. $R(\bar{E})$ is the expected number of new local entities produced by a "typical" (that is, sampled randomly from $b_{\bar{E}}$) local entity during its lifetime. We thus arrive at the following necessary condition for a nontrivial steady state:

$$R(\bar{E}) = 1. \tag{6.20}$$

Let $\bar{E}$ satisfy (6.20) and consider at a certain instant of time a collection of newborn local entities distributed according to $b_{\bar{E}}$. τ time units later their state distribution will be $T_{\bar{E}}(\tau)b_{\bar{E}}$. Notice that since the environment is now constant we use the time invariant version (6.10) of the next state operator. Since the distribution $b_{\bar{E}}$ of states at colonization does not change with time the measure $\bar{m}$ defined by

$$\bar{m} = \int_0^\infty T_{\bar{E}}(\tau)\, b_{\bar{E}}d\tau \tag{6.21}$$

satisfies

$$S(t)m = T_{\bar{E}}^c(t)\bar{m} \text{ for all } t \geq 0.$$

For a rigorous proof of this result we again refer to Diekmann *et al.* (1995b). So the pair $(m, \bar{E})$ is a steady state provided Eq. (6.17) holds. Taking (6.21) into account we see that the condition takes the form

$$\bar{E} = F\left(\int_\Omega \gamma(x, \bar{E}) \int_0^\infty T_{\bar{E}}(\tau)b_{\bar{E}}d\tau)(dx)\right). \tag{6.22}$$

Being an eigenvector corresponding to a simple eigenvalue, $b_{\bar{E}}$ is of course determined only up to an arbitrary multiplicative constant. If this constant can be chosen such that (6.22) holds, then the solution $\bar{E}$ of (6.20) and the corresponding $\bar{m}$ defined by (6.21) indeed form a steady state. If F is *linear* the arbitrary constant can always be adjusted such that (6.22) holds so in that case (6.20) is also a sufficient condition for a steady state. However, in general the equilibrium environment has to be determined from both (6.20) and (6.21).

C. Examples

1. The Levins Model

In the Levins model the basic local entity is the local population, and since the model is unstructured all populations have the same state. The local state space Ω thus consists of a single point. In particular, Λ and u do not depend on x. Since the colonization rate depends on the fraction of empty patches, we choose this fraction as the environmental state and denote it by $E(t)$. The expected number of new local populations (divided by the total number of patches) produced by a population extant at time t during the time interval $[t, t + s)$ given the course of the environment is thus

$$\Lambda_{\bar{E}}(t)([0, s)) = \int_t^{t+s} \beta E(\tau)e^{-\mu\tau}d\tau.$$

For time-independent environments the corresponding operator $W_{\bar{E}}(s)$ is simply the number

$$W_{\bar{E}}(s) = \beta\bar{E} \int_0^s e^{-\mu\tau}d\tau = \frac{\beta}{\mu}\bar{E}(1 - e^{-\mu s})$$

considered as a linear operator on the one-dimensional space **R**. In particular,

$$R(\bar{E}) = W_{\bar{E}}(\infty) = \frac{\beta}{\mu}\bar{E}.$$

The virgin environment is $\bar{E} = 1$; that is, all patches are empty, and we arrive at the well-known threshold condition

$$R_0 = \frac{\beta}{\mu} > 1 \tag{6.23}$$

for metapopulation persistence. When (6.23) holds, the unique nontrivial equilibrium is obtained from

$$R(\bar{E}) = \frac{\beta}{\mu}\bar{E} = 1.$$

2. A Simple Structured Model

In this section we show in detail how the model treated in Section IV.B can be put into the present framework. We use the same notation as in Section IV.B.

The metapopulation state is a measure m on the local state space $\Omega = \{x_1, x_2\}$ which can be represented by a vector (m_1, m_2) with m_i denoting the fraction of patches with local population of size x_i, $(i = 1, 2)$.

We choose

$$E_1 = \alpha_1 m_1 + \alpha_2 m_2 \tag{6.24}$$

and

$$E_2 = 1 - (m_1 + m_2) \tag{6.25}$$

as environmental variables. Note that E_1 describes the effect of immigration upon local population growth and that E_2 is simply the fraction of empty patches. The transition rate from x_1 to x_2 is thus $\gamma_{12} + E_1$ and a population in state x_i is expected to colonize empty patches at a rate $\beta_i E_2$.

Using these environmental variables and the assumption stated in Section IV.B it is a straightforward exercise to write down explicit expressions for Λ_E and u_E. Since we are mainly interested in metapopulation extinction and steady states we shall only calculate $\Lambda_{\bar{E}} = ([0, \infty) \times \omega)$ for constant environments $\bar{E} = (\bar{E}_1, \bar{E}_2)$.

The expected number of local populations produced by a local population during its sojourn in x_1 is

$$\frac{\beta_1 \bar{E}_2}{\mu_1 + \gamma_{12} + \bar{E}_1} \tag{6.26}$$

and during its sojourn in x_2 it is

$$\frac{\beta_2 \bar{E}_2}{\mu_2 + \gamma_{21}}. \tag{6.27}$$

The probability that the transition from x_1 to x_2 occurs before extinction is

$$p_{\bar{E}1} = \frac{\gamma_{12} + \bar{E}_1}{\mu_1 + \gamma_{12} + \bar{E}_1} \tag{6.28}$$

and the corresponding probability for the transition from x_2 to x_1 is

$$q = \frac{\gamma_{21}}{\mu_2 + \gamma_{21}}. \tag{6.29}$$

Consider a local population which initially (at time 0) has state x_1. The probability that it will enter state x_2 exactly n times is

$$p_{\bar{E}_1}^n q^{n-1}(1 - p_{\bar{E}_1}q) \tag{6.30}$$

and the expected number of entries in x_2 is thus

$$\frac{p_{\bar{E}_1}}{1 - p_{\bar{E}_1}q}. \tag{6.31}$$

Similarly, the expected number of sojourns in state x_1 is

$$\frac{1}{1 - p_{\bar{E}_1}q}. \tag{6.32}$$

Combining (6.26), (6.27), (6.31) and (6.32) one finds the expected number of new populations produced by a local population initially in state x_1 during its entire lifetime. Since all new populations are small (have state x_1) $\Lambda_{\bar{E}}(0, x_1)$ can conveniently be represented as a vector with zero second component:

$$\Lambda_{\bar{E}}(0, x_1)([0, \infty) \times \cdot)$$

$$= \begin{pmatrix} \dfrac{1}{1 - p_{\bar{E}_1}q} \dfrac{\beta_1 \bar{E}_2}{\mu_1 + \gamma_{12} + \bar{E}_1} + \dfrac{p_{\bar{E}_1}}{1 - p_{\bar{E}_1}q} \dfrac{\beta_2 \bar{E}_2}{\mu_2 + \gamma_{21}} \\ 0 \end{pmatrix}. \tag{6.33}$$

In a completely analogous way we obtain

$$\Lambda_{\bar{E}}(0, x_2)([0, \infty) \times \cdot)$$

$$= \begin{pmatrix} \dfrac{q}{1 - p_{\bar{E}_1}q} \dfrac{\beta_1 \bar{E}_2}{\mu_1 + \gamma_{12} + \bar{E}_1} + \dfrac{1}{1 - p_{\bar{E}_1}} \dfrac{\beta_2 \bar{E}_2}{\mu_2 + \gamma_{21}} \\ 0 \end{pmatrix}. \tag{6.34}$$

The corresponding operator $W_{\bar{E}}(\infty)$ at the metapopulation level can thus be represented by a 2 by 2 matrix with the vectors (6.33) and (6.34) as columns. The spectral radius (largest eigenvalue) of $W_{\bar{E}}(\infty)$ is

$$R(\bar{E}) = \frac{1}{1 - p_{\bar{E}_1}q} \frac{\beta_1 \bar{E}_2}{\mu_1 + \gamma_{12} + \bar{E}_1} + \frac{p_{\bar{E}_1}}{1 - p_{\bar{E}_1}q} \frac{\beta_2 \bar{E}_2}{\mu_2 + \gamma_{21}}. \tag{6.35}$$

The virgin environment corresponding to no local populations is given by $\bar{E}_1 = 0$, $\bar{E}_2 = 1$ and thus

$$
\begin{aligned}
R_0 &= \frac{1}{1 - (\gamma_{12}/(\mu_1 + \gamma_{12}))(\gamma_{21}/(\mu_2 + \gamma_{21}))} \frac{\beta_1}{\mu_1 + \gamma_{12}} \\
&+ \frac{(\gamma_{12}/(\mu_1 + \gamma_{12}))}{1 - (\gamma_{12}/(\mu_1 + \gamma_{12}))(\gamma_{21}/(\mu_2 + \gamma_{21}))} \frac{\beta_2}{\mu_2 + \gamma_{21}} \\
&= \frac{(\mu_2 + \gamma_{21})\beta_1 + \gamma_{12}\beta_2}{(\mu_1 + \gamma_{12})(\mu_2 + \gamma_{21}) + \gamma_{12}\gamma_{21}} \\
&= \frac{(1 + \gamma_{21}/\mu_2)(\beta_1/\mu_1) + (\gamma_{12}/\mu_1)(\beta_2/\mu_2)}{1 + \gamma_{21}/\mu_2 + \gamma_{12}/\mu_1}.
\end{aligned}
$$

From this expression for R_0 we see that the value of R_0 depends essentially on four parameter combinations, namely β_1/μ_1, β_2/μ_2, γ_{12}/μ_1, and γ_{21}/μ_2. R_0 is obviously increasing in β_1/μ_1 and β_2/μ_2. For most natural metapopulations one would assume that $\beta_2/\mu_2 > \beta_1/\mu_1$; that is, larger populations are less vulnerable to extinctions and exert a greater colonization pressure on empty patches than small populations, and then R_0 is also increasing in γ_{12} but decreasing in γ_{21}. However, if $\beta_2/\mu_2 < \beta_1/\mu_1$, then R_0 is decreasing in γ_{12} and increasing in γ_{21}. This result is in concordance with biological intuition. We emphasize that the approach employed in this paper allowed us to find the value of R_0 and in particular the threshold criterion $R_0 \geq 1$ for metapopulation persistence directly from the biological interpretation of the parameters. To arrive at the same result using the classical approach based on differential equations (Hanski, 1985) one would have to calculate all the eigenvalues of a matrix and determine the largest of them. The model treated by Hanski (1985) is only two-dimensional and this can be readily done but for higher dimensional systems the task is extremely tedious and sometimes even impossible. We point out that if $\beta_2/\mu_2 = \beta_1/\mu_1$, that is, if there is no difference in colonization capacity between the two size classes, then the threshold criterion $R_0 > 1$ reduces to the usual condition $\beta/\mu > 1$ for the Levins model.

We now proceed to look at nontrivial steady states $\bar{E} = (\bar{E}_1, \bar{E}_2)$. The eigenvector corresponding to the eigenvalue $R(\bar{E})$ has the form

$$b_{\bar{E}} = \begin{pmatrix} a \\ 0 \end{pmatrix}. \tag{6.36}$$

The interpretation of (6.36) is simply that all populations in newly colonized patches belong to size class x_1. It is now a straightforward task to calculate the steady metapopulation state $\bar{m} = (\bar{m}_1, \bar{m}_2)$ from (6.21). We emphasize that to apply this formula one need not evaluate the next state operator $T_{\bar{E}}$ since only its integral from 0 to ∞ is relevant and this can be calculated directly. The result is

$$\begin{pmatrix} \bar{m}_1 \\ \bar{m}_2 \end{pmatrix} = \frac{c_1}{c_2} \begin{pmatrix} a \\ 0 \end{pmatrix} + \frac{c_1}{c_2^2} \begin{pmatrix} -(\gamma_{12} + \bar{E}_1 + \mu_1)a \\ \gamma_{12} \end{pmatrix}, \tag{6.37}$$

where

$$c_1 = \gamma_{21} + \mu_2 + \gamma_{12} + \bar{E}_1 + \mu_1,$$
$$c_2 = (\gamma_{12} + \bar{E}_1)\mu_2 + \gamma_{21}\mu_1 + \mu_1\mu_2.$$

Substituting (6.37) into (6.24) and (6.25) we obtain together with the condition $R(\bar{E}) = 1$, with $R(\bar{E})$ given by (6.35), three equations in three unknown $(\bar{E}_1, \bar{E}_2,$ and a). Hanski (1985; see also Hastings, 1991) found that for certain parameter values there are two nontrivial steady states. This simple structured model thus predicts a fundamentally different qualitative behavior than Levins's model.

3. A Model with Continuous Local State

In this section we shall analyze a model originally described in terms of differential equations by Gyllenberg and Hanski (1992). Local population size is considered as a continuous variable and dispersion is modeled explicitly.

In this model, there are two basic local entities, a local population and a disperser. The size of a local population is denoted by x and it has the range [0, ∞). The state of a disperser is denoted by the symbol d. The local state space is thus $\Omega = \{d\} \times [0, \infty)$. We assume that local populations have an intrinsic density-dependent growth rate $g(x)$ which is due to local births and deaths. A local population with state $x \in [0, \infty)$ produces dispersers (emigrants) at a rate $\gamma(x)$. Dispersers enter a patch at a rate α. If this patch happens to be empty then the disperser dies immediately. If, on the other hand, it is occupied the disperser immigrates into the existing local population. The net growth of a local population is therefore modeled by the following ordinary differential equation:

$$\frac{dx}{dt} = g(x) - \gamma(x) + \alpha E_1(t). \tag{6.38}$$

Here $E_1(t)$ denotes the number of dispersers per patch at time t. Let $X_{E_1}(t, x; s)$ be the solution of (6.38) at time $t + s$ given the value $x \in [0, \infty)$ at time t. Dispersers disappear either because they die (at a rate ν) or because they enter a patch. The lifetime of a disperser is therefore exponentially distributed with parameter $\alpha + \nu$. Local populations go extinct as a result of local "disasters," which we assume to occur at the density-dependent rate $\mu(x)$. It follows that

$$u_{E_1}(t, x; s) = \begin{cases} \exp(-\int_0^s \mu(X_{E_1}(t, x; \tau)) \, d\tau) \, \delta_{X_{E_1}(t, x; s)} & \text{if } x \in [0, \infty), \\ \exp(-(\alpha + \nu)s) \, \delta_d & \text{if } x = d. \end{cases} \tag{6.39}$$

Here δ_x denotes the point mass concentrated at x.

The model includes two forms of reproduction by the local entities: local populations produce dispersers (emigration) and dispersers produce new local populations (colonization). We model colonization in the spirit of the Levins model by assuming that the rate at which a disperser colonizes an empty patch is proportional (with constant β) to the number of empty patches. We emphasize that in this model colonization is *not* the result of one disperser arriving at an empty patch (as mentioned above such dispersers are assumed to suffer sudden death) but dispersers colonize empty patches for instance by producing offspring that can initiate new local populations. The disperser itself does not enter the colonized patch but continues its life as a disperser. One disperser may thus very well colonize several patches during its lifetime. We realize that the true colonization mechanism may be very different in many real systems, but we have chosen this model since it allows for empty patches which do occur in nature. A model very similar to ours with a more realistic colonization mechanism but not allowing for empty patches has been analyzed by Val *et al.* (1995).

Denoting the fraction of empty patches by E_2 and assuming that the population of a newly colonized patch has size 0 we can now write down the colonization kernel as follows:

$$\Lambda_E(t, x)([0, s) \times \cdot)$$

$$= \begin{cases} \int \int_0^s \gamma(X_{E_1}(t, x; \sigma)) \exp(-\int_0^\sigma \mu(X_{E_1}(t, x; \tau))\, d\tau)\, d\sigma\, \delta_d & \text{if } x \in [0, \infty), \\ \int_0^s \beta \exp(-(\alpha + \nu))\sigma E_2(\sigma) d\sigma\, \delta_0 & \text{if } x = d. \end{cases}$$

$$(6.40)$$

We observe that the range of $\Lambda_E(t, x)([0, s) \times \cdot)$ is the two-dimensional subspace M_b spanned by δ_0 and δ_d, which reflects the fact that new local entities can be in either of two local states: d in case of a disperser and 0 in case of a local population. It follows that to find the eigenvalues of $W_{\bar E}(\infty)$ we need only to look at its restriction to M_b. The first component of an element in M_b refers to dispersers and the second to local populations.

Let $\bar E = (\bar E_1, \bar E_2)$ be a constant environment. The restriction of $W_{\bar E}(\infty)$ to M_b can be represented as a 2×2 matrix

$$\begin{pmatrix} 0 & \mathscr{E}(\bar E_1) \\ \beta \bar E_2/(\alpha + \nu) & 0 \end{pmatrix}, \tag{6.41}$$

where

$$\mathscr{E}(\bar E_1) = \int \gamma(x)\psi_{\bar E_1}(x)\, dx,$$

$$\psi_{\bar E_1}(x) = \frac{1}{g(x) - \gamma(x) + \alpha\bar E_1} \exp\left(-\int_0^x \frac{\mu(\xi)}{g(\xi) - \gamma(\xi) + \alpha\bar E_1}\, d\xi\right).$$

The elements of the matrix (6.41) have clear and important biological interpretations. $\mathscr{E}(\bar E_1)$ is the expected number of dispersers produced by a local population in a newly colonized patch during its lifetime. The element in the lower left corner is the expected number of patches colonized by a disperser. The matrix (6.41) has two eigenvalues: $+$ and $-$ the square root of the product of the nonzero elements of the matrix. The spectral radius $R(\bar{E})$ is thus *not* a dominant eigenvalue and there will be no convergence toward a stable distribution at the generation level. This is exactly as it should be: local populations produce dispersers and vice versa, and thus if the initial metapopulation consists entirely of dispersers (or of local populations), every second generation will consist entirely of dispersers and every second of local populations. However, since the lifetime of local entities is distributed, the metapopulation will converge toward a stable distribution in *real time*.

The condition (6.20) for a nontrivial steady state now takes the form

$$\frac{\beta}{\alpha + \nu} \bar{E}_2 \mathscr{E}(\bar{E}_1) = 1. \tag{6.42}$$

Recalling the interpretation of the elements of the matrix (6.41), we infer that Eq. (6.42) states that at steady state every local population exactly replaces itself.

Let $(b^{(1)}, b^{(2)})$ be an eigenvector corresponding to $R(\bar{E})$, that is, to the positive eigenvalue of matrix (6.41). Then

$$b^{(2)} = \sqrt{\frac{\beta \bar{E}_2/(\alpha + \nu)}{\mathscr{E}(\bar{E}_1)}} b^{(1)}. \tag{6.43}$$

Let $\bar{m} = (\bar{m}^{(1)}, \bar{m}^{(2)})$ denote the steady metapopulation state. Again the first component refers to dispersers and the second to local populations. Applying (6.21) we find

$$\bar{m}^{(1)} = \frac{1}{\alpha + \nu} b^{(1)} \tag{6.44}$$

$$\bar{m}^{(2)}(dx) = \psi_{\bar{E}_1}(x)dx \, b^{(2)}. \tag{6.45}$$

Using the definitions of E_1 and E_2, which have been stated verbally above and which mathematically have the form (6.22), we obtain

$$\bar{E}_1 = \frac{1}{\alpha + \nu} b^{(1)} \tag{6.46}$$

$$\bar{E}_2 = 1 - l(\bar{E}_1)b^{(2)}, \tag{6.47}$$

where

$$l(\bar{E}_1) = \int \psi_{\bar{E}_1}(x)dx$$

is the expected life-time of a local population. Equation (6.46) is an analog of the well-known relation in epidemiology: the prevalence of a disease equals the incidence rate times the average duration of the disease.

The system (6.42), (6.43), (6.46) and (6.47) is a system of four equations in four unknowns. Solving for $b^{(1)}$ and $b^{(2)}$ one obtains

$$b^{(1)} = \beta \int \gamma(x)\psi_{\bar{E}_1}(x)\, dx \bar{E}_1 \bar{E}_2, \tag{6.48}$$

$$b^{(2)} = \beta \bar{E}_1 \bar{E}_2 \tag{6.49}$$

and inserting (6.49) into (6.47) one gets

$$1 - \bar{E}_2 = \beta \bar{E}_1 \bar{E}_2 l(\bar{E}_1). \tag{6.50}$$

Eliminating $\bar{E}_2$ from (6.42) and (6.50) one finds the relation

$$\beta \left(\frac{1}{\alpha + \nu}\, \mathscr{E}(\bar{E}_1) - \bar{E}_1 l(\bar{E}_1) \right) = 1. \tag{6.51}$$

Once $\bar{E}_1$ has been solved from (6.51), $\bar{E}_2$ is obtained from (6.50) and finally the equilibrium size distribution of local populations from (6.45).

The virgin environment is given by $\bar{E}_1 = 0$, $\bar{E}_2 = 1$ and thus the trivial steady state corresponding to metapopulation extinction is stable as long as

$$R_0^2 = \frac{\beta}{\alpha + \nu}\, \mathscr{E}(0) < 1. \tag{6.52}$$

Since $\bar{E}_1 = 0$ satisfies (6.51) if $R_0 = 1$ we see that the branch of nontrivial steady states given by (6.51) bifurcates from the trivial solution at $R_0 = 1$. The bifurcation can be both supercritical and subcritical. By differentiating (6.51) implicitly with respect to the bifurcation parameter one can find out which case occurs. To illustrate this, let us take β as bifurcation parameter. Now, if

$$\frac{1}{\alpha + \nu}\, \mathscr{E}'(0) < l(0), \tag{6.53}$$

then the bifurcation is supercritical (Fig. 1), and if the reverse inequality holds in (6.53), then the bifurcation is subcritical (Fig. 2). The situation described in Fig. 1 is qualitatively identical to the prediction of the Levins model, whereas the pattern in Fig. 2 is a consequence of the effect of migration upon local dynamics. To understand this phenomenon better, let us have a closer look at the condition (6.53). The derivative $\mathscr{E}'(0)$ is a measure of how fast the number of dispersers produced by a local population increases as the number of dispersers increases from zero. As zero dispersers have no effect upon local dynamics, $\mathscr{E}'(0)$ can be interpreted as a measure of the impact of migration on local dynamics. If this impact is small, the model gives qualitatively the same prediction as the Levins model (where the impact is zero), but if it is sufficiently large (larger than the threshold $(\alpha + \nu)\, l(0)$), then there are multiple nontrivial steady states for certain values of the colonization parameter β.

Hanski and Gyllenberg (1993) considered an extension of the above model, in which each patch was assumed to have a fixed quality affecting local dynamics as well as extinction and colonization. This quality may be for instance patch size. This model can be written in the general framework described here. The basic local entity is either a disperser or an occupied patch structured by patch quality and the size of a local population. The analysis of Hanski and Gyllenberg (1993) showed that this model has a much richer structure than the one described here with the possibility of bifurcations shown in Fig. 3 among others.

Two-Species Metapopulation Models

Sean Nee   Robert M. May

Michael P. Hassell

I. INTRODUCTION

In this chapter we primarily discuss two-species metapopulation models although, for some topics, it is natural also to refer to results for single-species models and models with more than two species, and we shall do so. In Section II, we generalize the single-species Levins model (Hanski, this volume) to include the three simplest ecological relationships between two species coexisting as metapopulations: competition, predation, and mutualism. In Section III we focus primarily on predator–prey relationships in spatially explicit metapopulation models. The types of models discussed in this chapter have been studied over the years from a large variety of perspectives and interests, but in order to present a thematically unified discussion, we will describe these models from the point of view of the consequences of changes in the amount of suitable habitat on the abundances of the species and, ultimately, on their persistence. This question is not only topical, but one which we believe will become increasingly prominent in metapopulation studies. We will see that, for very simple models of each of the three relationships, the consequences are surprising indeed.

Devastation on the scale of the rain forests, for example, has the obvious consequence that vast numbers of species will be extinguished as their habitat is destroyed in its entirety. More subtle effects arise in less extreme circumstances,

such as the management of the old growth forests of the American Pacific Northwest or the ebb and flow of habitat types in Europe as a result of changing policies
on agricultural subsidies. The U.S. government has recently abandoned its policy
of "protect one, abandon ten," whereby national parks and nature reserves remain
as isolated islands in a sea of unlimited environmental devastation (Babbitt, 1995).
The new goal of landscape management is to maintain diverse communities and
the quality of habitat patches that still remain. In the models of habitat destruction
that we will study in this chapter, we deliberately assume that patches that remain
are of the same quality as before. Hence, we ignore such phenomena as "edge
effects" and other inevitable consequences of destruction and fragmentation
which are biologically important (e.g., Robinson *et al.*, 1995; Wiens this volume).
We will see that changes in the amount of habitat per se, even in the absence of
any other effects, can have surprising consequences.

The abundance and persistence of many species may be largely controlled
by a single limiting resource which is often related to the amount of suitable
habitat that is available. For example, we can imagine the abundance of a territorial species as being determined by the availability of suitable breeding territories, the abundance of a predator determined by prey availability, or the abundance of a disease organism determined by the fraction of a host population that
is unvaccinated. Many other things, such as life-history factors, will interact with
resource availability to generate the actual abundance, but such factors can be
treated as given, constants, while the availability of the limiting resource is altered
by changing the amount of habitat.

For such simply defined species, an important insight to emerge from metapopulation analysis is that a species may be unable to persist, even in the presence
of suitable habitat, if local extinction rates are greater than colonization rates
(Hanski, this volume). This was first realized in epidemiology, which is the most
well-developed metapopulation theory—a host is a "patch" of suitable habitat
for a disease organism, infection is "colonization," host recovery or death is "extinction." A cornerstone of epidemiology is the threshold theory of Kermack and
McKendrick (1927), which demonstrates the existence of a minimum number of
susceptible individuals required for a disease to achieve an epidemic outbreak in
a community. Furthermore, epidemiology has long known that it is not necessary
to destroy all the habitat of a species in order to eradicate it (vaccination programs are wanton acts of environmental vandalism from the point of view of a
disease organism). Smallpox would still exist outside a containment laboratory
in Atlanta if it was necessary to achieve the impossibility of 100% vaccination
coverage to eradicate it. Contrary to the prevailing opinion of the time, Ross
(1909) demonstrated that it was not necessary to eliminate mosquitoes entirely in
order to eradicate malaria but that, instead, there was a threshold ratio of mosquito
density to human density below which malaria could not persist (Heesterbeek,
1992). In a pathbreaking metapopulation analysis of the Northern spotted owl,
Lande (1988a) deduced that the planned level of destruction of breeding territo-

ries, the patches in his model, would entirely eradicate the owl, although the advocates of the plan thought that enough territories would be left to maintain a viable population.

The amount of habitat destruction that can result in the eradication of a species can be surprisingly small, and the estimation of "eradication thresholds" is clearly a valuable goal. For a single-species metapopulation, the Levins model suggests a simple estimate of the minimum number of patches required for the metapopulation to persist (Kareiva and Wennergren, 1995; Hanski *et al.*, 1996b): it is simply the number of patches which are observed to be unoccupied when the system is at a dynamical equilibrium between colonization and extinction. Inspired by epidemiological arguments (Anderson and May, 1991), Lawton *et al.* (1994) presented a simple and general derivation of this result. Nee (1994) observed that the result could be biologically generalized further: a simple estimate of the eradication threshold for a population, or metapopulation, is the unused amount of its limiting resource at equilibrium (patches of suitable habitat being the limiting resource for a metapopulation).

Such an estimate provides a starting point for the development of a deeper understanding by crystallizing in its derivation some important assumptions which can be relaxed for further analysis. For example, an important assumption is that the individual members of the population, or patches of the metapopulation, affect each other only indirectly, through the consumption of the limiting resource (see Section II.B for a discussion of this in the particular context of predator–prey relationships). Biologically, this assumption can be violated in many ways, with important implications for the estimate of the eradication threshold: Lande (1987), for example, studies the Allee effect and Hanski (1996b) studies the "rescue effect." In Section II.B, we acknowledge the implications of the possible ratio dependence of trophic interactions on the eradication threshold estimate. The simple estimate is also based on an entirely deterministic model, and the implications of stochasticity are discussed by Lande (1988a) and Hanski (1996b). Frankly pilfering epidemiological work, Lawton *et al.* (1994) discuss how several other factors which may be of real-life importance render the simple estimate either an over- or an underestimate.

It is possible that the simple estimate, treated as a rough empirical estimate, may be very useful in situations where our knowledge is grossly insufficient for an estimate of all the life history parameters and spatial complexities required for a detailed analysis (Anderson and May, 1991; Lawton *et al.*, 1994). However, then there arises the general question of how actually to determine the unused amount of limiting resource or suitable habitat, and an innovative approach to this question is presented by Doncaster *et al.* (1996).

The assumption that the metapopulation is actually at equilibrium when we come to estimate the eradication threshold is clearly of great importance (Lande, 1988a; Lawton *et al.*, 1994; Hanski, 1996b). However, it is of historic interest to observe that the first to suggest in the epidemiological context that the unused amount of limiting resource can be used to estimate the eradication threshold appears to have been Smith (1970), who was considering the *non*equilibrium situation of episodic epidemics. He observed that if, after an epidemic of yellow fever, for example, has died out one then observes that, say, 20% of the population is still susceptible, then this suggests that the disease could not establish itself in a population that consisted of only 20% susceptible individuals. The subsequent development of the subject begins from equilibrium arguments appropriate for endemic diseases (Anderson and May, 1991).

As we will see, habitat destruction may have qualitatively new consequences when we come to consider species which are enmeshed in an intricate web of ecological relationships, one which does not lend itself to caricaturing the abundance of the species as being determined by a single limiting resource. We will see that habitat destruction may actually be of *benefit* to inferior competitors and that small reductions in the amount of suitable habitat may catastrophically consign mutualist associations to oblivion. The analysis of two-species models is just the first step in developing an understanding of the consequences of habitat change for the tangled bank of Nature.

The models of Section II retain the Levins formalism of patch models (Hanski, this volume), so we do not explicitly model local dynamics (see Gyllenberg *et al.*, this volume), nor do we have an explicit spatial structure (see Hanski, this volume). Local populations do not have stable equilibria but, as in the Levins model, they go extinct at constant rates. Furthermore, asynchronous patch dynamics are "built in" to the structure of the models. In all the models studied in this section, for those values of the colonization and extinction parameters for which a nontrivial metapopulation equilibrium exists, this equilibrium is at least locally stable. Therefore, these more complex, two-species metapopulations, like their single-species counterparts, can persist even though local extinctions are inevitable. Because of the lack of explicit local dynamics and the built-in asynchrony, these models are not suitable for asking questions about possible interactions between local dynamics and metapopulation dynamics.

In Section III we turn to models which have both explicit spatial structure and explicit local patch dynamics. Questions of possible interactions between local and metapopulation dynamics may now come to the fore. We first ask: does it make any difference to the local stability of population equilibria if they are connected to other populations, at the same equilibria, by migration? As we will see, the answer is no for a broad class of models, and this is true for single-species as well as for multispecies metapopulation models. We then go on to discuss the rich spatial phenomena exhibited by predator–prey systems with unstable local dynamics and will see that, in this case, the effects of habitat destruction on metapopulation persistence depend on the dynamic geometry of the system. The models of Section II allow us to investigate changes only in the total amount of habitat. The spatially explicit models of Section III allow us to investigate the consequences of different *patterns* of habitat destruction.

II. TWO-SPECIES PATCH MODELS

The two-species patch models we will study in this section all have the same structure. At any particular time, patches of suitable habitat are either empty or occupied by one or the other, or both, of the species. The rates at which one species colonizes other patches, and goes locally extinct, are influenced by the presence or absence of the other species. We alter the total number of patches in the network and observe the consequences for equilibrium species abundances.

A. Competition

Consider two competing species utilizing the same habitat (resource) patches. One species is competitively superior to the other, and we make the simplifying assumption that the superior competitor, species A, completely excludes the inferior competitor, species B, from patches which it occupies. The inferior competitor can nevertheless coexist with the superior species at the landscape level because it has either a higher colonization rate—it is a "fugitive" or "weed"—or a lower local extinction rate, or both. Denote the proportion of empty patches in a landscape by x, patches occupied by species A by y, and patches occupied by species B by z. Species A and B have colonization rate parameters c_A and c_B respectively, and extinction parameters e_A and e_B. A simple generalization of the Levins model incorporating our assumptions is (Nee and May, 1992)

$$\frac{dx}{dt} = -c_A xy + e_A y - c_B xz + e_B z,$$

$$\frac{dy}{dt} = c_A y(x + z) - e_A y, \qquad (1)$$

$$\frac{dz}{dt} = c_B zx - e_B z - c_A zy.$$

Because the inferior competitor is "invisible" to the superior species, the dynamics of the superior competitor are described by the standard Levins model (Hanski, this volume). In the pristine world, $x + y + z = 1$. As patches are destroyed, $x + y + z = h$, where h is the fraction of suitable patches that remain in the landscape. It is easy to mentally switch between numbers and proportions, and one can describe the conclusions of this, and subsequent models, using the words interchangeably. If, for example, N_0 is the number of patches in the pristine state, then xN_0 is the number of patches which are empty, although not destroyed, yN_0 is the number of patches occupied by species A, which are, necessarily, not destroyed, and so on. Because $x + y + z = h$, the system is actually two-dimensional, as are the others we will study below, but we write down all three equations for completeness.

The nontrivial equilibrium solution of this system, found by setting Eqs. (1) equal to zero and solving, is

$$x^* = \frac{1}{c_B}(hc_A - e_A + e_B),$$

$$y^* = h - \frac{e_A}{c_A}, \tag{2}$$

$$z^* = \frac{e_A(c_A + c_B)}{c_A c_B} - \frac{e_B}{c_B} - \frac{hc_A}{c_B},$$

where the asterisk denotes the proportion of patches at equilibrium (there is, of course, always the trivial solution $x^* = h$, $y^* = z^* = 0$). Feasible equilibria are globally stable. If $c_A/e_A > 1$, the superior competitor persists in the landscape and a necessary condition for the inferior competitor to exist is

$$\frac{c_B}{e_B} > \frac{c_A}{e_A}. \tag{3}$$

To understand this expression, imagine a pristine world unoccupied by either species until a local population of the inferior competitor is established on one of the patches. The left-hand side of expression (3) is, simply, the average number of new patches that this patch would colonize before experiencing local extinction; in the language of epidemiology, it is the "R_0" of the inferior competitor. The right-hand side is the R_0 of the superior competitor. Hence, a necessary condition for the inferior competitor to exist is that it have a higher R_0. It is worth emphasizing that inferior competitors do *not* require a higher colonization rate to persist and this is discussed further in Nee and May (1992).

The effect of habitat destruction, modeled by lowering h, on the equilibrium abundances of the two species is illustrated in Fig. 1. Even though the inferior competitor can persist only by virtue of colonizing empty patches, nevertheless the effect of habitat destruction is to *increase* the abundance of the inferior competitor: this is a result of the decrease in the abundance of the superior competitor across the landscape. Hence, in this model, habitat destruction results in the competitive release of inferior competitors. Once the superior competitor has disappeared (at the eradication threshold level of destruction, $h = e_A/c_A$), the abundance of the inferior competitor then starts to decline with increasing destruction. Hence, simply changing the amount of suitable habitat, without any changes in the quality of the remaining habitat, is expected to increase the regional abundance of "weedy" species, to the extent that inferior competitors persist by virtue of a higher colonization rate. The same qualitative result is observed in a spatially explicit analog of this model, in which colonization is purely local (Dytham, 1995), and in a model which incorporates the many other "real life ingredients"

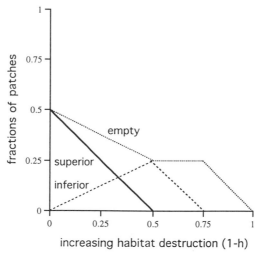

FIGURE 1 The equilibria of competition model (1), with $e_A = e_B$, $e_A/c_A = 0.5$, and $e_B/c_B = 0.25$. It is natural to view these equilibria as functions of increasing levels of habitat destruction; hence we plot them as functions of $1 - h$.

which are contained in Hanski's (1994a) "incidence function" approach to metapopulation models (Moilanen and Hanski, 1995).

This model, as all the others we will describe, has multiple interpretations. For example, interested in the co-occurence of *Daphnia* water fleas in rock pools on islands, Hanski and Ranta (1983) studied a three-species competition model in which, in our terms, h does not reflect habitat destruction but instead describes the different numbers of rock pools on islands of different sizes. For the *Daphnia* system, see also Bengtsson (1991).

It is important to keep in mind that we are making statements about equilibrium abundances. Destroying habitat does not instantaneously exert its effects, just as a population whose per-capita growth rate is reduced below one does not instantaneously go extinct, but dwindles to oblivion at a rate determined by its life-history parameters. So, for example, a level of habitat destruction sufficient to ultimately eliminate both competitors may seem to have no visible effect at all over the observational time span of individual observers and scores of contemporary species may actually be the "living dead" (Hanski *et al.,* 1996b). This fact has recently been given the vivid tag "the extinction debt," reflecting a growing interest in "transient" dynamics, i.e., the behavior and appearance of systems away from their equilibria (Tilman *et al.,* 1994). Such transient phenomena can be very long-lasting, especially in spatially structured systems, and bear no resemblance to the ultimate equilibrium state (e.g., Hastings and Higgins, 1994). However, even in simple systems it is important to emphasize that what we observe today, or even this decade, may not give a true picture of the next century, even if there are no further habitat alterations (Heywood *et al.*, 1994; Hanski *et al.*, 1996b). Furthermore, studies of coral reefs, an important class of metapopulations, have indicated that major disturbances occur sufficiently frequently to prevent the systems ever attaining their equilibrium compositions (Tanner *et al.*, 1994).

The model we have just described is the simplest extension of the Levins model to incorporate competitive relationships. Related models of varying degrees of complexity have been studied over the years and we will briefly describe a few of these. The purpose of this discussion is to illustrate the range of questions that can be addressed, and the range of principles that can be illustrated, by such models. Their ultimate origin lies in the easy impression that one can form that in terrestrial plant and marine communities there do not seem to be enough niche dimensions to allow the large number of species we observe to satisfy Gause's exclusion principle. This impression may or may not be mistaken (e.g., Knowlton and Jackson, 1994; Silvertown, 1987) but, in any case, it draws attention to other important factors potentially mediating coexistence. Hutchinson (1951), without an explicit quantitative model, drew attention to the fact that species can coexist, even while using the same resource, if they differ in their competitive and migration abilities. Recently, Tilman (1994) has emphasised this point in the context of terrestrial plant communities. Skellam (1951), with plants in mind, analyzed a quantitative model of competing species coexisting as metapopulations and generalized the analysis to a landscape with patches of different quality, as did Horn and MacArthur (1972). (Skellam's paper also analyzed what are now called "edge effects" and "source–sink" systems!) Skellam's analysis was motivated by what appears to have been a topical question at the time, namely why some plant species seem to thrive in "unpromising situations." Slatkin (1974) used a two-species metapopulation model to examine whether there is a metapopulation analog of the "priority effect" that one observes in the simple Lotka–Volterra model of two competing species, whereby the species to "arrive first" on a landscape excludes the other. There is not, in his model (but see Hanski, 1995). However, the analysis of Horn and MacArthur, which allowed for different types of habitat patches, did discover the possibility of alternative, stable communities depending on initial abundances (see also Case, 1991). Slatkin also enquired into the effects of changing the extinction rate parameters on the abundances of the competitors, inspired by ideas of predator-mediated coexistence, and found that, indeed, elevating the patch extinction rate could allow coexistence which was not previously possible. Hastings (1980) asked the same question in a multispecies version of Slatkin's model, inspired by speculation about the role of disturbance in maintaining coral reef diversity. He obtained the interesting result that, as the local extinction rate is changed, the number of species that can coexist does not change in a simple fashion, but can rise and fall several times. [The otherwise excellent review by Hastings and Harrison (1994) contains one error: model (1) is not based on Hastings (1980).] Most recently, multispecies models have been studied from the point of view of the evolution of virulence and of species diversity (Nowak and May, 1994; Tilman *et al.*, 1994).

B. Predation

In some circumstances, because of either overexploitation or unstable local dynamics (Section III), specialist predators may drive local populations of their prey extinct and, consequently, themselves as well. Nevertheless, the two species can persist as metapopulations over a landscape as long as local dynamics are not synchronous. We now examine a simple model of such metapopulations, understanding that the predator–prey relationship has a broad biological meaning including, in addition to the conventional relationship, plant–herbivore, host–parasite, and host–parasitoid relationships. In the absence of the predator, patches of prey ("victims") suffer local extinction at a rate e_v, which may be very small. Predators can colonize only patches containing prey, and patches containing both predators and prey go extinct at a rate e_p. Colonization parameters of prey-only patches and patches containing both predators and prey are c_v and c_p, respectively. Denoting the proportion of empty patches by x, prey-only patches by y, and patches containing both predators and prey by z, a simple model of this system is (May, 1994):

$$\frac{dx}{dt} = e_v y + e_p z - c_v xy.$$

$$\frac{dy}{dt} = c_v xy - c_p yz - e_v y, \tag{4}$$

$$\frac{dz}{dt} = c_p yz - e_p z.$$

Notice that we assume that prey in predator/prey patches (fraction z) do not colonize empty patches. Relaxing this assumption would complicate the analysis without introducing any interesting new features.

This system has the following nontrivial equilibrium solution

$$x^* = h - y^* - z^*,$$

$$y^* = \frac{e_p}{c_p}, \tag{5}$$

$$z^* = \frac{c_v}{c_v + c_p} \left(h - \frac{e_p}{c_p} - \frac{e_v}{c_v} \right).$$

Feasible equilibria are globally stable. The effect of habitat destruction on the abundances of predators and prey is illustrated in Fig. 2. The first thing we notice is that habitat destruction has no effect on the equilibrium number of the prey-only patches until it reaches such a level as to extinguish the predators

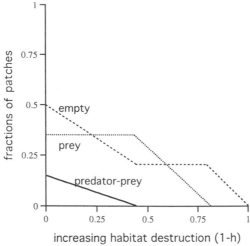

FIGURE 2   The equilibria of predation model (4), shown as functions of increasing levels of habitat destruction, $1 - h$, for parameter combinations $e_p/c_p = 0.35$, $e_v/c_v = 0.2$, and $c_v/c_p = 0.5$.

entirely. Second, the predator abundance declines with increasing habitat destruction in spite of the fact that there is no change in the prey abundance.

The result that habitat destruction does not affect the equilibrium abundance of prey, as long as predators can still persist, is not some uninteresting oddity generated by the simplicity of the model but is, in fact, true of a very broad class of models. The following result is well known in many ecological contexts, such as the effect of prey productivity on predator–prey abundances (e.g., Arditi and Ginzburg, 1989), and we simply rephrase it here for the context of habitat destruction. Consider the following general model for predator–prey, p–v, dynamics:

$$\frac{dv}{dt} = F(v, p, K, .),$$

$$\frac{dp}{dt} = pG(v, .).$$

(6)

This can be construed as either a model of predator–prey dynamics in a single, homogenously mixed population or a model of predator–prey patch dynamics, such as we have just considered. K is the carrying capacity of the prey, i.e., the abundance the prey, or prey patches, would achieve in the absence of predation. The dot denotes other parameters. K may itself be a function of other parameters, depending on one's choice of model. The functions F and G can be chosen in accord with any model of functional responses, migration regimes, life-histories, and so on. Indicating equilibrium abundances with an asterisk, equilibrium (v^*, p^*) satisfies $G(v^*, .) = 0$ and $F(v^*, p^*, K, .) = 0$. Since v^* is determined by the first of these equations, changing K, by habitat destruction, only affects p^*, leaving $v*$ unaffected. The only restrictive assumption is that the predator's per capita population growth rate, G, is independent of overall predator density, i.e., the model is "prey-dependent" (e.g., Arditi and Ginzburg, 1989). If the function G(v,.) is replaced by G(v, p,.), so the predators are now "interferential" rather than "laissez-faire" (May, 1976b), the argument breaks down and equilibrium prey abundances *are* affected by habitat destruction in ways that may, in principle, be very complex. Biologically, laissez-faire predators affect each other only indirectly, through their depletion of prey, or prey patches.

Although prey-dependent models of laissez-faire predators have dominated theoretical ecology, Arditi and Ginzburg (1989) have argued that a class of models of interferential predation may be generally superior. In particular, they advocate "ratio-dependent" models, in which v and p enter the general function G above as functions of their ratio v/p. This suggestion is based on both theoretical and empirical arguments and is the subject of lively controversy (see McCarthy *et al.,* 1995, and references therein). An entirely analogous, but uncontroversial, distinction between prey-dependent and ratio-dependent models exists in epidemiology: the former are considered appropriate for, for example, aerosol transmitted diseases such as measles, whereas the latter are more suitable for sexually transmitted and vector-borne disease (e.g., Thrall *et al.,* 1993).

Whether predators (or predator patches in metapopulation models) are laissez-faire or interferential has important implications for the consequences of habitat change on abundance and persistence. To illustrate this, we will see that the eradication threshold of the predators can, in principle, be readily estimated if they are laissez-faire, but not if they are interferential. It follows from the general arguments following model (6) that if the carrying capacity of the prey, K, is reduced by habitat destruction to the abundance of prey patches that we see today in the presence of predation, then the predator population will go extinct. This follows from the fact that the prey equilibrium is then the same as the carrying capacity of the landscape—hence, no predators. We can check this general result with the simple model (4). In the absence of predators, the system reduces to a Levins model for the prey, which informs us that the carrying capacity of the landscape is $h - e_v/c_v$. To find the eradication threshold of the predators, i.e., that value of h, h_{erad}, which extinguishes the predator metapopulation, the above arguments lead us to the equation

$$h_{erad} - \frac{e_v}{c_v} = v* = \frac{e_p}{c_p}, \tag{7}$$

which we solve to find

$$h_{erad} = \frac{e_p}{c_p} + \frac{e_v}{c_v}, \tag{8}$$

in agreement with Eq. (5c).

As observed in the Introduction, as long as organisms in general are laissez-faire, affecting each other only indirectly through their "consumption" of a "limiting resource," such as their occupation of suitable breeding territories, or, in this case, prey patches, then the simplest estimate of their eradication threshold is just the unused amount of that limiting resource (Anderson and May, 1991; Lawton *et al.,* 1994; Nee, 1994).

C. Mutualism

To construct a simple and illustrative model of mutualism, we imagine two species, one of which can survive for some time on a habitat patch, but requires the other species for its migration into new patches, whereas the other species depends on the presence of the first one for both survival and reproduction. One biological relationship of this type is the one between a plant species and a specialist seed disperser or pollinator. A less well-known inspiration for the model is the coviruses of plants (Bruening, 1977). These RNA viruses get their name from the fact that no single virus particle contains all the information necessary for a complete cycle of infection. For example, there are two tobacco rattle virus particles, a long and a short one. The long particle carries the gene for the replicase, while the short particle carries the gene for both particles' coat protein. The long particle's RNA can multiply in a plant on its own, but ultimately it requires the presence of the short particle for encapsulation and transmission to a new plant. Nee and Maynard Smith (1990) argue that this strange state of affairs arose by a process of mutual parasitism.

To provide a mnemonic for the subscripts, we will refer to plants and dispersers. As before, x refers to the proportion of empty patches, y refers to the proportion of patches occupied by the plant only, and z to the proportion of patches occupied by both plant and disperser. The subscripts p and d refer to these latter two patch types, respectively. Our assumptions lead to the model

$$\frac{dx}{dt} = e_p y + e_d z - c_p z x,$$

$$\frac{dy}{dt} = c_p z x - e_p y - c_d z y, \tag{9}$$

$$\frac{dz}{dt} = c_d z y - e_d z.$$

This has the nontrivial solution

$$x^* = h - y^* - z^*,$$

$$y^* = \frac{e_d}{c_d}, \tag{10}$$

$$z^* = \frac{1}{2}\left(\alpha \pm \sqrt{\alpha^2 - 4\beta}\right),$$

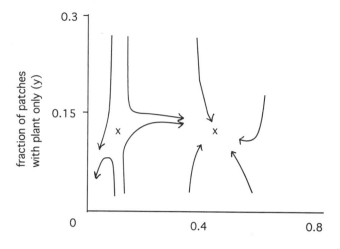

fraction of patches with
both plant and disperser (z)

FIGURE 3   Trajectories in the phase space of model (9) with $c_p - c_d - 4$, $e_d - 0.5$, $e_p = 1.5$, and $h = 0.8$. The symbol X marks the locations of the two equilibria: the one on the left is a saddle point and the one on the right is locally stable.

where

$$\alpha = h - \frac{e_d(c_p + c_d)}{c_p c_d},$$

$$\beta = \frac{e_p e_d}{c_p c_d}.$$

$$(11)$$

The eradication threshold, h_{erad}, is found to be

$$h_{erad} = \frac{e_d(c_p + c_d)}{c_p c_d} + 2 \sqrt{\frac{e_p e_d}{c_p c_d}}. \qquad (12)$$

Above h_{erad}, there are two equilibria that differ in the abundance of patches with both plants and dispersers, z^*. Local stability analysis shows that the larger of the two is a stable equilibrium, whereas the smaller one is a saddle point. Figure 3 illustrates the trajectories of the system in the vicinity of these two points for one particular choice of parameter values. (There is also, of course, the trivial solution $x^* = h$, $y^* = z^* = 0$.)

Figure 4 shows the equilibria as functions of increasing habitat destruction. When habitat destruction approaches close to the eradication threshold, there remains a large metapopulation of mutualists across the landscape at the stable equilibrium. However, at the eradication threshold, the two equilibria defined in Eqs. (10) collide and annihilate each other, leaving only the trivial equilibrium

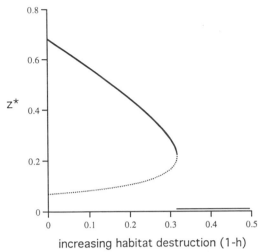

FIGURE 4   The two equilibria z^* of mutualism model (9), shown as functions of increasing levels of habitat destruction, $1 - h$. The solid line is the stable equilibrium. The colonization and extinction parameters are the same as in Fig. 3. y^* (not shown) remains at 0.125, up to the eradication threshold level of destruction. Thereafter, $y^* = z^* = 0$ and $x^* = h$.

of metapopulation extinction. This is an example of what is described in mathematics as a "catastrophe," an appropriate term in the conservation context. To describe this result more vividly, a perfectly viable association of mutualists living in great abundance across a large region can be completely destroyed by the construction of just one more shopping mall. This is vivid, but not entirely realistic. If the system was that close to the eradication threshold, then chance effects, not included in the simple deterministic model, and the existence of the saddle point would combine to create a serious threat to the persistence of the mutualists, rendering the system vulnerable to extinction if chance events move the patch abundances into a region of Fig. 3 which sweeps the above-threshold metapopulation to the alternative stable state of oblivion.

A sudden, large catastrophic change in the fate of the metapopulation as the result of a tiny change in the amount of suitable habitat may seem like a peculiar and unfamiliar outcome. One can become more comfortable with this result by considering the same phenomenon in a simpler and more familiar context. Consider a simple model of population dynamics in which a single population undergoes logistic growth described by the two parameters r and K, the per-capita growth rate and the carrying capacity, respectively. As long as $r > 1$, the stable equilibrium is a population of size K. However, lower r ever so slightly below 1 and the only equilibrium is extinction.

Armstrong (1987) noted that although there were metapopulation models of competition and predation, there were none of mutualism, and he constructed model (9) with $h = 1$ to fill the gap. Hastings and Wolin (1989) also studied a mutualism model to conclude that mutualist metapopulation systems always have a stable equilibrium, which they claim contrasts with mutualism models that do not incorporate spatial structure. A more general review of models, including metapopulations of humans and schistosomes, with two alternative stable states, "thresholds" and "breakpoints," can be found in May (1977). For a specific example, of alternative stable states, see Gyllenberg et al. (this volume).

III. SPATIALLY EXPLICIT METAPOPULATIONS

In this section we will emphasize the qualitatively new features that can arise in metapopulation models that are spatially explicit. We do this quite generally for a variety of interactions, including single-species, competing species, and predator–prey systems, but we will dwell in more detail on the latter. Most of the models assume that the habitat takes the form of a grid or lattice of "cells" containing local populations with discrete generations (for a discussion of various spatially explicit metapopulation models, see Hanski and Simberloff, this volume). Demographic parameters define population growth within, and migration of individuals between, cells. Such models with discrete time and space but continuous population state have been dubbed "coupled map lattices" (Kaneko, 1992, 1993; Solé et al., 1992) and, because of their complexity, have mainly been explored by numerical simulations. In all the examples below, the following rules apply. Within each generation there are two distinct phases: (1) a period when the local population reproduces and matures, and (2) a distinct migration stage when some mixing between local populations occurs. Reproduction can thus occur following, or prior to, migration, but not at the same time [this would require careful formulation in a model to keep track of those individuals that migrated and those that remained developing within the patch (Hassell et al., 1995)].

The models in this section have explicit local dynamics. They are, therefore, suitable for the study of the effects of migration on the dynamics of local populations in a metapopulation. This is where we begin.

A. Local Stability

Although most of the theory of population dynamics has concentrated on isolated populations with no interchange of individuals between other populations in the region, this does not mean that spatial dynamics have been neglected in population ecology; far from it. However, the emphasis has been primarily on the effects of the spatial distribution of individuals within a single patchy habitat. It is implicitly assumed in this work that the dispersing stages mix thoroughly before redistribution within the habitat according to specified behavioral or statistical rules. This implies that the individuals are able to disperse widely across the entire habitat which has, in turn, implications for the spatial scale that is appropriate for the study. The general conclusion from this body of work, whether it involves single species (e.g., de Jong, 1979; Hassell and May, 1985), competing species (e.g., Shorrocks *et al.,* 1979; Atkinson and Shorrocks, 1981; de Jong, 1981; Hanski, 1981, 1983; Ives and May, 1985), predator–prey interactions (e.g., Hassell May, 1973; Chesson and Murdoch, 1986; Pacala *et al.,* 1990; Hassell *et al,* 1991b; Rohani *et al.,* 1994), or disease–host interactions, is that spatial variation in the risk of mortality enhances population stability. Because of the assumptions made about migration, no patch or grouping of individuals can have any degree of independent temporal dynamics from generation to generation. By shifting our spatial scale upward to that of a metapopulation, asynchronous dynamics become a possibility.

We commence with a very simple case of a single species reproducing and competing for resources in a metapopulation. The environment is made up of uniform, discrete habitats or patches arranged in a regular grid in which live local populations of herbivoros insects. The insects have discrete generations, and in each generation some of the adult females disperse to neighboring populations. Following the migration stage, the females oviposit and the larvae that subsequently emerge compete for resources as a function of the density within their local population. Such a patchy environment can be conveniently modeled as a two-dimensional arena in which the local populations are distributed among a square grid of cells. In each generation there is a migration phase, in which a fraction of the adult insects leave the patch from which they emerged and move to neighboring cells. We first focus on the conditions for stability in such a system, when all the local populations move to a common, stable equilibrium, and then in the following section examine the more complex dynamics that can occur when the local populations are unstable.

We first assume that the local population dynamics within a single habitat are based on a familiar single-species model for intraspecific competition,

$$N' = \lambda N(1 + aN)^{-b}, \tag{13}$$

where N' and N are the population sizes in successive generations, λ is the finite rate of increase and a and b are constants defining the density dependent survival (Hassell, 1975; Hassell *et al.,* 1976; de Jong, 1979). The stability properties of Eq. (13) depend solely on the parameters b and λ and are described in Hassell (1975). We now ask the question "To what extent are these stability properties altered if the model is extended to a metapopulation by linking a number of these local populations, all with identical parameters, by limited migration?"

In a recent paper, Bascompte and Solé (1994) have explored such a metapopulation model in which local populations with dynamics described by Eq. (13) are linked by diffusive migration to their four nearest neighbors. Their results are surprising: as migration rate is increased, the dynamics become increasingly *unstable,* and thus increasingly diverge from those of the nonspatial, homogeneous model (13). This is counterintuitive, since one would expect increasing migration to link more effectively the separate local populations and so bring the properties of the spatially structured and homogeneous models closer together (Ruxton, 1994; Hassell *et al.,* 1995). The explanation lies in the biologically implausible way that Bascompte and Solé formulated migration within their model. Couched as a discrete analog of a reaction–diffusion equation, their model fails properly to segregate the processes of survival and migration, and as a result, the same individuals may simultaneously fail to survive and yet disperse (Hassell *et al.*, 1995).

If this problem is avoided by segregating competition and migration (for example, larvae that compete for resources and adults that disperse), completely different conclusions can be drawn: spatial structure now has *no effect* on the stability properties of the system. More specifically, if we assume periodic boundary conditions and migration of the form that a fraction μ of the emerging adults within a habitat disperse by moving with equal probability to one of the eight surrounding habitats, and hence a fraction $1 - \mu$ remain behind, it can be shown analytically that the local stability boundaries of this metapopulation are identical with those for a single local population (Rohani *et al.*, 1996). This is true for all migration rates, μ, and for all grid sizes. It does not depend on the number or location of the habitats to which the dispersing individuals move; all that is required is for the patterns of migration to be the same for all cells. Indeed, the result is also independent of the details of the within-habitat density dependence provided it takes the form f(N). This result—that the metapopulation and local populations have the same stability properties—is much broader. It applies equally to comparable models for interspecific and predator–prey interactions (Rohani *et al.*, 1996).

For a broad class of models, therefore, the introduction of spatial structure has no affect on the local stability properties of the systems. This result makes sense intuitively, provided that the environment is uniform, so that at equilibrium, all local populations have the same density. Thus, at equilibrium, migration to and from local populations is in balance and does not alter the equilibrium properties of the local populations. A number of factors will, of course, confound this simple conclusion. Most obviously, a spatially heterogeneous environment is bound to introduce different dynamics dependent on the nature of the spatial unevenness. However, even in homogeneous environments, moving to a metapopulation may change the stability properties under some conditions. Vance (1984) has discussed a range of single-species models in which migration between habitats often stabilizes, but occasionally destabilizes, the population as a whole; Reeve (1988) has shown for host–parasitoid models that stability is reduced by the interaction between migration and density dependent host rates of increase; and Hastings (1992) has explored age-structured metapopulations where strong levels of density dependence and asymmetric migration between age-classes is destabilizing.

Another class of studies is, in a limited sense, the converse of those discussed above and examines whether immigration has a stabilizing effect on unstable local population dynamics, in particular, chaotic dynamics. The general conclusion is that immigration readily turns chaos into periodic dynamics (e.g., Gonzalez-Andujar and Perry, 1995; Hastings, 1993). It is now understood that chaos is often a structurally unstable feature of dynamical systems that follow the period doubling route to chaos, easily abolished via period doubling reversals in the face of perturbations like immigration (Stone, 1993), although it may be a more robust feature of systems that approach chaos by other routes (Rohani and Miramontes, 1995).

B. Complex Spatial Dynamics

Moving beyond the region of local stability, spatially explicit metapopulations may show strikingly novel dynamics. Broadly, such metapopulations are characterized by unstable local populations tending to fluctuate asynchronously, and by the metapopulation as a whole tending to persist much more readily than in the comparable spatially homogeneous model. Such behavior has been documented for metapopulations of single species (Bascompte and Solé, 1994; Hassell et al., 1995), for competing species (Solé et al., 1992; Halley et al., 1994) and for various predator–prey systems (Taylor, 1988, 1990, 1991). Here, we concentrate on just one kind of interaction, between hosts and parasitoids, for which these dynamics have been thoroughly displayed (Hassell et al., 1991a, 1994; Comins et al., 1992).

Insect host–parasitoid systems are characterized by the adult female parasitoids being the only "searching" stage and laying their eggs on, in, or near the hosts that they encounter; these hosts are then subsequently killed by the feeding larvae (Askew, 1971; Godfray, 1994). This feature of having reproduction defined directly by parasitism makes host–parasitoid associations particularly simple and convenient models of predator–prey systems in general.

We begin with a general model for the interaction between an insect host and its specialist parasitoid in a completely homogeneous environment (Hassell, 1978),

$$N' = \lambda N f(P)$$
$$P' = cN[1 - f(P)], \tag{14}$$

where N', P', N and P are the host and parasitoid populations, respectively, in successive generations, λ is the host rate of increase, as before, $f(P)$ represents the probability of a host escaping parasitism (assumed here, for simplicity, only to depend on parasitoid density) and c is the average number of adult female parasitoids emerging from a parasitized host (henceforth assumed to be one). The dynamics of this model have been explored using a wide range of expressions for the different parameters (Hassell, 1978). The model will be unstable unless (1) the parasitoids attack hosts sufficiently nonrandomly (Pacala et al., 1990; Hassell et al., 1991b), (2) λ is sufficiently density dependent (e.g., Beddington et al., 1975; May et al., 1981), or (3) c is density dependent (e.g., Hassell, 1980; Hassell et al., 1983).

To extend this to a metapopulation, we assume the same environment as before, but now the local populations of hosts are attacked by parasitoids. In each generation the dynamics consist of two phases. First, a reproduction-and-parasitism phase in which hosts and parasitoids interact within individual patches according to Eqs. (14). The second phase is a migration phase where a fraction of the emerging adult hosts, μ_N, and a fraction of emerging adult female parasitoids, μ_P, in each patch redistribute themselves to the eight immediate neighboring patches. This migration is assumed to be spatially symmetrical for both species, but the fraction of dispersers is species-dependent. In most previous host–parasitoid studies, these dispersing individuals have been distributed over all other patches according to some specified behavioral or statistical rules (e.g., Hassell, 1978; Chesson and Murdoch, 1986; Pacala *et al.,* 1990; Hassell *et al.,* 1991b). Here, however, rather than entering a "pool" for such global migration (i.e., a global fornicatorium in the sky), the dispersing hosts and parasitoids move outward, using the same rule as above, to colonize equally the eight nearest neighbors of the patch from which they emerged (slightly different assumptions may be necessary along the boundaries depending on whether cyclic, absorbing, or reflective boundary conditions are used; the choice of condition has little effect on the outcome, provided the arenas are not very small).

The whole system is thus described by the following set of equations:

$$N_i = J_i \mathrm{f}(P_i)$$

$$Q_i = cJ_i[1 - f(P_i)] = cJ_i - cN_i \tag{15}$$

$$J_i' = \lambda[N_i(1 - \mu_N) + \mu_N\{N_i\}]$$

$$P_i' = Q_i(1 - \mu_P) + \mu_P\{Q_i\}.$$

Here N_i and J_i are, respectively, the adult and juvenile hosts in patch i, Q_i is the newly emerging parasitoids in patch i, and P_i is the postmigration population of parasitoids in patch i which search for host larvae to parasitize. The curly brackets represent incoming individuals, obtained as appropriate sums over the relevant patches. The function for parasitism is given by the unstable Nicholson and Bailey term, $\mathrm{f}(P_t) = \exp(aP_t)$, so a single isolated population is unstable with rapidly expanding oscillations although the metapopulation as a whole may be persistent.

Persistence in this model is associated with some striking spatial patterns of local population abundances, which have been labeled as "spatial chaos," "spirals," and "crystal lattices" (Hassell *et al.,* 1991b; Comins *et al.,* 1992). Figure 5 shows the approximate boundaries for these different patterns in relation to the host and parasitoid migration rates and for a chosen value of λ and an arena width of $n = 30$. The *spiral structures* are characterized by the local population densities forming spiral waves which rotate in either direction around almost immobile focal points. The phase-space dynamics of each patch form a close approximation to a fixed track, even though no exact repetition occurs. These patterns are apparently chaotic, since the position and number of focal points vary

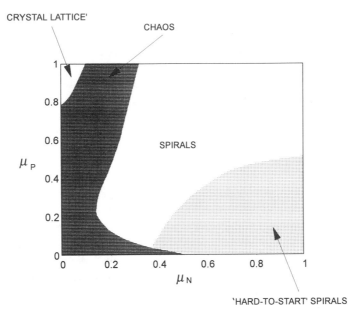

FIGURE 5   Dependence of the persistent spatial pattern on μ_N and μ_P for arena width of 30 and $\lambda = 2$. The boundaries are obtained by simulation and are approximate. The single hatched area indicates the region in which the spatial pattern is chaotic; the region marked as "hard-to-start spirals" represents parameter combinations for which the persistent spiral pattern is unlikely to be established by starting the simulation with a single nonempty patch. Spirals may be established in these cases by starting with a lower μ_N and increasing it after 50 to 100 generations. Metapopulation extinction occurs for some combinations with very small μ_N or μ_P; this area is imperceptible in the figures (after Hassell *et al.*, 1991a).

slowly with time in nonrepeating patterns. The combined metapopulation exhibits what appear to be stable limit cycles (Fig. 6a). *Spatial chaos* is characterized by the host and parasitoid population densities fluctuating from patch-to-patch with no long-term spatial organization. Randomly oriented wave fronts are observed, but each persists only briefly. The total metapopulation generally remains within narrow bounds, but occasional large excursions are observed (Fig. 6b). Despite the lack of recognizable structure, the populations appear to coexist indefinitely (as long as the arena is sufficiently large). Finally, the rather extreme combination of very low host migration and very high parasitoid migration gives persistent *crystal lattice-like* structures, in which relatively high density patches occur at a spacing of approximately two grid units, and the metapopulation as a whole is stable (Fig. 6c).

The entire metapopulation may go extinct in this model for several reasons. First, the total area may be too small (see next section). Second, the starting conditions for the simulation may be unfavorable for persistence. For example, in the region described as "hard to start spirals" in Fig. 5, persistence is impossible if the simulations are started from a single nonempty cell. Once the populations

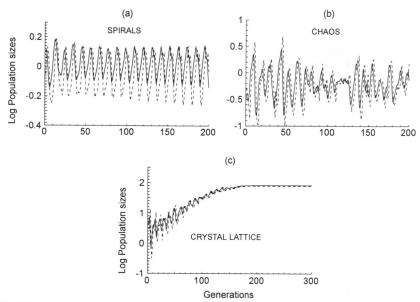

FIGURE 6   Typical time series of average population size corresponding to the three classes of spatial behavior shown in Fig. 5. (a) Stable cycles where spatial spirals occur, (b) chaotic population dynamics where spatial chaos occurs, and (c) a stable equilibrium where the "crystal lattice-like" spatial pattern is observed (after Hassell *et al.*, 1991a).

are initiated, however, by simultaneous colonization of many cells, persistence always occurs. Third, metapopulation extinction may arise from intrinsic dynamic instability. This region is small and in Fig. 5 is restricted to parameter combinations in which either μ_N or μ_P is very small. Finally, we note that persistence remains possible for a wide range of values of the host rate of increase (λ), ranging from close to unity to very large. The principal effects of increasing λ are to favor the formation of spirals (rather than spatial chaos) at low host migration rates and to reduce the spatial scale of the persisting spirals.

C. Habitat Destruction and Spatial Dynamics

The metapopulations described in the previous subsection persist readily for a wide range of host and parasitoid demographic parameters. An important additional requirement was a sufficiently large number of local populations (e.g., grid side length = 30). Any reduction of the grid size (Comins *et al.*, 1992) or fragmentation of the habitat (Hassell *et al.*, 1993) runs the risk of disrupting the dynamics of the metapopulation as a whole, either by reducing the number of local populations below a critical level required for the combined metapopulation to persist or by interfering with the migration required to link the unstable local populations.

Habitat destruction has generally the dual effect of reducing the amount of

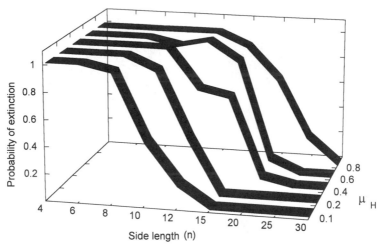

FIGURE 7 Extinction probability in relation to the numbers of patches in a square grid of side length n and the fraction of hosts migrating to neighboring patches (μ_N) ($\mu_P = 0.89$, $\lambda = 2$). Extinction is measured as the proportion of 50 replicates failing to persist over 2000 generations. Each replicate is started by setting a nonzero population density in only the third patch from the left in the top row. The same 50 pairs of initial host and parasitoid densities are used for all the parameter combinations. Local extinction occurs by numeric underflow (densities less than about 10^{-45}); however, the results are robust when local extinction thresholds for both host and parasitoid are modeled explicitly (after Hassell *et al.*, 1991a).

habitat and restricting the opportunities for migration between habitat fragments. Within the host–parasitoid metapopulation outlined above, it is clear that the probability of long-term persistence decreases as the number of patches is reduced (Fig. 7). The extent of this effect depends to a large extent on the characteristic spatial scale of the dynamics. Thus with parameter combinations producing "crystal lattice" patterns, the overall populations can persist in a stable interaction even with very small grids of $n = 2$. At the other extreme, interactions in which large-scale spirals occur are especially vulnerable to shrinking grid sizes, while interactions producing chaotic spatial patterns are intermediate between the two. This trend is also clear from Fig. 7, which represents a slice across Fig. 5. Within the region of chaos all interactions persist for side lengths of 15 and above, whereas within the region producing spirals the probability of extinction increases so that with the relatively large spirals generated with $\mu_N = 0.8$ some extinctions occurred for all simulations with $n < 30$ and there was no persistence at all with $n < 15$. Failure to persist in small arenas is thus associated with insufficient space in which to fit a self-maintaining pattern. These general trends remain true for different values of λ and also for increased migration distances (Comins *et al.*, 1992).

Regions of suitable habitat may become partially subdivided by inhospitable corridors that restrict movement within the overall area. Mankind's ever-increasing network of roads must have this effect on the habitats through which they plunge. By pushing ecological communities closer to the limits of their range, climate change is also likely to produce similar effects in which habitats shrink in a patchy way, leaving pockets with limited connections for the species within them. To illustrate the possible effects of such disturbance, let us modify the environment of 30×30 patches explored above by imposing barriers of one patch width with varying numbers of "gaps" for movement between the subareas (Hassell *et al.*, 1993). Two conclusions stand out. First, as noted above, interactions with characteristic spatial dynamics on a large scale are by far the most easily disrupted. For example, an interaction with a single persisting spiral filling the 30×30 arena always becomes extinct when a barrier bisects the habitat, while interactions with small-scale spirals or chaotic spatial dynamics can persist much more easily as the habitat is disrupted. In short, habitat subdivision affects species persistence in a way that is strongly influenced by the characteristic scale of the spatial dynamics.

D. Multispecies Systems

The two-species host–parasitoid models of the previous sections lay bare some interesting dynamical properties of metapopulations. Important questions remain, however, in our understanding of how metapopulation dynamics may affect community structure (Holt, this volume). Here, we examine the specific case of how spatial processes may influence the coexistence of three-species host–parasitoid systems (Hassell *et al.*, 1994; H. N. Comins and M. P. Hassell, unpublished) that are straightforward extensions of Eqs. (15), in which the third species may be another host, another parasitoid, or a hyperparasitoid.

The results are very similar for the different systems: a third species can coexist stably within the spatial dynamics (spiral waves or chaos) generated by an existing two-species, host–parasitoid interaction, provided that it is relatively sessile compared to its competitor. Coexistence thus depends upon a kind of fugitive coexistence (Hutchinson, 1951; Levins and Culver, 1971; Horn and MacArthur, 1971; Hanski and Ranta, 1983; Nee and May, 1992, 1994; Hanski and Zhang, 1993). For example, in the two-parasitoid–one-host system, coexistence occurs most easily when the two parasitoid species have very different migration rates, provided that low migration is matched by high within-patch searching efficiency, and vice versa. Similarly, in the case of two-host–one-parasitoid interactions, coexistence occurs readily when the two host species have very different migration rates and the relatively immobile species has either the larger rate of local-population increase or is less susceptible to parasitism. Finally, in the host–parasitoid–hyperparasitoid system, coexistence demands that the hyperparasitoid has a higher searching efficiency than the parasitoid and a much lower migration rate. That the hyperparasitoid should have the higher searching efficiency is also in accord with the conclusions from nonspatial models of host–parasitoid–hyperparasitoid interaction (Beddington and Hammond, 1977; Hassell, 1979; May and Hassell, 1981).

An interesting additional point is that coexistence in these models tends to

a   b   c

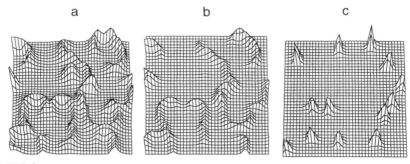

FIGURE 8   Maps of the spatial density distribution (with linear scales) of (a) hosts, (b) highly dispersive parasitoids, and (c) relatively sedentary parasitoids, in a snapshot from a one-host and two-parasitoid simulation with $\lambda = 2$, $\mu_N = 0.5$, $\mu_{P1} = 0.5$, $\mu_{P2} = 0.05$. The three grids should be mentally superimposed to perceive the relationships between the densities of the three species. Spiral foci exist at the ends of the "mountain ranges" in the left-hand figure (excluding ends at the edges of the grid). In the time evolution of the system the "mountain ridges" are the peaks of population density waves and are thus in continuous motion. The peaks or foci, by contrast, remain in almost exactly the same place, for indefinitely long times (after Hassell *et al.*, 1994, where further details are given).

be associated with some degree of self-organizing spatial separation between the competing species. This is best seen when the spatial dynamics show clear spirals. In the case of two competing parasitoids with very different migration rates, the relatively immobile species tends to be confined to the central foci of the spirals, where it is the most abundant species, and the highly dispersive species occupies the remainder of the "trailing arm" of the spirals, as shown in Fig. 8 (H. N. Comins and M. P. Hassell, unpublished). Since the foci of the spirals are relatively static in these models, the less mobile species appears to occur only in isolated, small "islands" within the habitat, much as if these were pockets of favorable habitat. As the migration rates become less divergent between the species, the niche of the less dispersive species spreads further into the arm of the spirals. Such spatial segregation of the competing species, purely as a consequence of the dynamics, is an intriguing property of these spatial models.

IV. CONCLUSION

Theoretical studies of spatially distributed populations with restricted migration between patches have revealed a fundamental tendency for populations to become spatially organized into spiral or chaotic patterns of local abundance. Spirals are associated with cycles in average population size over time, while chaotic patterns lead to time series that are also chaotic. A necessary condition for these patterns are unstable local dynamics, but the results persist with very low host rates of increase, with very low migration rates, and even if a small minority of adults disperse much more widely. Several interesting features follow on from these patterns, such as (1) the spatial segregation of competing species

described in the previous section and (2) the relative non-invasibility of populations showing spiral waves (Boerlijst *et al.*, 1993). Theory is far ahead of experiment and observation in this instance. While direct observation of these kinds of
spatial dynamics in the field presents enormous logistical problems, it may be
possible to determine properties of the population density distributions which are
diagnostic of spirals or spatial chaos (for example, particular patterns of delayed
covariance). Work of this kind would be most welcome to facilitate bridging the
gap between theory and empirical results.

7

From Metapopulation Dynamics to Community Structure
Some Consequences of Spatial Heterogeneity

Robert D. Holt

I. INTRODUCTION

The most fundamental structural properties of a local community are the number and relative abundances of its member species and the pattern of their dynamical interactions (Roughgarden and Diamond, 1986). The history of community ecology largely revolves around variations on a small number of perennial themes, including: (1) the relationship between species diversity and environmental heterogeneity (e.g., resource diversity or disturbance regimes; Chesson, 1986; Huston, 1994), (2) the implications of direct and indirect interactions for community structure (e.g., dynamical constraints on food chain length; Pimm, 1982; Schoener, 1993; Wootton, 1994), and (3) historical contingency, such as multiple stable states (e.g., priority effects in competition).

A consideration of spatial dynamics can enrich all these traditional themes in community ecology. The insight that local colonizations and extinctions determine local community structure was first articulated in the 1960s in the theory of island biogeography (MacArthur and Wilson, 1967). This seminal work was soon complemented by analyses of the effects on species interactions of patch dynamics and spatial fluxes in mosaic landscapes (e.g., Levins and Culver, 1971; Horn and MacArthur, 1972; Levin, 1974; Whittaker and Levin, 1977; Holt, 1985), a line of thinking which in recent years has crystallized into a rich body of theory under the rubric of "metapopulation dynamics" (Gilpin and Hanski, 1991).

If a "metapopulation" is defined to be a set of local populations coupled by dispersal (Hanski, 1991), a "metacommunity" may be defined simply as a set of local communities in different locations, coupled by dispersal of one or more of their constituent members (Gilpin and Hanski, 1991, p. 9). At present, there is an explosion of interest in the consequences of spatial dynamics for single-species dynamics (Hanski, this volume), interactions between species (e.g., Bengtsson, 1991; Hassell *et al.*, 1994; Kareiva and Wennergren, 1995; Nee *et al.*, this volume), and, more broadly, the structure of entire ecological communities (e.g., Case, 1991; Nee and May, 1992; Tilman, 1994; Caswell and Cohen, 1993; Holt, 1993).

My aim in this chapter is not to provide a synoptic overview of all pertinent work on the implications of metapopulation dynamics for community structure. Instead, I use variants of standard metapopulation model to examine several interlinked questions in community ecology which have not to date been examined in depth, but deserve further attention: (1) How does landscape heterogeneity influence the composition of local communities? (2) Can metapopulation dynamics constrain food web structure, for instance the average length of food chains? (3) When do indirect interactions constrain community membership at the level of entire landscapes? In this chapter, I use straightforward extensions of standard metapopulation models (e.g., Hanski, 1991) to examine these questions. My focus is on theory development and the articulation of hypotheses which warrant empirical scrutiny.

II. EFFECTS OF LANDSCAPE HETEROGENEITY ON LOCAL COMMUNITY COMPOSITION

Imagine a landscape that has been colonized over an evolutionary time scale from a larger species pool. For simplicity, I consider first a noninteractive community (i.e., no interspecific competition or predation) and examine the influence of heterogeneity at the landscape level on local community structure. Most metapopulation models to date have assumed that the patches comprising the metapopulation are physically homogeneous [though Horn and MacArthur (1972) and Hanski (1992b, 1995) do consider habitat heterogeneity in the context of interspecies competition]. Yet, in practice, large areas almost always encompass spatially heterogeneous local conditions (Williamson, 1981; Holt, 1992). Such regional heterogeneity can influence local community structure in a variety of ways, particularly if rarer species are considered.

Species abundance distributions typically reveal that a substantial fraction of species in local communities consists of rare species (Gaston, 1994). Surveys conducted at multiple sites, replicated over time, often show that many rare species display a pattern of local extinctions and recolonizations. For instance, in the Eastern Wood study of a bird community discussed by Williamson (1981, pp. 93–100), 28 of 44 recorded species went locally extinct at least once in the 26 years of the study. For some of these species, "extinctions" may be recorded because the site provided only a small sample drawn from populations experiencing the landscape at a coarser spatial scale (J. Bengtsson, personal communication). For other species, the Eastern Wood population may be part of a classical metapopulation, in which a balance between colonization and extinction across the landscape permits regional persistence, despite the ephemeral occurrence of populations in local communities (Hanski, this volume).

However, Harrison (1994b; see also Harrison and Taylor, this volume; Schoener, 1991) has argued that species which in a particular patch network show frequent extinctions and colonizations, may actually have a few persistent populations, which permit overall persistence. Moreover, a local population that never goes extinct may nonetheless prove to be a sink population, maintained by a regular flow of individuals from self-sustaining source populations (Shmida and Ellner, 1984; Holt, 1985, 1993; Pulliam, 1988; S. Hubbell, personal communication).

Local species richness thus may be enhanced if, at the landscape scale, habitat heterogeneity provides each species with some habitat patches permitting long-term persistence. Guaranteed local survival in some habitats allows a species to be present (e.g., as rare transients or sink populations) over a much broader range of habitats. To examine this effect in more detail, it is useful to consider a metapopulation model that incorporates habitat heterogeneity in colonization and extinction rates.

A. A Metapopulation Model for a Heterogeneous Landscape

Assume that the landscape consists of a large number of habitat patches, of which a fraction H are suitable for the community. For simplicity, I will consider that just three habitat types are present: patches of two distinct habitat types, potentially occupied by species in the community, embedded in a third matrix habitat, unsuitable for any of them. Let h_i be the fraction of habitat patches of type i. Necessarily, $h_1 + h_2 = H \leq 1$. Some species in the community may be habitat specialists on just habitat 1, others specialists on habitat 2, and yet others may be habitat generalists, able to use both habitat types (possibly to different degrees).

Let p_i denote the *fraction* of habitat patches of type i occupied by a focal species. The total occupancy of this species over the entire landscape is $p = p_1 + p_2$. Let e_i be the rate of extinction of the focal species in patches of habitat type i and c_{ij} the rate of colonization of type i patches due to emigration from patches of type j $(i, j = 1, 2)$. The following model describes dynamics of the total metapopulation:

$$\frac{dp_1}{dt} = (c_{11}p_1 + c_{12}p_2)(h_1 - p_1) - e_1p_1$$

$$\frac{dp_2}{dt} = (c_{21}p_1 + c_{22}p_2)(h_2 - p_2) - e_2p_2$$

(1)

In a metapopulation with homogeneous habitat patches, colonization and extinction rates should be independent of patch type, so $c_{ij} = c$ and $e_i = e$. Model (1) then reduces to the usual form, $dp/dt = cp(H - p) - ep$ (Levins, 1969a; Hanski, 1991, this volume).

Let us consider first a species specialized to just habitat i. By assumption, this species cannot occupy habitat j at all, hence $p_j = 0$. When rare, the species increases at a rate (per occupied patch) of $c_{ii}h_i - e_i$ and equilibrates with a fraction $\hat{p} = h_i - e_i/c_{ii}$ of the landscape occupied. If we define the "conditional incidence" (Holt, 1993) I_i of species i to be the probability that it occupies a patch, given that the patch is suitable for it, then at equilibrium $I_i = 1 - e_i/h_i c_{ii}$. The species will persist in the landscape (without repeated invasion from an external source pool) only if $c_{ii}h_i > e_i$.

Assume that in the regional source pool, species specializing to the two habitats are equally common, and that each ensemble of habitat specialists can be described with the same bivariate frequency distribution of colonization and extinction rates. Further, assume that in the landscape habitat 1 is the commoner, i.e., $h_1 > h_2$. It follows that the rarer habitat (habitat type 2) should have fewer species that are habitat specialists. Moreover, those specialist species which are present should on average have lower overall occupancies, compared with species specialized to the commoner habitat.

If two species have equal conditional incidences, but differ in their habitat specialization, the species specializing in the rarer habitat must have a higher colonization rate or a lower extinction rate. Extinction and colonization rates reflect many aspects of individual and population ecology, such as life history responses to disturbance, temporal dynamics in local population abundance, resource specialization, resistance to resident predators, and so forth. Hence, there should be systematic differences in entire suites of ecologically relevant traits between ensembles of species specialized to rare, as opposed to common, habitats.

To go further with this line of reasoning, one would need to specify statistical distributions for the parameters e and c among specialist species in the species pool. This would be an interesting exercise, but at the present juncture premature, given the paucity of data on these parameters at the level of entire guilds or communities. The above theoretical results provide testable hypotheses for future comparative community studies of rare and common habitats. I now turn to a comparision of habitat specialists and generalists.

B. Habitat Specialists and Generalists

The cross-habitat colonization terms in model (1) represent a kind of landscape "mutualism": the incidence of a species in one habitat type may be enhanced because the species is present in another habitat as well. If a species can colonize patches of a second habitat type, without reducing its rate of colonization of patches of the first habitat type, it obviously should be able to persist better in a heterogeneous landscape. Moreover, a habitat generalist may be a member of the local community in a particular habitat type, though this species would disappear in a homogeneous landscape consisting entirely of just that habitat type. The above model permits a closer analysis of these effects.

To determine whether or not a species can persist, one examines its rate of increase when it is rare (i.e., at low occupancy). If a species increases when rare, it will persist, whereas if it decreases when rare, it is vulnerable to extinction. When a species is rare across both habitat types, we can approximate the above model with a pair of linear differential equations. The initial growth rate of the species when rare is given by the dominant eigenvalue of this simpler model,

$$\lambda(h_1, h_2) = \tfrac{1}{2}[\lambda_1 + \lambda_2 + \sqrt{(\lambda_1 - \lambda_2)^2 + 4\,c_{12}c_{21}h_1h_2}], \tag{2}$$

where

$$\lambda_i = c_{ii}h_i - e_i \tag{3}$$

is the rate of metapopulation growth when habitat type i alone is available in the landscape.

If h_1 is fixed and $c_{12}c_{21} > 0$, λ increases with h_2. Thus, the ability of a species to utilize a second habitat may facilitate its persistence in a heterogeneous landscape.

Consider the special case of $c_{11}c_{22} = c_{12}c_{21}$. This constraint on parameter values could arise in two biologically distinct ways, each quite plausible in different circumstances:

1. Colonization could be determined entirely by the site of colonization (i.e., $c_{11} = c_{12}$ and $c_{22} = c_{12}$). For instance, the presence or absence of a particular mortality factor, say a natural enemy, could influence the likelihood of local colonization. If the natural enemy is found predictably in some habitats, but not others, this should lead to spatial heterogeneity in colonization rates.

2. Colonization rates of empty patches could be determined entirely by the site of origination for dispersers (i.e., $c_{11} = c_{21}$, and $c_{22} = c_{12}$). For example, the two habitat types could differ in the local average abundances achieved by a species. If individuals emigrate at a constant per capita rate, the habitat type with larger populations will exert a disproportionate effect on the colonization of empty patches.

In this special case, the combination of parameters defining the sign of the growth rate when the species is rare is given by the following expression:

$$G = \frac{c_{22}h_2}{e_2} + \frac{c_{11}h_1}{e_1}.$$

When $G < 1$, the metapopulation declines toward extinction; conversely, when $G > 1$, the metapopulation grows when it is scarce in the landscape.

If each $c_{ii}h_i/e_i > 1$, the species could persist in either habitat alone. If each $c_{ii}h_i/e_i < 1$, but $G > 1$, a species can persist in the entire landscape, even though it cannot persist in any single habitat alone. At equilibrium in model (1), for habitat i

$$0 = [c_{ii}p_i^*(h_i - p_i^* - e_i)] + c_{ij}p_j^*(h_i - p_i^*). \tag{4}$$

If the cross-habitat parameters c_{ij} are positive, both $p_i^* > 0$. The bracketed term on the left of (4) is zero at $\hat{p}_i = h_i - e_i/c_{ii}$ and is negative for larger p_i. Because the right-hand term is positive, the bracketed term must be negative; hence $p_i^* > \hat{p}_i$. Thus, a species' incidence in one habitat type is enhanced because of spatial coupling with the second habitat type. Habitat generalization permits a species to be present with a higher than expected incidence in each habitat, given colonization and extinction rates for each habitat in isolation.

Model (1) illustrates how "spillover" between habitats can enrich local communities. Assume that habitat 2 is a "black-hole" sink (Holt and Gaines, 1992), which can be colonized but does not provide colonists for either habitat type (for concreteness, one can imagine that population densities are very low in habitat 2, so these populations provide negligible sources for colonists). Hence, c_{11} and $c_{21} > 0$, but $c_{22} = c_{12} = 0$. The incidence of the species in habitat 2 is $I_2 = p_2^*/h_2 = c_{21}p_1^*/(e_2 + c_{21}p_1^*)$, which increases with p_1^*. If H is fixed, then $p_1^* = H - h_2 - e_1/c_{11}$, which decreases linearly with h_2. Hence, the incidence (probability of occurrence, per patch) of this spillover species increases in habitat 2, the less frequent is this habitat type, relative to the frequency of the habitat type that actually sustains a viable metapopulation.

Species with high colonization or low extinction rates in their preferred habitat should exhibit a high occupancy in this habitat and can secondarily have a high incidence in habitats where they cannot persist. Such spillover effects should be most noticeable in rare habitats, involving in particular those species with high occupancy in frequent habitat types.

In some circumstances, utilizing a second habitat may permit a species to persist in a landscape even if there is no colonization among patches of the second habitat type. For instance, imagine that patches of habitat 2 are overdispersed, sufficiently far apart that $c_{22} = 0$, and furthermore that the species cannot persist in habitat 1 alone. The condition for such a species to persist in the entire landscape is

$$\frac{c_{11}h_1}{e_1} < 1 < \frac{c_{11}h_1}{e_1} + \frac{h_1 h_2 c_{12} c_{21}}{e_1 e_2}. \tag{5}$$

The right-hand inequality is always met if e_2 is sufficiently small. Sparse habitats with low local extinction rates can have a large effect on the overall persistence of a species, even if the geometry of the landscape does not permit such habitats to sustain the species on their own. In effect, colonization of sparse but low-extinction habitat patches provides a kind of "spatial storage effect" (Holt, 1992), amplifying colonization rates overall in the more widespread habitat.

This two-habitat metapopulation model leads to several interesting and testable conclusions. In a heterogeneous landscape:

1. Habitat specialists will be disproportionately common in those habitats that are most common in the landscape.

2. Some generalists may persist in the landscape precisely because they can exploit a range of habitat types.

3. Species which can persist in one habitat can thereby incidentally occupy other habitats, enriching those local communities. This spillover effect should be particularly important in defining the community membership in sparser habitats and be characterized by species with high occupancies in commoner habitats.

4. Specialists on rare habitats should have unusually low extinction rates, or high colonization rates, relative to the entire ensemble of species in the landscape (including both specialists on common habitats and habitat generalists). This implies a systematic bias at the community level in entire suites of ecological factors correlated with local extinction or colonization rates.

The above model deliberately ignored species interactions. Yet, in practice, habitat suitability for a given species and its local colonization and extinction rates may be largely determined by interactions with other species. Several authors have considered metapopulation models for species interactions in homogeneous landscapes (e.g., see Nee *et al.*, this volume) and have examined competitive interactions in heterogeneous metapopulations (Horn and MacArthur, 1972; Hanski, 1992b). In the remainder of this paper, I consider some implications of trophic interactions in a heterogeneous metapopulation, using natural extensions of the above model.

III. METAPOPULATION DYNAMICS OF FOOD CHAINS

The simplest trophic interaction is the one between a specialist predator and its prey, and the simplest food web is an unbranched chain of trophic specialists. Here I first consider a metapopulation model for a three-level food chain. A food chain describes a set of tight sequential dependencies among species. In many circumstances, it is reasonable to expect that such sequential trophic dependency will lead to nested distributional patterns, in which a given species will be necessarily absent in a patch if its required prey population is absent (Holt, 1993, 1995).

Let the state of a patch be identified by the length of the food chain it contains, such that "0" denotes an empty patch, "1" a patch with just the basal prey species, "2" a patch with both the basal prey and an intermediate predator, and "3" a patch with both these plus a top predator. The fraction of patches found in state i is denoted by p_i. We assume that the basal species in the food chain is a habitat specialist and that its required habitat occupies a fraction $h < 1$ of available patches in the landscape. The following model describes colonization–extinction dynamics in this metapopulation:

$$\frac{dp_1}{dt} = (c_{01}p_1 + c'_{01}p_2 + c''_{01}p_3)(h - p_1 - p_2 - p_3)$$
$$- (c_{12}p_2 + c''_{12}p_3)p_1 + e_{31}p_3 + e_{21}p_2 - e_{10}p_1, \tag{6}$$

$$\frac{dp_2}{dt} = (c_{12}p_2 + c''_{12}p_3)p_1 - (e_{20} + e_{21})p_2 + e_{32}p_3 - c_{23}p_3p_2,$$

$$\frac{dp_3}{dt} = c_{23}p_3p_2 - (e_{30} + e_{31} + e_{32})p_3.$$

For clarity, the order of the subscripts for the colonization and extinction coefficients indicates the direction of flow among states (read them from left to right). Thus, the c_{ij}'s denote the rate at which colonization transforms patches from state i to state j; the e_{ij}'s likewise set the rates of extinction, changing patches from state i to state j.

In the basal prey equation, the parameter c'_{01} arises because empty patches can be colonized by prey originating from patches with both the basal prey and the intermediate predator. Likewise, c''_{01} denotes colonization of empty patches by basal prey emigrating from patches with both the intermediate and top predators (as well as the basal prey), and c''_{12} describes colonization of prey patches by intermediate predators dispersing from patches with the full food chain. If these parameters are positive, colonization dynamics at lower trophic levels involves habitat heterogeneity, comparable in spirit to model (1) (although such heterogeneity is not a fixed landscape feature, but instead emerges as a dynamical feature of the trophic interactions).

The most important assumption made in the above model is that the food chain builds up via sequential colonization (see, e.g., Glasser, 1982), and that if a prey population goes extinct in a patch, so does any predator directly or indirectly supported by that prey. Within the confines of these key premises, a wide range of assumptions about local dynamics can be embodied in the model.

It is useful to examine the properties of this model by building it up from its base. The basal species, on its own, satisfies the standard metapopulation model

$$\frac{dp_1}{dt} = c_{01}p_1(h - p_1) - e_{10}p_1.$$

At equilibrium, $p'_1 = h - e_{10}/c_{01}$. The basal prey species persists provided

$$h > \frac{e_{10}}{c_{01}}. \tag{7}$$

This inequality also ensures that the basal species increases when rare.

A. Two Trophic Levels

When the intermediate predator is also present, the model takes the form

$$\frac{dp_1}{dt} = (c_{01}p_1 + c'_{01}p_2)(h - p_1 - p_2)$$

$$- c_{12}p_2p_1 + e_{21}p_2 - e_{10}p_1 \tag{8}$$

$$\frac{dp_2}{dt} = c_{12}p_2p_1 - (e_{20} + e_{21})p_2.$$

The model resembles a standard predator–model, but with the crucial difference that the predator patches also contain prey and can therefore contribute to the rate of generation of new prey patches, either by prey colonization of empty patches, or by predator extinctions unaccompanied by prey extinctions. For simplicity, we will assume that the predator goes extinct locally only if the prey also goes extinct (i.e., $e_{21} = 0$), so here I consider only the former effect.

There are two kinds of effects a specialist predator may have on its prey in this model: it may alter the prey extinction rate, or it may change the rate of prey colonization of empty patches. In general, these effects could be either positive or negative:

1. Biogeographic "Donor Control"

Some predators may have negligible effects on local prey dynamics and so are unlikely to alter prey colonization or extinction rates, i.e., $e_{10} = e_{20}$ and $c_{01} = c'_{01}$. This is "donor control" (DeAngelis, 1992) in a spatial context: prey dynamics may constrain the distribution of the predator, without reciprocal effects by the predator on its prey.

2. Increased Prey Extinction

The scenario that has received by far the most attention in the literature on local predator–prey interactions is the one in which predators reduce prey abundances so greatly that both populations face enhanced extinction risks (i.e., $e_{10} < e_{20}$) (e.g., Gilpin, 1975; Taylor, 1991; Hassell et al., 1992). Even in the absence of any effect of the predator on prey abundance in "typical" years, the predator may heighten the risk of prey extinction during episodes of disturbance, reduced prey resources, or extreme climatic events. Even in predator–prey models with stable equilibria bounded well away from zero, following large perturbations there can be transient phases at low densities, greatly increasing the likelihood of local extinction for both species (R. D. Holt, unpublished results).

3. Decreased Prey Extinction

In a wide range of circumstances, predators can reduce the magnitude of fluctuations in prey abundances (May, 1973b; Rosenzweig, 1973) or even increase average prey abundances (Abrams, 1992). For instance, if prey respond behaviorally to predators by reduced exploitation of their own resources, over-exploitation may be less likely in the presence of a predator. Sih *et al.* (1985) reported a surprising number of cases in which removal of a predator led to a decrease in the abundance of the focal prey. Many of these cases seem to involve indirect interactions in multispecies assemblages (e.g., competitive interactions among prey, held in check by generalist predators), but it is not clear that all do. In cases where a predator enhances the mean abundance or reduces the temporal variability of its prey, it is conceivable that $e_{10} > e_{20}$.

4. Decreased Prey Colonization

If local prey densities are greatly reduced by predation, the flux of dispersers available for colonizing empty patches is likely to be reduced and, hence, we could expect that $c_{01} > c'_{01}$.

5. Increased Prey Colonization

If predators increase local prey density, as noted above, then predators may also indirectly facilitate prey colonization. Alternatively, if prey differentially disperse in response to perceived increases in the local risk of predation, rates of prey emigration may be higher from patches with predators than from patches without predators. In such cases, one might expect that $c'_{01} < c'_{01}$.

In general, I suspect that scenarios 1, 2, and 4 are more likely than either 3 or 5. In comparisons among systems, however, it is important to keep in mind the potential for "counterintuitive" effects of predators on prey extinction or colonization rates.

As noted above, when the prey occurs alone it equilibrates at $p'_1 = h - e_{10}/c_{01}$. The predator can increase when rare provided that $c_{12}p_1 - e_{20} > 0$, or

$$h > \frac{e_{10}}{c_{01}} + \frac{e_{20}}{c_{12}}. \tag{9}$$

This simple result has several implications. First, as May (1994) notes, the requirement for persistence of a specialist predator in a metapopulation is more stringent than the requirement for the persistence of its prey (compare conditions (9) and (7). Specialist predators are not likely to persist in rare habitats, unless they have very high colonization rates or very low extinction rates. Predators which increase the extinction rate of their prey are particularly unlikely to persist in rare habitats.

As a limiting case, consider a donor-controlled system (i.e., $e_{20} = e_{10}$) in which the predator colonization rate is a times that of its prey. In this case, the predator can increase when rare provided $a > (1 - I)/I$, where I denotes the equilibrial incidence of the prey when alone. A predator specializing on a prey with low incidence (i.e., $I \ll 1$) must have a much higher colonization rate than that of its prey. Hence, in a heterogeneous landscape, specialist predator–prey interactions should occur disproportionately in the more widespread habitat types. Those specialist interactions which do occur in rare habitats should involve predators which either have unusually high colonization rates, or little effect on prey extinction rates, or involve prey which themselves are habitat generalists.

The above conclusions were drawn from the condition for the predator to increase when rare, a condition which provides a criterion for robust persistence in the face of large perturbations in the fraction of patches occupied by either species. The above model also defines a joint equilibrium for the predator and prey with both present at positive occupancies, as follows (recall we are assuming that $e_{21} = 0$):

$$p_1^* = e_{20}/c_{12},$$
$$p_2^* = \tfrac{1}{2}[A \pm \sqrt{A^2 + 4B}],$$

where (10)

$$A \equiv h - p_1^* \left(1 + \frac{c_{01}}{c'_{01}} + \frac{c_{12}}{c'_{01}}\right),$$

$$B \equiv c_{01} \frac{p_1^*}{c'_{01}} \left(h - p_1^* - \frac{e_{10}}{c_{01}}\right).$$

If both $A < 0$ and $B < 0$, then $p_2^* < 0$, and hence a joint equilibrium with both species present in positive numbers does not exist. If $B > 0$, then the positive branch in the above solution leads to a unique positive equilibrium. The condition that $B > 0$ is equivalent to the condition for invasion by the predator, when the prey is at a predator-free equilibrium. The condition for $B > 0$ can be expressed as $p_1^* < p_1'$. Thus, when the predator invades, the system settles into a unique equilibrium in which the prey is reduced to a lower occupancy than when alone.

If $B < 0$, the predator cannot increase when rare. However, a joint equilibrium may nonetheless exist if $A > 0$, (the larger branch in the above solution). When this occurs, the system exhibits alternative, locally stable states, one with, and one without, the specialist predator. Moreover, the equilibrium with the predator present has the prey at a higher occupancy than when the prey is present alone.

However, this outcome is impossible if $c'_{01} < c_{01}$, and $e_{20} > e_{10}$. For alternative equilibria to exist in this predator–prey system, the predator must either enhance the prey colonization rate, or reduce the prey extinction rate, or both. As noted above, there are reasonable circumstances leading to such counterintuitive effects of predation on prey dynamics. When such effects are present, it is feasible for the metacommunity to exist in alternative stable states.

B. Three Trophic Levels

Let us then return to the full food chain model, Eq. (6). Rather than attempt a full analysis of this model here, I will simply touch on some interesting limiting cases. Consider a system which is donor-controlled at each level and where predators in a patch face extinction only when its prey goes extinct, but predators do not affect prey extinction rates (i.e., $c_{01} = c'_{01} \equiv c$, $e_{30} = e_{20} \equiv e$, and $e_{31} = e_{21} = 0$). Without the top predator, the intermediate predator has an equilibrial occupancy of

$$p_2^* = h - \frac{e}{c} - \frac{e}{c_{11}}.$$

The top predator invades if $p_2^* > e/c_{23}$, or

$$h > \frac{e}{c} + \frac{e}{c_{12}} + \frac{e}{c_{23}} \tag{11}$$

Alternatively, if the intermediate predator shuts down prey emigration (i.e., $c'_{01} = 0$), and $e_{21} = e_{31} = e_{32} = 0$, the top predator increases when rare provided

$$h > \frac{e_{10}}{c_{01}} + \frac{e_{20}}{c_{12}} + \left(\frac{c_{12} + c_{01}}{c_{01}}\right)\left(\frac{e_{30}}{c_{23}}\right). \tag{12}$$

For both special cases, the condition for invasion by the top predator is more stringent than the condition for invasion by the intermediate predator (compare conditions (11) or (12) to (4)). A habitat that is too rare to sustain the intermediate predator will not contain the top predator, either. However, more common habitats may be able to sustain a specialist intermediate predator, but not a similarly specialized top predator. Thus, if there are constraints on species' colonization abilities, long food chains composed of trophic specialists are not likely to characterize rare habitats (see also Schoener, 1989).

The basic conclusion that emerges from this model is that metapopulation dynamics can constrain the length of specialist food chains, particularly in heterogeneous landscapes where the basal species is specialized to a rare habitat. Trophic specialization on such species automatically forces habitat specialization on species of higher trophic rank, which thereby experience all the spatial constraints on the distribution of the lower-ranked species, compounded by additional limitations of their own (Holt, 1993).

The full model can admit alternative stable landscape states. For instance, a top predator may be able to stabilize an intrinsically unstable interaction between an intermediate predator and its own prey (May, 1973b; Rosenzweig, 1973), thus extinction rates may be low for patches with the full chain. If the landscape initially has all species in all patches, then it may persist in this state because of low extinction rates. However, if the system starts with just the intermediate predator and its own prey, the intermediate predator may go extinct because of highly unstable local dynamics. In this case, obviously the top predator cannot invade, because its own prey is absent. Thus, the landscape may either have just the prey alone or the entire food chain.

IV. APPARENT COMPETITION IN METACOMMUNITIES

Food chains are useful starting points for examining the implications of meta-population dynamics for community structure, but most natural food webs are much more complex, because there are typically multiple species on each trophic level and complex linkage patterns across levels (e.g., Polis, 1991). I will next examine the potential for strong indirect interactions arising in metapopulations, constraining species membership in local communities. In standard food web models, species richness at intermediate trophic levels is often limited by a combination of two mechanisms: exploitative competition (via effects of species at these levels on abundances of lower trophic levels) and apparent competition (via effects on the abundance of higher trophic levels) (Pimm, 1982).

Consider a landscape of patches of two habitat types, each containing a single habitat specialist. A generalist predator which can exploit both prey species in the two habitats is present. If predators can colonize patches of both habitat types from patches of either type, the dynamics of the two prey species are indirectly linked. If predators increase prey extinction rates or depress prey colonization rates, it may be possible for one prey species to exclude indirectly the other species, in effect by providing a reservoir maintaining a resident predator population (Holt and Lawton, 1994)

To explore the potential for apparent competition in a landscape context consider the following model, which splices the forms of model (1) (metapopulation dynamics in a heterogeneous landscape) and model (7) (predator–prey metapopulation dynamics):

$$\frac{dp_1}{dt} = c_1 p_1 (h_1 - p_1 - q_1) - e_1 p_1 - p_1 (c_{11} q_1 + c_{12} q_2)$$

$$\frac{dp_2}{dt} = c_2 p_2 (h_2 - p_2 - q_2) - e_2 p_2 - p_2 (c_{21} q_1 + c_{22} q_2)$$

$$\frac{dq_1}{dt} = p_1 (c_{11} q_1 + c_{12} q_2) - e_{1q} q_1$$

$$\frac{dq_2}{dt} = p_2 (c_{21} q_1 + c_{22} q_2) - e_{2q} q_2.$$

$$(13)$$

Here, the p_i are the fraction of patches of type i occupied by prey i, and the q_i are the fraction of such patches occupied by both this prey and the generalist predator. In the absence of the predator, each prey obeys a standard metapopulation model, in which c_i and e_i are respectively the colonization and extinction parameters of the prey species specialized to habitat i (which occupies a fraction h_i of the landscape). The quantities c_{ij} scale the rate of colonization of prey patches of type i by generalist predators dispersing from type j patches (as in model 1). The predator and prey go extinct in each habitat at a habitat-specific rate e_{iq}.

As in the food chain model (6), we assume here sequential colonization, so that predators do not colonize a patch until it is occupied by a suitable prey population. However, model (13) deals with only a subset of the predator–prey interactions feasible in model (6). In particular, prey colonization occurs only from patches in which the predator is absent. (Permitting colonization from patches with predators would make an already parameter-rich model even more complicated, so I defer until future work consideration of such more general models.) Predators over-exploit their prey, coupling predator to prey extinctions.

When prey i is at equilibrium and alone, it occupies a fraction $p_i = h_i - e_i/c_i$ of the landscape. The predator, when rare and invading a landscape with prey i only present a equilibrium, grows at an instantaneous rate $\lambda_i = c_{ii}p_i - e_{iq}$. If both prey are present at equilibrium, then expression (2) defines the initial growth rate of the predator population, as a function of its growth rate λ_i in each habitat, considered separately. All the above conclusions about the consequences of habitat generalization on persistence and equilibrial incidence carry over to a trophic generalist that encounters different prey in different habitats, including prey species unable to sustain the predator population by themselves, and so forth. However, the present system is dynamically much more complex than was possible in model (1), because the prey have their own colonizationextinction dynamics, constraining the predator's dynamics, and the predator can in turn drive prey extinctions.

Consider a system in which prey 1 and the predator are present at their respective equilibrial occupancies:

$$p_1^* = \frac{e_{1q}}{c_{11}}, q_1^* = \frac{c_1(h_1 - e_{1q}/c_{11}) - e_1}{c_1 + c_{11}}.$$

Prey species 2 can increase when rare provided

$$\frac{1}{p_2}\frac{dp_2}{dt} = c_2h_2 - e_2 - c_{21}q_1^* > 0.$$

The analogous equilibrial occupancies and criterion for invasion by species 1 are given by transposing the indices 1 and 2 in the above expressions. The prey species may coexist at the landscape level if both invasion criteria are satisfied simultaneously.

The resident prey indirectly reduces the rate of invasion by a second prey species, because it sustains a predator metapopulation which can invade patches once they contain the invading prey. This indirect inhibitory effect, called apparent competition (Holt, 1977), arises even though the two prey species never co-occur within any given habitat patch. Such apparent competition raises the possibility of exclusion due to shared predation in a metacommunity.

A limiting case of the above model suffices to illustrate the potential for exclusion by apparent competition. For simplicity, assume that there are no solo prey extinctions (i.e., $e_i = 0$), that the predator colonizes much more rapidly than it goes extinct, and that the predator colonizes the two habitats indiscriminately (i.e., $c_{ij} \equiv c_q$). With these assumptions, the criterion for invasion by prey 2, given that prey 1 occurs at equilibrium with the predator, is approximately

$$\frac{c_2 h_2}{c_1 h_1} > \frac{c_q}{c_1 + c_q}.$$

Similarly, the criterion for invasion by prey 1 is

$$\frac{c_2 + c_q}{c_q} > \frac{c_2 h_2}{c_1 h_1}.$$

If $c_q \ll c_i$ ($i = 1, 2$), both inequalities will usually be satisfied, and hence the two prey species should be able to coexist in the landscape. By contrast, when $c_q \gg c_i$ ($i = 1, 2$), then one of the two inequalities will not hold. In this case, the prey species with higher $c_i h_i$ can increase when rare and the other prey species is common, whereas the alternative prey cannot reciprocally increase when rare.

The model shows that given a predator which is both a habitat generalist and a trophic generalist, alternative prey species specialized to different habitats may indirectly interact via predator colonization of prey patches—apparent competition (Holt, 1977, 1984; Holt and Lawton, 1994) at the landscape level. If such predators are effective colonizers and can induce local prey extinctions, one prey species restricted to the community in one habitat can indirectly exclude another prey species in a different local community.

The potential for prey exclusion via metacommunity dynamics raises an interesting methodological dilemma. Given such exclusion, a survey of seemingly suitable but empty habitat patches will not reveal the cause of absence—generalist predators, which can colonize only after the missing prey has invaded. The usual sort of descriptive surveys may completely miss the dynamical cause for species exclusion from a heterogenous landscape.

A criterion for dominance in apparent competition is given by the compound parameter $c_i h_i$. Prey species with a low value of this quantity are particularly vulnerable to exclusion by shared predation. Prey specialized to rare habitats (low h_i) are more likely to be excluded by predators sustained by prey inhabiting more widespread habitats. Likewise, prey species which are poor colonists (low c_i) are more prone to exclusion by apparent competition. A low c_i may reflect either poor individual dispersal abilities or low local prey population sizes.

I have previously (Holt, 1984) analyzed a one-predator, two-prey species model in which each prey was specialized to a different habitat. This model explicitly tracks abundances in each habitat (unlike (13)) and assumes density-independent predator dispersal (leading to a source–sink population structure). Such dispersal permits prey to experience apparent competition, despite habitat segregation. The prey species with lower intrinsic growth rate is vulnerable to exclusion by the alternative prey, and the likelihood of such exclusion increases with increasing predator dispersal.

These earlier results are consistent with the conclusions drawn above for shared predation in a metacommunity. Given low inherent extinction rates, the "intrinsic growth rate" of a prey metapopulation is its rate of colonization, which is $c_i h_i$. This compound parameter determines prey community composition, just as the usual intrinsic growth rate does in determining dominance in within-patch apparent competition (Holt, 1984; Holt and Lawton, 1993).

CONCLUSIONS

Classical metapopulation theory assumes that landscapes are comprised of a large number of patches available for colonization. Most models assume that the patches are physically homogeneous. Yet in natural landscapes, metapopulations are likely to span a wide range of local environmental conditions. In this chapter, I have used variants of the Levins metapopulation model to examine some potential consequences for community structure of habitat heterogeneity.

These theoretical results suggest that sparse habitats in a heterogeneous landscape are likely to sustain a biased array of species, including habitat specialists with unusually high colonization or low extinction rates and habitat generalists sustained via spillover from more abundant habitats.

Trophic specialization leads to a kind of magnification of these effects, so that each additional level must satisfy increasingly stringent criteria for persistence. One broad implication of this result is that metacommunity dynamics automatically tends to constrain food chain length.

Trophic generalization leads to an avenue for indirect interactions among alternative prey species. If alternative prey species are habitat specialists, but a predator is a habitat generalist, predator colonization can couple the dynamics of these prey species. This gives rise to apparent competition at the metacommunity level, which in some circumstances can lead to the exclusion of prey species that are poor colonists, or are specialized to rare habitat types.

The ideas presented here provide a first pass through the potential implications of habitat heterogeneity for metacommunity dynamics and structure. One promising direction for future work will be in developing spatially explicit models (Kareiva and Wennergren 1995; Nee *et al.*, this volume) with limited dispersal and various patterns of spatial heterogeneity. My expectation though, is that the general conclusions reached here will prove robust.

ACKNOWLEDGMENTS

I thank Ilkka Hanski and Jan Bengtsson for very thoughtful reviews of the manuscript and the National Science Foundation for financial support.

8 Genetic Effective Size of a Metapopulation

Philip W. Hedrick   *Michael E. Gilpin*

Population structure has long been recognized as having a major influence on the maintenance and loss of genetic variation and has been the topic of extensive research in population genetics (e.g., Wright, 1978; Slatkin, 1985, 1987). Generally, investigation of the impact of population structure on genetic variation has assumed that subpopulation sizes remain constant over time, i.e., subpopulations do not go extinct. It has been shown that if a population exhibits metapopulation dynamics, i.e., patches in which subpopulations exist become unoccupied because of local extinction, that many of the generalizations of earlier studies of population structure may not hold (Slatkin, 1977; Maruyama and Kimura, 1980; Wade and McCauley, 1988; Gilpin, 1991).

The amount of genetic variation in a population is generally determined using the measure heterozygosity because of both its biological importance (individuals are either heterozygotes or homozygotes) and the extensive theory that predicts heterozygosity levels due to various evolutionary factors. In the present context, we are concerned with two different aspects of heterozygosity: the average level of heterozygosity in a subpopulation or a metapopulation and the spatial distribution of heterozygosity due to the structure of the population. Generally, the steady-state values of these heterozygosity values are of interest to evolutionary genetics, while changes, particularly losses, in heterozygosity are of particular importance to conservation biology.

Both the steady-state levels and changes in heterozygosity are governed by the effective size of the population. In general, the effective population size corrects census (or breeding) population number to account for a variety of (mainly demographic) real world considerations such as the sex ratio of breeding individuals and life history characteristics (e.g., Lande and Barrowclough, 1987; Caballero, 1994). The effective population size is usually calculated for a group of individuals with given particular demographic properties in which there is random mating (although other mating structures have also been examined, Caballero, 1994). The effective population size is generally defined as the size of an ideal population that results in a given variance in allele frequency or amount of inbreeding. However, because of our interest in the level of genetic variation, we will estimate the effective population size in an ideal population that results in a given loss of heterozygosity (this has been termed the eigenvalue effective population size by Ewens, 1989). This effective size can be estimated either for subpopulations of a metapopulation or for total metapopulation composed of a group of subpopulations.

Population dynamics similar to that in a theoretical metapopulation in natural populations are not uncommon (Harrison and Taylor, this volume) and there do appear to be particular instances in which habitats are fragmented that metapopulation dynamics is an appropriate description of the population structure at a regional level (e.g., Hanski *et al.*, 1995b; Thomas and Hanski, this volume). As a result, there has been increasing interest in the impact of metapopulation structure on genetic variation in endangered species and other organisms that exist in extremely fragmented habitats, either of natural or human causation (Hastings and Harrison, 1994).

I. AN EXAMPLE

Before examining the specific effects of metapopulation dynamics on effective population size, it is useful to give an heuristic example to demonstrate how metapopulation dynamics can influence the maintenance of genetic variation. Gilpin (1991) gave a simple simulation example in which he assumed that there are three subpopulations or patches in the metapopulation, each with an effective (and census) population size of 500. All the subpopulations were initiated with a high level of heterozygosity.

The important sequence of events in this simulation starts in generation 48 (see Fig. 1) when patch 2 goes extinct and is recolonized from patch 3 with a consequent reduction in heterozygosity. This loss occurs because it is assumed that recolonization is by only two individuals, e.g., a fertilized female. The next significant event is when empty patch 1 is recolonized from patch 2 with a founder population having no genetic variation. Finally, when patch 2 goes extinct in

PATCH

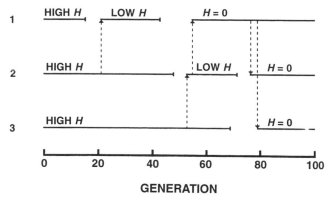

FIGURE 1   The level of heterozygosity (*H*) over time in a simulation of a population existing in three patches (after Gilpin, 1991). The short vertical bars on the right-hand end of horizontal lines indicate extinctions in a patch and the arrows indicate recolonization.

generation 71, the metapopulation has no variation although there are still 500 individuals remaining in patch 1. All of these individuals can be traced back to some individuals in patch 3 before generation 51. Gilpin (1991) termed this process through which metapopulation dynamics reduces genetic variation the coalesence of the metapopulation, i.e., the loss of genetic variation being traced back to a few individuals that are the ancestors of all the individuals in the present metapopulation.

This example illustrates an extreme case in which the loss of genetic variation in the metapopulation can be dramatically lower than that expected from a population the size of the average census number in the system. Gilpin (1991) found that in general the most dramatic lowering of genetic variation occurred for extinction and recolonization values at which the average number of occupied patches was low enough that the metapopulation itself was in danger of extinction. However, genetic variation may be of secondary interest in metapopulations with high extinction expectation so we will examine metapopulations that include more patches and with a balance of local extinction and recolonization rates that makes the likelihood of extinction low.

II. METHODS

One methical approach used to examine metapopulations in a population genetics context is that of Slatkin (1977), Maruyama and Kimura (1980), Wade and McCauley (1988), Ewens (1989), and Barton and Whitlock (this volume). These authors use an infinite (or finite) patch, spatially implicit, Levins (1970) metapopulation structure in which there is instant recolonization of empty patches. Thus, some constant fraction of the local populations go extinct each generation, all of which are immediately recolonized by some number of colonists which then, during the time step (or over time, Barton and Whitlock, this volume), grow up to the local carrying capacity.

Our model, on the other hand, is expanded from the earlier approach of Gilpin (1991) as introduced above and differs from the approaches of Slatkin (1977) and others in several ways. First, our model has a finite number of patches, each of which can support a local population, but which can be empty for a number of time steps (see examples in Figs. 1 and 2). Second, we decouple gene flow from the number of colonists, a possibility suggested by Wade and McCauley (1988). Third, the time that a local population remains extinct is governed both by the colonization probability and also by the number of extant source patches. Finally, we examine the influence of metapopulation dynamics on genetic variation separate from genetic drift within patches by assuming an infinite population size within a patch. While the previous approaches can be approximated analytically, our approach appears to be tractable only using computer simulation (however, see Whitlock and Barton, 1996). Further, while the general behavior of the two approaches are similar, in some cases they appear to yield quantitatively different answers.

Because there are a number of parameters that can influence the effective population size of a metapopulation, and because of the complicated nature of the interaction of stochastic processes within and between patches in our model, we will use computer simulation to understand the process of heterozygosity loss and estimate the effective size in a metapopulation. We will check the simulations through the use of analytical approximations for the behavior of single patches. Our approach will be to assume some standard conditions and then sequentially alter these various parameters and assumptions to determine their effects.

A. Description of Parameters

First, let us assume that the metapopulation is divided up into N_p patches, each of local population size K. Within each patch, random mating is assumed. We will examine the changes in heterozygosity for a single locus with two alleles, both of which have an initial frequency of 0.5 in all patches.

Following Levins (1970) we assume a colonization rate (probability) of c and extinction rate e and based on occupancy of all other patches in the metapopulation. The variable $p*$ gives the observed fraction of patches in the metapopulation that are occupied at any one time. We assume that the actual probability of colonization to an unoccupied patch is $c* = cp*$ so that if some of the patches are not occupied, the rate of colonization is lowered because the pool of potential colonizers is reduced. For $c* < e$, the metapopulation will go extinct. For a metapopulation with a finite number of patches, extinction is possible even for $c* > e$, much in the same way that a small population can go extinct from demographic stochasticity even with the individual birth rate greater than the individual death rate. Note that this is a spatially unstructured model, essentially equivalent to the original model of Levins (1970).

A patch is assumed to be colonized by N_f founders randomly chosen from a given occupied patch, termed the propagule-pool model or randomly chosen from all occupied patches, called the migrant-pool model (Slatkin, 1977). After recolonization, it is assumed that the subpopulation expands in one generation to its carrying capacity, K. The individuals in the following generation are randomly drawn from the parental allele frequency pool to simulate genetic drift within a patch. After evaluating the influence of genetic drift in local populations, to determine the impact of metapopulation dynamics independent of genetic drift within a patch, we will assume that the number of individuals within a patch is infinite. When the population size is assumed to be infinite within the patch, then there is no change in allele frequency from genetic drift and the only change within a patch occurs from gene flow when it is present. When there is gene flow, each generation a proportion m of the individuals in a given patch comes from another given occupied patch, making the total amount of gene flow into a patch per generation, mN_p^*, where N_p^* is the number of occupied patches.

B. Estimation of Effective Metapopulation Size and Other Values

To estimate the effective population size, N_e, the relationship which gives the change in heterozygosity between consecutive generations,

$$\overline{H}_{t+1} = \overline{H}_t\left(1 - \frac{1}{2N_e}\right), \tag{1a}$$

is used where $\overline{H}_t$ and $\overline{H}_{t+1}$ are the mean heterozygosities over all occupied patches and over replicate computer simulations in two consecutive generations t and $t + 1$, respectively (e.g., Hedrick, 1985). N_e is the effective population size that results in the given amount of loss of heterozygosity between the two generations. Therefore, an estimate of the effective population size is

$$N_e = \frac{\overline{H}_t}{2(\overline{H}_t - \overline{H}_{t+1})}. \tag{1b}$$

If we assume that the average heterozygosity within a subpopulation (or patch) is $\overline{H}_S$, then the estimate of the average effective subpopulation size is

$$N_{e(S)} = \frac{\overline{H}_{t(S)}}{2(\overline{H}_{t(S)} - \overline{H}_{t+1(S)})}. \tag{2a}$$

Likewise if $\overline{H}_T$ is the average heterozygosity in the total metapopulation (calculated from the global allele frequency in the metapopulation), then the estimated effective population size for the metapopulation is

$$N_{e(T)} = \frac{\overline{H}_{t(T)}}{2(\overline{H}_{t(T)} - \overline{H}_{t+1(T)})}. \tag{2b}$$

To estimate heterozygosity, a given metapopulation simulation was run for $2/e + 25$ generations. The first $2/e$ generations were used to allow the metapopulation dynamics to become stabilized (the expectation is that approximately 90% of the patches would have had one or more extinctions during this period, $1 - (1 - e)^{(2/e)}$), and the heterozygosities of the last 25 pairs of consecutive generations were used in the estimation. A number of preliminary runs demonstrated that the decay of heterozygosity had stabilized for this period (see also Fig. 3) and yet there was still enough heterozygosity remaining to give an estimate of the effective population size. For each parameter set, the mean heterozygosity in a given generation was the result of 1000 independent replicate simulations (excluding generations in which individual simulations had metapopulation heterozygosity values of zero). From these heterozygosities, 25 effective population size values (generations $2/e + 1$, $2/e + 2$, . . . $2/e + 25$) were calculated and these were averaged to give the estimate of the effective population sizes for the patches and the metapopulation, making each an average of approximately 25,000 values.

In addition, the extent of diversity among the subpopulations was measured using (Nei, 1987).

$$F_{ST} = \frac{H_T - H_S}{H_T}. \tag{3}$$

The average value of F_{ST} was still changing (increasing) during the period for which the heterozygosity and effective population sizes were calculated so its value was used only to determine the relative differences between the effects of various parameters.

At any one time in a metapopulation with an infinite number of patches, an average of $p = 1 - e/c$ of the patches are expected to be occupied (e.g., Levins, 1970). Because the metapopulations that we are examining here have a finite number of patches, there is stochastic variation in the number and fraction of occupied patches. The average or expected census number of individuals in a metapopulation with a finite number of patches should be approximately

$$N' = pKN_p. \tag{4}$$

Because our colonization rate is not constant but is a function of the number of patches occupied and we assume that extinction and colonization occur consecutively and not simultaneously, we have actually calculated the average census number, N. To determine the difference between the census number and the effective population size of the metapopulation, we can calculate the ratio $N_{e(T)}/N$ which should be close to unity if the two values are similar and near zero if the effective population size is much lower than the census number.

As a standard set of parameters, it was assumed that $c = 0.2$, $e = 0.05$, and $N_p = 10$. Given these colonization and extinction rates, and with this number of patches, from empirical estimates we found that slightly more than 70% of the patches were occupied on average at any one time and the probability of metapopulation extinction, all patches being unoccupied, over the first 50 generations was quite small (approximately 0.2%).

The effective population size within a subpopulation can be estimated using an analytical approximation to check the simulations in the following manner. The mean number of generations for a turnover, an extinction, and a recolonization for a given subpopulation, is approximately

$$t = \frac{1}{e} + \frac{1}{c} - 1, \tag{5}$$

which is the expected time to a subpopulation extinction ($1/e$) plus the expected time to a recolonization ($1/c$) minus 1 (because recolonization can occur in the same generation immediately after extinction has occurred). The heterozygosity after t generations is then

$$H_t = H_0\left(1 - \frac{1}{2N_f}\right)\left(1 - \frac{1}{2N}\right)^{t-1}, \tag{6a}$$

where N is the number of individuals in the subpopulation (progeny are drawn randomly so that the effective size within a patch should be equal to the census size). Given that the expected effective population size within the subpopulation is $N'_{e(S)}$, then

$$H_t = H_0\left(1 - \frac{1}{2N'_{e(S)}}\right)^{t}. \tag{6b}$$

By substitution, then

$$N'_{e(S)} = \frac{1/2}{1 - [(1 - 1/(2N_f))(1 - (1/(2N)))^{t-1}]^{1/t}}. \tag{7}$$

III. RESULTS

There are several factors that should influence the effective size of a metapopulation, namely, the carrying capacity of a patch, the rates of extinction and colonization, the number and source of founders, the number of local patches, and the rate of gene flow between patches. After briefly discussing several examples of the general patterns of the results, we will investigate these parameters individually in a sensitivity analysis to determine what impact they have on the effective population size and its relationship to the census number.

A. General Patterns

Before we discuss the influence of particular parameters, it is useful to visualize the general pattern of the results. As an introduction to the general pattern of results found in the following simulations, Fig. 2 gives a graphical

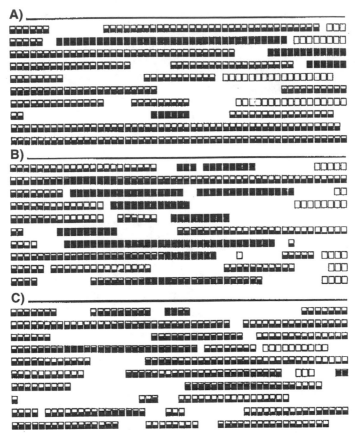

FIGURE 2 The allele frequencies, represented by the amount of shading in a box, in the 10 patches of a metapopulation over 50 generations. Gaps in a horizontal series of boxes represent unoccupied patches. In all three simulations, $c = 0.2$ and $e = 0.05$. The other parameters are (A) $K = \infty$, $m = 0.0$, and $N_f = 1$; (B) $K = 50$, $m = 0.02$, and $N_f = 1$; and (C) $K = 50$, $m = 0.1$, and $N_f = 5$.

summary from three representative simulations. In each case, there are 10 initially occupied patches, all having an initial frequency of 0.5 for both alleles and having colonization and extinction probabilities of 0.2 and 0.05, respectively. The simulation runs for 50 generations from left to right with occupied patches shown by boxes and extinct patches indicated by gaps of varying numbers of generations.

After the first few generations, approximately 7 of the 10 patches are occupied at any one time, as expected. The allele frequency in a local population, indicated by the fraction of the box that is shaded, is influenced differentially here by genetic drift, gene flow, and colonization. In these examples, at the end of the simulation the middle example is nearly fixed for one allele while the top one is split between subpopulations that have high frequencies of one allele or the other, and the bottom simulation has most subpopulations polymorphic for the two alleles. These results are generally consistent with the expectations from higher gene flow in the bottom example, the lower number of founders in the middle one, and the infinite carrying capacity in the top simulation.

Because we are not assuming that there is some force to supplement or sustain heterozygosity, such as mutation, balancing selection, or external gene flow inputs, genetic variation in the metapopulation will ultimately fall to zero. Trivially, it falls to zero if the metapopulation goes extinct. However, we have carried out most of our simulations with values of $c = 4e$ and $N_p = 10$, so that the probability of metapopulation extinction is very low in the short term.

Figure 3 illustrates the pattern of H in the metapopulation and F_{ST} over 50 replicate simulations with $N_p = 10$, $c = 0.1$, $e = 0.025$, $K = 50$, $m = 0.02$, and $N_f = 1$. Over the first $1 - (1 - e)^{2/e}$ or 90 generations of the simulation, the distribution of both H and F_{ST} widens and then appears to reach a quasi-steady state. In the last half of the generations, both H and F_{ST} seem more or less uniformly spread between zero and some upper value. However, the average level of heterozygosity is declining at a rate characteristic of the parameters of the simulation. There is also a probability of leakage each generation into the absorbing state of $H = 0$ and $F_{ST} = 0$ in which the amount of variation in the metapopulation is zero.

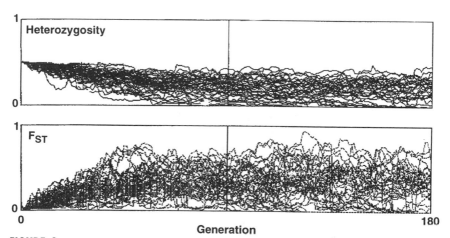

FIGURE 3   The distribution of 50 replicate simulations of metapopulation for H (top) and F_{ST} (bottom) when there are 10 patches, $K = 50$, $c = 0.1$, $e = 0.025$, $m = 0.02$, and $N_f = 1$. The simulation was run for 180 generations, twice the approximate time for the initially colonized patches, to have a 90% probability of extinction.

B. Carrying Capacity of a Patch

First, let us examine the effect of different numbers of individuals or carrying capacity, K, in the patches, a value that is central to the conclusions of Gilpin (1991). Table I gives the estimated effective population size for different patch sizes ranging from 25 to ∞. Notice that for the smallest patch size, the effective population size for the whole metapopulation is 38.2, 21.9% of the average census number, N. The effective population size within a patch is even reduced to 16.4 from the carrying capacity value of 25. Even for this small census number in a patch, the effective population of the metapopulation is about the census number for one and a half occupied patches. Because, on average, approximately 70% of the patches were occupied, an actual census of the number of patches occupied and the effective population size within a patch would give a very distorted picture of the actual effective population size of the metapopulation.

As the patch size increases, the effective size of the metapopulation does not increase very much and asymptotes fairly quickly; it is only 64.2 when the number in the patch is infinite. In other words, even though the census number within a patch may be very large, this has very little influence on the loss of heterozygosity in the metapopulation (and the estimated effective metapopulation size) which is primarily determined by the metapopulation dynamics. Notice also that the estimated effective population size within a patch is about 40 to 50% of that in the total metapopulation.

In the last column, the values of F_{ST} are given for the different K values. For the smaller carrying capacities, the values are extremely high, reflecting the influence of genetic drift within the patches as well as the extinction–recolonization process. For $K = \infty$, F_{ST} reflects the differentiation that occurs only by the extinction–recolonization process and this alone can generate substantial differentiation among patches.

The second column of Table I gives the expected analytical values for the subpopulation effective size from expression (7). The observed simulations are very close to the analytical expectations for effective population size values, supporting the validity of the methodology used here. In the sixth column of Table I, the expected census number is given using expression (4). The observed average census number for the simulations, given in the fifth column, is about 5% lower for all levels of the carrying capacity, reflecting the lower colonization rates resulting from less than 100% patch occupancy.

C. Rates of Colonization and Extinction

For metapopulation existence when there are an infinite number of patches, the rate of colonization must be larger than the rate of extinction. When there are a finite number of patches, the rate of colonization must be several times that of the extinction rate for long-term metapopulation existence (the exact ratio is

TABLE I The Estimated Effective Population Size within a Patch, $N_{e(S)}$, and for the Total Metapopulation, $N_{e(T)}$, for Different Local Carrying Capacities (K) Given That $c = 0.2$, $e = 0.05$, and $N_p = 10$[a]

K	$N_{e(S)}$	$N'_{e(S)}$	$N_{e(T)}$	N	N'	$N_{e(T)}/N$	F_{ST}
25	16.4	16.2	38.2	174.4	187.5	0.219	0.678
50	22.4	23.4	46.4	345.9	375	0.134	0.519
100	30.4	30.0	64.2	690.2	750	0.093	0.424
200	36.0	35.0	66.7	1392.9	1500	0.048	0.370
∞	40.7	41.9	76.6	∞	∞	0.000	0.312

[a] F_{ST} gives the diversity of allele frequencies over the subpopulations and N is the average observed census population size for the metapopulation. $N'_{e(S)}$ and N' are the approximate values given by expressions (7) and (4).

higher when the absolute levels of these parameters is higher and the number of patches is smaller). As a result, we have calculated effective population sizes for an array of colonization and extinction rates that permit metapopulation existence for a substantial time period (Table II).

First, notice that the rate of colonization has very little influence on the effective population sizes, either within a patch or for the total metapopulation,

TABLE II The Estimated Effective Population Size within a Patch, $N_{e(S)}$, the Metapopulation Effective Size, $N_{e(T)}$, and the Average Observed Occupancy Rates (p^*) of a Patch for a Range of Extinction and Colonization Rates When $K = \infty$ and $N_p = 10$

		e			
	c	0.025	0.05	0.1	0.2
$N_{e(S)}$	0.8	80.3	40.0	19.9	9.8
	0.4	74.6	38.6	20.6	
	0.2	81.2	40.7		
	0.1	87.6			
$N_{e(T)}$	0.8	205.1	98.4	41.9	14.4
	0.4	187.1	84.8	35.3	
	0.2	186.1	76.6		
	0.1	179.4			
p^*	0.8	0.989	0.976	0.936	0.794
	0.4	0.952	0.893	0.751	
	0.2	0.881	0.732		
	0.1	0.723			

with levels between 0.1 and 0.8 having nearly the same effective size. On the other hand, as the extinction rate increases, both effective population sizes are greatly reduced. For example with $c = 0.8$, the effective metapopulation size is reduced from 205.1 to 14.4 as e increases from 0.025 to 0.2.

In the bottom of the table, the observed occupancy rates for a patch are given. On the diagonals which have approximately the same occupancy rates, as the absolute values of the colonization and extinction rates are increased, the effective population is greatly decreased. For example, with occupancy rates around 75%, the effective metapopulation size is reduced from 179.4 to 14.4 as the absolute values of the colonization and extinction rates increase eightfold. The explanation for these results is that the level of extinction governs the number of turnovers (extinction and then recolonization) because the rate of colonization is high relative to that of extinction. This high turnover results in a large reduction in heterozygosity and consequently a reduction in the effective population size.

D. Number of Founders

Recolonization of an empty patch in a metapopulation model generally is assumed to occur from a limited number of founder individuals. Above we have assumed that they were all drawn randomly from a given patch [as in the propagule-pool model of Slatkin (1977)]. On the other hand, the founders can be drawn randomly from all occupied patches [similar to the migrant-pool model of Slatkin (1977)]. As shown in Table III, when the number of founders is small for the propagule-pool model, then the effective population size is greatly reduced whether the subpopulation is finite (here $K = 50$) or infinite. For example, when the number of founders is increased fourfold from 2 to 8, $c = 0.2$, $e = 0.05$, and the subpopulation size is infinite, the effective metapopulation size is increased approximately fourfold from 76.6 to 292.7. When the subpopulation is 50, the increase in the number of founders has a substantial influence but has a much smaller effect. For example, increasing the number of founders from 2 to 8 result in 63% increase, 46.4 to 75.5, in effective metapopulation for $c = 0.2$ and $e = 0.05$.

Increasing the number of founders also reduces the value of F_{ST}, particularly when $K = \infty$. For both combinations of colonization and extinction rates given here when the carrying capacity in infinite, a fourfold increase in the number of founders results in a nearly fourfold reduction in F_{ST}.

Estimated effective sizes and F_{ST} are given for the migrant-pool model in Table III where the N_f values are indicated by an asterisk. When the founders are drawn from the migrant pool and other parameters are equivalent, the effective population sizes are larger and the F_{ST} values lower than for the propagule-pool model. For example, when two founders are drawn randomly from all occupied patches rather than from one patch and the carrying capacity is infinite, the metapopulation effective size is 112.6 rather 76.6, about 50% larger.

TABLE III   The Estimated Effective Population Sizes and F_{ST} for Different Founder Numbers, N_f, when $N_p = 10$, $c = 0.2$ and $e = 0.05$ or $c = 0.4$ and $e = 0.1$, and $K = 50$ or ∞^a

c	e	K	N_f	$N_{e(S)}$	$N_{e(T)}$	F_{ST}
0.2	0.05	50	2	22.4	46.4	0.519
			4	31.5	59.9	0.400
			8	38.9	75.5	0.347
		∞	2	40.7	76.6	0.312
			4	81.9	152.7	0.166
			8	167.9	292.7	0.082
		∞	2*	78.9	112.6	0.217
			4*	173.1	222.9	0.096
			8*	408.3	487.9	0.046
0.4	0.1	50	2	15.3	27.0	0.474
			4	22.6	37.5	0.323
			8	32.8	52.2	0.238
		∞	2	20.6	35.3	0.354
			4	41.3	64.2	0.190
			8	78.8	122.1	0.098
		∞	2*	46.1	59.9	0.228
			4*	105.0	123.0	0.101
			8*	204.7	232.3	0.048

^{a} If N_f is indicated by an asterisk, then founders are randomly drawn from occupied patches.

E. Numbers of Subpopulations

Table IV gives the effective population sizes when the number of subpopulations ranges from 5 to 40. For all different numbers of subpopulations, the effective size within a subpopulation does not change and remains at approximately 40. The effective size of the metapopulation is approximately doubled as the number of subpopulations is doubled from 10 to 20 and from 20 to 40 and increases from 57.2 to 306.1 as the number of subpopulations is increased eightfold from 5 to 40. As a result, the ratio of the effective subpopulation size to the effective metapopulation size becomes less as the number of subpopulations increases. The diversity among subpopulations increases as the number of subpopulations increases. This suggests that with smaller numbers of patches in the metapopulation they have more similar allele frequencies because of a higher connectedness while with more patches in the metapopulation some patches are quite unconnected to each other.

TABLE IV   The Estimated Effective Population Sizes for Different Numbers of Subpopulations When $K = \infty$, $c = 0.2$, and $e = 0.05$

N_p	$N_{e(S)}$	$N_{e(T)}$	$N_{e(S)}/N_{e(T)}$	F_{ST}
5	43.3	57.2	0.756	0.166
10	40.7	76.6	0.531	0.312
20	39.4	148.8	0.265	0.386
40	39.5	306.1	0.124	0.418

F. Rate of Gene Flow

We have assumed until now that there is no gene flow between subpopulations and that the subpopulations are only connected because patches that have become extinct are recolonized from other patches. Table V gives the estimated effective population sizes when the level of gene flow is increased from 0.0 to 0.02 from each occupied patch. First, notice that the effective metapopulation size is increased from 76.6 to 219.7 as m increases. This result is somewhat counterintuitive because it is generally assumed that a strongly subdivided population, one with low rates of gene flow, would retain more overall genetic variation. In fact and as expected, when m is low, the level of F_{ST} is the largest and F_{ST} declines as m increases. In addition, the average effective subpopulation size is nearly as large as the effective metapopulation size as m gets to 0.005 or larger. The basis for these results is that the effective metapopulation size is being driven by the metapopulation dynamics making it very low and gene flow does somewhat overcome the loss of heterozygosity by restoring genetic variation into patches that have lost genetic variation. However, because all the subpopulations now are

TABLE V   The Estimated Effective Population Sizes for Different Levels of Gene Flow between Patches When $K = \infty$, $c = 0.2$, $e = 0.05$, and $N_p = 10$

m	$N_{e(S)}$	$N_{e(T)}$	$N_{e(S)}/N_{e(T)}$	F_{ST}
0.00	40.7	76.6	0.531	0.312
0.00125	66.5	95.1	0.699	0.224
0.0025	88.9	115.7	0.769	0.167
0.005	125.8	140.1	0.898	0.114
0.01	156.7	172.4	0.909	0.069
0.02	217.7	219.7	0.991	0.040

connected by the gene flow, the rate of loss of heterozygosity is the same for the total metapopulation and the separate subpopulations.

IV. CONCLUSIONS

We have determined the relationship between effective size of a metapopulation and the parameters that govern the dynamics of the metapopulation: the number of patches, the local extinction and recolonization rates, the local carrying capacity, the number of founders that recolonize a metapopulation, and the rate of gene flow between extant patches. Prediction of the effective metapopulation size, based on these parameters, may allow understanding of the retention of heterozygosity in spatially structured populations and should be of great value in conservation planning.

We have investigated in a sensitivity analysis the influence of these parameters on the effective metapopulation size and the relationship of the effective size to the census number. Under the assumptions of our basic model as the carrying capacity is increased, the effective metapopulation size is increased but asymptotes at a value less than 100 even when the carrying size in each patch is assumed to be infinite. In this case, the census number may not be at all related to the effective size because the effective size is governed by metapopulation dynamics.

Because the census size increases with an increase in K, the ratio of the effective metapopulation size to the census number drops quickly as the census size increases. This point was emphasized by Gilpin (1991) when he commented "that the ability of a metapopulation to retain genetic variation, which may be defined as proportional to its so-called effective population size . . . can be one to two orders of magnitude lower than the maximum total number of individuals in the system." Obviously the size of this effect can be large, but to be one order of magnitude lower, K must be around 100 (probably high for many vertebrates but low for many invertebrates or plants).

This effective subpopulation size could in theory be determined by genetic estimates of effective population size [changes in allele frequency (Waples, 1989) or the amount of linkage disequilibrium (Waples, 1991)] but would not be obvious from demographic estimates of effective population size. In other words, it would be difficult to assess the effective size of the metapopulation by determining the effective size within several patches because the metapopulation effective size is governed by the extinction and recolonization dynamics.

The main factors determining the low effective metapopulation size are the rate of extinction of patches and the number and type of founders recolonizing empty patches. Overall, the effective metapopulation size is increased if the rate of extinction (or the rate of turnover, extinction and subsequent recolonization) is reduced, the number of local populations increased, the numbers of founders is increased, and the rate of gene flow is increased. The increase in effective size with an increase in gene flow is unexpected but is due to gene flow countering the influence extinction–recolonization dynamics which causes the metapopulation to coalesce at a heterozygosity of zero. Gene flow acts to restore variation to patches with zero heterozygosity, thereby reducing the rate of heterozygosity loss and increasing the estimate of effective metapopulation size.

As expected intuitively, the effective metapopulation size is larger when the founders follow the migrant-pool model than for the propagule-pool model. Whitlock and McCauley (1990) generalized these two models to allow a proportion of the founders from the migrant-pool and a proportion from the propagule-pool and found intermediate combinations more closely mimicked the migrant-pool results. On the other hand, if there were spatial structure in the metapopulation with a higher probability of colonization from closer subpopulations, then the founders may more closely approach the propagule model because most founders would come from one or a few neighboring subpopulations (Wade and McCauley, 1988).

There are several aspects of the model which would make it somewhat more realistic or other factors that we did not discuss here. First, rather than allowing the population to grow to carrying capacity, it can be allowed to grow at some rate to the carrying capacity. When this is done, the effective size of the population is even lower than what is given in Table I because more genetic drift occurs during the growth period than at carrying capacity. Second, the carrying capacity of the metapopulation could be kept constant but partitioned between different numbers of patches (a contrast similar to the single, or few, large populations or many small populations idea for reserves). When more patches are present given the same total carrying capacity, the effective metapopulation size is somewhat lower than if the same number of individuals are spread across a few patches. Finally, several other factors, such variation of the parameters in space or time, correlation in space or time of rates of extinction or colonization, and extinction being a function of the genetic constitution, were not investigated.

In traditional studies of population structure, there has been an attempt to summarize the general impact of population structure using the number of migrants between subpopulations per generation parameter. In general, if the number of migrants is greater than unity, then there is little substructuring found while if the number of migrants is greater than unity, then the subpopulations may greatly differ in allele frequencies. On the other hand, with a metapopulation structure the amount of gene flow does not give the complete picture of population differentiation. For example, even with substantial gene flow, a metapopulation can have high diversity over patches because of the extensive impact of frequent extinction and recolonization.

Overall, a metapopulation in which all local populations suffer extinction and recolonization tends to reduce greatly genetic variation, with the effective population size being dramatically lower than what one would estimate from the average census number or the sum of the effective population sizes within each patch. We see two major lessons from these findings, one for evolutionary biology and one for conservation biology. First, in an evolutionary perspective given the observed high levels of heterozygosity for allozymes and other molecular markers in most species, it appears unlikely that most species spend their entire evolutionary history in a Levinesque metapopulation configuration. This is in general support of the observational conclusion of Harrison and Taylor (this volume). On the other hand, the results from our simulations might explain the difference between the inferred effective population sizes and local population numbers of well studied, high density species, such as *Drosophila,* which seem to have local and regional densities several orders of magnitude higher than their effective population size inferred from the level of genetic variation.

Second, with increasing anthropogenic fragmentation of the natural landscape, it may be that unnatural metapopulations are being formed that very well might be subjected to the forces similar to those modeled in our simulations. Because many human impacts are fairly recent, these fragmented populations might be in the initial phase of heterozygosity loss. In particular, conservation biologists should be cognizant of the genetic implications of metapopulation dynamics and realize that heterozygosity may be lost much faster than predicted from either census numbers or traditional estimates of effective population size.

As an example of an application to conservation, Pimm *et al.* (1989) and Gilpin (1991) have suggested that metapopulation dynamics, rather than a population bottleneck (O'Brien *et al.,* 1983), may have been important in reducing the level of genetic variation in the cheetah. The relatively normal levels of genetic variation recently found for minisatellites and microsatellites in the cheetah (Menotti-Raymond and O'Brien, 1993, 1995) appear to be consistent with regeneration of variation at these highly mutable loci since the bottleneck (Menotti-Raymond and O'Brien, 1993). On the other hand, the higher level of variation for these more highly mutable loci is also consistent with equilibrium levels predicted by neutrality theory given a low effective metapopulation size as determined from the present study (Hedrick, 1996).

ACKNOWLEDGMENTS

We appreciate the comments of Nick Barton and Ilkka Hanski on an earlier draft of the manuscript. P.W.H. was partially supported for this research by NSF.

The Evolution of Metapopulations

N. H. Barton   *Michael C. Whitlock*

I. INTRODUCTION

Natural populations differ from the simplest models in ways which can significantly affect their evolution. Real populations are rarely all of the same size; the rates of migration into and out of populations vary in space and time; some populations go extinct, and new ones are established, while all populations fluctuate in size. Furthermore, the genetic properties of real species are not like those assumed in simple models. Alleles are exposed to a wide variety of selection, mutation rarely creates novel genotypes with each mutation event, generations overlap, and environments vary from place to place. Evolution in a metapopulation can be substantially different from the predictions of single-population models and, indeed, very different from the simplest models of subdivided species.

Most species inhabit a wide geographic range and are subject to random drift in small local populations. Nevertheless, spatial subdivision and stochastic perturbations do not necessarily have a significant effect on evolution. Species might adapt as new alleles spread through the regions in which they are favored; they might split into new, reproductively isolated, species if adaptations established in different places turn out to be incompatible with each other. This is essentially the view held by R. A. Fisher (1930); as we show below, it implies that population structure and gene interactions can be ignored. In contrast, Sewall Wright (1932) argued that a "shifting balance" between random drift and natural selection is necessary to establish new combinations of genes. In a similar vein, Mayr (1942) argued that gene flow prevents divergence in the main part of a species' range and that speciation depends on random shifts in isolated founder populations (Mayr, 1982; Provine, 1986).

In this chapter, we summarize the present state of these contrasting views, by asking what effect population structure might have on adaptation and speciation. Neutral variation is irrelevant to these processes, and so we must understand the effects of migration and random drift on selected genes. To date, most theory makes the unrealistic assumption that species are divided into local populations of constant size and with constant migration rates. We pay particular attention to the consequences of random changes in population structure and, in particular, to local extinctions and recolonizations. We will see that if all genes affect fitness in the same way, regardless of the geographic location or genetic background, then population structure has no qualitative effect: it merely amplifies random drift and hence interferes with the efficacy of selection. Population structure has a more fundamental effect if there are interactions between genes (i.e., dominance, epistasis, or frequency dependence, allowing alternative "adaptive peaks") or if the environment is heterogeneous (so that the local population can adapt to local conditions). The crucial issue, then, is whether interactions between genes and with the local environment are extensive enough for evolution to depend strongly on population structure.

There is a close analogy between the subdivision of a population of genes across space and across diverse genetic backgrounds. A gene can find itself associated with a genetic background which is at a selective advantage and will thereby increase by "hitch-hiking." Similarly, a gene can find itself in an area in which the local population is expanding and hence can gain an advantage through "spatial hitch-hiking." Recombination transfers genes between genetic backgrounds, just as migration moves genes from place to place; both processes reduce the random variation in fitness induced by the geographic or genetic background. Also, genes can be adapted both to the place in which they find themselves and to the genetic background in which they are embedded. At present, considerable effort is being devoted to the study of hitch-hiking; many of the techniques and concepts in this area carry over to understanding the effects of spatial population structure (cf. Barton, 1994, 1995).

We begin by summarizing the effects of population subdivision on neutral variation, a relatively straightforward body of theory. We then consider the simplest kinds of natural selection, in which alleles are either uniformly favorable or deleterious; again, the results are straightforward. Population structure has more complex effects when different alleles are favored in different places; we show how gene flow and population turnover impede adaptation to a heterogeneous environment. Finally, we consider the most complex case, in which genes interact with each other, so that populations can evolve toward alternative stable states.

The effects of gene interaction in a structured population are crucial to an understanding of both speciation and adaptation via Wright's (1931, 1932) "shifting balance" process.

II. NEUTRAL VARIATION IN METAPOPULATIONS

A. Models with Stable Local Populations

Sewall Wright established many of the basic results of single-locus population genetic theory (Wright, 1931, 1932). In particular, he described a model of multiple populations which has come to be known as the "island model," a series of identical populations of size N individuals, with a fraction m of each population emigrating and being replaced by immigrants drawn at random from the common pool of migrants. Such a metapopulation will eventually reach an equilibrium genetic variance among local populations, expressed by the standardized coefficient F_{ST}:

$$F_{ST} \approx \frac{1}{4Nm + 1}. \tag{1}$$

(F_{ST} is, for a two-allele system, the variance among populations in allele frequency divided by $p(1 - p)$, where p is the allele frequency in the entire metapopulation.)

Other traditional descriptions of population structure include stepping-stone models (Kimura and Weiss, 1964; Maruyama, 1971, 1972), where populations exchange migrants only with adjacent populations. These models approach the geographic structure of real metapopulations, where local populations usually do *not* exchange migrants equally with all others (Hanski, this volume; see Maruyama, 1971, 1972; Hartl and Clark, 1989; Slatkin, 1993, for more details).

B. Models with Extinction/Recolonization

Many of the same factors which contribute to the partitioning of variance in the island model are also those which control the evolution of variance among metapopulations with unstable local dynamics and hence extinctions and recolonizations. Essentially, small population size increases the variance among populations and migration decreases it. Small population size allows genetic drift (and therefore more variance among populations), but because this scales with the reciprocal of population size, there is a nonlinearity between population size and the amount of variance created among populations. Thus, variable population sizes increase the amount of genetic differentiation among populations relative to the case where the same number of individuals were more evenly distributed among populations (see Whitlock, 1992a).

Extinction of local populations allows the variance among populations to increase in most circumstances (Slatkin, 1977; Wade and McCauley, 1988; Whitlock and McCauley, 1990; McCauley 1991, 1993; Whitlock *et al.*, 1993; Wade *et al.*, 1994), because extinction usually implies that new populations are being formed somewhere, either in the same habitat patches or elsewhere. Colonization is associated with founder effect; the size of a new local population is unlikely to be as large as the equilibrium size of a local population. It is possible for extinction and colonization to reduce differentiation among populations, but only if genes move predominantly when new populations are formed. Extinction and recolonization will increase the genetic differentiation among populations if

$$k < \frac{2Nm}{1 - \phi} + \frac{1}{2}, \qquad (2)$$

where k is the number of individuals which colonize new patches, N is the size of extant populations, m is the migration rate among populations, and ϕ represents the probability that two colonizing genes come from the same source population (Whitlock and McCauley, 1990). Note that in most cases this condition is met; unless the process of colonization is substantially different from the process of migration among extant populations, k will be approximately equal to Nm (both of which represent the number of individuals coming into a patch in one generation). The genetic differentiation of several species has been studied in relation to local extinctions (see McCauley, 1993; Giles and Goudet, this volume).

Similarly, whenever different populations have different demographic parameters, the variance in frequency of neutral genes across populations is affected. When populations split, the variance among them is increased, because of the drift allowed by reduced population sizes (Smouse *et al.*, 1981; Whitlock, 1994). When the populations split along family lines, the effect is greatly enhanced (Smouse *et al.*, 1981; Whitlock, 1994). The population structure of the Yanomama people (Smouse *et al.*, 1981) clearly demonstrates this effect.

Migration decreases the variance among populations, by mixing and homogenizing gene frequencies. The pattern of migration is critically important: spatially variable migration patterns result in a much higher degree of population differentiation than the mean migration rate would indicate (Whitlock, 1992b). This effect arises because some demes will have low immigration rates and can therefore drift especially strongly. Drift and migration affect population structure in a very nonlinear way, and hence mean differentiation is enhanced by the inclusion of relatively isolated populations. As will be seen below, the effects of demographic variability are even stronger when evolution depends on interactions between genes.

An important special case of spatial variation in migration rates is a type of metapopulation where some populations consistently send out more migrants than they receive and others receive more migrants than they put out (a "source-sink" metapopulation; Hanski and Simberloff, this volume). The sinks would have smaller (or zero) population sizes without migration from the sources. The differentiation among sink populations is likely to be much greater than that among source populations, but sink populations are not likely to diverge much from the source populations nearby (see Whitlock, 1996). The main implication from these models is that, while populations in nature can express large degrees of variance as usually measured, this may in fact mean very little in terms of the evolutionary potential of the metapopulation. If the sink populations are contributing much to the variance among populations, but have a low probability of persisting to contribute to future generations, then the evolutionary events which occur in such sinks have little relevance. (See the sections below on local adaptations and the shifting balance.)

Temporal variability of migration rate is less important than the above factors in affecting the differentiation among populations in neutral genes. The effective migration rate is decreased by a factor equal to half the temporal variance in migration rate ($m_e = \bar{m} - (\bar{m}^2 + \sigma_m^2)/2$), which will be a fairly small effect even with quite large coefficients of variation (Sved and Latter, 1977; Nagylaki, 1979; Whitlock, 1992b). However, temporal variation can become very important in predicting the evolutionary consequences of population structure for selected genes (see Moore and Tonsor, 1994, and below).

C. The Effective Size of a Metapopulation

One of the most important evolutionary properties of a species is its variance effective size. This is the size of an ideal population in which drift causes the genetic variance to decline at the same rate as in the population in question. (An "ideal" population is one in which individuals mate at random and have equal chances of reproductive success). The effective size helps predict the ability of the population to fix favorable mutations, the probability of fixation of deleterious alleles, the amount of standing variation available to selection for adaptation, and other important parameters (see, for example, the discussion in Crow and Kimura, 1970). A species which is subdivided into local populations has a different effective size than an undivided species with the same number of individuals (Slatkin, 1977; Maruyama and Kimura, 1980; Chesser *et al.,* 1993; Whitlock and Barton, 1995); therefore it is of interest to discover the specific implications of population subdivision for effective size.

In general, three factors affect variance effective population size, and each of them can be expressed in many ways. These are the census population size (i.e., the total number of individuals in the metapopulation), the variance in reproductive success among individuals (which can be caused by unequal sex ratios, variance in any fitness component, etc.), and nonrandom mating. Furthermore, correlations between these factors are important; correlation across generations in reproductive success of related individuals reduces effective size.

Subdivision also affects the ecological properties of a species. The total number of individuals in a subdivided species is likely to be much less than that in an undivided species. Subdivision is often the result of habitat destruction and fragmentation, which reduces the resource available to support the species. Moreover, smaller populations have a higher rate of extinction than large ones, so that many small habitat fragments are occupied only part of the time (Hanski, this volume). By the first principle above, as the census population size is reduced, the effective population size is also reduced.

Second, population subdivision can affect the variance in reproductive success. Higher variance in reproductive success translates into a lower effective population size, essentially because some individuals are contributing much to the next generation and there is hence a higher chance that a random pair of alleles will be identical. In the island model, with constant local population size and migration rate, the effective size of the metapopulation is actually increased relative to an undivided population of the same census size, by a factor of $1/(1 - F_{ST})$. This is because in the standard island model, the reproductive success of each population is fixed. Hence, the fitnesses of individuals within populations are negatively correlated. Thus, as the alleles within populations become more closely related, the variance in average reproductive success of each allele becomes smaller, and the effective metapopulation size grows. In contrast, if each population is allowed to have a different mean fitness, as for example in the natural case when local extinctions are allowed, the variance in reproductive success can be greatly increased by the clumping of individuals into local populations, and the effective size of the whole metapopulation can be greatly decreased (Hedrick and Gilpin, this volume). A more general model of effective size is complex (see Whitlock and Barton, 1996) but can be illustrated with a special case. If the census size of all populations is N, but their output varies, then the overall effective size, N_e, is

$$N_e = \frac{Nn}{(1 + V_B(R))(1 - F_{ST}) + 2NF_{ST}V_B(R)}. \tag{3}$$

Here there are n local populations, $V_B(R)$ is the variance in fitness among populations, and F_{ST} is the standardized variance across populations (Wright, 1969; Weir and Cockerham, 1984). Each population is assumed to have the same size, and migration is not geographically structured (Whitlock and Barton, 1996, describe more general cases). F_{ST} increases with decreasing local population size and decreasing migration rate; if F_{ST} is large and the variance in reproductive success among populations is small, then subdivision increases effective size. On the other hand, if populations vary in fitness, then N_e will be much less than it would be without subdivision (see Fig. 1 for an example with local extinction).

Finally, N_e is affected by nonrandom mating. In the equation above, the degree to which mating is nonrandom due to population subdivision is described by F_{ST}. Local mating can increase or decrease effective size, depending on the distribution of reproductive success.

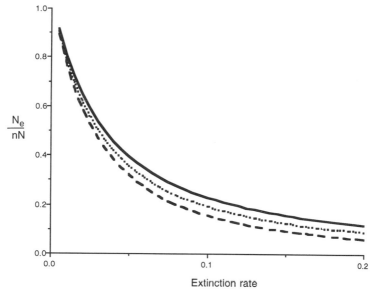

FIGURE 1   Increasing local extinction rates reduce the effective population size of the species. $N = 50$, $k = 2$, and $m = 0.1$. The upper continuous line is for unrelated colonists ($\phi = 0$), the dotted line is for $\phi = 0.5$, and the dashed line is for genetically identical colonists ($\phi = 1$). (From Whitlock and Barton (1996, Fig. 2).)

III. SELECTION IN METAPOPULATIONS

A. The Establishment of Favorable Alleles

The establishment of a new and favorable allele involves two stages: first, the increase from a single mutant copy to appreciable frequency around its original location, and second, its spread through the whole population. Even if a new mutant has an appreciable advantage, it is unlikely to be fixed: most copies will be lost by chance in the first few generations. With panmixis, the probability of fixation is just twice the selective advantage of the heterozygote ($2s$, for $s \ll 1$; Fisher, 1922; Haldane, 1927). Remarkably, this probability is not affected by spatial subdivision, provided that population sizes are constant and there is symmetric migration (Maruyama, 1970). Hence, the rate at which a species accumulates favorable alleles is not affected by subdivision and equals $4Ns$, where N is the effective size of the whole population. By analogy, the probability of fixation is not altered by balancing selection at genes linked to the gene in question, provided that the balanced polymorphisms do not fluctuate in frequency.

Random extinctions and recolonizations introduce an additional source of random drift and hence reduce the fixation probability: an advantageous allele may be lost when the population in which it arises goes extinct. In the island model, where migrants are drawn at a rate m from a common pool, the fixation probability can be calculated (Tachida and Iizuka, 1991; Barton, 1993). If colonists are drawn from many populations, the fixation probability cannot be reduced by more than half, to s instead of $2s$. However, if colonists are derived from a single ancestral gene (as may be if there is severe inbreeding at foundation of new populations), extinction can have a much greater effect (Fig. 2). The chance that a weakly favored allele will be fixed is reduced from $2s$ to $2s/[(1 + \lambda/m)(1 + 2N\lambda)]$, where λ is the rate of extinction and N the deme size (assumed to be constant; Barton, 1993). Thus, if the rate of extinction is higher than the rate of migration ($\lambda \gg m$), and faster than random drift ($\lambda \gg 1/2N$), the reduction can be severe; a species consisting of n demes will accumulate favorable mutations at a rate $2sn\mu m/\lambda^2$, which is independent of deme size and inversely proportional to the square of the extinction rate.

The effect of extinction on fixation probability is not the same as its effect on the effective size of the metapopulation ($N_e/N = (1 + 2N\lambda + 4Nm)/(4N(m + \lambda))$; Slatkin, 1977), and so extinction could interfere substantially with adaptation without having an appreciable effect on the diversity of neutral markers (provided that local population sizes are very small). However, abundant species may not be limited by the establishment of new mutations. Adaptation to changed circumstances may depend on alleles that previously occurred at appreciable frequency, as with quantitative traits of high heritability and as may be the case for industrial melanism in the peppered moth (Lees, 1981). Even if new mutations are required, these may occur so often as to cause negligible delay. The rate of mutation to any new protein differing by one amino-acid is $\approx 10^{-9}$, and so all such mutations occur at least every generation if the whole population is larger than $\approx 10^9$. For example, warfarin resistance in rats arose independently at least twice in Britain within 10 years of the application of this poison (Drum-

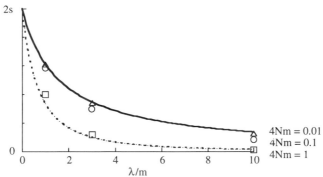

FIGURE 2   For a given ratio between extinction and migration (λ/m), the fixation probability decreases as migration and extinction rates increase. The solid curve gives the limit of low migration and extinction rates ($2s/(1 + \lambda/m)$), while the dotted curves give the limit of weak selection ($2s/[(1 + \lambda/m)(1 + 2N\lambda)]$). The symbols give exact numerical solutions. Colonists are derived from a single gene; $4Ns = 1$. (From Fig. 2 of Barton (1993).)

mond, 1966). Thus, the loss of favorable alleles caused by random extinction in a metapopulation may not have significant long-term consequences.

Once an allele is established in large numbers, it is almost certain to spread through the whole population. In the island model, it will increase exponentially, as $\bar{p}/\bar{q} \approx \exp(t\, s(1 - F_{ST}))$, where $\bar{p}$ is its average frequency, s its selective advantage, and t the time. The rate of increase is reduced by a factor $1 - F_{ST}$ because selection can act only on that fraction of genetic variance that is found within populations (see below). This result applies even if population sizes vary randomly, as in the usual metapopulation models; since F_{ST} is usually small, subdivision is unlikely to delay the spread of favorable alleles by much. When migration distances are restricted, the way in which an allele spreads depends critically on the pattern of migration (Mollison, 1977). If the number of long-range migrants falls away exponentially or faster, then the allele spreads as a smooth wave, with speed $\sigma\sqrt{2s}$, where σ is the standard deviation of distance between parent and offspring (Fisher, 1937). However, if migration is more leptokurtic than this, spread is faster and messier: single genes jump ahead and establish themselves, while others leapfrog past (see Shaw, 1995). While luck is needed to observe this process as it occurs, it may leave a trace in the pattern of secondary contact. If two populations advance to meet each other in smooth waves, they will become separated by a sharp genetic boundary, but if they spread through sporadic advances, they will eventually be demarcated by a broad cline of variable shape (Nichols and Hewitt, 1994).

B. Deleterious Mutations

In higher organisms, the net rate of mutation to deleterious alleles is high enough to cause a substantial "mutation load" (Kondrashov, 1988). In a sexual population, this manifests itself in three ways. First, when an equilibrium is reached between mutation and selection, mean fitness is reduced. If the effects of mutations are multiplicative, fitness is reduced by a factor $\exp(-U)$, where U is the total rate of mutation across the diploid genome. If additional mutations reduce fitness by a larger factor ("synergistic epistasis"), this load may be reduced (Kondrashov, 1984).

Second, mutations may be fixed in the whole population. This process may lead to a catastrophic decline in fitness in small populations ("mutational meltdown"; Lynch et al., 1993), but may also destroy adaptations based on very weakly selected alleles in even large populations. For example, codon usage is biased toward the most efficiently translated codon, but the bias is not complete; this suggests that the less favorable codon has often been fixed by drift (Bulmer, 1991). Finally, deleterious mutations interfere with selection at linked loci. In an asexual population, weakly favored alleles can be fixed only if they arise in the genetic background carrying no deleterious mutations, which greatly reduces their chances (Peck, 1994). Even in a sexual population, however, deleterious muta-

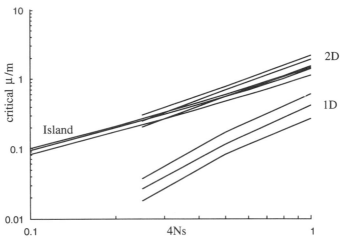

FIGURE 3   The critical ratio between mutation and migration $(\mu/m)_{\text{crit}}$, plotted against $4Ns$. If the mutation rate rises above this value, the deleterious allele is fixed. Each set of three lines is for (from bottom to top) $\mu/s = 0.2, 0.1, 0.05$. Values for the island model are calculated analytically, while those for one- and two-dimensional stepping stone models are taken from simulations of 401 populations (in 1D) or of 31×31 populations (in 2D), each with $2N = 20$ haploid individuals.

tions can substantially reduce fixation probabilities (Charlesworth, 1994a; Barton, 1995).

Does spatial subdivision accentuate these various deleterious effects of mutation? One can show by a simple argument that for arbitrary population structures, fitness in a mutation-selection balance is reduced from $\exp(-U)$ to $\exp(-U/(1 - F_{\text{ST}}))$. To see this, denote the average frequency of the deleterious allele by $\bar{p}$, and the frequency in the ith deme by p_i. Then

$$\Delta p_i = \mu q_i - sp_i q_i + \zeta_i + m(\hat{p}_i - p_i), \tag{4a}$$

Here, $\hat{p}_i$ is the frequency of the allele among immigrants to deme i, and m is the migration rate. ζ_i is the perturbation due to random drift. Taking the average across demes gives

$$\Delta \bar{p} = \mu \bar{q} - s\bar{p}\bar{q}(1 - F_{\text{ST}}) + \bar{\zeta}, \tag{4b}$$

$\bar{\zeta}$ is the fluctuation in average allele frequency due to drift, which is on average zero. F_{ST} is the standardized variance in allele frequency across demes. The fluctuations due to drift average out, and the effect of migration disappears if we make only the assumption that migration does not create or destroy alleles. Taking expectations and solving for equilibrium gives $\bar{p} = \mu/[s(1 - F_{\text{ST}})]$. Since F_{ST} is usually small, this remarkably general result implies that subdivision is unlikely to much increase the mutation load. Strictly speaking, Eq. (4b) applies with F_{ST} denoting the standardized variance of deleterious alleles, which may differ from that for neutral markers. Simulations of a variety of population structures show that when migration rates are sufficiently low relative to mutation, deleterious alleles can fix in some demes, leading to high F_{ST}'s and a high load. When μ/m is above a critical ratio, deleterious alleles fix everywhere. However, this requires very low migration rates (Fig. 3).

By assuming that fluctuations within populations occur on a much shorter time scale than changes in the overall frequency (reasonable if there are very many populations and selection is weak), one can show that the probability of fixation of a deleterious allele is $2s(1 - F_{ST})(N_e/N)/(\exp(4N_e s(1 - F_{ST})) - 1)$, where N_e is the effective size of the whole population, and N is its census size. Though fluctuations in population size (and in particular, local extinction) can reduce the effective size below the census size (see above), this argument shows that deleterious alleles are likely to be fixed in the whole species only if selection is very weak (say, $4N_e s(1 - F_{ST}) < 10$). Population subdivision therefore does not greatly increase the chance loss of adaptations.

IV. ADAPTATION TO LOCAL CONDITIONS

Spatial subdivision has potentially important consequences when selection varies from place to place. There has long been controversy over whether geographic isolation is required for the divergence of lineages. Darwin and Wallace held that populations could adapt to gradually varying conditions across their range and that this could lead to speciation. In contrast, Wagner emphasized the importance of barriers to migration in allowing speciation (see Mayr, 1982, p. 562). More recently, Ehrlich and Raven (1969) argued that because genes diffuse slowly through most species, gene flow cannot be a significant factor impeding divergence. Population genetic theory strongly supports this view, at least if one considers adaptations which can be built up by individual alleles or by continuous changes in quantitative traits. We consider the complications of gene interactions in the next section.

Whether populations can adapt to local conditions depends essentially on the relative rates of gene flow and selection. The simplest case is where an allele is favored in a population, and has selective advantage s in the heterozygote. The allele can be established despite gene flow, provided that the rate of immigration of individuals carrying the alternative allele is lower than the rate of selection ($m < s$; Haldane, 1931). One can generalize this to the island model, where the allele has advantage s in a fraction α of demes and disadvantage $-\beta s$ in the remainder. The allele can be established if migration is lower than the critical value $m_{crit} < \beta s/(\beta(1 - \alpha) - \alpha)$, or alternatively, if it is favored in at least a fraction $\alpha_{crit} = (1 - s/m)/(1 + 1/\beta)$ of populations. The allele can always be established if $m < s$ (heavy curves in Fig. 4a).

If genes diffuse at a rate s across a continuous habitat, then an allele can

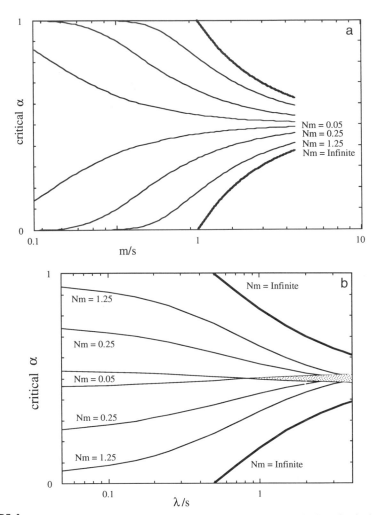

FIGURE 4 The critical proportion of populations, α, required to allow local adaptation in the island model. In a proportion α of populations, an allele is favored by additive selection s, while in the remainder, it is at a disadvantage $-\beta s$ ($\beta = 1$ in this example; fitnesses are $1:1 + s:1 + 2s$ or $1:1 - s:1 - 2s$). (a) Results with no extinction, plotted against the relative rates of migration and selection, m/s. The heavy curves show the deterministic limit, in which drift is negligible. The lower heavy curve shows the proportion of populations in which allele A is favored that is required for it to invade ($\alpha_{crit} = (1 - s/m)/(1 + 1/\beta)$). The upper heavy curve shows the corresponding criterion for the other allele to invade. Thus, both can be maintained for α between the two curves; polymorphism and local adaptation is always possible if $s > m$ (i.e. $m/s < 1$). The three pairs of light curves show the critical values for $Nm = 1.25$, 0.25, and 0.05: as the number of migrants decreases, conditions for polymorphism become more restrictive. (b) Corresponding conditions with random extinction at a rate λ, plotted against the relative rates of extinction and selection, λ/s. Colonization involves severe inbreeding, so that colonists are derived from one gene ($\phi = 1$). These are calculated exactly, from the transition matrix. Migration is fixed at $m = 0.025$ and selection at $s = 0.05$. The heavy curves give the deterministic limit ($\alpha_{crit} = (1 - s/(\lambda + m))/(1 + 1/\beta)$), while the three light curves give the limits for $Nm = 1.25$, 0.25, and 0.05 ($2N = 100$, 20, 4, respectively). The shaded region to the right, which lies between the curves for $Nm = 0.05$, shows the region in which neither allele can invade, so that the population evolves to fixation for one or the other.

become established provided that it is favored in a sufficiently large region, wider than a characteristic scale $l = \sigma/\sqrt{2s}$ (Slatkin, 1973). In a linear habitat, if the allele is at an advantage s in some region and a disadvantage $-\beta s$ elsewhere, then it can be established if the region is wider than $2\tan^{-1}(\sqrt{\beta})l$ (Nagylaki, 1975). New mutations have a probability of fixation approaching $2s$ if they arise within the appropriate region, but the probability falls to zero outside this region (Barton, 1987). A similar condition applies to adaptations based on quantitative traits; here, the characteristic scale is $\sigma\sqrt{V_s/2V_g}$, where V_s is a measure of the strength of stabilizing selection, and V_g is the additive genetic variance (Slatkin, 1978b). Since the dispersal range σ is typically much smaller than the species' range, weak selection can allow a species to adapt to diverse local conditions. This can be seen directly. For example, the grass *Agrostis tenuis* has evolved tolerance to heavy metals on mines only a few meters across (MacNair, 1987). Similarly, narrow clines ≈ 10 km wide separate races of *Heliconius* butterflies adapted to different mimicry rings (Mallet, 1993). Such examples leave little doubt that gene flow need not prevent divergence over very short scales.

A. The Effect of Population Turnover on Local Adaptation

How much do the local extinctions characteristic of metapopulation dynamics impede adaptation? Extinctions have two effects: they introduce an additional source of random drift, making selection less efficient; and extinctions increase gene flow, because colonists may be derived from different habitats. We can illustrate these effects using the simple island model introduced above, in which an allele is favored in a fraction α of populations; this fraction must be greater than some critical threshold for it to increase from low frequency. Similarly, α must be below a second threshold if the other allele is to invade; if both conditions are met, a polymorphism can be maintained, and there is some degree of local adaptation (Fig. 4a). Random drift reduces the conditions under which both alleles can be maintained; if Nm is much less than ≈ 1, then variation is likely to be lost, even if migration is weaker than selection ($m \ll s$). However, as we argued above, the low values of F_{ST} usually observed imply fairly high Nm, suggesting that local adaptation is not usually much impeded by drift. Our model includes no explicit spatial structure. However, the effect of random drift on neutral variation in a two-dimensional population also depends on Nm (or equivalently, on Wright's 1943 neighborhood size), so that the arguments may carry over to this more realistic case (Malécot, 1948, 1969).

We can now ask how extinctions, at a rate λ, affect the conditions for local adaptation. We need only consider alleles which are very rare in the migrant pool. Consider first the deterministic case ($Nm \gg 1$). where habitat patches are immediately recolonized at frequency $\bar{p}$, and suppose for the moment that selection is stronger than migration ($s \gg m$). In a fraction α of populations, the allele increases as $\bar{p}\ \exp(st)$, while in remainder, it decreases at a rate $\exp(-\beta st)$. The probability that the populations will survive for time t is $\approx \exp(-\lambda t)$; averaging over this distribution, the expected contribution to the migrant pool is $\alpha\lambda\bar{p}/(\lambda - s) + (1 - \alpha)\lambda\bar{p}/(\lambda + \beta s)$. If this value is greater than the present frequency, $\bar{p}$, then the allele will increase; hence, the threshold is $\alpha_{crit} = (1 - s/\lambda)/(1 + 1/\beta)$. If selection is faster than extinction ($s > \lambda$), then the allele is likely to reach very high frequency before the population goes extinct, and so invasion is possible for arbitrarily small α. The argument is easily extended to include the effects of migration, which reduces the rate of increase from low frequency to $\exp((s - m)t)$, leading to $\alpha_{crit} = (1 - s/(\lambda + m))/(1 + 1/\beta)$. Thus, continual migration and sporadic extinction have the same effect, both tending to increase the effective rate of gene flow. The allele can always be established if its advantage outweighs their combined effects ($s > (\lambda + m)$); otherwise, it can be sustained only if it is favored in a sufficiently large fraction of the metapopulation (heavy curves in Fig. 4b).

This argument applies only when populations are large enough to make drift negligible even during colonization. It is hard to find analytic results which incorporate the effects of random drift, because it is then likely that the allele will drift to high frequency in some populations, even when it is rare in the metapopulation as a whole. Analysis of selection, migration, extinction, and drift become intractably complicated when alleles become so common that nonlinear interactions are important. However, one can solve the model exactly, by calculating the transition matrix for each population—that is, the probability that a deme carrying i copies of the allele will carry j copies in the next generation, taking into account migration, selection, extinction, and drift. This shows that extinction can have surprising consequences when the number of migrants is small. As in Fig. 4a, Fig. 4b shows a pair of curves for each value of Nm; these give the critical proportions of populations (α) required to allow either allele to invade. To the left, the region between the curves shows the conditions under which both alleles can invade from low frequency, allowing an adaptive polymorphism. As the rate of extinction increases, polymorphism becomes less likely (right of Fig. 4b), as expected from the deterministic limit. As the rate of extinction increases further, the curves cross, so that when the proportions of the two kinds of habitat are equal ($\alpha \approx 0.5$; shaded region on right of Fig. 4b for $Nm = 0.05$), neither allele can invade. This implies that the metapopulation can fix for one or the other allele, depending on starting conditions.

B. Limits to Local Adaptation

It is striking that during drastic changes in climate, the distribution of many species shifts over large distances, apparently without much change in the range of environmental conditions which they experience (e.g., Atkinson et al., 1987). Species may fail to evolve further under natural selection, because they lack the necessary genetic variation, perhaps reaching some physical limit or becoming trapped in a coevolutionary race with competing species (Stenseth and Maynard

Smith, 1984). Another possibility is that gene flow from the more populated areas prevents adaptation at the edge of a species' range, thereby preventing advance into regions with different conditions from the ones within the occupied range.

The arguments so far have assumed that population size is independent of genotype and habitat. However, since small populations are likely to receive most immigrants from large populations, their ability to adapt may be substantially limited. An example is given by Dias *et al.* (1996), who show that populations of the great tit living in evergreen habitat on the island of Corsica lay their eggs later than do those living in deciduous habitat on the mainland, so that fledglings coincide with the later profusion of insects emerging on this habitat. However, tits living in the minority evergreen habitat on the mainland lay their eggs earlier than appears optimal. There is evidence of a high rate of gene flow into the evergreen habitat, since strong linkage disequilibrium is found between microsatellite allele combinations characteristic of populations in the majority habitat.

The potential power of gene flow to prevent adaptation at the edge of the range can be illustrated by a simple model of a quantitative trait. First, take the pattern of population density as given. The mean value of the trait then changes according to

$$\frac{\partial z}{\partial t} = \frac{\sigma^2}{2}\frac{\partial^2 z}{\partial x^2} + \sigma^2\frac{\partial \log(N)}{\partial x}\frac{\partial z}{\partial x} + \frac{V_g}{2}\frac{\partial \log(\overline{W})}{\partial z}. \tag{5}$$

The first term represents the diffusion of genes from place to place at a rate σ^2 (the standard deviation of distance between parent and offspring), while the third represents the tendency of natural selection to increase mean fitness (Pease *et al.*, 1989; Garcia-Ramos and Kirkpatrick, 1995). The second term gives the effects of a gradient in population density, which tends to increase the trait mean, z, if it is higher where density is higher ($\partial \log(N)/\partial x > 0$ and $\partial z/\partial x > 0$). To find the relative importance of regions with different density, define a quantity H which combines the effects of gene flow and natural selection[1] and which depends on the whole field of trait means $z(x, t)$

$$H[z(x, t)] = \int_{-\infty}^{\infty} N^2\left[\frac{V_g}{2}\log(\overline{W}) - \frac{\sigma^2}{4}\left(\frac{\partial z}{\partial x}\right)^2\right] dx. \tag{6a}$$

One can show, using Eq. (5), that H always increases:

$$\frac{\partial H}{\partial t} = \int_{-\infty}^{\infty} N^2\left(\frac{\partial z}{\partial t}\right)^2 dx. \tag{6b}$$

[1] This is a generalization of Eq. (2) of Rouhani and Barton (1987b) to allow for varying density; a similar expression can be found for allele frequency change, which is also weighted in proportion to N^2 (Barton and Hewitt, 1989).

Thus, H tends to a maximum, which is a compromise between the tendency of natural selection to increase mean fitness and of gene flow to reduce the steepness of clines, $(\partial z/\partial x)^2$. The important point here is that the effects are weighted in proportion to the square of population density, N^2: in this sense, a region with 10 times greater density has a 100-fold greater influence on the effects of gene flow and selection.

Slatkin (1973) and Nagylaki (1975) have used diffusion equations like Eq. (5) to study the effects of asymmetric gene flow in swamping adaptations based on single loci. Here, we illustrate the effect by using a simple model of a quantitative trait. Assume that the trait has constant genetic variance and is under stabilizing selection, with an optimum which changes linearly along a cline. The mean fitness of the population is then $\overline{W} = \exp(-(z - \beta x)^2/2V_s)$, where z is the mean of the trait, V_s is a measure of the strength of selection, and β is the rate of change of the optimum in space. Genetic variation around the optimum reduces mean fitness by $L = V_g/2V_s$, a measure of the load on the population due to stabilizing selection. For simplicity, we ignore the environmental variance. If population sizes were uniform, there would be a simple cline, with the mean tracking the optimum over an indefinite range ($z = \beta x$; Felsenstein, 1977; Slatkin, 1978a). However, if the density falls away as a Gaussian with width w (i.e., $N = N_0 \exp(-x^2/2w^2)$), the mean cannot track the changing optimum as effectively, because gene flow from the abundant region which is adapted to an optimum at zero tends to impede adaptation to the different conditions in less dense regions (Garcia-Ramos and Kirkpatrick, 1996). The solution which maximizes H is $z = \beta/[1 + (2V_s/V_g)(\sigma/w)^2] = \beta/[1 + L(\sigma/w)^2]$. This has a lower slope, reflecting a failure to adapt outside the region of high density. The effect depends on the load due to stabilizing selection (L) and on the distance over which density declines, relative to the dispersal range (σ/w).

So far, we have taken the population density as given. If density decreases with mean fitness, then there is a positive feedback, such that populations which are less well adapted become smaller, making it harder for them to resist gene flow, and so further reducing their mean fitness and their density (Kirkpatrick and Barton, 1996). Suppose that numbers are related to mean fitness through $N = N_0\overline{W}^\gamma$; $\gamma = 0$ implies soft selection, while large γ implies that the population size is very sensitive to mean fitness. One solution is perfect adaptation: $z = \beta x$, in which case numbers are constant, the cline is linear, and gene flow (the first two terms in Eq. (5)) has no effect. This solution is always stable to small perturbations, since the second term is proportional to $(z - \beta x)^2$. However, if γ is large enough, there can be another equilibrium, with adaptation only within a limited region. This second solution is $z = \omega x$, where $\omega = (\beta/2)(1 + \sqrt{(1 - (\beta_c/\beta)^2)}) < \beta$, and the dimensionless parameter $\beta_c^2 = 2V_g/\sigma^2\gamma$. This solution is neutrally stable: the population settles to be adapted to some arbitrary z, but perturbations can shift it to the left or right. For this possibility to exist, $\beta > \beta_c$ where $\beta_c^2 = 2V_g/\sigma^2\gamma$—that is, the change in optimum must be rapid (β large), and the population sensitive to $\overline{W}$ (γ large). This scheme may not be plausible

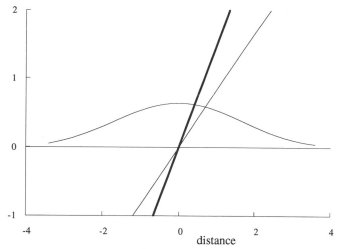

FIGURE 5   The collapse in adaptation at the edge of a species' range. The graph show numerical solutions to a model of population density, N, and the mean of a quantitative trait, z, across a one-dimensional habitat. Individuals reproduce and disperse continuously in time, with a net growth rate $r(1 - N/K) - (z - z_{opt})^2/2V_s$, and stabilizing selection toward an optimum that changes at a rate β in space ($z_{opt} - \beta x$; heavy line). Assuming that the trait z is normally distributed with mean $\bar{z}$ and variance V_g leads to equations $\partial N/\partial t - (\sigma^2/2)\partial^2 N/\partial x^2 + rN(1 - N/K) - (V_g/2V_s)N - N(\bar{z} - \beta x)^2/2$, $\partial \bar{z}/\partial t = (\sigma^2/2)\partial^2 \bar{z}/\partial x^2 + \sigma^2 \partial \bar{z}/\partial x \log(N)/\partial x - (\bar{z} - \beta x)(V_g/2V_s)$. Scaling, the only two parameters are $A = (V_g/rV_s) = 2L/r$, $B = (\beta\sigma/r)/\sqrt{2V_s}$. For this example, $A = 1$, $B = 1.5$. Collapse in adaptation is possible if $B\sqrt{2} > A$, or $(\beta\sigma/\sqrt{2V_g}) > \sqrt{8L}$. The figure shows the equilibrium numbers (bell-shaped curve) and trait mean (light line).

for one trait, since it requires that the strength of density-dependent regulation, γ, be lower than $(2V_g/\sigma^2\beta^2)$, which is the change in optimum over one dispersal range, measured in genetic standard deviations. However, if the optimum for many traits changes in the same region, this constraint becomes much less restrictive, suggesting that multivariate adaptations in marginal populations may collapse in the face of gene flow.

This model is simplified by the assumption that population density and gene flow are directly related ($N = N_0\bar{W}^\gamma$). However, a model which follows the joint change in density and in the trait behaves in a similar way (Fig. 5), with adaptation collapsing outside some arbitrary region if the optimum changes rapidly enough. There are analogous results for discrete loci. Suppose that there are two alleles, P and Q, with fitness of P given by $W_P = r(1 - N/K) - r\Omega_P(x)$, and similarly for Q. The functions Ω_P, Ω_Q represent additional death rates, over and above the density-dependent term, which differ slightly for the two alleles. With no migration, a population fixed for Q would equilibrate at $N = K(1 - \Omega_Q x)$ for $\Omega_Q(x) < 1$ and would go extinct if $\Omega_Q(x) > 1$. Ω_Q is chosen to be large enough that the population declines to zero to the left (Ω_Q large for $x \ll 0$). In the bulk of the species' range, allele P has disadvantage s ($\Omega_Q > \Omega_P$ for $x > 0$), but has a selective advantage S at the edge (($\Omega_P - \Omega_Q$) $\approx S$ for $x < 0$). For this case, one can show that if selection is weak ($\Omega_P \approx \Omega_Q$), there is a critical value of s, above which allele P cannot be fixed, even though it is favored at the edge of the range and would extend the range if it could be established. Crucially, this critical value is small ($s_{crit} \approx S^2$), so that an allele must have a very weak disadvantage in the bulk of the range if it is to be established by selection in a sparse and limited region.

Thus far, we have discussed adaptations based on one gene or on one quantitative trait. In this simplest of cases, gene flow only prevents adaptation if it exceeds some critical value: the outcome depends on the relative strengths of selection and gene flow. Divergence may usually occur in this way, and new species may arise as a by-product if the adaptations themselves cause reproductive isolation. For example, races of *Rhagoletis* flies adapted to different host plants mate on the fruit from which they emerged, and also emerge at different times, leading to partial isolation (Feder *et al.*, 1990a,b). In the monkey flower *Mimulus guttatus*, the allele responsible for resistance to heavy metals interacts with another gene to cause hybrid sterility (MacNair and Cumbes, 1989). In the next section, we consider the more complex issues which arise when certain *combinations* of genes are required for adaptation. In this case, it may be that populations must be reproductively isolated before they can adapt—either because of some physical barrier to gene flow, or because of genetic differences that impede interbreeding. Population structure then interacts with selection in a more complex way, and the processes of adaptation and speciation are closely intertwined.

V. SPECIATION AND THE "SHIFTING BALANCE"

Natural selection may cause populations to evolve toward alternative stable states. For example, heterozygotes between different chromosome arrangements may not pair and segregate properly in meiosis, leading to partial sterility (White, 1973). This kind of selection against heterozygotes leads to fixation of one or the other type. Thus, while species often contain several chromosome arrangements, these are usually found in different places, forming "chromosome races" separated by narrow clines. Other kinds of selection can lead to alternative equilibria. *Heliconius* butterflies are distasteful and have evolved conspicuous colour patterns to advertise their distastefulness to predators. In any one area, there is strong selection for convergence to a common pattern ("Müllerian mimicry"), but populations in different areas have established different patterns (Turner, 1981). Where these pattern races meet, they are separated by narrow clines which are maintained by selection against heterozygotes, against recombined genotypes, and against rare alleles (Mallet, 1993).

In general, multiple stable states are likely, since different gene combinations may often fulfill the same function. Sewall Wright (1932) introduced an influential metaphor for thinking about these multiple states, the "adaptive landscape." This is best defined as a graph of mean fitness against the state of the population, which can be described by allele frequencies or by the means of quantitative traits (see Provine, 1986). Natural selection tends to increase mean fitness, and so populations evolve toward the nearest "adaptive peak." However, this may not be the global optimum, in which case adaptation will be impeded, because populations cannot evolve toward the global optimum through a sequence of changes, each favored by selection. This problem was, of course, a major concern for Darwin, most notably in his discussion of the evolution of the eye (Darwin, 1859, Chapter 6). Multiple stable states are also involved in speciation: hybrids between populations at different peaks are less fit, and conversely, most models of reproductive isolation lead to multiple equilibria (Barton and Charlesworth, 1984).

Wright (1932) proposed that species may most efficiently adapt by means of a "shifting balance" between evolutionary forces. He divided this process into three phases. In the first, random fluctuations (due, for example, to sampling drift) cause local populations ("demes") to move into the domain of attraction of new adaptive peaks. In the second phase, selection within populations takes them to the new peaks. Finally, different adaptive peaks compete with each other, by a variety of processes, such that "fitter" peaks tend to spread through the whole species. This third phase involves an element of group selection. For example, adaptive peaks may spread if they increase the size of the local population, or the number of emigrants. However, selection between adaptive peaks is not opposed by selection between individuals, and so this component of the "shifting balance" is more plausible than the more familiar models in which group selection is opposed by individual selection (e.g., Kimura, 1985; Nunney, 1985; Wilson, 1987).

The theory underlying the first two phases of the "shifting balance" is well established, while the third phase has only recently received detailed attention. Unfortunately, evidence on the actual significance of the shifting balance for adaptation and speciation is sparse. We first summarize the existing understanding of the shifting balance model and then concentrate on the specific issue of how local extinctions and recolonizations (or more generally, variation in population structure) affect the process.

A. The Classical Shifting Balance Model

The first requirement is that alternative stable states—roughly speaking, adaptive peaks—exist. This is plausible on both theoretical and empirical grounds. Most multilocus models admit many equilibria, whether fixed or polymorphic (Feldman, 1989). Wright (1935) analyzed what is perhaps the simplest case, in which stabilizing selection acts on an additive quantitative trait. Here, many combinations of genes lead to the same phenotype, and so each can be established by selection. If mutation maintains variation around the optimal phenotype, the number of stable equilibria increases still further (Barton, 1986). More generally, if fitnesses are randomly assigned to genotypes, there are large numbers of local adaptive peaks, which are reached in only a few steps from a randomly chosen starting point and may be well below the global optimum (Kauffman and Levin, 1987). The most obvious evidence is that populations which hybridize and yet remain genetically distinct even when living in the same environment have presumably reached different adaptive peaks (Harrison, 1993).

Unfortunately, while the existence of multiple adaptive peaks is an essential precondition for the shifting balance process, it does not show that the divergence of populations onto different stable states has occurred in opposition to natural selection. Natural selection can cause populations to evolve from some ancestral state to a variety of fitter states, which may turn out to be incompatible with each other. For example, Robertsonian fusions between chromosome arms may cause little or no meiotic nondisjunction in the heterozygotes with the ancestral unfused chromosomes. However, there may be severe sterility in the heterozygote between different fusions, since many chromosomes may attempt to pair (Bickham and Baker, 1986; Searle, 1986). More generally, different mutations may arise and be established by natural selection in different places, and may turn out to be incompatible with each other when they meet (Bengtsson and Christiansen, 1983). It is plausible that much reproductive isolation evolves in this way, without the need for any peak shifts in opposition to selection (Orr, 1995). One should note, however, that while this avoids the first phase of the shifting balance, incompatible gene combinations may still compete with each other in the third phase (below).

The question of whether different populations are often at different adaptive peaks can be answered, at least in principle, by measuring the extent and scale of "outbreeding depression," in which crosses between populations lead to reduced fitness. There has been surprisingly little work along these lines. Recent experiments with plants have suggested significant reduction in hybrid fitness over short scales (Waser and Price, 1994; Burt, 1995). On the other hand, crosses between the much more divergent taxa involved in hybrid zones have given equivocal evidence of reduced hybrid fitness and suggest a more important role for adaptation to different environments (Arnold and Hodges, 1995).

Most theoretical attention has focused on the probability that random drift will establish a new adaptive peak in opposition to selection. Wright (1941) showed that the chance of a new mutation with disadvantage s in the heterozygote being established in a deme of size N is proportional to $\exp(-Ns)$; the diffusion approximation for this probability, which is reasonably accurate even for small N, is $\sqrt{s/N\pi}\exp(-Ns)$ (Lande, 1979; Hedrick, 1981). This conclusion extends to a wide class of models which can be described by drift and selection across an adaptive landscape: the probability of a peak shift is in general proportional to $\overline{W}^{2N}$, where $\overline{W}$ is the mean fitness of the population in the adaptive valley, compared to the original adaptive peak (Barton and Rouhani, 1987). Thus, peak shifts must either occur in a very small population or involve very slight reduction in mean fitness if they are to occur with reasonable probability. Migration into the population impedes peak shifts; for example, with selection against heterozygotes, Nm immigrants per generation reduce the probability by a factor 2^{-4Nm} (Lande, 1979; Barton and Rouhani, 1991).

The chances of a peak shift can be greatly increased by a severe population bottleneck, as for example during the founding of a new population. During the brief period of small population size, selection is negligible relative to random drift, and the occurence of a peak shift depends mainly on the chance that the population will drift across the "adaptive valley" (Rouhani and Barton, 1987a). For an additive quantitative trait, the variance of the population mean is $2FV_g$, where F is the net reduction in heterozygosity and V_g the initial genetic variance (Barton and Charlesworth, 1984). Thus, with severe inbreeding, and high heritability, a shift of a few phenotypic standard deviations is not unlikely. However, if substantial reproductive isolation is to arise during a founder effect, there must have been substantial variation in the initial population, and this variation must have been subject to selection. In a variety of models, the expected isolation produced by a founder event is proportional to the standing genetic load due to this variation (Barton, 1989).

Models of peak shifts in a single population address only the first two phases of the shifting balance process. A full description of the entire process must include the spread of a new adaptive peak through a metapopulation. The simplest case is the island model, where Wright's (1937) formula for the equilibrium between migration, selection, mutation, and drift allows a complete analysis. In a metapopulation consisting of a large number of populations of size N, allele frequency is distributed as $\psi(p) \approx p^{4Nm\bar{p}+4N\mu-1}q^{4Nm\bar{q}+4N\nu-1}\overline{W}^{2N}$, where $\bar{p}$ is the allele frequency in the migrant pool, and μ, ν the mutation rates from Q to P and vice versa. (There is a similar formula for the distribution of a quantitative trait). This distribution itself determines the composition of the migrant pool ($\bar{p} = \int_0^1 p\psi(p)\,dp$), giving an equation which can be solved numerically (Barton and Rouhani, 1993). For small numbers of migrants, populations shift independently of each other. Shifts are more likely toward the fitter peak than away from it, and so the stochastic equilibrium is biased toward the fitter peak. Because shifts are more likely to be to whichever state is commoner in the whole population, there is a positive feedback which increases the bias as the number of migrants increases (left of Fig. 6). However, when the number of migrants is greater than some critical value ($Nm > Nm_{crit} \approx 1$), a rare adaptive peak cannot spread in the face of migration from populations at the commoner peak even if it confers greater fitness. There are then two stable states for the whole metapopulation, and the global optimum cannot be reached (right of Fig. 6). Adaptation is thus most efficient when the number of migrants is just below the critical value, since the bias in favor of the fitter peak is then greatest. This bias can be large, even when the difference in fitness between the two peaks is small. If one adaptive peak also increases the population size or the number of emigrants, then group selection assists its spread. However, this is a weak effect, of second order in selection (Rouhani and Barton, 1993).

Analysis of a population structured in two dimensions is more difficult, but leads to qualitatively similar conclusions. A new adaptive peak can be established by chance provided that the number of migrants (or equivalently, the neighborhood size) is small; there is no requirement for strict geographic isolation. The probability of a shift is given by the chance that the new adaptive peak is established in an area large enough that its advantage over the old peak outweighs the swamping effect of gene flow, allowing it to spread through the whole population. In a continuous habitat with density ρ and dispersal rate σ^2 this probability is proportional to $\exp(- C\, Nb/\alpha)$, where $0 < \alpha < 1$ is a dimensionless measure of the asymmetry between the peaks, and Nb is Wright's neighborhood size, $4\pi\rho\sigma^2$ (or $4\pi Nm$ in a stepping-stone model). This applies for both quantitative traits under disruptive selection (Rouhani and Barton, 1987b) and selection against heterozygotes (Barton and Rouhani, 1991). Just as in the island model, there can be a very strong bias in favor of even a slightly fitter peak; indeed, in an infinite two-dimensional habitat, it is impossible for an inferior peak to be fixed. However, the chance of a shift depends only weakly on the strength of selection: a strongly selected shift is less likely to occur by chance over a given area, but need only spread over a smaller area to overcome gene flow from outside.

Wright believed that fitter peaks would spread because the populations in which they are established would send out more migrants, and hence would inevitably pull neighboring populations into the same state. This kind of deterministic spread occurs in the models discussed above, where a new peak sweeps through a continuous habitat (Rouhani and Barton, 1987b; Barton and Rouhani, 1991). In the island models (Barton and Rouhani, 1993; Rouhani and Barton, 1993), shifts are stochastic, but the evolution of the whole metapopulation is also deterministic. Again, there is an advantage to those populations which send out more migrants. However, in both cases the spread is primarily driven by selection

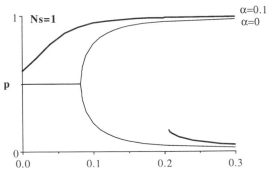

FIGURE 6   The overall mean allele frequency ($\langle p \rangle = \bar{p}$), as a function of the number of migrants (Nm), for selection against heterozygotes; $N\mu = 0.01$, $Ns = 1$; calculated from Eq. (23a) of Barton and Rouhani (1993). Fitnesses of (QQ, PQ, PP) are $1 : 1 - s + \alpha s : 1 + 2\alpha s$. The light curve gives the symmetric case, where the critical number of migrants is $Nm_{crit} = 0.088$. The heavy curve is for asymmetry $\alpha = 0.1$; the critical number of migrants is then $Nm_{crit} = 0.237$. Even this slight difference in fitness between the two homozygotes leads to a strong bias in favor of the fitter peak if Nm is just less than Nm_{crit} (upper heavy curve).

between individuals, rather than by excess emigration. Unless population density and migration rate depend very strongly on genotype, the latter has negligible effect (see Crow *et al.,* 1990; Barton, 1992). On this view of the third phase, shifts are likely to occur only in regions with low Nm and yet must presumably spread through the whole species, into regions with large Nm. This requires either that Nm varies through time or that it varies gradually across the species range, such that asymmetric gene flow from the more abundant regions does not swamp peak shifts that occur at the margins.

The variations in population structure which are central to the idea of a metapopulation facilitate the shifting balance process. However, they also introduce a random element which reduces the bias in favor of the fitter peak. The shifting balance may be important in the evolution of reproductive isolation, though it is hard to judge whether populations reached different adaptive peaks through random drift acting in opposition to selection, or through selection alone. It seems less likely that the shifting balance contributes significantly to adaptation: species would need to be divided into very many different adaptive peaks, and peaks involving different sets of interacting genes would need to be able to spread independently of each other.

B. Extinction/Recolonization and the Shifting Balance

The theory for a subdivided population with constant population size and migration rate shows that the rate of peak shifts is not directly proportional to the variance of fluctuations due to drift, or to the variance in allele frequencies across populations. Thus, taking the average variance across populations, as one might when defining "effective population size," does not tell us the chance that a population will shift from one peak to another. We will demonstrate this more clearly with a specific example.

Consider the probability of shifting from one adaptive peak to another, when the adaptive landscape is determined by the phenotype of one particular trait. Assume that there are two adaptive peaks. The population starts at the lower peak and must cross the region of reduced mean fitness (the "adaptive valley").

The probability of transition from one peak to another has been given by Barton and Rouhani, (1993, Eq. 13), for populations of a given size and immigration rate, and for a given fitness function. Let us examine the probability of transition when all populations are initially at the same peak. Imagine a metapopulation with variable local population sizes, but with a constant *number* (Nm) of migrants coming into each population in each generation. With a constant number of migrants, the F_{ST} of any given class of populations will be approximately the same; if the populations are sufficiently old, we will assume that the genetic variance within demes is approximately constant. However, the probability of transition to new peaks is not at all constant (see Fig. 7). The smaller populations have a much higher probability of transition to new peaks.

The difficulty with this simple analysis is that even though the probability of

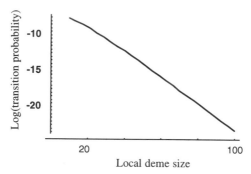

FIGURE 7   The probability of transitions between adaptive peaks as a function of population size. Smaller populations are much more likely to undergo peak shifts than are larger populations. These probabilities are given for a metapopulation in which all populations start at one peak and have a constant number of migrants coming into each deme each generation. The probabilities of transition are given in Barton and Rouhani (1993, Eq. 13a) based on the fitness function given in that paper. The parameter values used here (in the sense of Barton and Rouhani, 1993) are $s = 0.8$, $\Omega = 1$, $\alpha = 0$, $\nu = 0.3$, and $Nm = 2$.

forward transition from the old peak to new is higher for smaller populations, the contribution of those smaller populations to the gene pool is also smaller, and therefore the influence that they will have on the transitions of other demes is much smaller. Furthermore, the backward transition probability of smaller populations is also higher, so these new shifts are less stable.

In this kind of model, we can examine the probability of peak shifts directly. We know (from the arguments cited above) that the probability of a successful peak shift is proportional to $\overline{W}^{2N}$, with $\overline{W}$ being the fitness of the population in the adaptive valley. The mean probability of peak shifts across populations is therefore $\int \psi_i \overline{W}^{2N_i}$, where ψ_i is the frequency of populations of size N_i. This mean probability of a shift is thus *extremely* sensitive to small N values. For example, in a metapopulation with half its populations of size 10 and half of size 190, the arithmetic mean size is 100 and the harmonic mean is 19. However, the constant population size which has the same overall rate of shifting is about 13 if $\overline{W} = 0.8$. The probability of peak shifts is much more influenced by small population size than is reflected even in the harmonic mean.

How does local extinction affect the spread of new adaptive peaks in the shifting balance? In an elegant analysis, Lande (1979) showed that a new chromosome arrangement can spread if populations go extinct and are recolonized by colonists from a single deme, which is fixed for the new arrangement. If extinction and recolonization are random with respect to genotype, then the chance that a new underdominant mutation will be fixed through the whole metapopulation is equal to the chance that it is fixed within a single population. This is because (by analogy with the neutral theory of molecular evolution) all the populations in a species must trace back to one ancestral population, and the rate of evolution of the whole species must equal the rate of change of that one population.

In reality, new adaptive peaks may spread by migration between populations, and the process of extinction/recolonization may be nonrandom. Lande (1985) extended his analysis to include these effects and concluded that an adaptive peak is likely to gain a greater advantage from its stochastic spread between populations than from any increase in colonization or decrease in extinction which it causes. This conclusion clearly depends on the relative rates of migration and extinction: if almost all spread were by extinction and recolonization, then an adaptive peak which decreased extinction could gain a considerable advantage.

The other side of the coin, though, is that the smaller populations which are more likely to be able to drift through an adaptive valley are also subject to demographic stresses not experienced by larger populations. Therefore, larger populations usually have a higher probability of survival, and make a larger contribution to the migrant pool. This counteracts whatever increased fitness associated with a higher peak a small population may have found. A particular example of this is in source–sink metapopulations, where the smaller sink populations are more likely to experience genetic drift, but are also more likely to go extinct by demographic stochasticity or poor habitat quality, and they are also continually swamped by immigrants from the source population. Therefore, the distribution of population size alone, without considering correlated demographic parameters, is insufficient to predict the probability of evolution on a complex landscape.

At the same time, temporal variance in migration rates can *increase* the probability of peak shifts. One of the difficulties with the shifting balance model is that the probability of the first phase, where drift allows a population to drift to the domain of attraction of a new peak, is decreased by migration; but the probability of the third phase, where a new peak shift is exported to other populations, may be *increased* by higher migration rates. Temporally fluctuating migration rates may allow for both phases to occur successfully; phase one may occur during periods of low migration and phase three later when migration rates are higher (Moore and Tonsor, 1994).

Similar considerations apply in continuous habitats. The hybrid zones which separate different adaptive peaks are easily trapped by local barriers to gene flow (Barton, 1979). Thus, a fitter peak may be able to spread through the range of the species only if population structure fluctuates enough for the fitter peak to escape local barriers. This introduces a large element of chance into the outcome, since an adaptive peak which happens to be in a population that expands over a large area will gain a spurious advantage—a kind of spatial hitch-hiking. For example, the alpine grasshopper *Podisma pedesris* is divided into two chromosome races, which are separated by a narrow hybrid zone. One race is consistently more abundant than the other (Jackson, 1992), suggesting a fitness advantage; however, it cannot spread because the hybrid zone is trapped on the main ridge of the Alpes Martimes. The present distribution is likely to be determined more by historical patterns of recolonization after the last glaciation, rather than by the relative merits of the two karyotypes.

Most hybrid zones involve concordant changes in many characters. This is not an ascertainment bias, because contacts identified in one criterion (e.g., plumage or chromosome type) usually show extensive divergence in other genetic systems as well (Barton and Hewitt, 1985). This raises another difficulty for the shifting balance process. The expansion and contraction of local populations tends to bring unrelated differences together in complex hybrid zones; once together, both linkage disequilibria and the common influence of changing population structure tends to keep them together. It is therefore hard to see how a new peak could spread as a result of its own favorable effect on fitness, rather than as a result of its fortuitous association with other peak shifts. The problem is essentially that the shifting balance is an asexual process. Though Wright saw this as an advantage, in that it allows gene combinations to be kept together, it also makes it hard for the fittest gene combination to succeed.

C. Maintenance of Genetic Variation in a Metapopulation

Population structure is potentially important in the maintenance of genetic variation. Spatial heterogeneity in selection, coupled with limited migration, can substantially change allele frequencies from one site to another (see above). With some migration among sites, genetic variation is maintained both locally and globally.

Increased genetic variation can be maintained in a metapopulation even without spatial variation in selection. If there are multiple adaptive peaks, for example due to stabilising selection on a polygenic trait, then local populations can shift from the domain of alternative peaks, allowing genetic variation in allele frequencies (even if not in phenotypic states) to increase. Migration among these populations introduces locally unusual genotypes, whereby the genetic variation within populations can be greater than it would be without spatial population structure (see Goldstein and Holsinger, 1992; Barton and Rouhani, 1993). The following analysis shows how this is possible.

Alleles at loci under pure stabilizing selection are essentially under a one-locus selection function with heterozygote disadvantage. If stabilizing selection on a character is strong enough to substantially affect fitness, then the mean of that character in a population will be close to the most fit type. This implies that a substitution at one locus that, say, increases the character value will be compensated by a change at some other locus. We can therefore examine changes at single loci, using the strength of selection on the character and the number of loci which affect that character to predict the effective strength of selection against heterozygotes at one of the loci. (These assumptions have been tested by simulations of selection on quantitative traits in subdivided populations.)

Figure 8 shows some of the results. When selection is strong relative to some critical migration rate, the populations each go to adaptive peaks such that the frequency in the metapopulation of any given allele is 0.5 and each population is nearly fixed for one or the other of the alleles. As migration rate increases, the

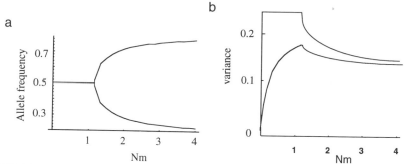

FIGURE 8   The maintenance of genetic variation in a subdivided population with stabilizing selection and mutation is approximated by a model of a single locus with heterozygote disadvantage. In these examples, each local population is composed of 10 diploid individuals with the strength of stabilizing selection (V_S/V_E) equal to 15. (a) The allele frequencies in the pool of migrants will be on average 0.5 until a critical value of Nm, after which a bifurcation occurs, when the allele frequencies become either greater than or less than a half. (b) The total variance in the metapopulation is defined by these allele frequencies, and is shown by the upper curve. It is at the maximum of 0.25 for Nm below the critical value and then decreases. The lower curve shows the genetic variance within populations. This reaches a maximum at the critical migration rate and then declines with increasing migration.

allele frequency in the metapopulation goes toward either 0 or 1 [with equal probability for these additively acting genes (see Fig 8a)]. Thus variation in the metapopulation is at a maximum anywhere below the critical migration rate (see Fig 8b). Variance within populations is more complex. With low migration rate, local populations are nearly fixed for one allele or the other; as migration rate increases, more alleles come into each population and increase local variation. Above the critical migration rate, however, the alleles which come into a population have a higher probability of being identical in effect to those already there, and the local genetic variation decreases again (see Fig. 8b). A shifting balance type of process for traits under stabilizing selection can maintain much more genetic variation than can mutation alone, if the number of migrants among populations is small enough. Variation in selection across populations or through time may cause populations to shift to different adaptive peaks even for large Nm, and hence may maintain variation in the same kind of way.

VI. DISCUSSION

As can be seen from many of the preceding sections, spatial population structure is potentially important to many different evolutionary processes. Unfortunately, different population structures which have similar properties from the point of view of one process can have very different properties for other processes. The obvious ways of measuring population structure, such as the effective sizes of local populations, the genetic variance among populations, or even some standardized measure like F_{ST} are thus inadequate. As shown in the previous section, the probability of peak shifts depends greatly on the demographic properties of a species, even when one holds F_{ST} constant. The effective size of a metapopulation depends on both F_{ST} and the distribution of population sizes (as well as on the distribution of population fitness as a function of population size, etc.). The probability of fixation of favorable alleles is not completely described by F_{ST} or by the effective size (Barton, 1993).

Different evolutionary processes must be considered separately in order to get the right answers. The distribution of neutral alleles in a metapopulation depends on random drift and migration, whereas local adaptation depends primarily on migration and selection. Wright's view of evolution involves a shifting balance between all three processes. It is critically important that studies of genetic metapopulation structure go beyond the usual measurement of F_{ST} or some such simple measure of the differentiation of neutral markers caused by sampling drift. It is essential that empirical studies are performed which give us a broader knowledge of the demographic and selective forces at work in determining the distribution of genotypes across populations. Moreover, as is clear from the section on peak shifts, we must learn more, at both the theoretical and the empirical levels, of what determines the overall distribution of genotypes across populations and not merely the mean and variance. Such studies are difficult but not impossible, and they are badly needed for a full picture of the causes and consequences of spatial population structure.

ACKNOWLEDGMENTS

We thank the Biotechnology and Biological Sciences Research Council and the Darwin Trust of Edinburgh, whose funds supported this work.

P A R T

METAPOPULATION PROCESSES

The principal processes in metapopulation dynamics are extinction, migration, and colonization, that is, establishment of new local populations. The key questions are how these processes jointly affect the dynamics and evolution of local populations and the entire metapopulation. The five chapters in this section extend the theoretical treatment of metapopulation ecology, genetics, and evolution in the previous section by a more detailed examination of particular processes, such as local population growth, temporal variance in growth, and the impact of immigration on local dynamics.

Much of the ecological interest in metapopulations stems from the observation that small local populations have a high risk of extinction, and hence the long-term persistence of a metapopulation consisting of extinction-prone small populations cannot possibly be understood at a smaller scale than the metapopulation. Foley replaces the verbal model of "small-populations-have-a-high-probability-of-extinction" with a series of mathematical models of stochastic population growth, which predict the expected time to population extinction under various scenarios, such as demographic stochasticity, environmental stochasticity, catastrophies, and their combinations. Though numerical simulations may produce more accurate results for well-studied populations, the insight provided by general mathematical models is invaluable. Foley also makes an effort to relate the models to real populations by discussing the often hard problems of parameter estimation. Foley identifies a number of issues that may be important for extinction but are not covered by the standard models. These additional factors include interspecific interactions, regional stochasticity, the form of density dependence, and so on. One additional problem is the "noise" due to slowly changing environmental conditions. Arguably, many if not most extinctions are due to "novel" events, which could not have been easily anticipated from short records of past population growth.

The calamitous process of extinction is countered, in metapopulations, by what Ims and Yoccoz term the transfer processes: emigration, migration, and colonization. Though the simple models of metapopulations and many field studies deal with only colonization of empty patches, it is well-understood that migration among existing populations may have significant consequences for both local dynamics and metapopulation dynamics in the broad sense (most chapters in Sections III and IV). Unfortunately, actual empirical measurement of the rates of emigration, immigration, and colonization are hampered by various difficulties, regardless of whether one uses indirect or direct observations or experiments. One of the greatest dilemmas is spatially and temporally varying rates of migration. Given all the variability at the individual level described by Ims and Yoccoz, it is a small wonder that any metapopulation models show any promise at all for predictive purposes. However, it may turn out that the details of migratory behavior matter little for predicting, for instance, the effect of distance on colonization, a population-level phenomenon. It is questionable whether an individual-based model would generally be more useful than a population-based model for predicting metapopulation dynamics, though a mechanistic understanding of say migration clearly requires an individual-level study. The lesson is not to funnel into one approach or another, but to realize that different approaches supply complementary information, and the clever biologists are the ones who succeed in forging a synthesis most helpful for a particular task.

Olivieri and Gouyon recognize that strong selective forces may operate in metapopulations on migration and other life-history traits. Migration has typically certain costs, in the form of mortality during the movement through hostile environments and due to possible relocation into less suitable localities than that of the natal population. Thus, genes promoting migration are generally selected against at the level of local population. However, this is only a part of the picture, as at the metapopulation level migration is selected for, which is most conspicuous when migrants succeed in establishing new populations. Olivieri and Gouyon term the action between these two antagonistic forces of selection the "metapopulation effect." This distinction is often seen as a conflict between individual and group selection, but as Olivieri and Gouyon remark, there is no conflict between two different types of selection, rather a hierarchy of selection components operating at different levels of the biological organization. Such issues apart, there is a substantial theoretical and empirical literature on factors affecting and scenarios of migration, which is reviewed in this chapter. One conclusion of potential importance to conservation of species in fragmented landscapes is that the evolutionarily stable migration rate is generally less than the rate that would maximize the size of the metapopulation (fraction of suitable habitat occupied at equilibrium). Taking into account that species have hardly had time to adapt to rapidly changing landscapes and that species may be even in an ecological nonequilibrium in increasingly fragmented landscapes (chapters by Hanski and by Nee, May, and Hassell), there appears to be substantial scope for managed reintroductions.

One of the more frequently used but often misunderstood concepts in the metapopulation literature is the rescue effect. As originally defined by Brown and Kodric-Brown, the rescue effect refers to reduced probability of population extinction due to immigration. In mechanistic terms, the rescue effect occurs because immigration props up population size, which decreases the risk of immediate extinction (Foley). The rescue effect was originally envisioned to occur in mainland–island metapopulations, but a similar effect may also occur in metapopulations without a mainland. In this case, existing populations provide a kind of mutual aid to each other. However, the complication here is that individuals that moved from population A to population B reduce the size of population A and hence potentially increase the risk of extinction of the source population (A) while decreasing the risk of extinction in the receiving population (B). In practice, the effect of population size on extinction is nonlinear, either because of some deterministic Allee effect, practically forcing the extinction of small populations, or because the consequences of stochasticity are most severe in small populations (Foley). Therefore migration from large to small populations may create a rescue effect at the metapopulation level. Stacey and Taper explore in their chapter the ecology of the rescue effect, in a range of taxa from butterflies to small mammals to amphibians. What they find is that migration among local populations is often so extensive that it is likely to affect local dynamics. This means, in terms of theory, that the kinds of structured metapopulation models described by Gyllenberg, Hanski, and Hastings in the previous section should be appropriate and, in terms of dynamics, that complex metapopulation dynamics with alternative stable states may be expected to occur.

The final chapter by Frank deals with host–parasite systems in a broad sense and their evolutionary dynamics in fragmented environments. If there are only small numbers of interacting host and parasite genotypes (or species), stable local dynamics may be expected. In contrast, with large potential numbers of types, inevitable extinctions due to small population size (of some types) open up possibilities for resistant hosts and virulent parasites, whose matching parasites and hosts went extinct, to dominate the local patch. A colonization by the missing type leads to another major perturbation, as the invader now has a strong selective advantage over the local types. Frank relates the theory to a range of observations on plant–pathogen and bacteria–phage systems, cytoplasmic male sterility in plants (a conflict between cytoplasmic and nuclear genes), disease resistance in vertebrates (allelic variation in the major histocompatibility complex), and genetic variance in plant resistance to herbivores. Under his hypothesis, which remains a challenge to test on a sufficiently large spatial scale, colonization–extinction dynamics not only maintain the ever-changing composition of local communities, but the turnover rates of allelic types are proportional to the number of potentially coexisting types.

10

Extinction Models for Local Populations

Patrick Foley

I. INTRODUCTION

The two underlying processes of metapopulation dynamics, local extinction and colonization, are subject to chance (Levins, 1970; Hanski and Gilpin, 1991). Life in a metapopulation is haphazard, and even when deterministic influences are at work, they may seem stochastic to an ecologist with limited information (Sugihara and May, 1990). This chapter reviews stochastic models of extinction within a local population. The main thread of the chapter is an analytic model of environmental stochasticity in which populations fluctuate between a ceiling and extinction. A diffusion analysis of environmental and demographic stochasticity is complemented by analytic and numerical investigations into the robustness of the analysis. Simulations in discrete time show that the diffusion analysis works (Foley, 1994; Hanski *et al.,* 1996a). The robustness investigations of this chapter show that the model can be extended to deal with many biological details. Although the model was developed to predict extinction rates, it speaks to the question of colonization rates also, since emigration is often a function of the population sizes whose fluctuations are analyzed by the same model.

Any mathematical model is bound to be a simplification of reality. This is a weakness (and virtue!) shared with verbal models and computer simulations. The beauty of a verbal model is that it sticks in your head and you can make use of it at will. However, verbal models of extinctions come down to this: small populations have higher extinction rates than large ones. The mathematical models presented in this chapter do better. They give expected times to extinction and population size thresholds for heightened vulnerability. They reveal the parameters we need to measure and the ones we can (at smaller risk) neglect. They allow comparisons between species and between landscapes. A computer simulation can make more accurate predictions since it can avoid many simplifying assumptions, but only if the biological parameters and processes are known in detail. Levins (1966) named three features of a good model: generality, realism, and accuracy. Verbal models of population dynamics are typically inaccurate, computer simulations too specific, and mathematical models unrealistic. Useful mathematical models in ecology are robust enough that slightly unreal assumptions produce only slightly inaccurate predictions (This virtue is sometimes called "the structural stability of the model"). The models we examine are fairly robust.

In view of the recent wave of discussion about individual-based models (Lomnicki, 1988; DeAngelis and Gross, 1992; Judson, 1994), the simple analytic approach of this chapter may appear anachronistic. Populations are not aggregations of identical atoms; sex, age, genotype, and history affect ecological dynamics, and local interactions surely influence competition and predation outcomes. When substantial resources are available, computer simulations based on individual interactions are particularly appropriate. For example, the Northern Spotted Owl *(Strix occidentalis caurina)* inspired a spatially explicit territory-based simulation in Pascal (McKelvey *et al.,* 1993) and a spatially explicit individual-based simulation in the object-oriented language Smalltalk (Foley *et al.,* 1993), each project costing over 2 person-years of effort. However, such models are expensive to build, hungry for parameter values, particular to the case at hand, not logically compelling, and hard to summarize in useful form. Some of these problems will be overcome with conceptual advances, but at present the (arguably) most illuminating analysis of Northern Spotted Owl dynamics is probably Lande's (1988a) simple demographic and metapopulation analysis. For insight, logic, and generality it is hard to beat mathematics (Maynard Smith, 1974), and mathematical models form a secure foundation for the construction of more complex analyses.

Most important for the theory of metapopulations, island biogeography, and landscape ecology, local extinction models allow theory (and computer simulation) at the higher level to proceed more quickly and gracefully. The first models of island biogeography used demographic stochasticity (MacArthur and Wilson, 1967) perhaps because it takes the form of the well-studied birth and death processes and it clearly affects small discrete populations. Many recent metapopulation models use catastrophic stochasticity (Ewens *et al.,* 1987; Hastings, 1990; Mangel and Tier, 1994), perhaps because the time scales of exponential population growth and catastrophes are different enough to permit analytic simplifications. I have focused on environmental stochasticity because it is persist-

ently important for small and large populations, it subsumes catastrophic stochasticity (at least in principle), and it can be empirically measured.

Most important for conservation biology, simple mathematical models allow
a ready comparison of conservation strategies. Optimization and cost–benefit
analyses depend on such comparisons. Programs such as RAMAS (Burgman *et
al.*, 1993) and VORTEX (Lacy and Kreeger, 1992) are good at including age
structure and other biological details, but they need extensive runs to cover much
parameter space. They can be made to do the job of comparing strategies, but at
some cost in time, generality, and conceptual clarity.

Most important for an understanding of natural populations, local extinction
models provide explanations about the persistent problems of ecology such as
population fluctuations, density dependence, and species–area curves. These explanations are testable against the observed patterns of biogeography and population time series data. This chapter makes some modest advances into this rich
and varied territory.

II. LOCAL POPULATIONS ARE VULNERABLE TO EXTINCTION DUE TO DEMOGRAPHIC AND ENVIRONMENTAL STOCHASTICITY

A population is extinct when it reaches the size of zero and it is doomed to
prompt extinction when the pooled reproductive value of all females reaches zero.
A single habitat patch in a fragmented landscape has limited resources which set
a limit to the possible size of the local population. The population size N (measured in number of females) thus lives on the interval that includes zero and this
population ceiling. Sooner or later any local population will go extinct due to the
stochastic nature of birth and death processes and to the temporal variation of the
environment. Each population is ultimately doomed; the challenge is to predict
when and understand why.

It is always possible to model population growth as

$$N(t + 1) = R(N, t) N(t), \tag{1}$$

where $R(N, t)$, the "fundamental net reproductive rate" (Begon *et al.*, 1990), is a
function of both N (thus providing density dependence) and t (permitting stochasticity). R is also a function of the age structure and genetic structure of the
population, but these effects are *usually* less important than the effects of N and
random effects which we attribute to t. The probability distribution of $R(N, t)$
becomes the critical feature directing the fate of the population. It is the job of
field ecology to describe typical probability distributions for R and the job of
theory to analyze the consequences. The equivalent challenge is to understand
the distribution of $r(N, t) = \log R(N, t)$, the realized per capita reproductive rate.
In the following, I will often shorten $R(N, t)$ to $R(t)$ or R, and I do the same to
$r(N, t)$ and $N(t)$. The natural logarithm will be used exclusively.

Deterministic models assume that R and r are constant over time. Two particular examples are the constant growth model,

$$r(N, t) = r_d, \tag{2}$$

for all N and t, except at the ceiling value K, and the Ricker model,

$$r(N, t) = r_d(1 - N/K), \tag{3}$$

which has properties similar to those of the logistic model (Burgman et al., 1993). Both models give simple extinction predictions: if $r_d < 0$, the population will go extinct. For model 1 the expected time to extinction T_e is given by

$$T_e = \frac{\log (N(0))}{r_d}. \tag{4}$$

These or similar deterministic models may often be appropriate for local populations, especially if their patches are sinks ($r_d < 0$, Pulliam, 1988) or the mainland ($r_d > 0$, MacArthur and Wilson, 1967). If every patch has negative r_d, unless, immigration can compensate (Hanski et al., 1996a), we should switch our study to a new metapopulation.

Stochastic models of extinction have been applied to local populations since MacArthur and Wilson (1963, 1967) developed their theory of island biogeography. MacArthur and Wilson employed demographic stochasticity in their analysis, assuming constant per capita birth and death rates, a population ceiling K, and population change to be a Poisson process. Demographic stochasticity is strongest in tiny populations and negligible compared to environmental stochasticity when population sizes reach the hundreds (Goodman, 1987a,b; Lande, 1993). Demographic stochasticity is analogous to random genetic drift in that it depends on the intrinsic uncertainty associated with an individual's reproduction and mortality, essentially a sampling problem for mother nature. Large populations average out birth and death across individuals and the sampling error (the variance of $r(t)$) is inversely proportional to N.

In a broad sense, environmental stochasticity refers to any randomness imposed by the environment on $r(t)$. It has become conventional in the conservation biology literature to distinguish between catastrophes and environmental stochasticity in the narrow sense (Shaffer, 1981). Catastrophe models typically assume a constant positive r (modified perhaps by demographic stochasticity) except when disaster hits (disease, new competitor, temporary habitat destruction), when r is negative and often large (Hanson and Tuckwell, 1981; Ewens et al., 1987; Mangel and Tier, 1993b; Lande, 1993).

Environmental stochasticity in the narrow sense can plausibly be modeled by assuming that $r(t)$ follows a gaussian distribution with mean r_d and variance v_r, i.e.,

$$r(t) \sim N(r_d, v_r), \tag{5}$$

which provides for good, bad, and indifferent years (Foley, 1994). Efforts to model and analyze environmental stochasticity have been extensive (Levins, 1969b; Richter-Dyn and Goel, 1972; May, 1973a; Ludwig, 1974, 1976; Turelli, 1977; Leigh, 1981; Nisbet and Gurney, 1982; Lande and Orzack, 1988; Dennis *et al.,* 1991). The early efforts were based on stochastic logistic models or models with mixed environmental/demographic stochasticity models, but these quickly run into intractable analysis. Lande and Orzack and Dennis *et al.* assumed no density dependence whatever. Recently Lande (1993), Foley (1994), and Middleton *et al.* (1996) have analyzed the simple ceiling model reviewed in this chapter and obtained succinct, useful results.

To compare the three modes of stochasticity, consider a model for the probability density of $r(N, t)$,

$$r(N, t) \sim \begin{cases} N(r_d, v_r + v_1/N) & \text{with probability } (1 - \lambda) \\ \log(1 - \delta) & \text{with probability } \lambda \end{cases}, \qquad (6)$$

where λ gives the probability of a catastrophic year, δ gives the proportion of the population destroyed in a catastrophe (Hanson and Tuckwell, 1981), and v_1 represents the variance in r due to demographic stochasticity in a single female. The notation "v_1" was first employed by Goodman (1987b, p. 219); Caughley (1994) calls this value V_{d1}. Under the usual assumptions v_1 should be equal to the sum of the mean per capita birth (b) and death (d) rates, while r_d is the difference $b - d$. Notice also that r_d is the expected value of ln R, $\mathbf{E}(\ln R)$, not $\ln(\mathbf{E} R)$. This distinction has caused alarm and confusion in the literature (see Lande, 1993, on Goodman, 1987b, or Turelli, 1977, p. 163, on Feldman and Roughgarden, 1975). This distinction between growth rates is clarified further by Caswell (1989, p. 213).

Equation (6) can only be an approximation, since real populations have discrete sizes, while Eqs. (1) and (6) assume a continuous state space for N. This approximation will do little damage as long as we make common sense interpretations of its consequences. For example, we must use $N = 1$ as a lower boundary for an extant population, not zero. The boundary behavior of the usual diffusions is peculiar at $N = 0$ and less problematic for $N = 1$ (see Gardiner, 1985, or Karlin and Taylor, 1981, for discussions of boundary behavior). Furthermore, when a population drops below one female, it is not viable in any biological sense. The upper boundary K is a "reflecting boundary." A good deal of subtle ecological thinking has been done on the boundary at infinity, but it hardly concerns real populations.

It is easy to simulate population growth using Eq.(6) using a computer language or a spreadsheet program. A population goes extinct when N goes below 1. When N goes above K, it is reflected below K. No analytic results can be hoped for in the general case, especially if r_d depends on N. If λ (or δ) and v_1 are small enough, then the pure environmental stochasticity model is an adequate approximation of the population dynamics, providing useful tools for the study of structured metapopulation models.

III. ENVIRONMENTAL STOCHASTICITY

A. Environmental Stochasticity in a Ceiling Model

The most practical approach to modeling environmental stochasticity is to try a diffusion approximation (Arnold, 1974; Turelli, 1977; Roughgarden, 1979; Gardiner, 1985). The main disadvantages of diffusion approximations are that they model a discrete state space and discrete time intervals with a continuous state space and continuous time and that there are subtle and debatable niceties to the pure mathematics of diffusions. The greatest advantages are that analytic results are usually easier to obtain and that the literature on applied diffusion theory is large, embracing Einstein's model of Brownian motion (1905) and recent results in the neutral and selectionist theories of molecular evolution (Kimura, 1983; Gillespie, 1991). Some of the following analysis can be found in more extended form in earlier papers (Foley, 1994; Hanski *et al.,* 1996a).

The simplest approach is to model the change in the natural logarithm of N denoted by n. The appropriate stochastic differential equation is

$$dn = r_d dt + \sqrt{v_r} dW(t). \tag{7}$$

The drift term r_d and the diffusion term v_r we have met. $W(t)$ is the standard Wiener or white noise process, $dW \sim N(0, dt)$. If v_r is very small, the solution to Eq. (7) approaches the deterministic result

$$N_t = N_0 e^{r_d t}. \tag{8}$$

A population ceiling at $N = K$ cuts off this exponential growth at the plateau of the carrying capacity. In the more familiar logistic growth model, population size can exceed the carrying capacity; in the ceiling model it cannot. As a technical point, since the diffusion has a constant diffusion term v_r, there is no distinction between the Ito and Stratonovich interpretations (Kloeden and Platen, 1992, p. 157). The logarithm of K ($=k$) is a reflecting boundary and 0 is an absorbing boundary for the diffusion of n.

The fullest knowledge that we can obtain about the future history of our population, which starts off at $n = n_0$, is the probability density $p(n, t)$. The density $p(n, t)$ satisfies the Kolmogorov forward and backward equations (Karlin and Taylor, 1981; Gardiner, 1985). The backward equation (KBE),

$$\frac{\partial p(n, t)}{\partial t} = \frac{v_r}{2} \frac{\partial^2 p(n, t)}{\partial n^2} + \frac{\partial p(n, t)}{\partial n}, \tag{9}$$

is especially useful in obtaining results about sojourn and extinction times (Foley, 1994; Hanski *et al.,* 1996a). Middleton *et al.* (1996) work with the forward equation (KFE, also known as the Fokker–Planck equation) to develop related results for population persistence time. Functions of the probability density also satisfy the forward and backward equations in appropriately modified forms. Some especially useful functions can be studied in this way, such as the probability of reaching carrying capacity before going extinct.

The expected time to extinction $T_e(n_0)$ satisfies the following ordinary differential equation which follows from KBE (Gardiner, 1985, p.138; Ewens, 1979, p. 120),

$$-1 = \frac{v_r}{2}\frac{d^2 T_e(n_0)}{dn_0^2} + r_d\frac{dT_e(n_0)}{dn_0}. \tag{10}$$

Since k is a reflecting boundary and 0 is an absorbing boundary for n, the boundary conditions are

$$T_e(0) = 0$$
$$\frac{dT_e(k)}{dn_0} = 0. \tag{11}$$

Equations (9) and (10) can be solved in several ways, but the sojourn time approach gives the most information about the population state before absorption (extinction) and has been popular with population geneticists. The sojourn time, $t(n; n_0)$, is the expected number of generations spent at size n before the population goes extinct, or to be more exact, $t(n; n_0)dn$ is the expected time spent in an interval of size dn around n (Maruyama, 1977; Ewens, 1979, pp.120–123). T_e is the sum of the sojourn times for all n up to k,

$$T_e(n_0) = \int_0^k t(n, n_0)\, dn, \tag{12}$$

where

$$t(n, n_0) = \frac{2}{v_r\psi(n)}\int_0^n \psi(x)\, dx \qquad \text{for } 0 < n < n_0 \tag{13}$$

$$t(n, n_0) = \frac{2}{v_r\psi(n)}\int_0^{n_0} \psi(x)\, dx \qquad \text{for } n_0 < n < k \tag{14}$$

$$\psi(x) = \exp\left(-2\int_0^x \frac{r_d}{v_r}dy\right). \tag{15}$$

If $r_d = 0$, T_e is easy to obtain:

$$\psi(x) = 1 \tag{16}$$

$$t(n; n_0) = 2n/v_r \qquad \text{for } 0 < n < n_0$$

$$t(n; n_0) = 2n_0/v_r \qquad \text{for } n_0 < n < k \tag{17}$$

$$T_e(n_0) = \frac{2n_0}{v_r}\left(k - \frac{n_0}{2}\right). \tag{18}$$

If r_d is not zero, define $s = r_d/v_r$ to get

$$\psi(x) = e^{-2sx} \tag{19}$$

$$S(n) = \frac{1 - e^{-2sn}}{2s} \qquad \text{(the scale measure which we discuss later)} \tag{20}$$

$$t(n; n_0) = \frac{1}{r_d}(e^{2sn} - 1) \qquad \text{for } 0 < n < n_0 \tag{21}$$

$$t(n; n_0) = \frac{e^{2sn}}{r_d}(1 - e^{-2sn_0}) \quad \text{for } n_0 < n < k$$

$$T_e(n_0) = \frac{1}{2sr_d}(e^{2sk}(1 - e^{-2sn_0}) - 2sn_0). \tag{22}$$

The assumption of most metapopulation models is that for a given habitat patch, the extinction rate is constant. This assumption does not quite hold for environmental or demographic stochasticity, since $T_e(n_0)$ is a function of the population size n_0. Thus extinction is not a Poisson process (Foley, 1994; Middleton *et al.*, 1996), though for practical purposes it comes close. Since

$$T_e(k) = \frac{k^2 e^{2sk} - 1 - 2sk}{v_r \quad 2s^2 k^2}, \tag{23}$$

the extinction rate $e(k)$ follows

$$e(k) = 1/T_e(K)$$

$$= \frac{v_r}{k_2 e^{2sk} - 1 - 2sk} \tag{24}$$

When sk is small, the extinction rate approaches v_r/k^2, as was noticed by Goodman (1987b). As sk becomes large, $e(k)$ approaches the form

$$e(k) \approx \frac{2r_d^2}{v_r}e^{-2sk}$$

$$= \frac{K^{-2s}}{2r_d s}$$

$$\propto K^{-2s}, \tag{25}$$

as was pointed out by Lande (1993). Hanski (1992a, 1994b) assumed the form of Eq. (25) in a series of papers on incidence functions in metapopulations (Hanski, 1992a, 1994b), and similar functions characterize species–area curves of the island biogeography literature (MacArthur and Wilson, 1967; MacArthur, 1972;

Brown and Gibson, 1983). In this literature K is implicitly or explicitly assumed to be proportional to A, the area of the habitat patch or the island.

B. Values of r_d, v_r, K, ρ, and v_l Estimated from Time Series Data

There are many unsolved problems in the estimation of extinction parameters. The simplest approach to estimating r_d is to take the mean value of $r(t)$ over a time series of population sizes (Dennis *et al.*, 1991). The technical scruple is that $r(t)$ is not drawn from a set of independent identically distributed random variables. The biological problem is that near the carrying capacity r values are bound to be underestimates of r_d. On the other hand, estimating r_d by regressing r on N (Roughgarden, 1979, p. 306; Dennis and Taper, 1994; J. E. Foley, P. Foley, and S. Torres, unpublished) will overestimate r_d if, as seems often true with territorial animals, density dependence is not linear but hits hardest close to K. Foley (1994) discusses the estimation of r_d, v_r, K, and ρ (the serial autocorrelation of environmental effects) and uses simple, robust but nonoptimal parameter estimators on time series including checkerspot butterflies, grizzly bears, mountain lions, and wolves. More rigorous estimation procedures would be welcome.

Table I surveys representative values of the extinction parameters. The r_{d0} estimator is the arithmetic mean of the r values. The r_{d1} estimator uses the mean r from the bottom three quartiles of population sizes (the idea is to avoid r values near the ceiling). The v_r estimator is the variance in r over the course of the time series. This estimate seems rather robust based on simulations, but it may be biased by error in census taking. Wolda (1978) also reports an abundance of v_r values (ranging from 0.018 to 0.642) for temperate and tropical insects. The estimate for ρ is just the serial autocorrelation between consecutive r values (Chatfield, 1989). K is taken as the maximum value in the time series of N, but other estimators may be preferable, especially if K itself varies from year to year or if outbreak populations temporarily shoot above the normal ceiling (Foley, 1994).

The extinction predictions are most sensitive to inaccuracy in estimating r_d, regrettably the parameter trickiest to estimate. Consider Menges' (1990) intensive study of 16 furbished lousewort *(Pedicularis furbishiae)* populations. His first year of data on λ (our R) is distinctly discouraging ($R = 0.68, 0.76, 0.64$, respectively) for all three of the populations studied (Menges, 1990, Table 5). The next 2 years of R values look better, partly because these three populations improve their growth rates ($R = 1.27, 1.03, 0.98$) and partly because he now has data for 13 mostly superior populations (they show a mean R of 1.18). Over all 3 years, an arithmetic mean of yearly mean $r(t)$ values gives a negative r_d estimate, perhaps because he studied worse populations in the first year. An unweighted mean of all r values would look much better for the lousewort, since Menges had more data for better years! If data from each population could be considered a realization of an identical random variable, much tighter confidence limits could be put around an r_d, but some habitat patches may be sinks.

TABLE I Estimated Values of Extinction Parameters

Species	T	r_{d0}	r_{d1}	v_r	ρ	K	k	$s_0 = r_{d0}/v_r$	$s_1 = r_{d1}/v_r$	Source
Euphydryas editha, JRC	26	0.002	0.307	1.46	−0.24	7259	8.89	0.00116	0.21027	Harrison *et al.* (1991)
Euphydryas editha, JRH	26	−0.052	0.126	0.84	−0.32	1998	7.6	−0.0619	0.15	Harrison *et al.* (1991)
Moth sp. 1, Rothamsted	13	0.048	0.094	0.564				0.08511	0.16667	I. Hanski (personal communication)
Moth sp. 2, Rothamsted	11	0.083	0.768	1.61				0.05155	0.47702	I. Hanski (personal communication)
Moth sp. 3, Rothamsted	10	0.154	0.509	2.59				0.05946	0.19653	I. Hanski (personal communication)
Moth sp. 4, Rothamsted	10	0.101	0.093	0.351				0.28775	0.26496	I. Hanski (personal communication)
Moth sp. 5, Rothamsted	9	0.117	0.369	0.462				0.25325	0.7987	I. Hanski (personal communication)
Metapeira orb spider, Bull Cay, Bahamas	4	0.545		0.971		234.5	5.46	0.56128		Schoener and Spiller (1995)
Metapeira orb spider, Longest Cay	4	0.508		3.025		172	5.15	0.16793		Schoener and Spiller (1995)
Metapeira orb spider, Dichotomous Cay	4	0.659		2.103		125.5	4.83	0.31336		Schoener and Spiller (1995)
Metapeira orb spider, Cay 405	4	−0.588		3.207		12	2.48	−0.1833		Schoener and Spiller (1995)
Furbished lousewort	4	−0.062		0.128				−0.4844		Menges (1990)[a]
Grizzly bear, Yellowstone	28	0.003	0.018	0.011	−0.46	57.97	4.06	0.28182	1.62727	Dennis *et al.* (1991)
Wolf, Alaska	11	0.03	0.175	0.18	−0.18	2592	7.86	0.16667	0.97222	Young and Goldman (1944)
Blackbird	16	0.17		0.174		1		0.97701		Diamond (1984)
Wheatear	16	0.08		0.487		1		0.16427		Diamond (1984)
Great tit Dutch	16	0.025	0.189	0.265		145.5	4.98	0.09434	0.71321	Perrins (1965)
Great tit Marley, UK	16	0.109	0.211	0.216		85.63	4.45	0.50463	0.97685	Perrins (1965); Lack (1966)
Great tit Dean, UK	16	0.007	0.251	0.345		75.19	4.32	0.02029	0.72754	Perrins (1965)
Tawny owl	12	0.053	0.007	0.004		32.14	3.47	13.25	1.75	Southern (1970)

[a] v_r Averaged over space and time for 16 populations

Or consider Lande's (1988a) survey of Northern Spotted Owl demography. His estimate $\lambda = 0.96 \pm 0.03$ (or equivalently $r_d = -0.04$) is neither significantly different from 1.0 ($r_d = 0.0$), which timber interests in the Pacific Northwest of North America would like to believe, nor from $r_d = -0.01$ obtained by longterm census data (Forsman *et al.*, 1984). Alvarez-Buylla and Slatkin (1991) provide a recent review of methods to set confidence limits around growth rate estimates obtained from typical age–structure models, but these models assume no density dependence.

In a metapopulations the r_d problem becomes one of distinguishing sinks from sources, a problem made more difficult by the complexities of migration (Watkinson and Sutherland, 1995). In general, the uncertainty of r_d is a vexing problem for deterministic analysis, which depends entirely upon r_d, and only slightly less problematic for the environmental stochasticity analysis, in which the random fluctuations measured by v_r can swamp the effect of r_d.

Two possible ways to estimate v_1 suggest themselves; either assume that $v_1 = b + d$ (Goodman, 1987b) or regress Var(r) on $1/N$. The second method demands a lot of data and has not been done to my knowledge. The first method assumes a standard birth–death process with independence across time and among individuals (Feller, 1971). The birth–death process theory assumes, for example, that competition does not operate and that individuals do not compensate for demographically bad periods (b is the *instantaneous* birth rate, d the corresponding death rate). Recall that a demographic accident is not necessarily a failure to accumulate energy reserves or to survive hardship. In many cases, a female who fails to reproduce early in the season may do so later or save the extra resources for next year. We know that such individual compensation exists, especially in perennial plants and large vertebrates; but we need to discover empirically how far it will go to diminish demographic stochasticity. Competitors may also take opportunistic advantage of the demographic accidents of their unfortunate neighbors. If no such compensation holds, then v_1 ranges from $|r_d|$ to several times $|r_d|$. Thus $\gamma = v_1/v_r$, the relative effect of demographic to environmental stochasticity for a single female, will range in practice from 0 to about 1.

A local population's extinction rate due to environmental stochasticity depends on all these parameters and the initial population size N_0, but a rough prediction can be obtained by assuming that r_d and ρ are close to zero and that N_0 is close to K. Then Eq. (18) implies that $T_e = (\log K)^2/v_r$. Extinction parameters for populations of Table 1 give T_e's ranging from 3010 years for the tawny owl near Oxford to 2 years for the Metapeira orb spider of Cay 405 in the Bahamas. The median value is 54 years for the great tit in Marley, England. So we see that typical local extinction rates, e, range from 0.0003 to 0.5 populations per year.

Healthy populations will usually have positive r_d values which can raise persistence times greatly. When sk is much greater than one, $T(K)$ becomes proportional to K^{2s} as shown in Eq. (25). The populations of Table 1 have somewhat intermediate sk values, with dynamics influenced substantially by both chance (v_r) and the upward pressure (r_d) of a positive population growth.

C. Environmental Stochasticity May Explain Species–Area Curves

MacArthur and Wilson (1967, p. 8) examined the species–area curves for several archipelagoes. The relationship between the number of species S on an island and island area A often fits the equation

$$S = cA^z, \tag{26}$$

where c depends on the taxon and the region under study, and z is a value that is consistently close to 0.3 for islands, but only 0.15 for continental habitat patches that are not very isolated (MacArthur and Wilson, 1967; Diamond and May, 1976). The number of species on an island depends on colonization rates and community structure, which affects extinction rates, but much of the community influence can be subsumed under its effects on r_d, K, and v_r.

Consider first the "disconnected community" in which species have no ecological interactions. If the species colonizing an island are independent, then community assembly is a simple process depending on constant colonization (m) and extinction (e) rates. With a continent to assure colonization, the probability of a particular species incidence on an island is (Hanski, 1992a).

$$p = \frac{m}{m + e}. \tag{27}$$

If m is small compared to e, that is if the islands are depauperate compared to the continent, then we have

$$p \approx mT_e, \tag{28}$$

where T_e depends on K which should be proportional to A. Referring to Eq. (25), we then obtain approximately

$$p = 2r_d smK^{2s}. \tag{29}$$

Since S, in this disconnected community, is just the sum of all persisting species from the continent which is a reservoir for S_{max} species, the expected number of species satisfies

$$ES = \sum_{i=1}^{S_{max}} p_i. \tag{30}$$

If we assume that K and s and m are similar for each species on the island, we obtain approximately

$$ES \approx 2r_d sm S_{max} K^{2s}, \tag{31}$$

which has the form of the standard species–area curve (4.1) with

$$z \approx 2s = 2\frac{r_d}{v_r}. \tag{32}$$

This result suggests that if environmental stochasticity sets extinction rates and if communities are disconnected, then typical s values are close to $s = 0.15$ on islands.

Consider next the "zero-sum community" of competitors who share a common carrying capacity K_{total}. Doubling the species number on an island might halve the effective carrying capacity for each species on average in this guild. Then

$$p \approx c \left(\frac{K_{total}}{S} \right)^{2s},$$ (33)

and on a typical island

$$S^{1+2s} \approx c S_{max} K^{2s}$$ (34)

$$S \approx (c S_{max})^{1/1+2s} K^{2s/1+2s},$$ (35)

again giving us the form of Eq. (26) if K is proportional to area. Given the two extremes of the disconnected and the zero-sum community, we get

$$2s \leq z \leq \frac{2s}{1 + 2s},$$ (36)

or equivalently

$$\frac{z}{2} \leq s \leq \frac{z}{2(1 - z)}.$$ (37)

which means that typically $0.15 \leq s \leq 0.22$ on islands. The power form of the species–area curve is consistent with the theory, and the calculated s values are plausible as can be seen from Table I.

On archipelagoes in "relaxation" with no migration among islands, the incidence patterns are entirely decided by extinction dynamics. For example, Brown's (1978) data on montane mammals of the Great Basin Ranges suggest similar values as the MacArthur and Wilson table. Montane birds show a lower z than do mammals, and Brown reasonably argues that this is due to the added influence of colonization. It is also possible that his birds may have lower s values, or less community connectivity.

The above ideas touch the surface of the literature on species–area relationships. For other insights see Diamond's (1975b, 1984), the Gilpin and Diamond (1976, 1981), and Hanski's (1992a, 1994b; this volume) series of papers on incidence functions. The form of Eq. (26) is traditionally explained in terms of Preston's (1962) canonical lognormal distribution for species abundances (MacArthur, 1972; May, 1975), but the canonical distribution still needs explanation. There may well be a deep connection between the theory of this section and Preston's observations.

IV. MODELS WITH DEMOGRAPHIC AND ENVIRONMENTAL STOCHASTICITY

A. Adding Demographic to Environmental Stochasticity

The environmental stochasticity models developed in the previous sections are mathematically tractable because the diffusion function v_r is constant. Small and large populations are all under the same hand of nature. To add demographic stochasticity to the model means uglier (at least more involved) mathematics (Leigh, 1981). The diffusion coefficient is now a function of population size, and I employ the Ito interpretation on Turelli's (1977) recommendation. Following (6), we must replace v_r by $v_r + v_1/N$. This leads to

$$\psi(x) = \exp\left(-2\int_0^x \frac{r_d}{v_r + v_1/N}\,dn\right)$$

$$= \exp\left(-2\frac{r_d}{v_r}\int_0^x \frac{1}{1 + (v_1/v_r)\,e^{-n}}\,dn\right)$$

$$= \exp\left(-2\frac{r_d}{v_r}[x + \log(1 + \gamma e^{-x}) - \log(1 + \gamma)]\right)$$

$$= \frac{(1 + \gamma)^{2s}}{(e^x + \gamma)^{2s}}, \tag{38}$$

where $\gamma = v_1/v_r$, $s = r_d/v_r$. The integral was evaluated using Gradshteyn and Ryzhik (1980, p. 92). Note that if γ is zero, $\psi(x)$ reverts to the earlier form in Eq. (19). The new $\psi(x)$ is apparently hard to integrate in closed form. Approximations are required for further exploration. Karlin and Taylor (1981, p. 194) discuss the use of the scale function

$$S(n) = \int_0^n \psi(x)\,dx \tag{39}$$

and the speed function

$$m(n) = \frac{1}{(v_r + v_1/\exp(n))\psi(n)} \tag{40}$$

to obtain more insight into the diffusion process. (Their formulas are more general than ours). As already suggested, Eq. (39) may not be integrated in closed form. From Eq. (24), we see

$$\psi(x) = e^{-2sx}\left(\frac{1 + \gamma e^{-x}}{1 + \gamma}\right)^{-2s}$$

$$\approx e^{-2sx}\left(1 + \frac{2s\gamma}{1 + \gamma}x + \frac{s\gamma(2s\gamma - 1)}{(1 + \gamma)^2}x^2\right), \tag{41}$$

which we get by expanding the second product around $x = 0$. Thus

$$S(n) \approx \frac{1 - e^{-2sn}}{2s} + \frac{\gamma}{1 + \gamma s} \frac{1}{s} \left[\frac{1}{2} - e^{-2sn} \left(sn + \frac{1}{2} \right) \right]$$

$$= \frac{1 + 2\gamma}{2s(1 + \gamma)} \left[1 - \left(1 + \frac{2s\gamma n}{1 + 2s\gamma} \right) e^{-2sn} \right]. \tag{42}$$

Notice that as n gets large, $S(n)$ approaches the value

$$S(\infty) = \frac{1 + 2\gamma}{2s(1 + \gamma)} \tag{43}$$

and that when n is very small

$$S(n) \approx n. \tag{44}$$

In fact the critical value of n turns out to be

$$n_c = S(\infty) = \frac{1 + 2\gamma}{2s(1 + \gamma)} \tag{45}$$

For n somewhat below n_c, $S(n)$ follows Eq. (44), and for n somewhat above it, $S(n) \approx S(\infty)$. This plateau-like shape of $S(n)$ is common to any model (demographic and/or environmental) when $r_d > 0$. This makes the critical value n_c (and N_c) of some interest.

The sojourn time near k starting from k is given by an equation which begins as a rewritten version of Eq. (13),

$$t(n; k) = 2m(n)S(n)$$

$$\approx \frac{ne^{2sn}}{v_r} \frac{2(1 + \gamma e^{-n})^{2s-1}}{(1 + \gamma)^{2s}} \qquad \text{for } 0 < n < n_c \tag{46}$$

$$\approx \frac{e^{2sn}}{v_r} \frac{1}{s} \frac{1 + 2\gamma(1 + \gamma e^{-n})^{2s-1}}{(1 + \gamma)^{2s+1}} \qquad \text{for } n_c < n < k, \tag{47}$$

which is continuous and piecewise differentiable. Since almost all of the sojourn time is concentrated above n_c if k is larger than n_c, we get

$$T_e(k) \approx \int_{n_c}^{k} t(n; k) \, dn \approx \frac{(e^k + \gamma)^{2s} - (e^{n_c} + \gamma)^{2s}}{2sr_d} \frac{(1 + 2\gamma)}{(1 + \gamma)^{2s+1}}, \tag{48}$$

which may be improved by adding in the integral of the sojourn time below n_c.

If r_d is zero, a simpler calculation than Eqs. (38)–(48) gives

$$T_e(k) = \frac{k^2}{v_r} \left(\frac{1}{1 + \gamma} \right). \tag{49}$$

This result shows the most exaggerated influence of demographic stochasticity on $T_e(k)$ since when s is large populations hover around K. When s is close to

TABLE II   Effects of s and γ on N_c, the Critical Population Size

s	γ							
	0	**0.2**	**0.4**	**0.6**	**0.8**	**1**	**1.2**	**1.4**
0.05	22026	116619	383518	936589	1875610	3269017	5150197	7521930
0.06	4160	16684	44994	94687	168896	268337	391910	537371
0.07	1265	4160	9737	18424	30256	44994	62253	81595
0.08	518	1468	3089	5398	8331	11790	15664	19847
0.09	259	653	1265	2077	3055	4160	5355	6610
0.1	148	341	619	968	1370	1808	2269	2743
0.15	28.0	48.9	72.7	97.8	123	148	173	196
0.2	12.2	18.5	24.9	31.1	37.0	42.5	47.6	52.4
0.3	5.3	7.0	8.5	9.9	11.1	12.2	13.1	14.0
0.4	3.5	4.3	5.0	5.6	6.1	6.5	6.9	7.2
0.5	2.7	3.2	3.6	4.0	4.2	4.5	4.7	4.9
0.6	2.3	2.6	2.9	3.1	3.3	3.5	3.6	3.7
0.7	2.0	2.3	2.5	2.7	2.8	2.9	3.0	3.1
0.8	1.9	2.1	2.2	2.4	2.5	2.6	2.6	2.7
0.9	1.7	1.9	2.0	2.1	2.2	2.3	2.4	2.4
1	1.6	1.8	1.9	2.0	2.1	2.1	2.2	2.2

zero, population sizes more frequently fall in the danger zone for demographic stochasticity. Still Eq. (49) provides insight into the effect of the parameter γ. If γ is 1 and s is close to zero (so that population sizes fluctuate into the demographic risk zone), extinction rates will be doubled.

Define a successful colonization as one that produces a population that reaches carrying capacity before it goes extinct. Then the probability of this success is

$$P_k(n) = \frac{S(n)}{S(k)}, \tag{50}$$

where, to be precise, $P_k(n)$ is the probability of reaching k before 0, starting at n (Ewens, 1979, p. 119; Karlin and Taylor, 1981, p. 195). Equations (43) and (44) then lead to two major insights. First, populations greater in size than N_c are fairly sure of success, and second, populations below N_c can expect success in proportion to log N.

Pure environmental stochasticity implies $\gamma = 0$, so Eq. (45) becomes

$$n_c = \frac{1}{2s}. \tag{51}$$

Table II gives critical values of N_c for ranges of typical s and γ values. N_c should be interpreted for a particular population in this way:

$$P_k(n_0) \approx \frac{n_0}{k} \quad \text{if } N_0 < K < N_c \tag{52}$$

$$P_k(n_0) \approx \frac{n_0}{n_c} \quad \text{if } N_0 < N_c < K \tag{53}$$

$$P_k(n_0) \approx 1 \quad \text{if } N_c < N_0 < K. \tag{54}$$

A nice feature of the critical population size concept is that it does not depend on K, but only on s and γ, facilitating comparisons among populations.

B. Demographic Stochasticity Alone

Models involving only demographic stochasticity are best analyzed with discrete models as done by MacArthur and Wilson (1967), Nisbet and Gurney (1982), Talent (1990), and Renshaw (1991). Real populations surely undergo some environmental stochasticity, so pure demographic stochasticity models are mainly of theoretical interest. Still for the sake of completeness and comparison, we employ our usual diffusion methods for the analysis. Assume that $v_r = 0$ and define $s_d = r_d/v_1$. Then

$$\psi(x) = e^{-2s_d(e^x - 1)} \tag{55}$$

$$S(n) = e^{2s_d}(E_1(-2s_d e^n) - E_1(-2s_d)) \tag{56}$$

$$S(n) \approx n \qquad \text{for } 0 < n < n_c \tag{57}$$

$$S(n) \approx e^{2s_d}E_1(2s_d)$$

$$\approx \ln\left(1 + \frac{\ln(2)}{2s_d}\right) \quad \text{for } n_c < n < k \tag{58}$$

$$t(n; n_0) = \frac{2}{v_1} e^n e^{2s_d(e^n - 1)} S(n) \quad \text{for } 0 < n < n_0 \tag{59}$$

$$t(n; n_0) = \frac{2}{v_1} e^n e^{2s_d(e^n - 1)} S(n_0) \quad \text{for } N_0 < n < k \tag{60}$$

$$t(n; k) \approx \frac{2}{v_1} e^n e^{2s_d e^n} E_1(2S_d)$$

$$t(n; k) = \frac{1}{2r_d} e^n e^{2s_d(e^n - 1)} \qquad \text{if } 1/2s_d < k. \tag{61}$$

The critical n_c is here

$$n_c \approx e^{2s_d} E_1 (2s_d)$$

$$\approx \ln \left(1 + \frac{\ln(2)}{2s_d} \right) \tag{62}$$

and the probability of success is given by

$$P_k(n) = S(n)/S(k)$$

$$\approx 4s_d(n - s_d n^2) \quad \text{for } 0 < n < n_c \tag{63}$$

as long as $1 < 2s_d dk$. Equations (56) and (58) use exponential integral functions, special functions of mathematical physics (Olver, 1974, p. 40). The sojourn time of Eq. (61) can be integrated to give

$$T_e(k) \approx \frac{e^{2s(e^k - 1)} - 1}{4r_d s_d} \quad \text{if } 1/2s_d < k. \tag{64}$$

V. ROBUSTNESS OF THE MODEL

A. Autocorrelated Environmental Stochasticity

Diffusion models of environmental stochasticity traditionally assume that year to year fluctuations in the environment are not autocorrelated. In the real world, this assumption fails. Foley (1994) derived a corrected value of v_r, the effective environmental variance v_{re}, that satisfies the equation

$$v_{re} = v_r \frac{1 + \rho}{1 - \rho}, \tag{65}$$

where ρ is the autocorrelation of environmental effects between consecutive years. An equivalent adjustment has been developed but rarely used in population genetics (Gillespie, 1991; Gillespie and Guess, 1978). Turelli (1977, p. 148) examined the convergence of stochastic processes with more general autocorrelation structure to white noise.

Equation (65) shows that positive autocorrelation leads to exaggerated swings (because one bad year is likely to be followed by another) and that negative autocorrelation dampens swings in r. Simulations confirm the utility of Eq. (65) (Foley, 1994). Formulas for extinction times, extinction probabilities and so forth can be adjusting for nonzero ρ by using v_{re} in place of v_r.

Many population size time series show slightly negative ρ values, more negative than can be accounted for by the bias of the standard estimator. Some of the negativity may be due to the inevitable bounciness near the ceiling and near extinction for data sets that have been selected to analyze. Some of the variance in apparent r may be attributable to error in estimating $N(t)$ and subsequent corrections. Some may be due to age–structure disequilibrium (but not in most insects!). This remains a statistical puzzle and perhaps a biological one also.

B. Logistic Density Dependence

The theory presented in this chapter assumes that K is a ceiling, a barrier above which a population cannot remain for long to wander. The ceiling assumption is about as realistic as the logistic assumption, and it is easier to do the math. A substantial body of verbal theory, simulations and field observations about "density vagueness" (Strong, 1986), "spreading the risk" (den Boer, 1968, 1981), and "density independence" (Andrewartha and Birch, 1954), mainly by insect population biologists, has driven a rebellion against the pervasive density dependence of the logistic model. Vertebrate ecologists have usually supported the logistic model, despite the territorial nature of many of their subjects. Territoriality may be expected to lead to strong density dependence only near the population ceiling. However, vertebrates rarely show the wild population fluctuations of invertebrates and thus appear to be regulated in a more consistent way. Nonetheless, most of this regulation may be at or very near the ceiling. If so, there may be less difference than there appears between vertebrate and invertebrate population regulation with respect to density dependence.

The classical logistic model of Verhulst,

$$dN = rN \left(1 - \frac{N}{K} \right),$$ (66)

shows awkward behavior in the vicinity of K. Populations above K are supposed to retreat toward K with the same force with which they approach K from below. It is possible to generalize the logistic (Hassell et al., 1976), but there is no consensus about the prevalent form of density dependence in real populations (Royama, 1992, discusses the possibilities at length), and it is difficult to detect density dependence from time series (Pollard et al., 1987; Vickery and Nudds, 1991; J. E. Foley, P. Foley, and S. Torres, unpublished). The logistic model becomes markedly peculiar if $r < 0$ and $N > K$. This peculiarity becomes especially significant in the analysis of environmental stochasticity because $r(t)$ will often fall below zero, even if N is above K.

For these reasons and for the sake of an intelligible comparison, the model of pervasive density dependence examined here includes these features: K remains a ceiling, and r_d is replaced by $r_d(1 - e^{n-k})$. In other words our model becomes a truncated, stochastic Ricker equation

$$dn = r_d (1 - e^{n-k}) dt + \sqrt{v_r} \, dW(t).$$ (67)

Compare this with Eqs. (3) and (7) which it generalizes. We do the usual diffusion analysis to get

$$\psi(x) = e^{-2sx}e^{2s(e^{x}-k-e^{-k})} \tag{68}$$

$$\frac{1 - e^{-2sn}}{2s} < S(n) < \frac{1 - e^{-2sn}}{2s} + \frac{ke^{-2sk}}{2}(e^{2s(1-e^{-k})} - 1). \tag{69}$$

Unfortunately $S(n)$ appears difficult to obtain in closed form. A good approximation and certainly a lower bound is the scale function for the case of no density dependence. The upper bound is obtained by using the value of $\psi(k)$ and the concavity of the exponential function. The sojourn function takes its lower bound and a good approximation in the form

$$t(n;k) \approx \frac{2}{v_r} e^{2sn}e^{-2s(e^{n}-k-e^{-k})} \frac{1 - e^{-2sn}}{2s}$$

$$= \frac{1}{r_d}(e^{2sn} - 1)e^{-2s(e^{n}-k-e^{-k})}. \tag{70}$$

This is also hard to integrate in closed form, but numerical approximations are easy. Remarkably, the ratio of $T_e(k)$ with the ceiling Ricker model to the $T_e(k)$ without the logistic term depends mainly on s and rather simply so. As a numerical rule of thumb,

$$\frac{T_e(k) \text{ Ricker}}{T_e(k)} \approx 1 - 0.7s \quad \text{for } s < 1. \tag{71}$$

This result is so consistent, it should be analytically obtainable. Simulations reveal that the stochastic Ricker model without a ceiling at K leads to a higher T_e than in Eq. (71) since sojourn times above n add to the integral, but as shown above, this model has some built in biological unreality.

The conclusion of this is analysis is that for small s values, extinction rates do not differ much between the ceiling model and the logistic model of density dependence.

C. Other Adjustments to the Model

Allee effects due to social aggregation requirements (Allee *et al.*, 1949, p. 393) pose few difficulties to the model. As discussed in Dennis *et al.* (1991) and Foley (1994), the lower population boundary changes from $n = 0$ to $n = n_a$, the critical minimal log population size that permits social aggregation. This is equivalent to lowering k by the amount n_a. Note that this adjustment may substantially diminish demographic stochasticity if n_a is large enough. All the demographic stochasticity may occur in populations too small to persist on deterministic grounds.

The ceiling K need not be constant. Roughgarden (1975) analyzed a logistic model with fluctuating K and found that populations would fluctuate with a time lag due to the effects of recent density dependence. If K represents a ceiling, this lag should be smaller, and if the fluctuations are small compared to those of r, the results of this chapter's model should still hold approximately using an effective K close to the arithmetic mean. If all of the environmental stochasticity is attributable to K rather than r, the most straightforward extinction model (ignoring demographic stochasticity) is to examine the distribution of K, looking for the probability that K drops below one. This approach assumes that we know the full probability distribution of K rather precisely, a doubtful proposition. In a sense, this would imply knowledge of the rate of catastrophes.

VI. CATASTROPHES AND GENETIC IMPOVERISHMENT

Catastrophes are extraordinary, almost unpredictable population threats. Typical models assume a catastrophe rate λ and a population mortality rate δ (Hanson and Tuckwell, 1981; Lande, 1993). Of course δ may follow some random distribution; it need not be constant (Mangel and Tier, 1993b, 1994). Extraordinary, unpredictable events are hard to study, especially when they include agents as diverse as disease outbreaks, hurricanes, superinvading organisms (predators or competitors), asteroids, widespread fires, and acid rain.

For a population to become extinct due to catastrophe, one of three scenarios seems necessary: (1) a clean sweep catastrophe, (2) a catastrophe sequence, or (3) catastrophes mixed with lesser problems. In the clean sweep scenario, everyone dies; i.e., δ is one. In the catastrophe sequence scenario, λ is high enough so that two or more catastrophes can occur in close sequence, before a population has time to get back to K. Perhaps most plausibly, in the mixed model, catastrophes bring the population down low enough for normal environmental stochasticity or demographic stochasticity to finish the job. The analysis of the clean sweep scenario is easy; the extinction rate e is exactly λ. Local dynamics do not matter. The mixed catastrophe scenario does not lend itself to simple analytic results, although Mangel and Tier (1993a) show how to construct generating matrices to obtain T_e. The catastrophe sequence scenario stretches the concept of catastrophe into a common occurrence; nonetheless theorists frequently resort to it (Hanson and Tuckwell, 1978, 1981; Lande, 1993). In comparing the relative importance of environmental stochasticity (in the narrow sense) and catastrophes, Lande (1993) concludes that it depends upon the parameter values. Unfortunately, little is known about real catastrophe parameter values.

To get a quantitative handle on catastrophes, let us consider disease outbreaks. Mortality due to disease is a normal part of the life of a population. Most variation in disease-based mortality will then fall under the category of environmental stochasticity (although disease often shows predictably periodic dynamics). However, the primary outbreak of a new viral or bacterial strain can often have severe effects. After such an outbreak, the relationship between the disease and the host usually becomes less virulent for a variety of reasons including host immunity, host evolution and disease agent evolution (Anderson and May, 1991;

TABLE III   Candidates for Catastrophes Caused by Disease in Natural Populations

Species	Disease	δ	λ	Source
Grey fox	Canine distemper	0.78	1	Davidson et al. (1992)
Wolf, Superior NF	Parvovirus	High	?	Mech and Goyal (1993)
Wolf pups	Parvovirus	0.75	0.44	Johnson et al. (1994)
Rabbit, Australia	Myxomatosis	0.99	?	Fenner and Myers (1978)
Lion, Serengeti	Canine distemper	0.25	?	Morell (1994)
Domestic dogs, Africa	Canine distemper	0.35	0.2	Alexander and Appel (1994)
Ringed seal, Lake Baikal	Phocid distemper	0.2	?	Grachev et al. (1989)
Cheetah, captive colony	FIP	0.42	?	O'Brien et al. (1985)
Ungulates, Africa	Rinderpest	0.9	?	McCallum and Dobson (1995)

Ewald, 1994). It is the primary outbreak that behaves like the catastrophe of theory. Such outbreaks can be rare, unpredictable, and highly virulent. Table III gives examples of high mortality induced by disease in natural populations. Some barely fit the catastrophe profile; most do not, because they have either high λ or low δ. Little progress has been made in the literature so far to estimate λ and δ, although the increasing interest in disease threats to conservation efforts (McCallum and Dobson, 1995) should change that. A particular disease may only sweep catastrophically through a population (or metapopulation) once, so that we need to estimate the pooled λ for all potential diseases. This kind of data is available only for humans and domesticated animals, but even the most substantial efforts to study the ecological effects of disease (Anderson and May, 1991) concentrate on specific diseases.

Disease-caused catastrophes interact in a complicated, poorly understood way with the genetic structure of host populations. It remains difficult to place a value on genetic variability although we know it has value. Survivorship and fertility may be depressed by inbreeding (Ralls et al., 1988), but populations can evolve to compensate for the problems (Falconer, 1981, p. 229). Furthermore *within* the individual host, the somatic genetic variability of the immune system defenses evolves in a fashion heterodox to population genetics. Franklin (1980) derived magic numbers for critical population sizes vulnerable to inbreeding depression and genetic impoverishment. However, the latter was based on a doubtful application of Lande's early quantitative genetics theory (Foley, 1992), and the former often fails due to the purging effects of natural selection against deleterious recessives. At present we have no magic numbers, and the convoluted relationship between the loss of genetic variability and catastrophic (or other) extinction remains mysterious. For conservation biology, the wisest strategy is to conserve founder genome equivalents (Lacy, 1989), but we have no real idea how valuable an extra genome equivalent is for a natural population.

Novel parasites, predators, competitors, and other living and evolving agents of catastrophe will surely test a population's ability to evolve, which is approximately proportional to genetic variability (Fisher, 1958). Therefore, with the possible exception of abiotic catastrophes, the genetic structure of a population is hard to ignore. Unfortunately we do not yet know how to use it. Nor do we have established methods for estimating the parameters λ and δ, so the application of catastrophic extinction theory to real populations remains problematic.

VII. IMPLICATIONS FOR METAPOPULATION DYNAMICS

A. Structured Metapopulations

Levins (1970) assumed a local population would usually be either at carrying capacity or extinct. This simplifies the math and approximates reality if successful new propagules rapidly approach carrying capacity and stay there. Structured metapopulation models take explicit account of population sizes within a patch, allowing them intermediate values (Hastings and Wolin, 1989; Gyllenberg and Hanski, 1992; Gyllenberg *et al.,* this volume). Populations subject to environmental stochasticity will fluctuate near the carrying capacity if r_d is positive. We can use the sojourn times to obtain the probability distribution of extant population sizes, the mean and the variance of N.

Define $f_p(n; n_0)$ as the probability density of n conditional on the population's persistence given a starting value of n_0. Similarly define the cumulative distribution function $F_p(n; n_0)$. Then from the definition of the sojourn time,

$$f_p(n; n_0) = \frac{t(n, n_0)}{T_e(n_0)}. \tag{72}$$

This density allows us to calculate the expectation and variance of n, $\mathbf{E}n$, and $\mathbf{V}n$. The two most interesting starting points n_0 are the carrying capacity k (since populations above n_c are effectively at k) and the propagule size at the establishment of a new population, a very small n_0. Foley (1994) and Hanski *et al.,* (1996a) derive $f_p(n; n_0)$, $\mathbf{E}n$ and $\mathbf{V}n$ for the environmental stochasticity model, and these results are given in Appendix 3.

Appendix 3 shows that if s is close to zero, a population will spend much of its time well below the carrying capacity whether we condition the density on a starting point of near zero or near k. The probability distribution of n is not gaussian, nor is N log normal, as would occur in the absence of boundaries in logistic or density-independent models (Lewontin and Cohen, 1969; Levins, 1969b; Ludwig, 1974; Nisbet and Gurney, 1982; Lande and Orzack, 1988; Tuljapurkar, 1990). In fact if s is zero, the conditional probability density of n is uniform, a result expected from the standard theory of random walks (Feller, 1971). The conditional probability density of N starting from a small propagule is then $1/(N \ln K)$; i.e., N spends most of its time closer to the extinction threshold than to the ceiling. This resembles the pattern of herbivorous insects capable of outbreaks (Ito, 1980).

One way to use the formulas of Appendix 3 in metapopulation models is to calculate a more realistic colonization rate based on the mean population size of extant populations (Hanski *et al.*, 1996a). A practical application of the formulas could include the rough estimation of the extinction parameters r_d and v_r from the empirical distribution of n. This would require either a long time series or observations from several uncorrelated habitat patches with estimatable carrying capacities.

Most interestingly, the sojourn times for population sizes permit a careful analysis of the erosion of genetic variability due to random genetic drift. Such an investigation has been undertaken with simulations by Gilpin (1991), Hedrick and Gilpin (this volume), and Ewens *et al.* (1987) for catastrophe driven systems. Analytic results for environmental and demographic stochasticity become possible, given the sojourn times of this chapter.

Population size affects molecular clock rates and genetic distances. While strictly neutral theory is independent of population size (Kimura, 1983), under nearly neutral models small populations evolve at faster rates, and much of molecular evolution is apparently nearly neutral (Ohta, 1972; Foley, 1987). Genetic distances between local populations will thus depend on population densities as well as on migration, so explicit models of local dynamics can be of some use in metapopulation genetics analysis for species with substantial population fluctuations.

B. Regional Stochasticity: Correlated Environmental Fluctuations

Demographic stochasticity is not spatially correlated (barring substantial migration between populations), but environmental stochasticity often is, and catastrophes due to weather, invaders, and disease outbreaks are likely to be. Most metapopulation theory assumes patchwise independence of extinction rates, and this theory is vulnerable to the charge of unreality. How much modification of the theory is needed remains an open question although simulations have explored some of the possibilities (Harrison and Quinn, 1989; Gilpin, 1990; Akçakaya and Ginzbug, 1991; Lahaye *et al.*, 1994). As Levins (1969a) argues from his diffusion analysis, correlated extinctions lead to lower average N values, and as the simulations show, shorter metapopulation persistence times. In the extreme case, local populations undergo uncorrelated extinctions, so that metapopulation persistence time, T_M, depends only on the analog to demographic stochasticity that Hanski (1992a) calls immigration–extinction stochasticity (Gurney and Nisbet, 1978; Nisbet and Gurney, 1982). At the other extreme, realized extinction rates are so perfectly correlated that $T_M = T_e$ for local populations. What about intermediate cases, and what are the patterns of correlation in actual landscapes?

The California Spotted Owl populations of southern California appear to form a metapopulation of upland forested habitat patches with substantial regional stochasticity (Hanski, 1992a) due to drought. Pairwise correlation estimates for rainfall in three habitat patches ranges from 0.811 to 0.896, and the previous year's rainfall explains 52% of the variation in owl fecundity in this region (Lahaye *et al.*, 1994). This level of environmental correlation will sharply lower the expected persistence time for spotted owls in the region. Den Boer (1981), Pollard (1991), C. D. Thomas (1991), Taylor (1986), and Hanski and Woiwod (1993) have documented numerous examples of spatially correlated population fluctuations in insects, much of it due to correlated environmental fluctuations. Regional stochasticity is a very real feature of metapopulations, population data and climatological data are piling up (e.g., Lamb, 1985), and it cries out for analysis.

Levins did not extend his 1969 analysis (1969a) to metapopulation extinction times, and Gurney and Nisbet (1978) Nisbet and Gurney (1982) analyzed the extreme case of immigration–extinction stochasticity without regional stochasticity. Regional stochasticity presents several entangled questions. How do correlated $r(t)$ values between two patches translate into correlated parametric extinction rates, $e(t)$? How does distance-based environmental correlation translate into the global environmental correlation of the metapopulation? How does a shared parametric extinction rate translate into a correlated realized extinction rate?

Consider a slight extension of Levin's (1969a) model, with H defined as the number of suitable habitat patches in a metapopulation, $Q(t)$ as the number of patches occupied at time t, $e(t)$ as the extinction rate at time t, and $m(t)$ as the colonization rate at time t. Then

$$\frac{dQ}{dt} = m(t)\, Q(t) \left(1 - \frac{Q(t)}{H} \right) - e(t)\, Q(t) \qquad (73)$$

may, if $m(t)$ is constant and $e(t) \sim N(E, V_e)$, be modeled by the stochastic differential equation

$$dq(t) = m(1 - e^{q(t)-h})\, dt - E + \sqrt{V_e}\, dW(t), \qquad (74)$$

where $q(t) = \log Q(t)$ and $h = \log H$. This equation is similar (but not identical in form) to those analyzed already in this chapter, and it can be extended similarly to include immigration–extinction stochasticity by changing V_e to $V_e + V_1/Q$, where V_1 represents the birth–death process stochasticity experienced by one population. The metapopulation is extinct when q drops below 0 and has a ceiling at $q = h$. Note that random movements during colonization, and the independence of local populations prevent some of the simplifications of the local analysis, such as the downplaying of logistic density dependence and demographic stochasticity. I am presently engaged in the analysis of these equations together with efforts to answer the other questions of this section (Foley, 1996).

Stochastic approaches such as random field theory (Vanmarcke, 1983) and point process theory (Cox and Isham, 1980; Thompson, 1988) and statistical techniques (Cressie, 1993) for dealing with spatial correlational patterns are becoming accessible to the ecological research community. Thus, much of the problem of regional stochasticity should soon yield to analysis and field work.

C. Scales for Growth, Extinction, and Colonization

The population size assumption of Levins, either $N = 0$ or $N = K$, is also somewhat weakened by an explicit stochastic analysis. Harmlessly, K should be replaced by $E_p(N)$. A new population can expect to reach this level in a time $T_s(n_0)$ given by

$$T_s(n_0) \approx \frac{\log E_p(N) - n_0}{r_d} \tag{75}$$

about 10 to 100 generations given the parameter values of Table I. Meanwhile expected extinction times for the same populations are on the order of 100 to the astronomical. However, many populations are going extinct at rates comparable to the rate they would go from N_0 to K. This potential overlap in scale is threatening to the Levins metapopulation model. If a metapopulation is to persist, high extinction rates need to be compensated for by high colonization rates. If colonization rates are high enough, then neighboring populations may begin to show coupled dynamics, and very complex, potentially chaotic and destabilizing dynamics may ensue as suggested by the literature on weakly isolated, spatially explicit, deterministic models (Hastings, 1990; Hastings and Higgins, 1994; Bascompte and Solé, 1994, 1995).

There are other effects of the local dynamics which are not inevitably harmful to the Levins metapopulation model. Consider the functional relationship between local population density and emigration. Emigration is likely to increase as population densities increase. This holds true for butterflies (Ehrlich, 1984) and vertebrates, but the relationship between density and emigration rate need not be linear (Hansson, 1991). Many species emigrate disproportionately at densities close to carrying capacity (Wynne-Edwards, 1962), so that a knowledge of this relationship and the probability density $f_p(n)$ become valuable. It is doubtful that modifications to the colonization assumptions of the Levins model will gracefully accomodate all natural dispersal/density relationships. Complex dynamics may result as suggested in the previous paragraph.

D. Metacommunity Dynamics

The single-species local population dynamics explored in this chapter have inevitable implications for community interactions at the local and the landscape level. Populations that fluctuate have erratic interactions with competitors, predators, prey, and mutualists. As has been argued by Andrewartha and Birch (1954,

p. 20), Wiens (1977), Strong (1986), and others, random populations lead to less deterministic and more diffuse competition. This makes competition harder to detect in the field, encouraging irresolvable controversy. However, stochasticity has real consequences for communities. In fluctuating environments, some resources often go unutilized. This may diminish competition and provide opportunities for invasion and evolution of new high-dispersal competitors. Or competition can become occasionally more intense, leading to transient bursts of natural selection for character displacement (Grant and Grant, 1989). Fluctuating predator–prey communities can lead to transient refugia for prey. Population fluctuations may erode mutualisms, since they depend on predictable reciprocation. The literature on these consequences of local stochasticity is vast (recent reviews include Chesson and Huntly, 1989; Chesson, 1990; Pimm, 1991). However, we do not know how much community disruption arises from particular forms of local stochasticity and extinction.

May (1973a), Chesson (1981, 1990), Caswell and Cohen (1993), and others have analyzed models of communities in flux, but most empirical work depends on comparative, macroecological approaches (Brown, 1995; but see Bengtsson and Milbrink, 1995, for a recent, fairly tight lab study with green algae and two *Daphnia* species). Hassell (1978), Hastings (1977, 1978, 1990), Kareiva (1989), Taylor (1990, 1991), and others have paid special attention to the enhancement of predator–prey stability by spatial and temporal heterogeneity although the distinction between metapopulation level processes and local processes has not always been empirically established (Taylor, 1990, 1991). The analysis of any ecological interaction can be improved only by an understanding of the local extinction dynamics of individual species, but a species embedded in the community, tightly coupled to other species, is likely to have distinctive dynamics that are poorly modeled by a single-species stochastic process. If most of v_r for a lynx population can be attributed to fluctuations in snowshoe hares (Royama, 1992), then a two-dimensional stochastic analysis is called for. The SIR models of epidemiology lead to local disease agent extinctions, partly due to the finite size of host populations, partly due to the predator–prey-like deterministic dynamics (Anderson and May, 1991; Mollison, 1995). Much of environmental "stochasticity" is presumably attributable to the deterministic, perhaps chaotic, dynamics of the community in which a species is embedded (Sugihara and May, 1990). It is not yet clear whether the "predictablility" of deterministic dynamics for diffuse interactions differs from the 'unpredictability' of stochastic dynamics when it comes to extinction.

The other major metapopulation process, immigration, may also depend on the presence of competing species (MacArthur and Wilson, 1967) or it may not (Simberloff, 1978b). Community interactions, single-species stochasticity, and colonization interplay sometimes tightly, sometimes diffusely in metapopulations, and the appropriate analytic tools vary likewise. However, these problems take us well beyond this chapter.

VIII. CONCLUSION

Local extinction can be plausibly modeled in many cases by a population undergoing environmental stochasticity, with a ceiling set by resource caps and a lower threshold set by the minimum number needed to successfully reproduce. The environmental stochasticity model presented here generates simple analytic predictions of extinction rates and population densities. These can be used in metapopulation models to improve our analysis of the fundamental processes of extinction and colonization. For extinction dynamics, demographic stochasticity may play a smaller role than has sometimes been assumed, but catastrophes remain hard to study empirically. It is easy to estimate the parameters of the environmental stochasticity mode, but surprisingly little work has been done estimating v_1, the fundamental parameter of demographic stochasticity, or δ and λ, the catastrophe parameters. We still know little about the effect of genetic impoverishment on the viability of a population.

The environmental stochasticity model has many largely unexplored implications for metapopulation dynamics. Emigration rates are influenced by local fluctuations, extinction rates become correlated across populations, the time scales of local growth and extinction may overlap. Most of these have been discussed, but not mathematically analyzed. Hopefully this chapter will provide tools with which to solve some of the outstanding problems of metapopulation dynamics, especially those related to colonization and regional stochasticity.

The processes that create environmental stochasticity are fairly well known in a general way, but the systematic quantitative partitioning of the environmental variance v_r into its components for a range of representative species would help to disentangle the classic problems of population regulation, competition intensity, density vagueness, regional stochasticity, and metapopulation persistence. Population fluctuations are probably better understood in light of the variance in r (e.g., Wolda, 1978) than in the variance of N or log N (e.g., Pimm, 1991). To understand the local extinction patterns of tightly coupled, nondiffuse ecological interactions (e.g., mountain lion and deer, infectious disease) will require two-dimensional stochastic analysis. The transition from Levins models to less-isolated patch models needs to be examined more carefully, since rescue effects become increasingly important in local persistence (Brown and Kodric-Brown, 1977; Hanski and Gyllenberg, 1993) and populations become more synchronous in their fluctuations. So there is plenty of work to do.

ACKNOWLEDGMENTS

Dr. Janet Foley (DVM, UC Davis) has been a persistent source of discussion about disease. Ilkka Hanski has kept me metathinking. The work of Levins, MacArthur, and Wilson continues to inspire.

APPENDIX 1

Symbols Used in This Paper

Symbol	Explanation
N	Population size measured in females; $n = \log_e(N)$
$R(N, t)$	Realized fundamental net reproductive rate: $N(t + 1) = N(t)\, R(N, t)$; $r = \log_e(R)$
K	Carrying capacity (in females) considered as a ceiling; $k = \log_e(K)$
r_d	The drift coefficient or deterministic component of r, the expected value of r
ρ	Serial autocorrelation in environmental effects between consecutive years
v_r	Variance in r due to environmental stochasticity
v_{re}	Effective environmental variance taking ρ into account: $v_{re} = v_r \dfrac{1 + \rho}{1 - \rho}$
v_1	Variance in r for one female due to demographic stochasticity
v	The diffusion coefficient, the variance in r: $v_r + v_1/N$
γ	v_1/v_r: relative importance of demographic and environmental stochasticity
s	r_d/v_r: strength of population growth (r_d) relative to environmental stochasticity
s_d	r_d/v_1: strength of population growth (r_d) relative to demographic stochasticity
λ	Rate at which catastrophes occur, a number close to zero
δ	Mean fraction of individuals killed in a catastrophe, a number close to one
$\psi(x)$	$\exp(-2 \int_0^x \dfrac{r_d}{v}\, dy)$, an unnamed, but useful function
$S(n)$	Scale measure of the diffusion: $\int_0^n \psi(x)\, dx$
$m(n)$	Speed measure of the diffusion: $\dfrac{1}{v(n)\psi(n)}$
$t(n; n_0)$	Sojourn time near n: $t(n; n_0) = 2m(n)S(n)$ for $n < n_0$, $t = 2m(n)S(n_0)$ for $n > n_0$
$T_e(n)$	Expected time to extinction starting at n: $T_e(n_0) = \int_0^k t(n, n_0)\, dn$
$P_k(n)$	Probability of reaching k before going extinct: $S(n)/S(k)$
$p(n, t)$	Probability density of $n(t)$: $\mathbf{E}n$ and $\mathbf{V}n$ are the expected value and variance, respectively, of n
N_c	The critical value for population success ($n_c = \log_e(N_c)$): $P_k(n_c) \approx P_k(k)$

APPENDIX 2

Main Results of Stochastic Extinction Theory with a Population Ceiling at K
No Stochasticity

$$T_e(n_0) \quad -\frac{n_0}{r_d} \quad \text{if } r_d < 0 \quad \text{otherwise } T_e(n_0) = \infty$$

Environmental Stochasticity (Exact Results Given Diffusion/Continuity Limitations)

$$T_e(n_0) \quad \frac{1}{2sr_d}\left(e^{2sk}(1-e^{-2sn_0})-2sn_0\right) \quad \frac{2n_0}{v_r}\left(k-\frac{n_0}{2}\right) \quad \text{if } r_d = 0$$

$$T_e(k) \quad \frac{1}{2sr_d}\left(e^{2sk}-1-2sk\right) \quad \frac{k^2}{v_r} \quad \text{if } r_d = 0$$

$$n_c \quad \frac{1}{2s}$$

Demographic Stochasticity (More Approximate, for $r_d \geq 0$)

$$T_e(n_0) \qquad\qquad\qquad\qquad \frac{n_0 e^k - e^{n_0}+1}{v_1} \quad \text{if } r_d = 0$$

$$T_e(k) \quad \frac{e^{2s(e^k-1)}-1}{4r_d s_d} \quad \text{if } 1 < 2s_d k \quad \frac{e^k(k-1)+1}{v_1} \quad \text{if } r_d = 0$$

$$n_c \quad e^{2s_d}E_1(2s_d) \approx \ln\left(1+\frac{\ln(2)}{2s_d}\right)$$

Environmental and Demographic Stochasticity (Also Approximate, for $r_d \geq 0$)

$$T_e(k) \quad \frac{(e^k+\gamma)^{2s}-(e^{n_c}+\gamma)^{2s}}{2sr_d} \frac{(1+2\gamma)}{(1+\gamma)^{2s+1}} \frac{k^2}{v_r}\left(\frac{1}{1+\gamma}\right) \quad \text{if } r_d = 0$$

$$n_c \quad \frac{1+2\gamma}{2s(1+\gamma)}$$

Environmental Autocorrelation

$$v_{re} \quad \frac{1+\rho}{1-\rho}v_r$$

Rule of Thumb: Effect of Logistic Density Dependence on Environmental Stochasticity

$$\frac{T_e(k) \text{ with logistic density dependence}}{T_e(k) \text{ without logistic density dependence}} \approx 1 - 0.7s \quad \text{for } 0 < s < 1$$

APPENDIX 3

Conditional Population Densities Predicted by Models with Environmental Stochasticity

In General

$$f_p(n; n_0) = \frac{t(n; n_0)}{T_e(n_0)} \qquad\qquad F_p(n; n_0) = \int_0^n f_p(x; n_0)\, dx$$

$$E_p(n) = \int_0^k n f_p(n; n_0)\, dn \qquad\qquad V_p(n) = \int_0^k (n - E_p(n))^2 f_p(n; n_0)\, dn$$

$$f_p(N; N_0) = \frac{f_p(\log(N); \log(N_0))}{N} \qquad \text{using the change of variable formula}$$

Starting from Small n_0 (Propagule Size)

$$f_p(n; k) \approx \frac{2s e^{2s(n-k)}}{1 - e^{-2sk}} \qquad\qquad f_p(n; k) = \frac{1}{k} \qquad\qquad\qquad \text{if } r_d = 0$$

$$E_p(n) = \frac{2sk e^{2sk} - e^{2sk} + 1}{2s(e^{2sk} - 1)} \qquad \approx \frac{k}{2}\left[1 + \frac{sk}{3}\right] \qquad\qquad \text{if } sk < 1$$

$$V_p(n) \qquad\qquad\qquad\qquad \approx \frac{k^2}{12}\left[1 + \frac{(sk)^2}{3}\right] \qquad \text{if } sk < 1$$

$$f_p(N) \approx \frac{2s N^{2s-1}}{K^{2s} - 1} \qquad\qquad f_p(N) \approx \frac{1}{kN} \qquad\qquad\qquad \text{if } r_d = 0$$

$$F_p(N) \approx \frac{N^{2s-1}}{K^{2s} - 1}$$

$$E_p(N) = \frac{2s(K^{2s+1} - 1)}{(2s + 1)(K^{2s} - 1)} \qquad \approx \frac{K}{k}[1 + sk] \qquad\qquad \text{if } sk < 1$$

$V_\mathrm{p}(N)$

$$= \frac{2s(2s+1)^2(K^{2s}-1)(K^{2s+2}-1) - 4s^2(2s+2)(K^{2s+1}-1)^2}{(2s+2)(2s+1)^2(K^{2s}-1)^2}$$

$$\approx \frac{K^2[1+sk]}{2k}\left(1 + \frac{2(1+sk)}{k}\right) \quad \text{if } sk < 1$$

Starting from k (or above n_c)

$$f_\mathrm{p}(n; k) = \frac{2s(e^{2sn}-1)}{e^{2sk}-2sk-1} \quad \text{for } 0 < n < k$$

$$f_p(n; k) = \frac{2n}{k^2} \quad \text{if } r_\mathrm{d} = 0$$

$$E_\mathrm{p}(n) = \frac{e^{2sk}(k-1/2s) - 1/2s - sk^2}{e^{2sk}-2sk-1}$$

$$\approx k\left[\frac{2}{3} + \frac{sk}{18}\right] \quad \text{if } sk < 4$$

$$V_\mathrm{p}(n) = \frac{e^{2sk}(k^2 - k/2s + 1/2s^2) - 1/2s^2 - 2sk^3/3}{e^{2sk}-2sk-1} - (E_p(n))^2$$

$$\approx k^2\left[\frac{1}{18} + \frac{sk}{135}\right] \quad \text{if } sk < 4$$

11

Studying Transfer Processes in Metapopulations

Emigration, Migration, and Colonization

Rolf A. Ims   *Nigel G. Yoccoz*

I. INTRODUCTION

The transfer of individuals across space is a key process in metapopulations. In fact, metapopulations are often seen as sets of local populations the very existence of which is dependent on mutual transfer (or exchange) of individuals (Hanski and Simberloff, this volume). Components of transfer rates such as emigration, migration (dispersal), and colonization become critical variables.

These variables must be understood and analyzed in quantitative terms for a better understanding of metapopulation dynamics. Unfortunately, there are often great difficulties associated with obtaining data on the transfer process in metapopulations. Events such as the emigration of individuals from habitat patches, the colonization of empty patches, and the behavior of migrating individuals are difficult to observe. For this reason, the transfer rate must often be inferred with indirect approaches based on rather stringent assumptions of uncertain validity.

In this chapter we will draw together the various empirical approaches to the study of emigration, migration, and colonization processes in a metapopulation context. In particular, we stress the limitations and advantages of different approaches and assess critically the lines of inferences that can reliably be drawn from them. Some of our conclusions may seem overly pessimistic: our knowledge is still very sparse, and there are severe limitations on what can be learned from the different approaches. Nonetheless, we are convinced that knowing these limitations allow us to devise better experimental and observational designs and methods of analysis.

II. TRANSFER OF INDIVIDUALS IN METAPOPULATIONS: DEFINING THE COMPONENTS

We will define *transfer* as the process by which an individual organism is brought (actively or passively) from one site to another. Often the transfer is associated with a change in life cycle phase, e.g., migration (natal dispersal) precedes reproduction in mammals and birds. Our definition is rather similar to the one used by students of social organization and demography, in particular of primate populations (e.g., Clutton-Brock, 1989), for which the transfer concept relates to the exchange of individuals between social groups.

Generally transfer of individuals can be viewed as a three-stage process (Fig. 1). It is initiated by an *emigration* stage in which the individual leaves its home site. The emigration event is followed by a *migration* stage which is the displacement process (movement) that brings the individual away from its home site. We will here use migration as synonymous to *dispersal* which is a term often favored by population ecologists to describe spatial displacement processes in settings other than metapopulations (see Wiens, this volume). The migration phase may or may not bring the individual to a new home site where it can settle. We will term this settlement stage *immigration* regardless of whether settlement takes place at a site already inhabited by conspecifics or at a site which is empty. Immigration in an empty patch followed by successful establishment of a new

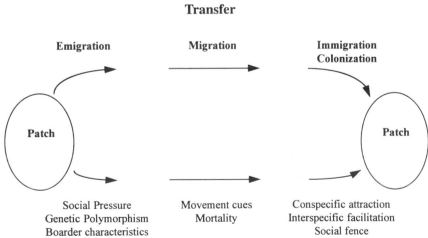

Transfer

Emigration   Migration   Immigration
Colonization

Patch   Patch

Social Pressure   Movement cues   Conspecific attraction
Genetic Polymorphism   Mortality   Interspecific facilitation
Boarder characteristics   Social fence

FIGURE 1   The stages of the transfer process in metapopulations and some of the causal mechanisms that may be involved at each stage.

population is *colonization*. Alternatively the migration stage may not lead to a settlement, if it is terminated by death. In a metapopulation context the scopes of these three general stages of the transfer process are somewhat restricted. Below we address the emigration, migration, and colonization/immigration as they pertain to the particular settings of metapopulations. We will be very brief on the ultimate and proximate mechanisms that may be involved in each stage as such considerations have been dealt with extensively in several recent reviews (for reviews on factors that trigger emigration and dispersal, see, e.g., Chepko-Sade and Halpin, 1987; Johnson and Gaines, 1990; Hansson, 1991; Stenseth and Lidicker, 1992a; Olivieri and Gouyon, this volume, the migration/dispersal process itself is dealt with in detail by, e.g., Ims, 1995; and regarding colonization/immigration, see, e.g., MacArthur and Wilson, 1967; Bazzaz, 1986; Ebenhard, 1991).

A. Emigration

Emigration is defined as leaving a local population whose spatial extent is determined by the extent of the respective habitat patch. The suitable habitat is typically restricted to discrete patches imbedded in a *matrix* (or *bath,* or *nonhabitat:* Levin, 1976; Czaran and Bartha, 1992), an area not suited for long-term survival and reproduction. Emigration events therefore requires crossing a boundary between habitat and nonhabitat. This boundary may be recognized as a distinct physical patch edge or it may be more like a gradient (Holland *et al.,* 1991). The sharpness of the border zone may affect the emigration rate from the patch, expecially if behavioral mechanisms are involved in the emigration process (Stamps *et al.,* 1987; Durelli *et al.,* 1990). There may, however, be some situation even in a metapopulation setting for which this habitat–nonhabitat dichotomy does not apply to the emigration stage. Sometimes organisms leave a local patch when it has been depleted for resources either by exploitation by the local population or by the independent dynamics of the patch itself (Whitlock, 1992b, Roff, 1994b). In this case the patch does not any more constitute habitat for reproduction and survival.

B. Migration

The migration stage involves displacements/movements in nonhabitat (matrix), often over long distances. The critical points here are the hazards associated with migration, as well as the specific movement trajectories. Both mortality and movement patterns may differ substantially from what may be typical inside habitat patches (Ims, 1995).

As a heritage from the original island-type models (MacArthur and Wilson, 1967; Levins, 1969a, 1970), present metapopulation studies even in terrestrial environments tend to treat the matrix as if it was water. However, terrestrial matrices are typically not as featureless and homogeneous as is often implicitly assumed (Wiens, this volume). Terrestrial matrices may contain physical and biological features which are distributed heterogeneously in space and time. Matrix features will to varying extents impede or enhance migration. This applies to all organisms, whether they migrate by active movements (most animals) or by passive displacements (wind dispersal of seeds). Migration–matrix heterogeneity interactions tend to make the density distribution of dispersing individuals a much more complex function than if it was determined solely by patch specific emigration rates together with the global mean and variance of the migration distances.

C. The Settlement Stage: Immigration and Colonization

Migrating organisms in metapopulations may or may not encounter a new habitat patch depending on the spatial distribution of patches (interpatch distances) in relation to species specific migration capacity and its interaction with matrix properties. Spatially isolated patches will often tend to be encountered less often than patches that are located close to other patches. For this reason patch-specific isolation indices, functions of distances to occupied neighboring patches weighted by the sizes of these populations, are used to describe encounter rates in spatially explicit metapopulation models (Hanski, 1994a, this volume). Note, however, that such encounter rates may not always decrease with increasing patch isolation. For example, if the migrating individual uses distant cues to orientate, isolated patches may attract a disproportionate number of migrants (Bell, 1991). Properties of the matrix may also be decisive. Certain physical features in the matrix may act as corridors and thus channel migrating individuals to certain patches on the expense of others (Ims, 1995).

For passively dispersing organisms, for example wind-dispersed seeds, the stochastic components of patch encounter rates tend to be large with the majority of dispersers not reaching any new habitat at all. The probability of a disperser reaching a new patch may be an important selective pressure shaping life history traits of organisms with a metapopulation structure. Organisms that disperse passively and thus have low patch encounter probabilities typically produce large numbers of offspring at the expense of the quality of each offspring. The opposite is true for actively migrating animals that are able to orient toward patches from some distance. In this case the encounter rate may be sufficiently high to permit the individual to sample several patches before settlement and thus to increase the likelihood of successful establishment. Patch or habitat selection involves decisions based on the assessment of environmental cues (patch characteristics) that reflect the quality of the patch. Apart from patch characteristics such as patch size and quality, the settlement may also depend on whether the patch is already occupied by conspecifics or not. Immigration into already established populations may be facilitated (by conspecific attraction or facilitation; Smith and Peacock, 1990; Wood and del Moral, 1987) or prohibited (social fences; Hestbeck, 1982) by the presence of conspecifics. Heterospecifics may similarly facilitate or impede immigration (Danielson and Gaines, 1987; Tilman, 1993). However, *colonization* of empty patches (no conspecifics present) must by definition be possible in a metapopulation context.

Colonization and immigration in metapopulations are not conceptually very different from the same processes under other settings. General theories of habitat selection such as Rosenzweig (1981) may be applicable to metapopulations studies (Danielson, 1992; Morris, 1992). However, generally one would expect that the stochastic components of habitat selection are greater in metapopulations than otherwise given the more restricted opportunities for organisms to get information on the availability of habitats being unpredictably scattered in time and space.

III. INDIRECT, GLOBAL APPROACHES TO PARAMETER ESTIMATION

Of the three transfer stages defined above, colonization has been of principal interest in both theoretical and empirical investigations of metapopulation dynamics. The reason for this is partly historical; colonization is the only stage that appears explicitly in Levins' metapopulation model (Levin, 1969a, 1970). Colonization rate has also remained the key transfer parameter in the most recent extensions of Levins's model (Hastings and Harrison, 1994).

Although colonization rate may occasionally be measured directly (Pokki, 1981; Solbreck, 1991; Whitlock, 1992b; Valone and Brown, 1995), in most cases it is an impossible task to estimate by direct observation how colonization rates change for instance as a function of interpatch distances. For both practical and conceptual reasons, colonization is the most difficult of the three transfer stages to study. The practical difficulty hinges on the fact that colonization events are typically rare and thus require studies of exceptionally large spatial and temporal extents (Wiens, 1989a). In contrast, emigration and migration are more local and continuous processes. The conceptual difficulties hinge on the fact that colonization is the most complex transfer stage, being a function of the two preceding stages (emigration and migration) of the transfer process.

Due to the great difficulties of obtaining direct estimates of colonization rates, most empirical metapopulation studies have employed indirect approaches. The simplest approach to getting an indirect global estimate of the colonization rate for a Levins-type metapopulation would be to solve the model for a known extinction rate, and assume that the metapopulation is at equilibrium. Unfortunately, direct estimates of extinction rates are at least as hard to obtain as estimates of colonization rate. Estimates of extinction rates require long-term data sets, which are available mainly for islands. Even then, available methods make homogeneity assumptions (temporal and spatial) which need to be tested before these methods can be used (Clark and Rosenzweig, 1994). A more realistic empirical approach may be to obtain estimates of the two other processes which are easier to study, the emigration rate and the distribution of migration distances. Such estimates would give a measure of the expected density distribution of migrating individuals in the matrix and hence the expected encounter rate of migrants with empty and occupied patches. Combined with the incidence function, describing the spatial distribution of extinct and extant patches in the landscape, the expected encounter rates may be used to derive colonization (and extinction) rates by numerical techniques. Recently Hanski et al. (1994) have used such an approach to develop spatially explicit metapopulation models (allowing for spatially variable population sizes and interpopulation distances in steady-state metapopulations) and estimate patch-specific colonization probabilities. Compared to the earlier, more general models (see Hanski, 1991, 1994a; Hanski and Gyllenberg, 1993; Hastings and Harrison, 1994), these new spatially explicit models represent a significant advance for predicting the equilibrium and transient dynamics of specific metapopulations. These approaches still go under the heading "global approaches" as the parameter estimates represent averages at the metapopulation level; the transfer parameters are not spatially explicit in the sense that per capita rates might vary according to patch sizes, matrix conditions, etc. This assumption about homogeneous processes may limit the applicability of this approach as predictive tools in real management situations.

Our greatest concern in this context relate to how the estimates are obtained. First of all, this approach requires a relatively large number of parameters to be estimated, most likely from the same data set. For example, in the model by Hanski et al. (1994) there were eight parameters of which five were related to patch (population) specific dynamics (α^{-1}, carrying capacity; r, density-independent growth rate; θ, a parameter measuring the strength of density dependence; and finally, s_d and s_e, two parameters measuring, respectively, demographic and environmental stochasticities in extinction rates). Three parameters describe the two transfer stages emigration and dispersal (c, constant emigration rate; τ, a parameter describing the migration distances; and μ, constant mortality rate during migration). While a few of the parameters were obtained by direct estimation based on specifically designed capture–recapture studies (i.e., emigration rate and dispersal distances), others were obtained by curve fitting (the five population specific parameters) or (subjective) adjustment procedures (mortality of migrants). This approach to deriving estimates of demographic parameters stands in contrast to the state of the art of recent demographic research which emphasizes (1) rigorous designs for sampling statistical populations (e.g., Burnham et al., 1987; Skalski and Robson, 1992), (2) analyses based on the formulation of alternative statistical models for estimating single demographic parameters and their statistical intervals (under the most appropriate statistical model) (Lebreton et al., 1992, 1993), and (3) tests of specific hypotheses deducing the outcome of interactions between mechanisms and processes. For example, in the case of the three transfer parameters of Hanski et al. (1994) it is not clear whether the data collected for estimation purposes was a representative sample for the whole metapopulation. Further, spatial constancy of the parameter estimates was assumed, but not tested. Finally, no statistical intervals were provided, nor could they have been as the parameters were not obtained independently of each other.

However, we will not dispute the fact that models such as those of Hanski (1994a) and Hanski *et al.* (1994) may be useful for predicting the dynamics of the specific metapopulation for which they were developed. In fact, sensitivity analyses showed that the predictions did not depend very much on the choice of specific values of the parameters determining transfer rates. However, this may imply that the dynamics of these butterfly metapopulations are mostly affected by other processes. The dynamics of other metapopulations may be more transfer rate dependent. In such cases the predictive power of models is likely to be much more dependent on precise and unbiased estimates of emigration, migration, and immigration/colonization rates, and knowledge about how these rates are shaped by various ecological factors. Clearly there is a need for assessing the reliability of empirical metapopulation models, in particular through forecasting, i.e., prediction under a different set of conditions (Conroy *et al.,* 1995). This has not been done to our knowledge, but would probably provide knowledge of the most critical parameters.

Another method of obtaining indirect, global estimates of transfer rates in metapopulations is to use population genetic methods to estimate gene flow. The estimation of gene flow between populations, as pioneered by Wright (reviewed in Wright, 1969), has been the subject of numerous studies. In particular, the use of Wright's F_{ST} statistics and spatial autocorrelation methods have been extensive (see Slatkin and Barton, 1989; Epperson, 1993). Such methods aim at estimating the product of N_e, the effective population size of each local population, and m, the immigration rate. These methods could therefore provide an indirect measure of one of the essential components of metapopulation dynamics, immigration, and colonization rates. However, such indirect methods are by themselves either unsuitable for metapopulations as they assume spatially homogeneous populations (Epperson, 1993) or they assume that the metapopulation is at a genetic and demographic equilibrium (Olivieri *et al.,* 1991; Slatkin, 1994, 1995).

When direct methods of estimating transfer rates and indirect methods based on gene flow have been used for the same species, they have often been found to give rather different answers; indirect methods typically indicating greater extent of dispersal than direct methods (Slatkin, 1987). For example, Slatkin (1987) described very limited dispersal distances for the checkerspot butterfly *Euphydryas editha,* using direct observations, but extensive gene flow using allele frequencies. Similar results have been found for, e.g., the Colombian ground squirrel, *Spermophilus columbianus* (Dobson, 1994), banner-tailed kangaroo rats *Dipodomys spectabilis* (Waser and Elliott, 1991), and grey-crowned babbler *Pomatostomus temporalis* (Edwards, 1993), all species for which direct methods indicated limited dispersal. Moreover, genetic differentiation within a metapopulation will be dependent on the origin (e.g., from the same or different patches) and nature (e.g., virgin or inseminated females) of the dispersers, and therefore the latter characteristics cannot be inferred without supplementary observations (McCauley, 1991). Slatkin (1994) suggested that the results of direct and indirect methods should be *combined* to determine whether the species is at genetic/demographic equilibrium and, more generally, the consistency of the different approaches.

We conclude that the existing indirect methods of obtaining metapopulation level estimates of demographic or genetic parameters will be themselves provide little opportunity for predicting present transfer rates between populations. However, the incidence function approach (Hanski, 1992a, 1994a, this volume) and the genetic methods (Hastings and Harrison, 1994; Morin *et al.*, 1994) may provide useful information on migration on longer time and larger spatial scales than direct methods could. The fact that metapopulation dynamics may change dramatically on a long time scale (Solbreck, 1991) shows that this is of great empirical interest.

IV. DIRECT, PATCH-SPECIFIC APPROACHES

The rich textures of real landscapes is playing a significant role for the resulting metapopulation dynamics (Wiens *et al.*, 1993). Habitat characteristics such as patch area and quality, spatial distribution of patches, and presence of dispersal corridors or barriers (see Ims, 1995, for a review) may influence local population dynamics as well as exchanges between patches. In the same way that demographic analyses of single populations have focused on the sensitivity of the population growth rate to the different demographic parameters (e.g., Eberhardt *et al.*, 1994), metapopulation studies should use similar sensitivity analyses to identify the critical parameters and assumptions (Alvarez-Buylla, 1994). The great difficulties involved in the analysis of the dynamics of single populations will surely propagate at the metapopulation level. We do not think the indirect approaches outlined above could be developed much further without focusing more on the specific mechanisms at the population level (Conroy *et al.*, 1995).

In addition to providing parameter estimates for predictive models, patch-specific studies may provide tests about common assumptions made in metapopulation models. For instance, patch specific studies may tell whether local populations are demographically self-sustained or not, i.e., whether they are sources or sinks in a deterministic sense, and whether exchange rates between patches are sufficiently low to justify the common assumption of independent dynamics of local populations (but see Gyllenberg *et al.*, this volume).

Below we review some of the most recent methods that can be used to obtain patch-specific estimates of emigration and immigration/colonization rate as well as methods of studying the spread of individuals (migration/dispersal) in the matrix. We suggest that the choice of methods must be guided by explicit hypotheses about which ecological mechanisms underlie these rates. Moreover, we emphasize that the methods must be sound with respect to general statistical principles for sampling and modeling. We focus mainly on observational methods since it appears to us that large-scale metapopulation processes are not often amenable to replicated, manipulative experiments. Experimental model system (EMS) studies

(Ims and Stenseth, 1989; Wiens *et al.*, 1993), which are feasible for certain "scaled-down" systems, may be important as a general method of exploring ecological mechanisms that may operate under various conditions (Ims *et al.*, 1993). EMS studies may also provide tools for validating new statistical methods (Yoccoz, 1994). For a summary of our survey, see Table I.

A. Studies of Migration

In plants, migration in metapopulations is generally restricted to seed dispersal, since vegetative reproduction will usually not allow for colonizing distant patches. The agents for seed dispersal are numerous, and their relative importance varies enormously both between and within species (e.g., Hughes *et al.*, 1994). However, we may distinguish between passive physical dispersal, such as wind dispersal and migration by animals, since they require different research approaches. Some animals, small insects in particular, and fungi disperse also passively (Shaw, 1995).

Wind dispersal is a physical process, and the seed shadow has therefore been analyzed recently using mechanistic (physical) models (e.g., Greene and Johnson, 1989a,b; Okubo and Levin, 1989; Andersen, 1991). Such models have the advantage of not being site- or species-specific, even if more empirical approaches, using for example physical models of seeds (Augspurger and Franson, 1987) are needed. While the average or mode of the dispersal distribution are usually rather well explained by such models (e.g., Okubo and Levin, 1989), the tail of long-distance dispersal is not well known, partly because of sampling problems, partly because of data lumping for long distances (Portnoy and Willson, 1993). Most seeds disperse short distances (a few meters) giving only small sample sizes for long distances (McEvoy and Cox, 1987). However, it is the long tail of the distribution which is of relevance to metapopulation dynamics. Local factors, such as landscape features influencing wind patterns, may greatly modify the distribution expected in more homogeneous environments (McEvoy and Cox, 1987). We are not aware of studies quantifying the role of landscape features for long distance seed dispersal, as most studies have been focusing on the colonization part of the transfer process (e.g., Wood and del Moral, 1987; Augspurger and Kitajima, 1992; Tilman, 1993).

Animal migration involving active movements is typically more complex than passive modes of dispersal. Still rather simple mechanistic models of vertebrate migration may show reasonably good fit to empirical data on migration distances [e.g., Buechner, 1987; Miller and Carroll, 1989; Caley, 1991; however, Porter and Dooley (1993) have shown that this may result from a sampling bias]. These mechanistic models are, however, inadequate in a metapopulation setting as the only parameter explicitly considered is a constant or monotonically changing probability of stopping (settling or dying) some distance from the home site in homogeneous landscape (Waser, 1985; Buechner, 1987; Caley, 1991; Miller and Carroll, 1989).

TABLE I Methodological Approaches to the Study of Transfer in Metapopulations

Approach	Spatial scale	Advantages	Inconvenients	References
Incidence function	Metapopulation	Only presence/absence data Large spatial and temporal scales	Large parameter uncertainty Mechanisms poorly known	Hanski (1992a, 1994a)
Gene flow	Metapopulation	Large spatial and temporal scales	Large parameter uncertainty Ecological mechanisms unknown	Slatkin (1994)
Experimental model system	Metapopulation: Small absolute spatial scale	Access to the mechanisms influencing the components of transfer	Artificial environments	Huffaker (1958); Forney and Gilpin (1989)
Capture–recapture studies	Metapopulation and patch	Estimation of patch-specific demography, population sizes and exchange rates	Require large sample sizes Components of transfer not separated	Brownie *et al.* (1992); Nichols *et al.* (1992)
Radiotelemetry	Metapopulation and patch	Allow study at the individual level	Small samples	White and Garrott (1990)
Experimental release in matrix	Metapopulation	Aim at dispersal stage	Origin of individuals	Harrison (1989)
Experimental release in patch	Patch	Aim at settlement stage	Origin of individuals	Augspurger and Kitajima (1992); Danielson and Gaines (1987)
Artificial empty patches	Metapopulation and patch	Aim at colonization; control of patch size and isolation	Artificial nature of patches	Schoener (1974); Whitlock (1992b)
Enclosed populations	Patch	Aim at emigration/colonization	Small scale Fence artifacts	Johnson and Gaines (1985, 1987); Valone and Brown (1995)

Mechanistic computer simulation models provide a more realistic approach to the study of migration in metapopulations. This approach involves computer simulations of movement patterns in spatially explicit mosaics consisting of lattices of suitable and nonsuitable habitat. The focus here is on the movement process itself and its interaction with environmental factors. In the simplest models the algorithms are usually based on simple or first-order correlated random walks (e.g., Gardner *et al.,* 1989; Wiens and Milne, 1989; Johnson *et al.,* 1992a,b). The cell size of the lattice corresponds to the grain size (the smallest spatial scale at which an organism recognizes spatial heterogeneity; Wiens, 1989a; Kotliar and Wiens, 1990). Spatial structures at larger scales emerge as clusters of cells. The clusters may or may not be connected (percolating networks; Gardner *et al.,* 1989; Johnson *et al.,* 1992a), depending on the fraction of the cells that has been designated as suitable area. Such spatial mosaics can be modeled to resemble the spatial structures of real matrix habitats (e.g., Johnson *et al.,* 1992b) or may be constructed by assigning a given proportion of suitable and nonsuitable cells at random (Gardner *et al.,* 1989).

The computer simulation approach to modeling individual movements may serve two related purposes. First, it may serve as a null model against which real movement patterns of individuals may be compared (e.g., Turchin, 1986). The second purpose is to check whether a random walk is a reasonable approximation of real movement behaviors and hence could be assumed in spatially explicit metapopulation models (Hanski, 1994a, Hanski *et al.,* 1994).

Insects and other ground living invertebrates moving slowly on the ground are particularly amenable for movement pathway studies because they are easy to observe such that the whole pathway of the moving individuals together with relevant parameters such as speed or step lengths (for discretized pathways) and tortousity can be measured and quantified. Such measurements may even to some extent be obtained by means of radiotelemetry (Andreassen *et al.,* 1993, 1996a, 1996b) or snow tracking (Tegelström and Hansson, 1987; Oksanen, 1993) for organisms that are hard to observe directly. In particular, radiotelemetric studies provide a way to measure long-distance dispersal, which may indicate the scale of specific metapopulations. Radiotelemetry also provides more than any other techniques opportunities to obtain direct estimates of rates and causes of mortality in the matrix (see below).

Wiens and colleagues have compared observed movement patterns of ground beetles and ants against the expectations from mechanistic simulation models (Wiens and Milne, 1989; Johnson *et al.,* 1992b; Crist *et al.,* 1992). Real pathways usually differ from the expectations based on first order correlated random walks models. This is also the case when the input parameters (turning angles and step lengths) (Crist *et al.,* 1992) and the spatial structure of the lattice (Johnson *et al.,* 1992b) are based on empirical data. Generally, real movements have a higher displacement rate and are less complex than simulated movements (Crist *et al.,* 1992). This is likely to be caused by the fact that most movement patterns have a directional bias (e.g., Kennedy, 1951; Johnson, 1969).

Not surprisingly, displacement rates of ground-living insects are dependent on the spatial structure of the vegetation as well as the particular species studied (Wiens and Milne, 1989; Crist *et al.*, 1992). Furthermore, the insects apparently respond to scale-dependent changes in the vegetation structure (Johnson *et al.*, 1992a). However, the most intriguing and potentially important result from these studies is that the structure of the pathways (measured by their fractal dimension) appears to be similar for different species, in different vegetation types and over the range of spatial scales being studied (Crist *et al.*, 1992). This gives hope for establishing certain generalizations about movement patterns that are valid across species, spatial, and temporal scales.

At the same time, the widespread capacity for spatial memory and orientation in many animals (Johnson, 1969; Danthanarayana, 1986; Bell, 1991) brings into question the value of this null model approach. Including more biologically plausible assumptions about movement decisions, for example elements of optimal search theory (see Bell, 1991), can make the simulation models more useful for improving our understanding of dispersal patterns in relation to spatial structures. Useful developments along such lines have been recently made (Kareiva and Odell, 1987; Turchin, 1987, 1991; Odendaal *et al.*, 1988; Crist *et al.*, 1992; Morris, 1993; Vail, 1993).

Migration may be risky, especially in metapopulation settings characterized by long-distance migration and long exposure times to unfavorable environmental conditions and biological enemies. Mortality during migration is likely to be a more significant modifier of the migration distance distribution for species in metapopulations than otherwise. To understand and predict the "filtering effect" of the migration stage (Lomolino, 1993) both on demographic and genetic processes in metapopulations, it is important to identify and quantify the importance of the different mortality agents. While the fate of a wind-dispersed seed may largely depend on chance events, whether it happens to be displaced to a suitable patch or not, mortality factors are likely to act more nonrandomly among actively migrating animals. Unfortunately, very little is actually known about the hazards involved in animal dispersal in general (Johnson and Gaines, 1990; Stenseth and Lidicker, 1992a).

Although the methodological difficulties of studying the causes of mortality during migration are formidable for most animals, recent advances in radiotelemetric techniques (Kenward, 1987; White and Garrott, 1990) bear promises for species large enough to carry transmitters. Studies such as those by Small *et al.* (1993), Larsen and Boutin (1994), Steen (1994), and van Vuren and Armitage (1994) give reasons for more optimistic expectations for the future than those conveyed by Johnson and Gaines (1990) in their general review of dispersal studies on birds and mammals. In particular, given a large sample of migrating radiotagged individuals, even rough mortality rate estimates may be obtained (unbiased, but with still large statistical intervals due to sample size limitations). Hazard rate models specifically tailored to data on the fates of radiotagged animals

(Pollock *et al.,* 1989) provide possibilities for stratification of the data with respect to characteristics of the dispersing animals (sex, age, etc.) as well as allowing one to include relevant environmental covariates such as matrix characteristics. Clearly, such analyses could contribute greatly to our understanding of the ecological processes determining transfer rates in metapopulations. The sample size requirements may still be beyond the logistic constraints of most research projects, and precise estimates of mortality rates of migrating individuals for most species may only be obtained from extensive, adequately designed and analyzed capture–recapture studies (see below).

B. Estimation of Transfer Rates Using Capture–Recapture Methods

Capturing, marking, and recapturing individuals has been used for a long time for estimating survival rates and population sizes (e.g., Fisher and Ford, 1947) as well as migration rates (Jackson, 1939). However, it is mainly since the development of the stochastic models of Cormack (1964), Jolly (1965), and Seber (1965) that the flexibility and usefulness of the models become clear (see Lebreton and North, 1992). Most models focused on the analysis of single populations, or the comparison of different populations (e.g., Burnham *et al.,* 1987; Lebreton *et al.,* 1992), with fewer developments regarding the estimation of exchanges between populations (e.g., Arnason, 1973). In the former models, survival rates are local survival rates in the sense that they measure the probability that an individual released at time t will be alive in the same population at time $t + 1$. These estimates confound mortality and emigration rates. Similarly, the recruitment rates will often confound immigration and reproduction *in situ*. We will below describe recent methods which have been proposed to disentangle these processes. We will focus on methods which do not assume any spatial or temporal homogeneity in the migration processes (e.g., Matsuda and Akamine, 1994), as we are indeed interested in the spatial and temporal variability of these processes.

1. Estimation of Immigration and Reproduction *in Situ* Using Capture–Recapture Studies of Single Populations

Nichols and Pollock (1990) assumed that the total adult population at time $t + 1$, $N(\text{ad}, t + 1)$, is made of three groups: adults that were present as adults in the population at time t, $N(\text{aa}, t + 1)$; adults that immigrated between time t and time $t + 1$, $N(\text{ai}, t + 1)$; adults which were present as trappable young in the population at time t, $N(\text{ay}, t + 1)$ (assuming that maturation time is equal to one time step). $N(\text{ai}, t + 1)$ could then be estimated using

$$N(\text{aa}, t + 1) = \phi_{\text{ad},t}\, N_c(\text{ad}, t),$$

$$N(\text{ay}, t + 1) = \phi_{y,t}\, N_c(\text{yo}, t),$$

$$N(\text{ai}, t + 1) = N_c(\text{ad}, t + 1) - N(\text{ay}, t + 1) - N(\text{aa}, t + 1),$$

where $\phi_{ad,t}(\phi_{y,t})$ are the adult (young) survival rates from time t to $t + 1$ estimated using open population models, and N_c denotes population sizes estimated using closed population models. This approach is also useful in documenting whether the population has a rate of increase which is larger or smaller than one, in the absence of immigration: we use the ratio $[N(aa, t + 1) + N(ay, t + 1)]/N_c(ad, t) = \phi_{ad,t} + \phi_{y,t} N_c(yo, t)/N_c(ad, t)$. The latter ratio represents the reproductive rate for the adults. It may be compared to more direct estimation, using, e.g., data on clutch size and proportion of breeders.

Another recent approach is provided by Pradel (1996). Instead of considering capture histories forward (i.e., from time t to $t + 1$), we could read it backward: "survival" rates would then be the probability that an animal caught at time $t + 1$ was present in the population at time t. Such probabilities are called seniority probabilities by Pradel (1996) and represents the resident fraction of the population. These probabilities could be estimated using similar approaches than are used for survival rates (Pollock *et al.*, 1990; Lebreton *et al.*, 1992). As population growth rates are related to both survival rates and seniority probabilities Pradel's approach could also be used to estimate population growth rates (including immigration), without relying on estimation of population sizes. The robustness of survival rate estimation (Lebreton *et al.*, 1992) may also apply to the estimation of seniority probabilities and therefore to the estimation of population growth rates. This approach would then be an alternative to the use of closed population models when the latter cannot be used.

Both methods present the advantage of making no specific assumptions about the dynamics of the population studied; they aim at estimating the rate of increase of the population, with and without immigration, which is one of the crucial parameters in metapopulation dynamics. However, as all statistical models, they rely on some assumptions, in particular about the growth rates of the young for Nichols and Pollock's approach (Yoccoz *et al.*, 1993) and about the absence of age-dependency in the Pradel's approach (i.e., parameters can only be time-dependent but not age-dependent; Pradel, 1996). Nevertheless, at least for the Nichols and Pollock's approach, appropriate study design could ensure that these assumptions are checked (using individual growth rates) and eventually modified (Yoccoz and Lambin, 1997).

2. Estimation of Immigration and Emigration between Populations Using Capture–Recapture Studies

Statistical models for estimating survival rates from capture–recapture data in open populations are rather robust to heterogeneity of capture probabilities or trap response, a common problem in many ecological studies. Survival rate estimation is therefore recommended for testing biological hypotheses instead of population size estimates (Lebreton *et al.*, 1992). Population size estimates are based on closed population models for which heterogeneity and trap response cannot be simply modeled (Pollock *et al.*, 1990; Yoccoz *et al.*, 1993).

There is of course no way that emigration rates could be estimated without trapping outside the study area. In contrast, immigration can be directly estimated using, for example, two different approaches (see above). One obvious way of estimating emigration rates has been to assume the equality of immigration and emigration rates (Waser *et al.*, 1994; see also Alberts and Altmann, 1995). In the context of metapopulation dynamics, the approach of Waser *et al.* (1994) seems to be of limited interest, since it is assuming that all populations are at demographic equilibrium. This assumption is often not validated using demographic data (survival and reproductive rates), and the accuracy of this approach is therefore difficult to assess. In practice, to estimate emigration rate, it is necessary to monitor several populations connected by movements of individuals. While this approach will provide us only with estimation of rates of exchanges (which result from emigration from a given population *and* mortality during the dispersal movement), it is a necessary first step in the direction of direct estimation of emigration rate.

Existing models for estimating exchange rates based on capture–recapture or sight–resighting data are special cases of multiple strata models (sensu Brownie *et al.*, 1992; Nichols *et al.*, 1992). In these models the strata correspond to different populations included in the study. These models provide an extension of usual capture–recapture models, where there is only one survival rate: the survival rate within the focal population. This approach has been successfully used for estimating exchange rates in geese (Hestbeck *et al.*, 1991), but as these authors recognize the method has limitations which can probably be overcome through appropriate study design.

The first limitation is due to the large number of parameters in the model, since we need to incorporate all pairwise exchange rates between populations. The parameters may also vary with time. For example, with five populations studied during 3 years, we have five local survival rates $+ 2 \times 10$ exchange rates per yearly interval, that is 50 survival parameters. If we add to this the recapture parameters, we end up with 60 parameters in the full model, which will obviously require a large data set to draw any reliable inferences. Moreover, it is assumed that the probability of survival for a given individual (within or in another population) was independent of its previous history (in particular whether it has previously moved or not). Obviously, if transfer between populations is a unique event in the individual lifetime, this assumption will not be satisfied. Brownie *et al.* (1992) have analyzed models where this assumption was relaxed (by including additional parameters), but as they pointed out, it resulted in models with a very large number of parameters, probably too large for most available data sets.

Our opinion is that this approach is necessary to get reliable estimates of exchange rates. However, it will probably be possible to implement it mainly for either experimental model systems (for which the number of parameters can be reduced by design: assuming equal exchange rates between identical patches) or species for which it is possible to obtain large samples (Hestbeck *et al.*, 1991).

V. EXPERIMENTAL APPROACHES TO STUDYING TRANSFER RATES IN METAPOPULATIONS

Experimental manipulations of populations have been used to identify some of the causal mechanisms affecting each of the components of transfer between populations, emigration, migration, and colonization. In the following, we will review the experimental approaches that have been used to explore these three components.

A. Emigration

Apart from a few studies in which emigration has been studied by direct observation of marked individuals leaving habitat patches (e.g., Lawrence, 1987b; Odendaal *et al.*, 1989), many experimental studies have relied upon certain devices to trap individuals attempting to leave the experimental populations. The use of fences (enclosures) has been extensive in vertebrate studies (Gill, 1978a; Johnson and Gaines, 1985, 1987; Danielson and Gaines, 1987; Bondrup-Nielsen, 1992), but also in a few studies on invertebrates (Hertzberg, 1996). Individuals attempting to leave the population are forced to move along fences and ultimately enter a fence trap (Johnson and Gaines, 1985, 1987; Bondrup-Nielsen, 1992; Aars *et al.*, 1995). Emigration rates are then measured by the proportion of individuals caught in the fence traps. However, with this method it may be problematic to distinguish between emigration and short-distance exploratory movements outside the habitat patch. This problem can be somewhat eased by increasing the spatial scale of the experiment, for instance by allowing for a quite large distance from the patch border to the fence (e.g., Hertzberg, 1996) or by using semipermeable border zones, for instance water (Ims, 1989).

Experimental studies on emigration have examined the effect of factors such as population density (e.g., Kareiva, 1985; Lawrence, 1987b), patch size (Kareiva, 1985; Turchin, 1986), patch shape (Harper *et al.*, 1993), and edge characteristics (Back, 1988b) on emigration rate. Not surprisingly, the experiments have shown that different species often respond differently to such factors depending on social structure (Wiens *et al.*, 1993) and perception of the "hardness" of patch edges (Stamps *et al.*, 1987; Ims, 1995).

B. Migration

The most common experimental approach to migration/dispersal studies is to release individuals on areas assumed to be matrix habitat for the species. One problem with this approach is that migration may be imposed on individuals not motivated to migrate. The migration stages may be easy to distinguish in some organisms (e.g., seed in plants or certain life stage in insects; Johnson, 1969), but there may be much individual variation in the migration tendency in other organisms (Swingland and Greenwood, 1983; Chepko-Sade and Halpin, 1987).

Combining studies on emigration (e.g., using fences) to define individuals prone to emigrate with relevant migration experiments might be a way of getting around the problem of studying migration on an inadequate sample of animals. The most common factor affecting migration in the release studies is the physical structure of the matrix habitat. For instance, Crist et al. (1992) manipulated the structure of the vegetation in their studies of movement patterns of three *Eloides* beetle species.

The possible role of linear habitats as dispersal corridors in highly fragmented landscapes (see Simberloff et al., 1992; Hobbs, 1992) has been addressed by a few experimental studies on small rodents (LaPolla and Barrett, 1993; Andreassen et al., 1996a,b). For example, Andreassen et al. (1996a) manipulated the width of dispersal corridors and found that root voles *Microtus oeconomus* showed highest transfer rates between patches connected with corridors of intermediate width due to both high emigration rates and straight line movement in the corridor. There is a great need for experimental studies on the effect of various corridor designs for the benefit of more effective conservation measures (Mann and Plummer, 1993). Inglis and Underwood (1992) give useful guidelines about the design of experiments addressing transfer rates in patch systems connected with corridors.

C. Colonization/Immigration

Without additional information it remains often an open question why a habitat patch which is seemingly suitable for a species is nonetheless empty. There are two possibilities. First, the patch may be suitable but is empty because the local population has gone extinct or because it is beyond the dispersal range from any extant population. Although observational studies may in some cases provide useful information (e.g., long-term studies may reveal colonization events; Valone and Brown, 1995), and incidence functions may indicate the migration ranges (Peltonen and Hanski, 1991), experimental studies are often required to determine under which conditions colonization/immigration onto a given patch is possible. Harrison (1989) used an experimental approach in which individually marked checkerspot butterflies were released in the matrix habitat at increasing distance from an empty habitat patch in order to estimate colonization probabilities as a function of the distance from the release point. Colonization rates may also be estimated by introducing artificial patches (proved to be suitable) within migration range from known local populations (Schoener, 1974; Whitlock, 1992b).

Second, the patch is not in fact suitable for colonization in its present state for the species. The most common approach to determining whether empty patches are suitable or not is translocation/transplantation experiment (e.g., Harrison, 1989; Massot et al., 1994). More information about which factors determine patch suitability for colonization have been obtained by experiments varying patch characteristics such as patch/island size (Schoener and Spiller, 1995), propagule size (Crowell, 1973; Ebenhard, 1987), the presence of predators (Schoener and Spiller, 1995), and conspecific and heterospecific residents in the patch (Danielson and Gaines, 1987; Tilman, 1993; Valone and Brown, 1995). All these factors have been shown to have effects on immigration rate and colonization success, but the strength, and sometimes even the sign, of the effect may vary between species. For example, the presence of conspecifics may decrease (intraspecific competition; Danielson and Gaines, 1987) or increase the success of immigrants (conspecific attraction; Stamps, 1991, or Allee-effect) and the same applies to the presence of heterospecifics (negative effects due to interspecific competion, Valone and Brown, 1995; Tilman, 1993; versus interspecific facilitation or nursing effect, Wood and del Moral, 1987). The outcome of transplantation experiments may depend on the time scale. It seems important to extend the experiment beyond the settlement phase as certain patch-specific factors may take a long time to exert their effects (e.g., Schoener and Spiller, 1995). Such experiments should pay more attention to characteristics of the translocated individuals as this may have profound influence on the success of the propagules (Bright and Morris, 1994).

VI. CONCLUSION

In this chapter we have focused on emigration, migration, and immigration/colonization as separate processes, which need to be approached empirically by different study designs. At the same time, the three transfer stages are clearly linked both methodologically and biologically. For example, emigration, dispersal, and immigration/colonization must be translated into transfer rates to be used in predictive models at the metapopulation level. Likewise, the three transfer processes are also linked to patch-specific demographic parameters (Stacey and Taper, this volume). For example, emigration rate may be a function of population density and growth rate (Stenseth, 1983), and immigration and colonization success may depend on local densities and number of colonists (Back, 1988a,b; Lawrence, 1987a; Augspurger and Kitajima, 1992; Stenseth and Lidicker, 1992a). Methods used for estimating population growth rates or density dependence will often be the same as the ones for estimating exchange rates (in particular capture–recapture methods). The choice of the most efficient study design should be guided by the critical parameters as well as how complementary information about these parameters may be obtained. We believe that no single approach will suffice if we are to advance our understanding of the transfer processes. A combination of individual-based approaches to study dispersal patterns, patch-specific demographic studies to estimate exchange rates between populations, as well as specific experiments for exploring the causal links between environmental factors and the focal processes will provide a robust framework for empirical transfer rate studies.

We have advocated the importance of patch-specific parameter estimates in the context of proper hypothesis testing, study design, and statistical modeling.

Global approaches to obtaining average values over the population of patches may or may not suffice to parameterize specific models. In any case, on their own, they are not likely to provide new insight into the role of transfer in metapopulations. Patch-specific estimates are needed to evaluate the assumptions underlying the global approaches to parameter estimation. For example, are unoccupied patches at all suitable? Are occupied patches sources or sinks in a deterministic sense? If so, what are the actual values of the rate of increase/decrease in the absence of emigration and immigration? Furthermore which factors determine patch-specific growth rates? The last question will require parameters that are related to the demographic processes (mortality and reproduction) rather than simple descriptions of the habitat characteristics and patch-specific densities. For example, both empirical and theoretical studies have shown that density as such may not be a good measure of habitat quality in a demographic sense (van Horne, 1983, Pulliam, 1988; Danielson, 1992; Kellner *et al.,* 1993).

It may be argued that we are asking for more than can be achieved in most field studied and that the "details" we ask for may be redundant and complicate rather than improve the matters. Estimating transfer rates will remain a crucial, but difficult task in metapopulation studies, because of the large sample size required. It is most probably through a combination of direct approaches in model systems and global approaches in natural metapopulations that further progress will be made.

ACKNOWLEDGMENTS

We thank K. Hertzberg, D. Hjermann, S. Mesnager, and N. Rioux for comments on the manuscript and Ilkka Hanski for his extensive editorial help. RAI was supported by UiO Support Programme/ELF Petroleum Norge AS and NGY by the "Programme Environnement du CNRS—Méthodes, Modèles, Théories."

12
Migration within Metapopulations
The Impact upon Local Population Dynamics

Peter B. Stacey   Veronica A. Johnson

Mark L. Taper

I. INTRODUCTION

The recent interest in metapopulation systems (e.g., Gilpin and Hanski, 1991, and references therein, Hanski and Simberloff, this volume) stems from the fact that members of a species usually are not distributed continuously in space, but are often clumped together as a result of variation in the geophysical and ecological characteristics of the landscape. The concentration of individuals within a particular area constitutes the local population, and it is at this level of organization that most behavioral, genetic and ecological interactions occur. In turn, local populations are generally separated from one another by more or less unsuitable habitats, where densities of the species are low or zero. Traditionally, most population and community level studies have focused on the analysis of individuals living within a specific area, with the assumption that events within the local population were both representative and generally sufficient to understand most important phenomena. However, almost all species have evolved mechanisms that enable individuals to cross unsuitable habitats at some stage of their life cycle, and thus most local populations of a species are potentially connected to other populations through dispersal and migration. Many studies have shown that even a very limited amount of migration can have a profound effect upon the recipient population. For example, on a genetic level, one or two migrants per generation will cause two otherwise isolated populations to behave as if mating between them is panmictic (Wright, 1951, 1978). On a demographic level, recent simulation models of multiple populations have shown that a surprisingly small number of immigrants per year (often between three and five adults: e.g., Stacey and Taper, 1992) will allow individual populations to persist in stochastic environments where they would otherwise quickly go extinct (e.g., Fahrig and Merriam, 1985; Gilpin, 1987; Hanski, 1991; Beier, 1993). These results indicate that a collection of populations that is connected together through migration can function at a higher level of organization, the metapopulation, and that these systems can exhibit important emergent phenomena that can be understood only by considering the entire system of populations as a whole.

A central issue in the analysis of metapopulations is the frequency of migration, or demographic connectivity, among component populations. For most analyses, the actual number of migrants that successfully move between two populations per breeding season or generation is the most important measure of the level of connectivity between them. Migration frequency will be a continuous variable and may range from zero, where populations are completely isolated from one another, to a value that may be nearly equal to the number of individuals in each unit, in which case the two units function as a single population. This emphasis on connectivity allows us to differentiate a metapopulation from other population complexes. Specifically, we consider a metapopulation to be a system of geographically or ecologically isolated populations *within which there is sufficient migration among populations to have a significant impact on either the demography or genetic structure of each component population.* As a result, in a metapopulation, the dynamics of each component population cannot be fully understood without reference to other populations and to the system as a whole. Thus, not all population complexes are metapopulations. Two populations may exist close together in space, but unless there is regular movement of individuals between them, the events in one population will not affect the other population, and they will not constitute a metapopulation. Similarly, if migration is so frequent that matings among members of the two "populations" are essentially random (i.e., the probability of an individual mating with any other individual in a sexually reproducing species is nearly $1/2(N_1 + N_2)$), then the system actually functions both genetically and demographically as a single population that occupies different habitat patches.

In addition to connectivity, the specific direction and timing of migration among component populations are important, since it can occur either before or after local population extinctions, and it may be either unidirectional or omnidirectional. For example, in many early models of metapopulations (e.g., Levins, 1970; see review by Hanski, 1991) migration from a source population to a target population occurs only after the target population goes extinct. In this situation, the metapopulation consists of units where each has a finite probability of extinction, $p(e)$, and (re)colonization, $p(c)$. Currently extant populations serve as sources of migrants that can, with a certain probability, reestablish extinct populations, and thus the overall system can persist longer than the same number of populations that are isolated from each other. The nontrivial equilibrium solution for the fraction of extant populations in this model is $P^* = 1 - e/c$. In this conceptualization, migration is important only as it impacts the frequency of recolonization, and the primary dynamic of the metapopulation system is the extinction/recolonization event. The identities, locations, and temporal histories of individual populations are not relevant, and the models are therefore "unstructured" in both time or space. In the field, the existence of periodic population extinction and recolonization events is the primary indicator that this type of metapopulation system is present (e.g., Smith, 1980; Menges, 1990).

A second group of early analyses examined individuals that occurred in patchy habitats (e.g., Roff, 1974; Chesson, 1981; Crowley, 1981). Most of these studies involve situations where migration rates (connectivity) among patches are high enough that the probability of individuals mating among patches is not very different from that of mating within patches. In such situations, the system is best considered as a single population, with spatial variation in the distribution of individuals.

More recently, attention has also been given to situations where connectivity is limited, but migration among populations occurs *prior to,* as well as after, an extinction event (e.g., Pulliam, 1988; Harrison, 1991; Hastings, 1991; Schoener, 1991; Stacey and Taper, 1992; Gyllenberg and Hanski, 1992; Hanski and Gyllenberg, 1993; see also Boorman and Levitt, 1973; Brown and Kodric-Brown, 1977). The source population can potentially affect the dynamics of the target population at any time and may provide sufficient new immigrants to "rescue" local populations before they go extinct. These types of metapopulations are more difficult to detect than those in which populations "wink off and on," because local extinctions may rarely occur under normal conditions. Individual elements in the system can in fact appear quite stable through time, and unless migration among the populations is disrupted or can be directly assessed (e.g., Stacey and Taper, 1992), the importance of metapopulation exchange may not be apparent. It is also more difficult to construct models of this type of metapopulation, because the system is structured in time and space and the dynamics of each population cannot be described in simple linear fashion using extinction and recolonization probabilities. Migration can occur at any time, and its frequency may depend upon the size and locations of both the source and the target populations. Because of the resulting complexity, these systems are best analyzed using either rather complex analytic models (e.g., Gyllenberg and Hanski, 1992; Hanski and Gyllenberg, 1993; Gyllenberg *et al.,* this volume) or simulation approaches (see below).

One special case in which migration can occur before extinction is the source–sink metapopulation, which exhibits similarities to the situations described by island biogeography (e.g., MacArthur and Wilson, 1967; Diamond and May, 1976). In these systems, one or more source populations, typically large in size or that occupy prime habitats, regularly produce an excess of individuals that disperse to smaller populations in less optimal habitat (sinks). The recipient target populations may have population growth rates that are consistently less than zero (or the net reproductive rate, R_o, is < 1); however, they are prevented from going extinct by the constant input of immigrants from more productive populations (where $R_o > 1$). This type of system can apply both to populations living on islands that are isolated from a mainland and surrounded by unsuitable water habitats (e.g., Pimm *et al.* 1988; Schoener, 1991) or to patches of varying size and quality in a fragmented landscape (e.g., Brown and Kodric-Brown, 1977; Pulliam, 1988; Hanski and Gyllenberg, 1993; Valone and Brown, 1995). The central characteristic of these systems is that the direction of migration remains consistent through time; some populations consistently produce an excess of individuals, while other populations are consistently rescued from extinction by immigration.

In this paper, we focus on the more general "rescue effect" metapopulation system, in which migration can prevent local population extinctions, but where there may be stochastic variation in both the rate and the direction of migration. Thus, the identities of the "source" and "sink" populations can change unpredictably through time (Harrison, 1991; Stacey and Taper, 1992). The shifting of sources and sinks is most likely to occur in situations in which local population growth rates are both highly variable and stochastic. In any particular year, reproduction and survivorship within one or more populations may be high, and thus those populations may produce a surplus of individuals and act as sources. In a following year the situation may reverse itself, and these populations may suffer declines to the point of accepting immigrants and becoming sinks. This system differs from source–sink models in that no one habitat occupied by the species is predictably better or more productive than any other occupied area.

Metapopulations (and structured metapopulation models) that involve stochastic variation in migration frequency and direction constitute the most inclusive case, and the extinction/recolonization and source/sink systems can be incorporated within them as special situations. Thus, within a particular complex of populations, the dynamics of very small and isolated units may be best understood using an extinction/recolonization approach, whereas larger populations that are closer together may interact through migration that changes in both direction and frequency through time (Harrison and Taylor, this volume; Thomas and Hanski, this volume). Which conceptualization is most useful will depend on the particular characteristics of the populations involved, and in many cases the most appropriate model may simply be a question of scale, depending upon both the dispersal abilities of the individual organisms and their tendency to move prior to reproduction (see below).

One of the important characteristics of all metapopulation systems is that the overall metapopulation can be much more stable than the component populations, because migration can buffer or "rescue" (Brown and Kodric-Brown, 1977) individual populations from negative stochastic and genetic events in their local environments (see also Leigh, 1981; Foley, 1994). This characteristic may be particularly relevant to conservation, since human activities often fragment once continuous habitats, creating artificial metapopulations, or disrupt the ability of individuals to disperse to different areas within natural metapopulations (e.g., Robinson *et al.,* 1995; Gibbs, 1993; McKelvey *et al.,* 1993; Ouborg, 1993). In fact, there is evidence (e.g., Harrison, 1991, and below) to suggest that metapopulations that include stochastic variation in migration and rescue effects may be common in many species that occupy fragmented environments. Unfortunately, because these systems can be difficult to recognize in nature because of the lack of regular extinction/recolonization events, many studies of metapopulations involving strong rescue effects have depended upon indirect methods and utilize either an analysis of genetic population structure to estimate previous levels of migration and gene flow (Olivieri *et al.,* 1990; Gilpin, 1991, pp. 17 – 38; Hastings and Harrison, 1994; Stacey and Johnson, 1996) or mathematical models to generate specific predictions that can be tested in the field. Below, we use a simulation model that includes stochastic variation in migration to illustrate how migration can affect both local population dynamics and the overall persistence of the metapopulation. We then examine some of the empirical evidence for metapopulation systems that may include strong rescue effects in several different taxonomic groups and offer suggestions for further research that would be useful in detecting and understanding these types of systems.

II. THE EFFECT OF STOCHASTIC VARIATION IN MIGRATION RATE ON METAPOPULATION DYNAMICS

Theoretically, there is considerable *a priori* reason to expect that many populations will be connected together by levels of migration that are sufficient to affect local population dynamics both prior to, as well as after, local extinction events. For example, as a result of conspecific attraction (e.g., Smith and Peacock, 1990; Ray *et al.,* 1991), dispersing individuals will often join an extant population when there is available breeding space, rather than wait to immigrate until the target population is extinct, as assumed under the most simple extinction/recolonization models. Metapopulation systems that involve significant migration before extinction (as opposed to only after extinction) can be difficult to detect in the field, simply because individual populations will appear to remain relatively stable through time. For example, Stacey and Taper (1992) examined the population dynamics of an isolated population of acorn woodpeckers *(Melanerpes formicivorus)* in Water Canyon, located in the mountains of central New Mexico, which had been extant in the area for at least 50 years. Detailed data on annual reproductive success and survival of juveniles and adults in the population had been collected from individually marked birds as part of a long-term study of cooperative breeding in this species (Stacey, 1979; Stacey and Ligon, 1987, 1991). Each demographic variable exhibited large annual variation, presumably as a result of stochastic events in the local environment, including rainfall and acorn production. Using data from the field study in a simple simulation model of single population growth through time, Stacey and Taper (1992) found that if this population was considered to be completely closed, without immigration, demographic stochasticity would lead to rapid extinction (usually < 20 years, depending on the particular assumptions about density dependence in growth rates in the model; see Fig. 1A). Extinction always occurred because, simply by chance, there would be a series of bad years for either reproduction or survival that would drive population numbers so low that they could not recover. However, allowing migration from other populations rapidly increased local persistence times, and even four immigrants per year (8% of population at carrying capacity) enabled the local population to continue for > 250 years (Fig. 1B). The number of immigrants required by the model to obtain long-term persistence was similar to the migration rate actually observed in the study population.

The study of Stacey and Taper (1992) suggests that the Water Canyon population of acorn woodpeckers is part of a larger metapopulation, whose existence and importance had not been suspected until there were sufficient data to conduct detailed analyses of local population dynamics and particularly the impact of stochastic variation in growth rates on persistence times. The local population by itself was not likely to continue for very long, but it was presumably rescued from extinction by regular migration from other populations. As a result, it appeared relatively stable through time even though its dynamics were affected in large measure by migration from other populations. Thus, while the single population model predicted relatively rapid extinction, placing it within a metapopulation complex with among-population migration predicted continued persistence.

This study illustrates how migration within a metapopulation can buffer individual populations from stochastic variation in growth rates that might otherwise lead to extinction. To explore this phenomenon further, we developed a more complex, spatially and temporally structured model that allows for migration in variable frequency and direction (depending on local growth rates at any particular point in time) and for immigration either before or after target populations go extinct. The details of this model will be discussed elsewhere (Taper and Stacey, 1997); here, we describe some of the results from the model to illustrate some of the important consequences of migration within metapopulations for individual population dynamics. Because of the potential complexities of individual population size trajectories within this type of stochastic system, we focus on the question of how long all populations within a metapopulation persist under different conditions and with different initial population sizes. The model is designed to apply primarily to vertebrate populations with distinct breeding periods; other simulation models that incorporate different assumptions are available (e.g., RAMAS; see Akçakaya and Ferson, 1992; Wu et al., 1993).

In this model, a predetermined number of local populations are established and allowed to grow or decline based upon an annual growth rate parameter (r) that has both a mean and a variance. Dispersal within the system is density dependent; populations that grow above a preset maximum carrying capacity (K)

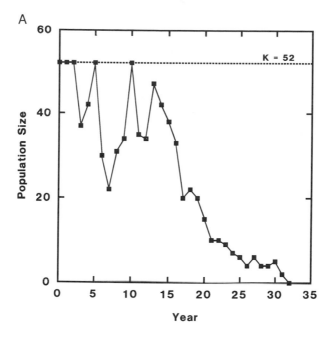

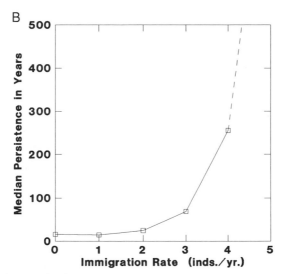

FIGURE 1   (A) An example of the annual changes in the size of a closed acorn woodpecker population from a typical computer simulation, using demographic parameters taken from data collected over a 10-year field study in Water Canyon, New Mexico. Simulations were started at the maximum population size ($K = 52$) and continued until the population declined to zero. Because the observed annual reproductive and survival rates were highly variable, most simulated populations went extinct in less than 20 years. (B) Increase in simulated persistence times of the same population when it is open and there is migration into the population with different annual frequencies (from Stacey and Taper, 1992).

produce emigrants that leave their home populations and encounter new populations with a transition probability t, where $t < 1$. If the new population is currently below its carrying capacity, it accepts an immigrant; if the population is at K, the immigrant moves again to another population. For each transition, the emigrant has a specified probability m of dying before it reaches a new population. The process is repeated until the individual becomes a member of a new population or dies. The size of each population is calculated after all births, deaths, and dispersal movements have occurred, and then the process repeated for another reproductive "year." Key parameters of the model that can be varied for each analysis include the number of component populations in the system, their initial sizes and carrying capacities, the mean and variance in growth rates of each population, and the correlation among the growth rates of all populations in the system. Metapopulation connectivity is varied by specifying the likelihood that a dispersing individual will encounter another population before it dies. If connectivity equals zero, all populations are isolated from each other and the system does not function as a metapopulation. Temporal structure is included by independently tracking the size of each population through each simulation "year." Spatial structure is incorporated into the model by specifying the connectivity of each pair of populations independently in a matrix such that pairs that are close together have high connectivity values, while those that are far apart have low values. However, in the results described here we considered the most simple case where individuals have equal probabilities of encountering any other population within the system after migration. Finally, the type of density dependence in migration frequency modeled here may be most typical of vertebrate populations, whereas many insects and other invertebrates may exhibit more density independent reproduction and, therefore, migration. As a result, this model may be less applicable to those taxa.

We are able to compare the dynamics of metapopulations where immigration occurs only after extinction with continuous migration by changing the conditions under which immigrants enter new populations. When modeling generalized rescue effect dynamics, immigrants enter a target population if the population size (n) is ≥ 0 and $< K$. When modeling strict extinction–recolonization dynamics, immigration to a population occurs only if $n = 0$ for that population. Local populations are declared extinct, at least temporarily, whenever population size declines below two (for sexually reproducing organisms); extinction of the entire system occurs when all local populations are extinct simultaneously. Simulations were repeated 200 times for each set of parameter values, and either the median or the geometric mean time to extinction for the entire metapopulation was determined.

Most metapopulation models have focused on determining the buffering effect that migration can have in either preventing extinction and/or allowing for reestablishment of a population after it has gone extinct. The threat of extinction is likely to be higher in small, fragmented populations (e.g., Gilpin, 1987; Goodman, 1987a, Pimm *et al.*, 1993) and where growth rates vary stochastically (Foley, 1994, this volume). Figure 2 illustrates the effects of environmental and demographic variation in our model, both with and without the buffering effect of migration. These graphs illustrate several points. First, even when the growth rate is slightly positive ($r = 0.001$), all populations eventually go extinct as a result of random fluctuations (if there were no variance in growth rates, all populations would continue to grow indefinitely, albeit slowly). Second, when there is no migration in the system and all populations are isolated from each other (i.e., there is no metapopulation structure), individuals concentrated in one or two large populations will persist longer than the same number of individuals divided up into more, but smaller, populations. However, when migration is allowed, an intermediate number of populations, each containing an intermediate number of individuals, last longer than either a few large populations or many small ones. This is an emergent property of the metapopulation structure that results from the buffering effects of among-population migration. Because of the stochastic variation in growth rates, when some populations are declining, others may be increasing at a rate sufficient to produce immigrants that can supplement growth in the declining populations. Thus, a connected group of populations can actually have a greater probability of persistence than a few large, but isolated, populations. Most important, these results illustrate how the dynamics of local populations can be strongly affected by metapopulation exchange and that the continued stability of each population cannot be understood without reference to the system as a whole.

Figure 3 illustrates the effect of the timing of migration and compares the persistence of metapopulations in which migration occurs only after extinction ($n = 0$), with the situation where immigration can occur at any time as long as $n < K$. All other parameters in the model are held constant. In both situations, an intermediate number of populations of intermediate size persist longer than either a few large populations or many small ones. However, both individual populations and the metapopulation system as a whole always last longer if migration can occur before extinction, because populations will be rescued before, as well as after, extinction events. The pronounced "advantage" of the strong rescue effects metapopulation system over the extinction–recolonization model suggests that many species in fragmented environments may have evolved highly efficient mechanisms for migration across unsuitable habitats. Without such mechanisms, many species may not be able to persist in such landscapes. For example, one species of particular interest and political importance in the United States is the spotted owl *(Strix occidentalis)*. Two of the subspecies, the Mexican *(S. o. lucida)* and the California *(S. o. occidentalis)* spotted owls, occur in naturally fragmented habitats in the mountain ranges in the American southwest and southern California, and a functional metapopulation structure with frequent migration among the populations may be necessary for the continued persistence of the species in this region (Lahaye *et al.,* 1994; Stacey, 1994). In contrast, the Northern subspecies *(S. o. caurina;* Lande, 1988a; McKelvey *et al.,* 1993), whose habitats were once primarily contiguous forests in the Pacific northwest, may not

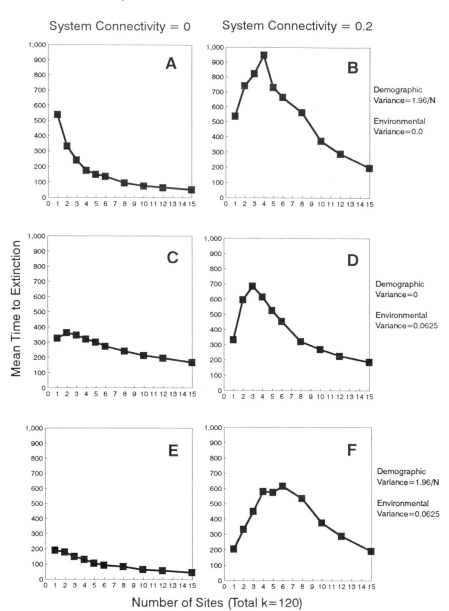

FIGURE 2   Results from a computer simulation model that illustrate how migration among populations in a metapopulation can "buffer" those populations against the effects of stochastic variation in annual growth rates and decrease the probability of extinction. Each graph gives the mean time to extinction in 200 computer simulation runs for systems containing the same total number of individuals divided into different numbers of patch sites or populations, each with a size equal to 120 divided by the number of patches (e.g., one population of 120 individuals, two populations of 60 individuals each, three populations of 40 individuals each, etc.). Shown here is the effect of different combinations of demographic and environmental variation on metapopulation persistence times (A, C, and E).

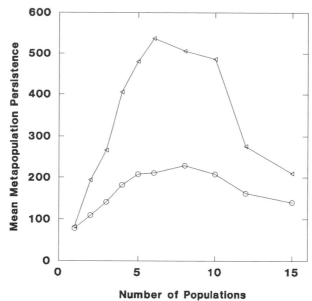

Number of Populations

FIGURE 3   Comparison of persistence times for two different types of metapopulation systems. A rescue effect model where migration can occur both before and after a population might go extinct (triangles). The more restrictive condition of the extinction/recolonization model where migration (or recolonization) occurs only after population extinction (circles). Results of simulations are given as in Fig. 2; all parameters except the timing of migration events were the same for each comparison. In each case, metapopulations that include migration before extinction persist longer than those systems that do not include the rescue effect.

have the ability to shift into a metapopulation structure with efficient migration once their habitat has been artificially fragmented by human activities, and the survival of this subspecies may be much more problematic.

   The buffering effect of the metapopulation depends on the ability of some populations to act as sources of migrants at the same time that other populations are declining and ready to accept those migrants. Because of this, migration will have little effect on individual persistence times if all populations are increasing or declining together (see also Harrison and Quinn, 1989; Gilpin, 1990; Burgman

FIGURE 2   (Continued)

Systems in which there is no migration among populations (between-population connectivity, measured as transition survival probability = 0), and therefore they do not function as metapopulations (B, D, and F). All of the same parameter values except that migration among populations is possible, and the connectivity or transition survival probability is 0.2 (20% of all dispersing individuals survive to immigrate into another population). Systems that include migration and function as metapopulations persist longer in all cases than do isolated populations as a result of the rescue effect (see text for details).

et al., 1993). Thus, the rescue effect is likely to be most important when growth rates of local populations in the system are either uncorrelated or negatively correlated. The impact of among-population correlations in growth rates on the persistence of our model metapopulation is shown in Fig. 4. These results suggest that metapopulations with strong rescue effects may be most common in situations where there is little correlation in the environments of different habitat fragments and therefore in the annual growth rates of the populations that occupy each of those fragments. Unfortunately, very few empirical studies have examined the degree of among population correlations in growth rates because it requires detailed and long-term demographic data gathered over an area large enough to include a number of separate populations. Studies of this nature are just beginning (Lahaye *et al.*, 1994; Dennis *et al.*, 1996; Martin *et al.*, in press; Stacey and Martin, 1997).

The results of both the simulation model presented here and earlier models indicate that migration within a metapopulation can have a major impact upon local population dynamics and increase the persistence times of both individual populations and the entire metapopulation system under conditions that may be relatively common in certain taxa and types of environments. Directly testing the predictions of these models, however, will be difficult in many cases because observing migration directly in the field requires identifying and following a large number of individuals whose migration paths may be unpredictable and for whom

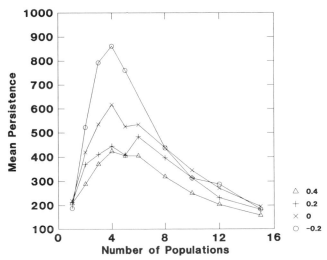

FIGURE 4   Effect of different degrees of among-population correlations (*rho*) in population growth rates on mean persistence times of the metapopulation. Results of simulations are given as in Fig. 2; all other parameters in the model were held constant for each simulation. Persistence times increase as the correlation in growth rates becomes more negative. Functionally, this represents a situation where some populations in the system are growing large enough to produce emigrants at the same time that other populations are declining to the point where they need immigrants to avoid extinction.

mortality rates may be high. As a result, much of the current evidence for the importance of migration in local population dynamics, and for the existence of metapopulations with strong rescue effects, is necessarily indirect. In the following section, we review selected studies to illustrate both where these systems are likely to occur and the evidence that is needed to detect them.

III. EMPIRICAL EVIDENCE

The importance of migration in population dynamics is indicated not only by theoretical studies; there is growing empirical evidence for the role of rescue effects in natural metapopulation systems. We have not attempted an exhaustive review of all possible natural metapopulations; rather, as examples, we focus on several taxonomic groups in which metapopulations are likely to occur and where a sufficient number of studies have been published to explore the nature of migration in these systems. These groups include butterflies, small mammals, and amphibians. Each group illustrates somewhat different patterns and suggests productive directions for future research.

First, it is important to establish the type of evidence needed to demonstrate the existence of a metapopulation with strong rescue effects mediated by migration. (1) As in all metapopulation systems, the species must inhabit distinct habitat patches surrounded by areas of unsuitable habitat. (2) There must be consistent movement between already occupied habitat patches such that the migration rates are relatively high compared to predictions from the simpler extinction/recolonization systems where migration occurs only after extinction. (3) Specific demographic and genetic effects of migration should be distinguishable at the local population level. For example, observed demographic parameters such as population size and density, age structure, sex ratio, and population growth rate should be different from those predicted for an isolated population. Similarly, genetic measures such as heterozygosity, inbreeding levels, and genetic differentiation should be different from those expected in an isolated population (e.g., Hastings and Harrison, 1994). Genetic parameters also may be used to distinguish among the different types of metapopulation systems. For example, Gilpin (1991, pp. 17–38) noted that low heterozygosity should be a feature of the extinction/recolonization metapopulations as a result of the repeated founder effects that occur with the recolonization of areas by small numbers of migrants. In contrast, heterozygosity should be higher in metapopulations with frequent migration as a result of high gene flow among component populations, and the system should act genetically more like a single large unit (Stacey and Johnson, 1996). (4) Finally, if metapopulation exchange is important, there should be greater persistence or stability of the local population as well as of the overall metapopulation, reflected in terms of decreased probability of extinction and dampened population size fluctuations relative to an isolated population in the same environment.

A. Butterflies

In one of the first detailed studies of the population dynamics of butterflies, Baker (1969) analyzed movements of adults out of natal patches and concluded that most species should exhibit low migration rates. Since then, many researchers have assumed that butterflies have extremely limited migration abilities and tend to exist in relatively isolated populations (Warren, 1987a,b; Baguette and Nève, 1994). Thus, when metapopulation theory emerged, butterflies seemed to be prime examples of extinction–recolonization metapopulations in which small, isolated populations could decline to extinction, and habitat patches could then be recolonized by rare long-distance migration events. Yet several lines of evidence suggest that migration among populations may be much more common in some taxa of butterflies than was originally thought (see also Thomas and Hanski, this volume), and long-distance migration may increase the persistence of local populations through both demographic and genetic effects.

1. Checkerspot and Fritillary Butterflies

Most checkerspot butterflies in the genus *Euphydryas* in North America appear to specialize on one or a few species of plants which occur in patchy distributions. For example, *Euphydryas gillettii*, a northern Rocky Mountain endemic, lays its eggs only on the black twinberry *(Lonicera involucrata)* which is restricted to wet, sunny, early successional patches (Williams, 1988; Debinski, 1994). A northeastern species, *E. phaeton*, is confined to wet, low areas that contain its primary food plant, turtlehead (*Chelone glabra;* Brussard and Vawter, 1975). As a result, many *Euphydryas* butterflies tend to occupy distinct habitat patches separated by areas of unsuitable habitat, predisposing them toward a metapopulation structure.

Unlike traditional extinction–recolonization metapopulations, relatively high rates of migration have been detected among existing populations, resulting in clear demographic and genetic effects. In a mark–recapture study of *E. anicia*, a species restricted to high peaks in the Rocky Mountains, White (1980) directly observed migration between two populations. In a single season, approximately 3% of the individuals in the source population successfully migrated to a new population. White remarked that such a rate should be high enough to eliminate genetic drift, minimizing genetic differentiation between populations. Indeed, at the time of publication, no significant genetic differences had been detected (White, 1980). Other studies have inferred relatively high migration rates from genetic data. In an early study of *E. phaeton*, Brussard and Vawter (1975) examined protein differentiation at seven allozyme loci in three local populations. They expected to find distinct genetic differentiation because each population had a small effective population size ($N_e \approx 20\text{--}200$) and no migration had been detected through a preliminary mark–recapture study. Yet only one of the seven protein loci showed any statistical difference between populations, and all populations exhibited high heterozygosity and intrapopulation genetic diversity.

Brussard and Vawter (1975) concluded that gene flow, and thus migration, must be occurring frequently enough to cause these genetic effects. Ehrlich (1983) reviewed data from various *Euphydryas* species and calculated that individual populations with effective population sizes (N_e) $\approx$ 30–50 that do not immediately rebound to higher levels will be unable to maintain themselves in isolation and inevitably decline to extinction. Thus, the apparent persistence of the small populations studied by Brussard and Vawter suggest that migration in *E. phaeton* prevents loss of heterozygosity and genetic differentiation, and can apparently rescue populations from extinction.

Similarly, in a more recent study of genetic differentiation in *E. gillettii,* Debinski (1994) expected to find high levels of differentiation, inbreeding, and homozygosity in local populations of Glacier National Park, because little migration had been directly detected in transplant experiments (Holdren and Ehrlich, 1981). She analyzed isozyme electrophoretic differences at 19 loci in two local populations and calculated unexpectedly low values of inbreeding and high values of heterozygosity ($F_{ST} = 0.041$; $H = 0.079, 0.084$). She calculated that Nm, the number of migrants between these populations per generation, must be as high as 5.80 in order to cause the observed genetic effects (Debinski, 1994). Although increased population stability or persistence was not directly inferred in this study, Ehrlich (1983) observed that genetic variability and polymorphism tend to be maintained in *Euphydryas* species despite large population size fluctuations. One possible explanation suggested by Debinski s (1994) results is that sufficient migration exists to preserve genetic variability and rescue small populations (Ehrlich, 1983), increasing the persistence of those populations as well as of the overall metapopulation.

Several studies of European butterfly species also provide evidence for metapopulations with strong rescue effects. The bog fritillary *(Proclossiana eunomia)* is found in natural wet meadows in Belgium, where its only larval food plant, *Polygonum bistorta* grows. Thus, this butterfly species also occupies distinct habitat patches surrounded by areas of unsuitable habitat. Using mark–recapture techniques, Baguette and Nève (1994) demonstrated that high levels of adult migration occur between these patches. At least 4.6% of all males and 10.5% of all females moved between populations. These researchers suggested that high levels of movement may lead to high levels of genetic variability, explaining why populations of the bog fritillary have persisted despite habitat loss (Baguette and Nève, 1994).

In a similar study in Britain, Warren (1987a) used mark–recapture experiments to examine mobility of the heath fritillary *(Mellicta athalia),* an endangered species that exists in small, discrete populations or colonies (Warren *et al., 1984).* He found that at least 1.7% of all males and 1.3% of all females moved between populations nearly 1 km distant from each other. No specific demographic or genetic effects were attributed to this substantial migration rate, but Warren (1987a,b) noted that these populations were unlikely to be genetically isolated from each other and that local population growth rates appeared to be uncorrelated. This suggests that the substantial migration observed may help dampen overall population size fluctuations through a rescue effect.

Thomas et al., (1992) performed an entirely different kind of study of four rare British butterfly species: *Plebejus argus, Hesperia comma, Thymelicus acteon,* and *M. athalia.* All of these skippers and fritillaries are restricted to distinct habitat patches due to very specific vegetation requirements for feeding and laying eggs. Rather than directly assessing migration, these researchers examined patterns of patch occupancy in the context of island biogeography theory. Using logistic regression, they analyzed presence and absence of each species in relation to patch size and isolation distance. They found that the presence of a local population was negatively correlated with patch isolation distance. Because isolation distance in turn should be negatively correlated with migration rate, populations within the local area presumably persisted because they were close enough to be in a metapopulation structure linked by migration (Thomas et al., 1992; Thomas and Hanski, this volume).

Hanski and his co-workers have conducted what is perhaps the most detailed analysis of metapopulation structure for any butterfly species to date with the Glanville fritillary *Melitaea cinxia* on a series of islands in southwest Finland (e.g., Hanski et al., 1994, 1995a,b). This and similar studies are summarized in Thomas and Hanski (this volume). Their study provides convincing evidence not only that a metapopulation structure exists in *M. cinxia,* but also that there is sufficient migration within the system to affect the dynamics of local populations, including their probability of extinction and subsequent recolonization.

All of these studies suggest a pattern of evidence for important rescue effects in butterfly metapopulations. Many species are clearly predisposed toward a metapopulation structure because they occupy distinct habitat patches as a result of specialized vegetation requirements during some stage of their life cycle. Though traditionally viewed primarily as extinction–recolonization metapopulations, many of these species show relatively high rates of migration detected directly through mark–recapture and suggested indirectly by genetic studies. Migration in butterflies has been connected to high genetic variability (suggesting a rescue effect rather than extinction/recolonization system), low genetic differentiation, and greater persistence of local populations. However, further research is needed to directly quantify the demographic effects of migration and demonstrate that high migration rates in various species are actually rescuing local populations from extinction.

B. Small Mammals

Considerable research has been directed toward studying migration in mammals (see Chepko-Sade and Halpin, 1987). These studies have rarely linked migration patterns to population dynamics and metapopulation systems, perhaps because the spatial structure of different local populations has not been studied in detail. Particular migration patterns have been connected to observed *genetic* population structure in small mammals, so many populations may exhibit metapopulation dynamics if it is assumed that genetic structure must be the result of spatial population structure (Lidicker and Patton, 1987). Yet Amarasekare (1994) notes that small mammals such as the banner-tailed kangaroo rat *(Dipodomys spectabilis)* can exhibit some spatial structure without occupying the discrete habitat patches required for metapopulation structure. Therefore, we consider only examples of possible metapopulations with strong rescue effects that satisfy the criterion of discrete populations, either as a result of natural habitat patchiness, social structure, or where the species habitat has been artificially fragmented into distinct patches by human activities.

1. Pikas—Natural Habitat Patchiness

A. Smith and his colleagues' research on pikas *(Ochotona princeps;* Smith, 1980, and this volume) provides a direct verification of the rescue effect in a natural metapopulation. Pika populations inhabit talus slopes surrounded by unsuitable sagebrush desert habitat. Smith tested the Brown and Kodric-Brown (1977) original prediction that immigration rate should increase with decreasing isolation distance, depressing rates of local population extinction. He censused pika populations in 1972 and again in 1977 to examine population persistence. Though extinction and recolonization were common on the most isolated habitat patches, migration was negatively correlated with isolation distance, apparently resulting in reduced extinction rates of the less isolated populations. In addition, some patches were too small to support breeding populations, yet they persisted presumably because migration rates were high enough to create a demographic rescue effect (Smith, 1980). In a detailed study of juvenile migration among the same populations, Peacock (1995) found that most young pikas tended to disperse to neighboring talus patches. However, multilocus DNA fingerprint data demonstrated that within-population band-sharing scores did not differ from what would be expected if individuals were assorting randomly among all patches in each generation. Migration among habitat fragments appeared to be important in maintaining genetic variation within the metapopulation and preventing the loss of heterozygosity in component populations. The mean heterozygosity in the fragmented system ($H = 0.736$) was essentially the same as that found in a continuous "mainland" population nearby ($H = 0.709$; Peacock, 1995). Together, these studies provide some of the best evidence available that apparent migration within mammalian metapopulations can maintain genetic diversity and rescue local populations from extinction.

2. Voles and Prairie Dogs—Social Systems That Create Spatial Structure

Stoddart (1970) reported on early research of water voles *(Arvicola terrestris)* in Scotland. These microtine rodents aggregate into spatially distinct populations along riverbanks, not because the suitable habitat is organized into discrete patches, but rather because each population represents a discrete social group and few individuals are ever found outside a social unit. The entire system exhibits migration patterns similar to those of rescue effect metapopulations based on distinct habitat patches. Stoddart conducted mark–recapture experiments and found that most movements were confined within a particular social group, but several female voles dispersed long distances and joined new social groups. Though further research is necessary to determine whether or not such migration could rescue populations from extinction, this study provides early evidence that social structure can cause spatial population structure and migration patterns consistent with theoretical rescue effect metapopulation structure.

Studies of the black-tailed prairie dog *(Cynomys ludovicianus)* provide more detailed and convincing evidence that metapopulations with strong rescue effects may be common among social mammals. Black-tailed prairie dog populations or colonies consist of several family groups called coteries. Populations remain spatially distinct from each other because of this rigid social structure (King, 1955; Smith, 1958). In addition, social groups actively alter the vegetation around colonies by clipping tall plant species to encourage the growth of favored food plants (Bonham and Lerwick, 1976), in effect creating distinct habitat patches across a relatively homogenous natural landscape. Garrett and Franklin (1988) used radio telemetry to assess migration of prairie dogs between these discrete populations. They detected consistent migration between colonies by male yearlings and a few adult females and verified that several individuals began breeding in new colonies after migration. Though the direct effects of migration were not monitored, Garrett and Franklin stated that their research supports the hypothesis that migration functions to minimize inbreeding. The regular influx of male immigrants should reduce inbreeding within local populations and elevate levels of heterozygosity, potentially rescuing populations from inbreeding depression and possible extinction (Hoogland, 1982). Indeed, Foltz and Hoogland (1983) examined four polymorphic loci and found a consistent excess of heterozygotes ($H = 0.068$) that they attributed in part to the high rate of male migration between colonies. This suggests the existence of a metapopulation structure and genetic rescue effects in black-tailed prairie dogs. However, Chesser (1983) found high levels of inbreeding ($F_{IS} = 0.3297$) and reduced heterozygosity in a separate study of seven polymorphic protein loci in black-tailed prairie dogs. Further research is clearly needed to determine accurately the patterns of migration and their effects in prairie dogs and other social mammals.

3. Mice, Chipmunks, Rats, and Squirrels—Rescue Effects in Fragmented Habitats

Forests that were once contiguous in the eastern United States and Canada are now highly fragmented as a result of agricultural and urban development. Farmlands often now contain distinct woodlots separated from each other by fields, buildings, and roads, thereby creating a mosaic of suitable habitat patches for many woodland species (MacArthur, 1972; Middleton and Merriam, 1981). White-footed mice *(Peromyscus leucopus)* have persisted in this network of woodlot patches and farm buildings connected by migration routes (Middleton and Merriam, 1981), creating a distinct spatial population structure. If each woodlot patch can be said to contain a distinct local population of white-footed mice, then both population persistence and migration between populations have been directly assessed for this species. Middleton and Merriam (1981) experimentally removed all white-footed mice from one farm woodlot to monitor immigration. They trapped a total of 54 individuals that immigrated into the woodlot and detected additional migration between three other woodlots and farm buildings included in the study. They also found that populations linked by these high levels of migration had higher growth rates than populations linked by lower levels of migration. Since winter populations can be as small as two individuals, rapid population growth in the spring (apparently facilitated by migration) could be critical for the persistence of local populations (Middleton and Merriam, 1981). Such clear evidence for rescue effect in the short term led Fahrig and Merriam (1985) to model migration and population structure in *P. leucopus* in order to examine patterns of long-term population persistence. The model predicted that population growth rates should be higher in populations connected by migration (as opposed to isolated populations), that winter population sizes should consequently be higher in these metapopulations, and that persistence of local populations and the overall metapopulation would depend on the rate of migration. They then used both tracking and trapping field data to verify that population growth rates were indeed higher in populations linked by migration (Fahrig and Merriam, 1985). Gottfried (1979) also examined population persistence of *P. leucopus* in relation to patch isolation distance. He found that populations in woodlots isolated by less than 0.5 km persisted, apparently because of high densities and high rates of migration.

Other small mammal species exhibit rescue effect metapopulation dynamics in artificially fragmented agricultural mosaics. Eastern chipmunks *(Tamias striatus)* breed in farm woodlots and have been trapped in high numbers migrating between woodlots and immigrating into patches where all resident chipmunks had been experimentally removed (Henderson *et al.*, 1985). Migration probably facilitates patch persistence, because populations that had declined to two to three nonbreeding individuals usually rebounded to become sizable breeding populations. They concluded that patterns of occupancy and the stability of chipmunk populations in the entire mosaic of woodlots depend on the rescue effect. Woodlot patches may be too small to support discrete local populations in a metapopulation structure; however, if rescue effect dynamics are occurring at this smaller scale, they also may be present at the metapopulation level.

One final set of evidence for rescue effect in small mammals comes from studies of Columbian ground squirrels *(Spermophilus columbianus)*. Before this century, ground squirrel habitat in western North America was virtually continuous, but agricultural development has since fragmented it, producing islands of habitat (Weddell, 1991). Boag and Murie (1981) studied migration of squirrels to and from a single colony in one habitat patch and estimated that 20% of all juveniles successfully migrated from the population each year. A smaller but detectable proportion actually became established breeders in other populations.

Boag and Murie also noted that local population fluctuations were relatively small compared to those of other species in the genus *Spermophilus,* suggesting that migration between populations may increase population stability. Weddell (1991) used radio telemetry to estimate a juvenile migration rate of 15%. In addition, patch occupancy was negatively correlated with isolation distance such that most occupied patches were isolated by less than 1 km. Weddell concluded that migration between habitat patches was critical for the continued persistence of local squirrel populations. He further emphasized the importance of migration by observing that connected populations persisted despite eradication attempts by local farmers. Though he does not use the term, it is clear that Weddell (1991) is describing a rescue effect metapopulation of Columbian ground squirrels.

The studies discussed above point out patterns of evidence for rescue effect metapopulations in small mammals and highlight gaps in our current knowledge. Many diverse species of small mammals may be predisposed toward metapopulations because they show spatial population structure as a result of sociality or habitat fragmentation. High rates of migration have been detected through tracking and trapping efforts, but more studies of migration in species that exhibit *natural* spatial structure are clearly needed. Migration has been linked to some demographic and genetic effects, but it has more frequently been implicated in local population and metapopulation persistence and stability through island biogeography studies. Such studies address the consequences of migration for metapopulation persistence, but the process by which that result is achieved, either through extinction – recolonization, rescue effect, or both, remains to be determined.

The previous examples also illustrate that the spatial structure and migration patterns of metapopulations with strong rescue effects can arise in several different ways. Natural habitat patchiness may be the least common way in which spatial structure is created for small mammals. Social structure also appears to encourage spatial aggregation and the threat of inbreeding may promote migration in social systems. Yet Lidicker and Patton (1987) found no consistent relationship between social systems and migration patterns, so such generalizations may be too simplistic. The most common way in which spatial structure is created appears to be habitat fragmentation. Weddell (1991) suggested that mammals in historically contiguous habitats never evolved strategies for very long-distance migration or persistence in small isolated populations, so these species may be unable to persist in human-created metapopulations. Metapopulations with strong rescue effects, in which local populations are close enough to accommodate short-distance migration, may be the inevitable outcome of habitat fragmentation for these species. Similarly, black-tailed prairie dogs apparently never disperse to unoccupied habitat patches (Garrett and Franklin, 1988), so they could not exhibit extinction/recolonization metapopulation dynamics. Conspecific attraction could promote higher rates of successful migration, leading to rescue effect metapopulation dynamics.

C. Amphibians

Wetland or pond habitats are naturally patchy, usually separated by terrestrial habitats that may be unsuitable for wetland-associated species, including many amphibians. Thus, many of these species may exhibit metapopulation dynamics. For example, Gibbs (1993) assumed a metapopulation mosaic to model the loss of small wetland patches and its effect on wetland animals. Using demographic data and observed migration distances from previously published empirical studies of wetland species, including some amphibians, he found that species with high migration rates appeared to be buffered from local extinctions by the rescue effect. The species that showed the highest probabilities of persistence were amphibians. The studies discussed below support Gibbs' (1993) theoretical evidence for the occurrence of rescue effect metapopulations in numerous wetland systems.

1. Red-Spotted Newts

In an early empirical investigation of metapopulations, Gill (1978a,b) studied the population dynamics of red-spotted newts *(Notophthalmus viridescens)*. These newts breed in mountain ponds in the eastern United States, traveling into surrounding terrestrial habitat only to enter winter hibernacula. Despite this distinct spatial structure, Gill (1978a,b) found evidence for a relatively high migration rate between ponds. Though he directly observed only one migration event, he inferred high migration from ponds off his study site, since approximately 50% of his breeding population was replaced each year by new recruits. In addition, juveniles in these populations pass through a distinctive "eft" life stage that is absent in coastal populations that do not exhibit metapopulation spatial structure. Gill (1978a) suggested that this life stage may be particularly adapted for migration in a metapopulation context. The direct effects of migration on local dynamics were not extensively assessed, but Gill (1978b) calculated very low inbreeding coefficients for each population as well as the overall metapopulation ($F = 0.0001$ to 0.0051). In addition, no local extinctions were observed during the 3-year study despite often negligible reproductive success, suggesting demographic rescue of local populations by large numbers of immigrants. Gill (1978a) concluded that the populations he studied were completely dependent on migration for persistence, and that most immigrants came from a shifting mosaic of "metapopulation centers" outside of his study area. He proceeded to model this metapopulation, showing that migration or colonization rates are so high that extinction almost never occurs and virtually all patches are occupied all the time, even if reproductive success is low. In current terminology, Gill's model would be considered a rescue effect metapopulation.

2. Natterjack Toads

Sinsch (1992) studied a similar system of natterjack toads *(Bufo calamita)* in Rhineland, Germany. These toads make more extensive use of terrestrial habitats than the red-spotted newts, but they are still restricted to breeding in ponds, and they concentrate their terrestrial activities near these breeding sites. Through direct observation, mark–recapture, and radio telemetry, Sinsch found that most adult female toads and some juveniles migrate between breeding ponds within seasons, while most males never leave their first breeding sites. During the 6-year study, one pond was consistently a source of migrants. Yet this system was not a traditional source–sink metapopulation because the other breeding sites shifted between being sources and sinks; therefore, the location of source populations moved through the mosaic of populations over time. Sinsch (1992) did not discuss the demographic consequences of frequent migration except to conclude that, as in the red-spotted newt, natterjack toad populations may persist, despite reproductive failure, because of immigration and the rescue effect. To assess the genetic consequences of high migration rates, he used allozyme electrophoresis to calculate genetic distances between the local populations. All calculated genetic distances were low (Nei's D = 0.0023 to 0.0646), though the most distant area showed greater genetic differentiation. Overall, Sinsch concluded that natterjack toads exhibit metapopulation dynamics similar to those modeled by Gill (1978a). Most sites are continually occupied, and local populations persist due to immigration and the rescue effect.

3. Pool Frogs

Along the Baltic coast of Sweden, pool frogs *(Rana lessonae)* occur in natural metapopulations, reproducing only in distinct water bodies. Sjögren (1991) and Sjögren Gulve (1994) studied extinction patterns in one frog metapopulation, incorporating island biogeography theory. Over a 6-year period, he found that populations isolated by greater than 1 km went extinct, while less isolated populations tended to persist. He attributed this persistence to the demographic and genetic consequences of migration, since most dispersing frogs travel less than 1 km and persisting populations probably received 2–15 immigrants per generation, as estimated from demographic data (Sjögren, 1991). Migration apparently had the demographic effect of replacing individuals lost to pike predation in heavily depredated ponds (Sjögren Gulve, 1994). Sjögren (1991) also examined the proportion of fertilized eggs in egg masses and determined that small local populations showed no evidence of inbreeding depression, a possible result of genetic rescue effect. Finally, immigration may have also increased population stability by mitigating population size fluctuations, which can be large in this species. Sjögren Gulve (1994) concluded that the metapopulation structure of pool frogs resembles the rescue effect metapopulation model proposed by Gill (1978a) for red-spotted newts.

Several researchers have found evidence for rescue effect in amphibian species. Since amphibians are restricted to breeding in water, they are predisposed toward spatial metapopulation structures. Migration has rarely been directly detected, but high rates of migration have been inferred from arrivals of new recruits and instances of population persistence despite reproductive failure. Genetic effects of migration, such as low inbreeding coefficients and lack of inbreeding depression, have been measured. Demographic effects are usually inferred from observed population persistence; therefore, persistence of local populations has been attributed to migration and the rescue effect. Population stability may also be enhanced by the rescue effect, because migration appears to dampen local population size fluctuations.

IV. CONCLUSIONS

The studies discussed above suggest that metapopulations with strong rescue effects may be more common than currently supposed (see also Harrison, 1991; Harrison and Taylor, this volume). In most cases, the research was not designed to provide evidence for a rescue effect, and some were published before the term was even introduced. Many organisms live in naturally fragmented environments, and areas of suitable habitat often will be too small to support viable populations over the long term, particularly in highly stochastic environments, if they are completely independent of other populations. Thus, it is not surprising that many species have developed the ability to disperse over considerable distances. Both our models and those of others clearly show that a collection of populations connected together through migration can persist longer than the same number of populations of identical sizes that are isolated from one another. Migration and metapopulation exchange will buffer individual populations against "bad" years, and they can allow species to persist in fragmented habitats where they might otherwise quickly go extinct. The ability to disperse successfully on a regular basis across unsuitable areas may act as a "filter" that determines the presence or absence of particular species in particular areas. As a result, the best way to look for metapopulations with strong rescue effects may be to examine habitat types, rather than particular taxa. Metapopulation systems are likely to be widespread throughout many different taxa, but all should be found within patchy habitats, because patchiness confers the spatial population structure that is a prerequisite of all metapopulation models. For example, the amphibians discussed above were first identified as potential metapopulation systems because they occupied spatially distinct ponds; therefore, it is likely that other amphibian species and other wetland-associated species will exhibit rescue effect metapopulation dynamics. Additional habitat types that should be examined include tide pools in the Pacific northwest, riparian areas throughout the western United States, and mountain tops in the desert southwest (Stacey and Johnson, 1996).

As is true of many biological phenomena that have a spatial component, a central issue in metapopulation structure is scale. For example, the distinction between one population that occupies a number of different patches and a metapopulation system that is composed of individual populations in discrete habitats, will not always be clear. One system will necessarily grade into the other because spatial distribution is a continuous rather than discrete variable. However, there may be a real difference in these systems, based upon the behavior of the organisms. In a patchy environment, individuals should have a relatively equal probability of breeding with other individuals both within the natal patch and in nearby patches; this probability in turn will be a function of the migration capabilities of the species. In a metapopulation consisting of individual populations, the individual will have a much *lower* probability of mating with an individual in another population than in their natal one. Mating probabilities are highly discontinuous. In terms of demographic models, this means that the growth of the population that occupies a collection of patches can be described as the sum of birth and deaths within that set of patches. In contrast, in a metapopulations system, we must describe both local population processes and migration frequency among populations. Because migration will often be independent of local demography, metapopulations will exhibit emergent properties (such as continued persistence through rescue effects) that are greater than the sum of its parts.

Similarly, the distinction between metapopulations where extinction and subsequent recolonization is the primary dynamic and systems where migration is frequent enough to prevent most local extinctions may also be primarily a matter of scale, both spatially and temporally. All local populations are likely to go extinct eventually, if they are observed for a long enough period. Similarly, many of the mammal studies discussed above (e.g., Gill, 1978a; Sjögren, 1991; Sinsch, 1992) have found that persistence of many small populations is a matter of their degree of isolation from other populations. Adjacent populations may last longer because they are part of a rescue effect metapopulation with frequent migration, where more distant populations are rescued less frequently and exhibit extinction/recolonization dynamics. Although these models may vary in some predictions about local population characteristics or processes (e.g., the level of genetic diversity within the metapopulation), they are not likely to be mutually exclusive in nature. Real world metapopulations can and apparently do exhibit more than one model dynamic. In all of the above examples, rescue effects seemed to occur in the central, least isolated populations while extinction/recolonization probably occurred among the peripheral, most isolated populations. The combination of strong rescue effects and extinction/recolonization dynamics may be common in nature, because migration rates should be negatively correlated with isolation distance (Brown and Kodric-Brown, 1977). Differences in migration rates may be the primary factor determining which metapopulation dynamic is most important, as there is no *a priori* reason to expect that dispersing individuals will settle only in habitats that are currently unoccupied. In each case, a central challenge in understanding these systems will be to determine the frequency and timing of among-population migration and its resulting impact upon local population dynamics.

Metapopulations can be more resistant to extinction than independent populations because the stochastic nature of variation in growth rates means that while some populations are declining, other elements in the system are likely to be increasing and producing potential immigrants (Goodman, 1987a). The ability of organisms to successfully migrate among populations is critical for this effect; even small increases in system connectivity result in large increases in persistence times. We thus repeat the conclusions reached by other authors in this volume. For natural metapopulations, any change in the environment (either natural or human caused) that makes migration more difficult may lead to the decline and possible extinction of both individual populations and the entire metapopulation. Management and preservation efforts usually focus on the single largest, or most representative, population or ecosystem. Smaller populations or habitats are often ignored and destroyed by development. Even though large populations may appear to be stable and self-sustaining, they may actually be highly dependent on other populations in a metapopulation system for their continued persistence (Stacey and Taper, 1992). Without immigration, these populations may collapse, even though their local habitat has been protected. Migration has long been considered to be important to prevent the negative effects of inbreeding in small populations; our analyses, and those of other researchers, emphasize that it may have even more immediate value in terms of demographic rescue. Recognition of the potential importance of metapopulation structure in the dynamics of species will make conservation efforts more challenging, but also more likely to succeed in the long run.

ACKNOWLEDGMENTS

Ric Roche helped with the code for the simulation model. We thank Elisabeth Ammon, Erik Doerr, Brenda Johnson, and Patricia Zenone for comments that substantially improved this work. P.B.S. is supported by NSF Grant DEB-9302247, V.A.J. by a University of Nevada Graduate Fellowship, and M.L.T. by EPA Grant CR-820086 and NSF Grant DEB9411770.

13

Evolution of Migration Rate and Other Traits

The Metapopulation Effect

Isabelle Olivieri   Pierre-Henri Gouyon

I. INTRODUCTION

Traditionally, population geneticists have put a strong emphasis on the evolutionary consequences of subdivision within a population (Wright, 1952, 1969; Malécot, 1948, 1969; Maruyama, 1970; Kimura and Maruyama, 1971; Nei, 1987). Population structure has been viewed as a way to produce spatial differentiation and founder effects through stochastic processes (e.g., Wright's shifting balance theory, 1969; Maruyama, 1970; Maruyama and Kimura, 1980). More recently, another class of models, hereafter called metapopulation evolutionary models, has focused on the evolutionary consequences of population extinctions and recolonizations (Van Valen, 1971). In these latter models, the emphasis is on the selective pressures created by population turnover within a metapopulation. The focus is on the evolution of particular traits (migration, sex-ratio, life-history traits, etc.) whose genetic determinism is usually unknown.

These two classes of models have generated two terminologies (see Hanski and Simberloff, this volume) that amount to the same object (i.e., a set of individuals subdivided into more or less ephemeral subsets): classical population genetics considered a population subdivided into subpopulations, or demes, or neighborhoods, while metapopulation evolution is about a metapopulation made of ephemeral local populations. These two approaches have not yet converged,

Metapopulation Biology
Copyright © 1997 by Academic Press, Inc. All rights of reproduction in any form reserved.

though some efforts in this direction are under way. Models of subdivided populations now include the effects of local extinctions and local dynamics (e.g., Barton and Whitlock, this volume; see also Slatkin, 1977; Slatkin and Wade, 1978; Lande, 1985a; Ohta, 1992; Michalakis and Olivieri, 1993; Nichols and Hewitt, 1994; see Hauffe and Searle, 1993, for an application of these concepts to Robertsonian fusion in mice). Metapopulation evolutionary models have so far been mostly used to study the evolution of migration rate (Van Valen, 1971; Comins *et al.*, 1980; Levin *et al.*, 1984; Olivieri and Gouyon, 1985; Venable, 1993; but see Caswell, 1982, for a study of the influence of demographic disequilibrium on life-history traits). In this chapter, we will focus on metapopulation evolutionary models. We wish to show that the demographic functioning of metapopulations creates intrinsic emergent properties that influence the evolution of major biological traits such as migration rate. We suggest that this effect, which we call the *metapopulation effect*, should be taken into account when studying the evolution of life-history traits and genetic systems in species which are composed of transient populations, where the unit of evolution is the metapopulation.

Migration plays a central role in metapopulation dynamics and evolution: it contributes to metapopulation spatial structure, local dynamics, and metapopulation evolution, as shown by other chapters of this volume (see also Venable, 1993). Migration is itself an evolving character, and many species exhibit distinct adaptations to migration (e.g., in plant species, pappi on seeds, spines on pods). The possible evolution of migration is testified to by the occurrence of polymorphisms for migratory behavior in numerous species of animals and plants (see examples of seed migration polymorphism in Venable (1979), Venable and Lawlor (1980), Olivieri *et al.* (1983), Olivieri and Berger (1985), Schmitt *et al.* (1985), and Clay (1982) and of wing dimorphism in insects in Southwood (1962), Dingle *et al.* (1980), Kaitala (1990), Ben-Shlomo *et al.* (1991), Karlson and Taylor (1992), Hairston (1993), and Tsuji *et al.* (1994); see Harrison (1980), Roff (1986), and Roderick and Caldwell (1992) for reviews of insects and Berthold and Pulido (1994) for a study of heritability of migratory activity in birds). We will not make a distinction between migration and dispersal and will consider both as undirected and irreversible movement away from the habitat patch of origin (e.g., strict dispersal as defined by den Boer (1990)): the migration rate of a genotype will be equal to the probability that an individual of that genotype leaves the patch in which it was born.

Why does migration evolve? At first sight, migration should be selected against. First, in most species, there is a risk of dying during migration, for instance because of predation. There may thus be an intrinsic cost to migration. Second, imagine a landscape covered with a suite of habitat patches, where some patches are suitable for a given species, while others are not. Suppose also that migration is restricted to within this landscape and is uniform across the patches. It is clear that migration may lead individuals to bad environments. The total number of emigrants (i.e., those who leave the patches) should be the same as or larger (if there is mortality during migration) than the number of immigrants (i.e., those who arrive in patches). Since only suitable patches (those inhabited by the species in question) produce emigrants, while all patches, including unsuitable ones, receive immigrants, it is clear that, in a suitable patch, the number of emigrants is on average larger (often much larger) than the number of immigrants. A gene enhancing migration should thus leave patches more often than it is reintroduced by immigration and should, on average, decrease in frequency within patches. Overall there exists therefore a selection for residency within each local population.

There are, however, also factors that select for migration among populations; these are factors acting at the among-population level: avoidance of sib competition, improved conditions elsewhere. Migration is especially favored when the spatio-temporal variability in population sizes is large (Levin *et al.*, 1984; Southwood, 1987; Venable and Brown, 1988). An extreme case of temporal variability in population sizes occurs if local populations go extinct. Johnson and Gaines (1990) recently reviewed the main models on the evolution of migration. Few of these models explicitly consider local extinctions, though such extreme variability is realistic in many cases, as shown in this volume. In many species, in fact, population extinction is unavoidable, either because of stochastic disturbances, or the ecological process of succession, or ultimately because of demographic stochasticity. Global persistence of any genotype in such species necessitates colonization after local extinctions. Each particular population will eventually go extinct and therefore only offspring which have emigrated will be able to reproduce. Contrary to the short-term cost of migration, there may thus be a long-term benefit to it.

Thus two opposing selection pressures, selection for migration during recolonization and selection against migration once a population has been established, act on the migration rate when local extinctions are the sole source of environmental variation. *These two antagonistic selective forces create a metapopulation effect.* This metapopulation effect was observed in the very first studies on the evolution of migration (Southwood, 1962, 1987; Van Valen, 1971; see also den Boer, 1990). Other traits apart from migration rate may also experience variable selection due to the metapopulation effect. Caswell (1982) has shown that, in a given population, the variability in population size would create a succession of episodes of *r* selection and *K* selection. In this paper, we will consider in more detail the evolution of life-history traits in a metapopulation. Moreover, we will show that selection on migration may interact with selection on other life-history traits, so that coevolution among characters may occur as a result of the two-level selective process.

II. HOW DOES A METAPOPULATION PERSIST IN A LANDSCAPE? AN EXAMPLE OF A MODEL WITH LOCAL DISTURBANCES AND SUCCESSIONAL PROCESSES

In this chapter, we present either published (Sections II–IV.A) or new (the rest of the chapter) results based on a deterministic model which has been described in Olivieri and Gouyon (1985) and in Olivieri *et al.* (1995). This model is based on a Leslie matrix approach, equivalent to the continous time model in Levin and Paine (1974), and the discrete version used by Paine and Levin (1981). We describe here the main assumptions of this model, extended to a species with overlapping generations. This model will be used in the following sections.

The metapopulation is assumed to evolve in a landscape made of an infinite number of patches, each containing at most one population (Fig. 1). A given patch may exist in any of states $0, 1, 2, ...,z$, in which state 0 represents an unoccupied patch as a result of extinction through disturbance, state $i (i \in \{1, z - 1\})$ a patch that has persisted through i consecutive seasons without an extinction event, and state z a late successional patch in which recruitment by the organism under study is prevented by later successional species. We assume that extinction through disturbance occurs at a rate which may depend on the number of seasons since the most recent extinction. A_i $(i \in \{0,z\})$ is the probability that a patch in state i is not disturbed (and proceeds to state $i + 1$ in the next season if $i < z$ or remains in state z if $i = z$). This holds for empty patches as well, so that the probability that a given disturbed patch is recolonized and produces propagules is equal to its probability of persistence A_0.

As shown in Olivieri and Gouyon (1985), Horvitz and Schemske (1986), Hastings and Wolin (1989), and Olivieri *et al.* (1995), these assumptions describe a Markovian process, such that, at equilibrium, one can express a stationary age distribution of patches V_i, as a function of A_i and z. When z is infinite and $A_i = A$ for all i, then $V = A = 1 - e$, where e is the local extinction rate and V, the proportion of occupied patches, is the sum of V_i over all i except $i = 0$. In all examples given in this paper, we assume that disturbance in different patches occurs independently. Moreover, we assumed that $A_i = A_1$ for all $0 < i < z$, but assume that the probability of recolonization of empty patches, A_0, and the probability of persistence of late successional patches, A_z, have distinct values.

At the beginning of the season, residents of a given patch reproduce. Adults

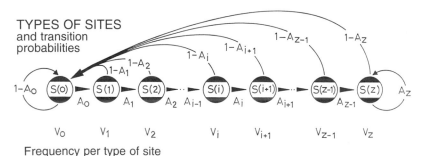

FIGURE 1 Metapopulation dynamics in a fragmented landscape. Each patch of age i can be either disturbed (with probability $1 - A_i$) or left undisturbed (with probability A_i) and then proceed to the following successional stage $i + 1$. (Reprinted with permission of University of Chicago Press from Olivieri *et al.*, 1995.)

either die after reproduction or persist. Juveniles either remain in the natal patch or disperse at some fixed, genetically based rate. A proportion q (equal to 1 in our simulations) of dispersing juveniles survive to form the migrant pool. Comins (1982) has shown that the ESS of migration rate in a two-dimensional stepping-stone model was very similar to that in an island model. In contrast, Lavorel *et al.* (1995) and Lavorel and Chesson (1996) have shown that under some conditions, the detailed spatial pattern of environmental variation has to be taken into account. In some species at least, however, an island model of migration appears to be a good approximation of the actual functioning of the metapopulation. In our model, we assume such an island model of migration. Each patch receives an equal fraction of the migrants, which then compete with local nondispersing juveniles. We assume local density-dependent regulation of juvenile viability and a patch age-independent carrying capacity equal to K, which is reached at a rate dependent on the parameter values. For a given number N of adults in a patch in season t, we first let adult survival to take place in a density-independent manner. If s is the constant adult survival rate from one season to the next, juveniles compete for the $K - Ns$ places left empty. If the number of competing juveniles is less than $K - Ns$, they all establish as adults in the following season. Otherwise, there are exactly $K - Ns$ juveniles that become adults at time $t + 1$, and the patch has reached the carrying capacity. All patches, including empty ones, receive migrants. Extinction or persistence of each population then occurs according to the local probabilities A_i. If there is no disturbance, migrants and resident propagules (if any) establish with equal competitive abilities. If there is disturbance, no individual establishes in the site.

The metapopulation must be regulated either at the patch (local) level or at the metapopulation (global) level. Without density-dependent regulation (see Hastings and Wolin (1989) for a model with global regulation and local exponential growth), either the metapopulation asymptotic growth-rate is greater than 1, in which case there is metapopulation explosion, or it is less than 1, in which case the metapopulation goes extinct. In the latter case, it is nevertheless possible that each extant population is in fact growing. The metapopulation structure itself does not create any overall regulation of the number of individuals (contrary to Wilson, 1973), but some kind of density dependence is necessary to prevent the metapopulation from demographic explosion (Hanski, 1990; Taylor, 1990). In our deterministic simulation model, a viable metapopulation reaches a demographic equilibrium through within-patch density-dependent regulation.

In a density-independent context, such as the one modeled by Horvitz and Schemske (1986) and Cipollini *et al.* (1994), the response of metapopulation growth rate to changes in life-history parameters provides the necessary information concerning selective pressures on those characters. However, density dependence is likely to affect the evolution of life-history traits, first, because it creates frequency dependency in the evolution of migration and, second, because it is unlikely that in species with overlapping generations all age classes would be equally sensitive to density. The Cipollini *et al.* model of patch dynamics thus cannot be really used to study adaptive evolution.

III. WITHIN-POPULATION SELECTION VERSUS COLONIZATION SELECTION: AN INSIGHT INTO THE METAPOPULATION EFFECT

A. Polymorphism for Migration

Van Valen (1971) suggested that migration polymorphisms could be maintained by opposing selection pressures within and between groups (populations). Using an analytical model for limiting cases (patch saturation at recolonization), and the more general numerical model described in the previous section, we have shown that two genotypes straddling the evolutionarily stable migration rate can indeed be maintained together in a stable pattern, as suggested by Van Valen (1971) and Roff (1975) and shown recently by Roff (1994b); (Olivieri *et al.*, 1995; see an example in Fig. 2). We will see in the next section that this polymorphism is in fact not evolutionarily stable.

At equilibrium, and in the cases in which a stable polymorphism could be

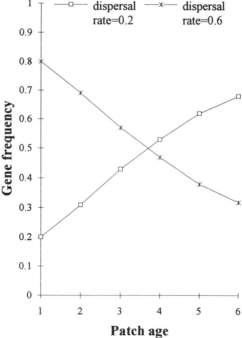

FIGURE 2   Equilibrium gene frequencies in local populations, as a function of population age. Two genotypes with migration rates of 0.2 and 0.6 are maintained in landscape L2 described Table I, with 2 = 7 instead of 3. Both genotypes have the same constant fecundity (10).

maintained, all patches of a given age had the same genotypic frequencies. These frequencies differed among patches of different ages. Figure 2 shows that the frequency of the genotype with the lowest migration rate increases with the age of the population. Therefore, each population is evolving toward fixation of the nonmigrating genotype. This pattern demonstrates that selection is indeed operating at opposite directions at the time of colonization (when migration is favored) and thereafter (when migration is selected against in each population).

We are aware of only three empirical studies of the relationship between successional stage and migration ability (Brown, 1985; Olivieri and Gouyon, 1985; Peroni, 1994). Only two of them (Olivieri and Gouyon, 1985; Peroni, 1994) consider within-species variability. Brown (1985) found that along plant succession, the proportion of Heteroptera species that were fully winged decreased from 95 to 65% (see her Fig. 2). The observed decrease was, however, not statistically significant. The thistles *Carduus pycnocephalus* and *C. tenuiflorus* show a pronounced seed heteromorphism: in each capitulum, inner seeds have a pappus and may be dispersed by wind, whereas outer seeds remain in the capitulum and fall on the ground with it (Olivieri *et al.,* 1983). By studying four populations of *C. pycnocephalus* and three populations of *C. tenuiflorus,* we found that, as predicted, the average proportion of seeds with a pappus (used for wind migration) decreased in both species along a successional gradient. The differences were slight but significant. More recently, Peroni (1994) tested the theoretical prediction in the red maple, *Acer rubrum,* by collecting samaras from five populations located in early successional environments and five populations located in late successional environments. She measured the wing loading ratio (samara mass/samara area), assumed to be inversely proportional to migration ability. She found that samaras from the early successional red maples showed slightly but significantly lower wing loading ratios than those from late successional environments, thus confirming the theoretical prediction. The evolutionary interpretation assumes that at least a part of the variation observed in natural populations is heritable. There are good reasons to believe that this is the case in the studies of Olivieri and Gouyon (1985) and Peroni (1994); see also Olivieri and Berger, (1985).

One might wonder whether all migration is the result of these two opposing selection pressures. Hamilton and May (1977) have shown that migration could evolve in a uniform population, because of sib competition and random choice of individuals forming the next generation. We have suggested (Olivieri and Gouyon, 1985) that at least some of long-distance migration is in fact a by-product of selection for short-distance, within-population migration. In several plant species, seeds (including those with migration apparatus) are more likely to remain close to the parent plant than to experience long-distance migration [e.g. in thistles (Oliveri and Gouyon, 1985) and in grey mangrove (Clarke, 1993)]. Peroni (1994) proposed that because rodent predation of red maple seeds is intense and seems to be density-dependent, moderate levels of migration within each population may be advantageous. She suggested that this might explain why migration ability in red maple seems greater than what would be expected from the sole consideration of local extinction probabilities. It could even be that the evolution of migration structures has little to do with migration itself. For instance, Tsuji *et al.* (1994) have proposed that wing dimorphism in *Cardiocondyla* ants was maintained through correlations between winglessness and emergence time. Motro (1991) and others (e.g., Holsinger, 1986) have suggested that migration and sexual dimorphism for migration may evolve as a way of avoiding inbreeding. Other reasons why migration may evolve can be found in Harrison (1980) and Johnson and Gaines (1990). We will discuss genetic correlations between migration and other traits in Section V.

B. Polymorphism vs ESS

Although polymorphisms such as the one shown in Fig. 2 are stable (i.e., gene frequencies return to some equilibrium values following perturbations away from these values), they are generally not evolutionarily stable states (*sensu* Maynard Smith, 1982). In all the cases we have studied (Olivieri *et al.*, 1995), there exists one evolutionarily stable migration rate (ESS), and the introduction of a genotype with a migration rate closer to the ESS than is any of the previously existing ones leads to the loss of at least one of the previous types. If the newly introduced genotype is the ESS itself, then all other, possibly coexisting types, are lost. In other studies, in which particular hypotheses were made about spatial heterogeneity of local carrying capacities, there was no ESS (McPeek and Holt, 1992; Cohen and Levin, 1991; Ludwig and Levin, 1991). For instance, Cohen and Levin found that in some situations, there may exist a strategy which may invade any other single type (evolutionarily compatible strategy), but which is also open to invasion by any type. The population is then likely to be polymorphic. It may also be that ESS are unreachable because of the genetic determinism. We have modeled a situation in which the migration rate among the offspring (proportion of those offspring which do migrate) is determined by the genotype of the mother. If a given mother is able to produce both offspring which migrate and offspring which do not, an intermediate migration rate can then be genetically determined through the mother genotype, so that mixed strategies can evolve. In some species, however, a given organism can produce only offspring which migrate or only offspring which do not. In other species, the migration behavior is determined by the genotype of the offspring; i.e., the character under selection is not the proportion of migrating offspring but the individual probability of migration. In both last cases, even though the genetic determinism of migration behavior might be polygenic, there may be no available intermediate strategy, and a polymorphism would be maintained between individuals who disperse and individuals who do not disperse (Roff, 1994b). For instance, according to Roff (1994a), wing histolysis in insects, which leads to winglessness, is a threshold character determined by the individual genotype, so that morphological variation is discontinuous, with no intermediates: individual probability of migration is either 0 or 1.

The heritability of migration tendency in insects is usually quite high (see

Roff, 1994a, for a review), suggesting that polymorphisms between winged and wingless genotypes result from balancing selection such as the one suggested in this chapter (though see Roff, 1990, for many other factors). Quite unexpectedly, however, Roff (1994a) has shown that when winglessness is a polygenic but threshold character, much of the variability in the trait is retained even in the presence of directional selection for one morph. The number of loci involved had very little influence on this result. Therefore, the high polymorphism level cannot be readily interpreted as a consequence of a balancing selection acting on migration behavior.

IV. DOES SELECTION ADJUST MIGRATION RATE AT THE METAPOPULATION LEVEL?

A. Influence of Landscape and Species Characteristics on ESS

Various factors influence the evolution of migration. Many authors have shown that local extinctions, by enhancing selection among populations, favor migration. For instance, according to Roderick and Caldwell (1992; see also Denno et al., 1991; Denno, 1994), frequent migration is common in insects in temporary habitats, where variation in local carrying capacity is large; whereas less migration and more winglessness occur in stable or isolated habitats. In an extensive review on the evolution of flightlessness in insects, Roff (1990) showed that flightlessness is associated with decreased environmental heterogeneity. His data broadly supported the prediction that flightlessness is also associated with habitat persistence, though he suggested that more quantitative data would be needed to fully support this prediction.

In the case when local extinctions are due to deterministic causes related to succession, for instance when a given population has a maximal lifespan due to the invasion of its patch by other species, we found (Olivieri et al., 1995) that the ESS varies in a nonmonotonic manner, first decreasing but then increasing with increasing population lifespan (Fig. 3). An increase in population lifespan increases the expected future lifetime of the home deme, and hence local selection against migration (Fig. 2) lasts longer. However, if population lifespan is very large, then the proportion of patches that have reached the end of succession is decreased, and hence the cost of migration due to falling in such patches is decreased.

Using numerical simulations, we found a quite strong effect of fecundity on the ES migration rate (Olivieri et al., 1995). Other things being equal, high fecundity species should disperse more offspring than low fecundity species. Under the assumption that in saturated patches juveniles may become established only when some adults die, the ES migration rate also increases with adult survival rate. This prediction is explained by the negative influence of adult survival on the establishment of offspring in the home deme. The main possibility for juveniles to establish is then to reach an empty site.

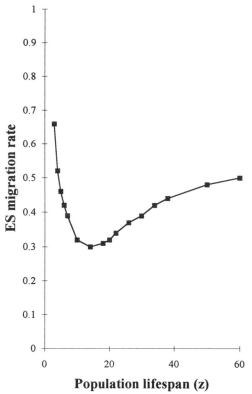

FIGURE 3   Evolutionarily stable migration rate as a function of population lifespan. Results based on iterations of the model in Fig. 1; with landscape L2 (Table I), fecundity of 10.

When the population lifespan is very long, and patches are saturated immediately at recolonization, making it more important to reach an unsaturated patch, the ES migration rate is given by

$$d^* = 1 \text{ if } A_0 \geq A_1$$

and

$$d^* = (1 - A_1 + A_0)/(1 - s A_1) \text{ if } A_0 < A_1,$$

where s is the adult survival rate, A_0 is the recolonization rate, and A_1 is the persistence probability of colonized patches. This result can be obtained from Eq. (14) in Olivieri *et al.* (1995) by assuming that z is infinite. The result shows that in the absence of succession (z large) and of mortality during migration ($q = 1$), maximal migration rate is selected for as soon as the probability A_0 of recolonization of empty patches is larger than the probability of persistence of established populations, A_1. The result also shows that the ES d^* increases with the adult survival rate s, the increase depending on the probability A_1. The ES migration rate is equal to 1 as soon as the survival rate is larger than $(A_1 - A_0)/A_1$, i.e., as soon as adult mortality rate is below A_0/A_1.

It is often concluded that high fecundities are associated with high migration rates in nature (the "colonizer" syndrome; Baker and Stebbins, 1965). For instance, in a survey of ecological characteristics of British angiosperms, Peat and Fitter (1994) found that species with wind-dispersed seeds produced more seeds than species with no specialized migration mechanism. Few studies have considered the association of high adult survival rates with migration. Some data appear to agree with our nonintuitive prediction that perennial species should be selected to have higher rate of long-distance migration than annuals. Venable and Levin (1983), in an extensive review of some 6000 species in the family Asteraceae, found that a significantly smaller percentage of annual than perennial species had migration structures that could affect medium and long-distance migration. They concluded that their data supported the idea that migration may be more important for perennial than for annual plants. Moreover, as we will show in the next section (V.B), there are interactions between life-history evolution and evolution of migration: we will suggest that when migration rate is increased, there may be selection for increased adult survival rate (detrimental to fecundity), thus reinforcing the effect just described.

B. Conditional Migration

The result shown in Fig. 2 raises a paradox. Intuitively, it would seem optimal to have low migration rate in the colonizing phase and to leave the patch when successional events are about to drive the population to extinction. What happens in our model is exactly the reverse. Newly formed populations send their progeny out while old ones keep their offspring in the patch where they have no future. This result makes sense, given that the colonizers must by definition be composed mainly of migrators. This is because until now, we have considered that a given genotype is characterized by a single migration rate (an unconditional migration strategy, McPeek and Holt, 1992). In reality, it might be that migration behavior varies with some environmental factors. For instance, it is well-known that in some species of mammals, emigration rate increases with density (Johnson and Gaines, 1990). In fishes, Okland et al. (1993) suggested that the migration decisions of brown trout and Atlantic salmon depend on individual growth rates. In birds, Pruett-Jones and Lewis (1990) studied the interaction between habitat limitation, habitat quality, and sex ratio in influencing migration decisions in a population of fairy-wrens. They suggested that young males delayed their migration in response to a limited number of mates and secondarily to habitat limitation. In insects, many studies have shown that environmental factors as well as wing or flight polymorphism influence behaviors related to migration (see Roderick and Caldwell, 1992, for several references). Turning to theoretical studies, Hastings (1992) considered age-dependent migration. He found that under some conditions (stronger density dependence in age classes which migrate less), chaotic population dynamics might appear. Such patterns are likely to influence the evolution of migration. Kaitala *et al.* (1989) showed that winglessness conditional on population density is an ESS when compared to an unconditional strategy in a model applied to the evolution of migration in a waterstrider. McPeek and Holt (1992), using a two-patch model, studied the influence of spatio-temporal variability in local carrying capacities on migration (thus phenotypically plastic migration, which they called a conditional migration strategy). They found that *"local population sizes and the proportions of local populations that disperse should be negatively correlated if populations are both at their evolutionary and demographic equilibria and if fitnesses are density-dependent."* Their model thus predicts that emigration from low density patches should be larger than emigration rates from high density patches, contrary to what is observed in mammals, for example. In our model, carrying capacity is the same for all patches. A strategy in which migration rate increases with population age or local density, should be advantageous: low-migration rate following arrival in an empty patch would facilitate colonization, whereas migration would be a better strategy when the patch is crowded. We may define a plastic strategy as one in which migration rate increases with population age. By running pairwise contests between nonplastic and plastic genotypes, we found that such plastic strategies would usually invade, provided that their average migration rate was not too different from the ES migration rate for a nonplastic strategy. Levin *et al.* (1984) reached similar conclusions.

C. Metapopulation Viability as a Function of Migration

One may wonder whether all migration rates allow the metapopulation persistence, and if not, how far from the optimum for the metapopulation can selection push the migration rate. Figure 4 shows the equilibrium metapopulation sizes for a single annual genotype as a function of the migration rate in the two landscapes L1 and L2 (see Table I). For a given landscape, there exists a limited range of migration rates that allow metapopulation persistence. Below a minimum value, the number of migrants establishing new populations is not sufficient to compensate for extinctions. Beyond a maximum value, the number of progeny remaining in a patch may not be sufficient to allow population establishment. For some parameter values corresponding to particularly favorable habitats (few end of succession sites) the maximal value may be equal to 1. The minimal value is always positive. Hanski *et al.* (1994) observed the same kind of pattern in a parameterized butterfly metapopulation model. Their model suggested that the migration rate of the butterfly *Melitaea cinxia* should lie between 10 and 40%, and indeed the observed migration rate, as measured from mark–recapture studies, was about 30%.

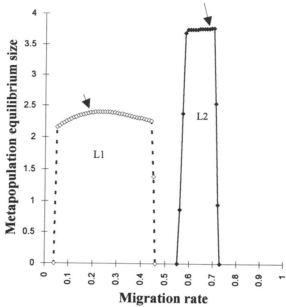

FIGURE 4   Metapopulation equilibrium size, as a function of migration rate, in L1 and L2 described in Table I, with fecundity of 5 (L1) and 7.5 (L2). The ES migration rate is indicated by an arrow. Size is expressed in Log(average number of individuals per patch × 1000)/log(10).

D. ESS versus Optimal Migration Rate

Migration is important for evolution (it ensures the genetic cohesion of a species and allows its demographic survival), and migration is an evolving character itself. However, what are the effects of a given migration rate on the survival of the metapopulation? Structured population genetics models (Wade and Mc-Cauley, 1988; Whitlock and McCauley, 1990) show that genetic differentiation among populations may be maximized by some intermediate migration rates.

TABLE I   Landscape Parameters Used in the Numerical Examples of the Model Described in This Chapter[a]

z	A_0	$A_i(i = 1..z - 1)$	A_z	z	K
L1	0.1	0.9	0.999	7	1000
L2	0.4	0.9	0.95	3	1000

[a] Landscape L1 is characterized by a longer population lifespan, but smaller recolonization rate and smaller disturbance rate of late successional (uncolonizable) sites (S_z sites). Carrying capacity K is the same in the two landscapes.

From a purely demographic perspective, one may define the metapopulation density as the average number of individuals per site. This is equivalent to the "site occupancy" used by Hamilton and May (1977) and Comins *et al.* (1980). One may then wonder how the metapopulation carrying capacity, that is, its density at demographic equilibrium, is influenced by the migration rate and define an *optimal* strategy (OS) as the migration rate which maximizes the metapopulation carrying capacity (Roff, 1975; Hanski *et al.*, 1994). This OS does not give any indication about the direction of evolution. However, knowledge of the optimum value of the migration rate could be of interest in management programs, if one could manipulate the level of exchanges between populations or the number of immigrants in newly founded populations. Several authors (Comins *et al.*, 1980; Motro, 1982) have shown that the OS is indeed different from the ESS. In the present model, we found that the OS was always larger than the ESS. Figure 4 shows two examples of the relation between migration rate and metapopulation equilibrium density; the ESS are indicated by the arrows. The ES migration rate always lies within the range that permits persistence: a strategy leading to metapopulation extinction cannot be evolutionarily stable.

Declines in the frequencies of migration enhancers in older populations may reduce the migration rate below the level considered optimal at the among-population level. Such within-population effect does not, however, depress the migration rate to the point leading to metapopulation extinction, because the "nonviable" genotypes disappear first, at least in our deterministic model. In a very different theoretical setting, Iwasa and Roughgarden (1986) were able to produce a model in which a species which could not invade an empty landscape could nonetheless invade if another species was present, and in some cases both species went subsequently extinct. Hanski and Gyllenberg (1993) and Hanski and Zhang (1993) have discussed the processes by which species can bifurcate into either a core or satellite distribution, occupying a large or small fraction of the suitable habitat, respectively.

An OS would be to retain just enough offspring in the local population to ensure maximum local population size, and to export all remaining propagules in search of other patches. Such a strategy is never evolutionarily stable, however, because "cheater" genotypes that would keep their offspring at home would increase in frequency within populations. The situation might be even worse if migration behavior is determined by the genotype of the offspring: as suggested by den Boer (1990), local selection will often favor the individuals which stay in the natal population, irrespective of whether this increases the chances of survival of the species (see also Motro, 1983; Roitberg and Mangel, 1993, for parent–offspring conflict in the evolution of migration). At the other extreme, genotypes that would never export offspring could never become established in other patches and would disappear following the extinction of their local population. Our results that the OS is above the ESS is in agreement with the numerical results of Roff (1975) and the analytical results of Comins *et al.* (1980). In contrast, the ES rate exceeds the optimal rate in the Hamilton and May (1977) sib competition model. In their model, each patch can support a single adult, with competition for the patch occurring among the nonmigrating offspring of the resident and the migrating offspring from other patches. Migrating offspring have a constant probability of survival (p) during migration. In their first model, Hamilton and May assumed that all patches were occupied. The ES migration rate was then equal to $1/(2-p)$. In their second model, they further assumed that some patches might remain vacant, because of demographic stochasticity. They then found that the ES migration rate was higher than in the first model, and they described the "optimum" migration rate as the one which maximized the proportion of patches occupied. They found that the optimal migration rate was lower than the ES migration rate. Hamilton and May (1977) suggested that Roff's (1975) assumption of several adults per patch increased selection favoring those remaining in the natal patch, causing the ES migration rate to fall below the optimal rate. Additionally, the possibility of environmentally caused local extinctions in the Comins *et al.* but not in the Hamilton and May sib competition models may promote migration as a mechanism for increasing total population size, causing the optimal migration rate to rise above the ES rate (Comins *et al.*, 1980, p. 214). In fact, if there are only a few individuals per patch on average and local extinctions are due to demographic stochasticity only, the optimal migration rate is always lower than the ES migration rate. In contrast, if local extinctions are mainly due to environmental stochasticity, the optimum migration rate is higher than the ES migration rate, whatever the number of individuals per patch.

There is not always an ES migration rate and an optimal rate to compare. First, an optimal migration rate exists only if there are local extinctions, otherwise the number of individuals in the metapopulation is constant. Second, as shown by Frank (1986), selection depends on the among-population variance in gene frequency. An ES migration rate thus exists only if there is some gene frequency variance among populations. If there is no mortality during migration, many individuals per patch (no demographic stochasticity), and no environmentally caused local extinctions, gene frequencies quickly homogenize across local populations in a constant environment, and hence neutral migration polymorphism is always maintained and there is no ES migration rate. The analytical expression for the ES migration rate under such assumptions is indeterminate (see, for instance, Comins *et al.*, 1980, p. 213, or Eqs. (B6)–(B8) with $q = 1$ in Olivieri *et al.*, 1995). If, however, there is some mortality during migration but no extinctions, migration is always selected against, and the ES migration rate is zero (Comins *et al.*, 1980, p. 213). With local extinctions due to environmental stochasticity, there is always some variance of gene frequencies among populations, and hence there is evolution of gene frequencies. In summary, an optimal migration rate exists only if there are some kind of local extinction. An ESS exists if there is some cost to migration, or if there are local extinctions, or both. In the former case, the ES migration rate is zero.

The fact that the metapopulation may not be viable for too low or too high migration rates has interesting consequences for population genetic models of subdivided populations with local extinctions (with rate e). These models usually assume recolonization rates of 1 or $1 - e$ (Whitlock and McCauley, 1990), that local carrying capacities are reached in a single generation, and very low effective migration rates between extant populations. In other words, these models assume that a very small number of emigrants is able to recolonize any number of patches. This is clearly impossible as long as one does not assume either a very efficient habitat choice by migrants, so that most of them arrive in unoccupied patches, or a very high fecundity together with a very low survival during migration, so that migrants have no chance to establish themselves in occupied patches, but saturate newly colonized patches in a single generation. In the model presented in this chapter, we have assumed that colonizers are simply migrants which, by chance, arrived in an empty patch. The number of recolonizers is then a function of the migration rate, whereas the rate of recolonisation is equal to A_0 as soon as the number of recolonizers is positive (viable metapopulation).

The result that the optimum migration rate is not equal to the ES migration rate, could possibly be used in the management of those species whose migration rate can be manipulated: since the optimum is usually higher than the ESS, we might enhance the conservation of a species by increasing the number of colonizers and migrants. Actually, an artificially increased migration rate (as long as it remains reasonnable) will probably be beneficial for the species in most cases, not only for the reason just given, but also because enhanced migration decreases inbreeding, a fact which might be desirable from a genetic point of view.

E. A Mixture of Two Landscapes, Each Sustaining a Viable Metapopulation, May Cause Metapopulation Extinction When Connected to Each Other

Until now, we have defined a landscape as a set of patches sharing common transition probabilities and demographic properties. Many landscapes, however, include different kinds of components, for instance rocky versus nonrocky patches. Landscape heterogeneity has been considered by several authors. For instance, Cohen and Levin (1991) have shown that spatial heterogeneity have some features in common with positive temporal correlations; they found that the optimal dispersal rate was in general decreased as compared to a homogeneous environement. We consider here a quite different model, in which we find that the ES dispersal rate may increase or decrease with increased landscape heterogeneity. Assume for instance a landscape made of the two sublandscapes described in Table I, with a proportion α of landscape 2 and a proportion $(1 - \alpha)$ of landscape 1. Figure 5 describes the metapopulation equilibrium size as a function of the migration rate within landscapes with α varying from 0 to 1. It can be seen that for some values of α, the metapopulation is not viable whatever the migration rate and that a low proportion of landscape 1 can have tremendous consequences for metapopulation viability. The sudden decline of the metapop-

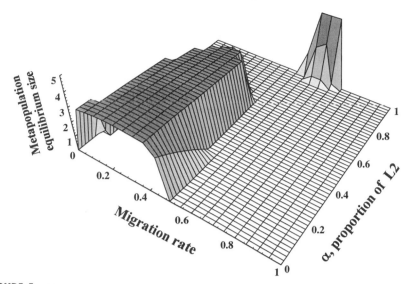

FIGURE 5   Metapopulation equilibrium size, in a landscape made of a sublandscape L2 (fraction α) and of sublandscape L1 (fraction $1 - \alpha$) (see Table I), as a function of migration rate and α. Fecundity was 5 in L1 and 7.5 in L2. (size is measured as in Fig. 4).

ulation size at the limits of the interval is a striking result which does not seem easy to explain intuitively.

Figure 6 describes the same pattern, for a different species or a different landscape, in which fecundity is slightly larger. Note that the sublandscape 2 (L2) is more favorable than the sublandscape 1 (L1), since metapopulation density is greater in L2 than in L1. However, when the two landscapes are mixed, L1 seems to influence more strongly than L2 the range of adapted migration rates that allow the persistence of the metapopulation. Implication of this result is that one should be cautious in interpreting source–sink systems. In particular, when the evolution of the migration rate of the entire metapopulation leads to a rate which is out of the range necessary for the existence in one of the sublandscapes, the set of populations in this sublandscape behaves like a sink. This does not mean that, were this sublandscape alone, it could not be favorable for the species, provided another migration rate was allowed to evolve. Evolution may not always be driven by adaptation to the potentially most productive environment.

The results shown in Figs. 5 and 6 have important consequences. Although a species may be able to adapt to two different landscapes, it may not be able to persist in a mixture of these two landscapes. This occurs when the ranges of migration rates allowing metapopulation persistence in each landscape do not overlap. In terms of land management, it follows that a change in part of a landscape (for example if a given proportion of landscape 1 is transformed into landscape 2) may lead to metapopulation extinction. This extinction does not result

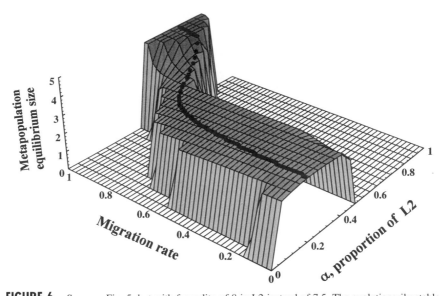

FIGURE 6   Same as Fig. 5, but with fecundity of 8 in L2 instead of 7.5. The evolutionarily stable migration rate is indicated as well.

from either a decrease in the availability of suitable patches or from an increase in the fragmentation of the landscape. The extinction is caused by there being no longer sufficient migration allowing the persistence of the metapopulation. None of the two landscapes is unsuitable, each of them being able to sustain a viable metapopulation; but the mixture is unsuitable. To clarify such situations further, let us imagine that a landscape starts with $\alpha = 0$ in Fig. 5, and imagine that the proportion of L2 is increased. For instance, assume that fertilizers are applied to some patches. As a consequence, fecundity in these patches is increased, but also the successional processes becomes faster, thereby decreasing the population life-span (z). We may then observe that the species disappears because of increased interspecific competition (lower z). In contrast, consider another landscape in which all patches receive fertilizers ($\alpha = 1$). Assume that some farmers decide to stop applying fertilizers to their soil. The species may go extinct, and we will think that this is because the environment has become poorer (fecundity has decreased). The explanation is proximally correct, but it accounts a part of the truth only. Increased competition in a rich environment may not be lethal for a species which can invade numerous new patches while being eliminated from old ones. Decreased fecundity may not be lethal if the species can stay longer in each patch. However, for all these conditions, there may or may not exist a migration rate allowing the metapopulation to persist. When such migration rates exist, the metapopulation may or may not persist, depending on its ability to adjust the migration rate fast enough. Figure 6 also gives the ES migration rate as a function of α, the proportion of landscape L2 in the mixture. It is striking that, for certain critical values of α, very small changes in α provoke large changes in the ES migration rate. There are thus certain mixtures of different kinds of habitats which are critical for the ability of the metapopulation to adapt its migration rate, given the available additive variance that might occur for this character.

The evolution of migration rate in particular landscapes is thus a critical characteristic of a metapopulation determining persistence or extinction. Extinction may result from landscape changes which lead to situations in which no migration rate allows persistence, or when changes in landscape structure are so fast that the available genetic variability does not allow adaptation. Moreover, as pointed out by Gomulkiewicz and Holt (1995), when a metapopulation adapts to a novel environment, its density may fall below a critically low level for a period of time during which the metapopulation is highly vulnerable to extinction.

Metapopulation adaptation can result in constant evolution toward new equilibrium values, which are adapted to (although not optimal in) novel landscapes. When a landscape is changing, conservation decisions on species management should depend on whether those new values are attainable or not. If they are not, management decisions should be mainly concerned with the landscape itself. If they are, it will not be sufficient, or even desirable, to reinforce or reintroduce populations without taking into consideration the consequences of such actions on the evolution of the species. For instance, let us imagine a plant species which is going extinct because its migration rate is too low with respect to increased habitat fragmentation. Conservation biologists could try to prevent extinction of such a species by reintroduction. For this purpose, it will be necessary to collect seeds and reproduce them. Care should be given not to favor the less dispersed seeds, which are possibly easier to collect: this would create a selection pressure against migration, and thus decrease metapopulation adaptation. If this point was not taken into account, the species could gain new individuals at the expense of its ability to maintain itself without human help. Den Boer (1990) suggested that the SLOSS discussion in Wilcox and Murphy (1985) on the design of nature reserves passed over the essential point: the differences in migration rate between species. We may go further and suggest that differences in migration rate between genotypes within a species may be a critical consideration in some conservation tasks.

V. THE METAPOPULATION EFFECT: OTHER TRAITS

We have called the antagonistic two-level selection observed on migration behavior as the *metapopulation effect*. We will now show that this effect can be observed for other traits as well, for instance life-history traits and properties of the genetic system.

A. Evolution of Dormancy and Diapause

Ellner and Shmida (1981) suggested that the desert plants of Israel have evolved a variety of migration-restricting seed-containers that protect the seed from predation and flooding, regulate the within-season timing of germination, and spread germination over several years. Yeaton and Bond (1991) suggested that differences in both migration in space and dormancy in seed bank could promote coexistence of two shrub species in a habitat disturbed by fire (see also Ellner, 1987; Shmida and Ellner, 1994). Recently, Tsuji *et al.* (1994) have suggested, using an ESS model, that a trade-off between dormancy and migration may evolve even in stable environments. Migration in space is one way of escaping local adversity. Dormancy (in plants) and diapause (in animals), by allowing escape in time, represent other ways of achieving a comparable result (Hairston, 1993; Ellner and Hairston, 1994). Models (Venable and Brown, 1988; Levin *et al.,* 1984; Cohen and Levin, 1991; I. Olivieri, unpublished) as well as observations (Venable and Lawlor, 1980; Olivieri *et al.,* 1983) demonstrate the antagonism between the two strategies, which are often two kinds of responses to the same sort of environmental variation. All models assuming only unconditional strategies predict that dispersed seeds should not be dormant. Nondispersed seeds might be dormant or nondormant.

The metapopulation effect described in the evolution of migration rate is also operating in the evolution of dormancy. Consider two competing strategies, with different dormancy rates of nondispersed seeds. Within a patch, those seeds which germinate will obviously be overrepresented in the adult population. Thus the gene frequency of an enhancer of dormancy decreases through time in the adult population. However, when conditions are such that a stable polymorphism is maintained between two germination strategies, we observe (unpublished simulation results) that the frequency of the more dormant genotype increases through time in the seed bank. In the absence of disturbance, this type will have little success since nondormant genotypes constantly produce seeds which occupy the available space. Following a complete disturbance in a given patch, no offspring are produced in the patch, and recolonization occurs either through migration or through germination from the seed bank. The dormant strategy is then overrepresented at recolonization. Just like with migration, dormancy is favored at the metapopulation level, while selected against at the population level. Although this process clearly occurs at the patch level, so that one actually need not consider several patches to study the evolution of dormancy, it would be difficult, from a genetic point of view, to consider the new population as the same as the one previously occupying the patch. We thus still consider that the process occurs at the metapopulation level, even though the metapopulation may occur in a single patch, successively occupied by different populations.

B. Evolution of Life Histories

Classical models of life-history evolution usually consider the asymptotic population growth rate as a measure of fitness in species with age-structure (Charlesworth, 1994b). This is reasonable only in the absence of frequency dependency

(Kawecki, 1993) and provided that populations ever reach this growth rate. In a metapopulation, new populations are regularly established and old ones go extinct. It is unlikely, in such a system, that the asymptotic growth rate calculated for a population could be of any relevance. Kawecki (1993) has shown that in a patchy environment, where individuals compete for resources within patches, increasing mortality always favors decreased allocation to late survival, if mortality is caused primarily by destruction of entire patches. This means that an increase in local extinction rate decreases evolutionarily stable adult survival. At equilibrium, the realized asymptotic growth rate of the metapopulation is one, and the metapopulation reaches a dynamic equilibrium, in which populations that have not yet reached the carrying capacity increase in size. In our particular model of local density dependence, while all populations undergo similar phases (density-independent growth followed by density-dependent regulation unless extinction occurs), these phases are not synchronized. At equilibrium, a stable age distribution is attained only at the metapopulation level, not necessarily at the population level. An example is given in Fig. 7. All populations of a given age i have the same age-class distribution of individuals, but the age-class distribution of a given population may vary between years.

Let us then ask the question of whether stable polymorphisms of genotypes with different life histories may be maintained, and what is the influence of population lifespan, fecundity, and local extinction rates on the evolution of life histories. For instance, compare three genotypes with similar migration rates of 0.5. Assume no senescence. Let genotype P be a typical perennial *(iteroparous)* genotype, with low annual net fecundity (four viable seeds per adult) and high adult survival rate (0.9). Let genotype I be intermediate, with net fecundity of 8,

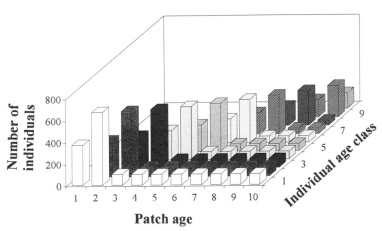

FIGURE 7 Age structure of individuals in each population age class at metapopulation equilibrium. The model was iterated with disturbance rates of landscape L2 (Table I) and population lifespan of $z = 10$, migration rate of 0.5, constant adult survival rate of 0.9, and annual fecundity of 2.

and adult survival rate of 0.45. Finally, let genotype A be annual (semelparous), with net fecundity of 12. The results are given Table II.

The results show that the longer the population lifespan, the more the perennial genotypes are selected for.

By dividing all fecundities by half (P, 2; I, 4; A, 6), interesting differences emerge (Table II). Notice that a polymorphism between annuals and perennials is possible (here for $z = 11$), indicating that disruptive selection on life histories may occur. When z is small, the annual genotype is favored. When z is high, the perennial type is favored. When z is intermediate, the intermediate I may be able to replace one or the other genotype in pairwise contests (here the intermediate wins against the annual for $z = 11$) and still be eliminated by natural selection when both other types are present. This example illustrates that dynamics of game theory models with more than two strategies may be complex (Maynard-Smith, 1982). Such disruptive selection occurs here because the empty patches are faster recolonized by the annual type, whereas the more perennial type is selected for in old populations; in both niches, the intermediate type does less well than the best adapted genotype. In this particular case, the dynamics are highly dependent on initial conditions, even though the only evolutionarily stable state is the occurrence of both annual and perennial types in frequencies of ca $2:3$. This is illustrated in Fig. 8, which shows that when the simulation is initiated with 100 seeds of each genotype in each empty patch, the annual genotype almost goes extinct because of competition with the intermediate genotype I, before I starts to decline because of competition with the perennial, allowing the annual type to increase again. The metapopulation processes may thus lead to particular and

TABLE II   Invading Genotype in Contests Involving Either the Three Types, A, I, P, or Pairs of Them, as a Function of Population Lifespan, z^a

| | Genotypes intially present | | | | | | | |
| | A, I, P | | I, P | | A, P | | A, I | |
z	1^b	$1/2^c$	1	1/2	1	1/2	1	1/2
7	A	A	I	I	A	A	A	A
8	I	A	I	I	A	A	I	A
9	I	I	I	I	A	A	I	I
10	P	I	P	I	P	A	I	I
11	P	A+P	P	P	P	A+P	I	I
12	P	P	P	P	P	P	I	I

^{a} A is annual, P is a perennial, I is intermediate.
^{b} Fec $= 1$; see text for fecundity and survival of each genotype.
^{c} Fec $= 1/2$; same as Fec $= 1$, but all fecundities were divided by 2.

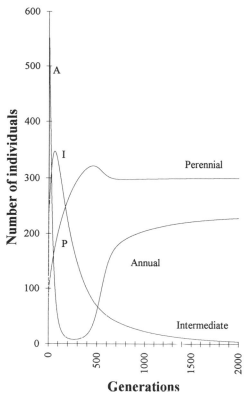

FIGURE 8   Evolution of the number of individuals of three genotypes with variable life-histories. The y-axis gives the numbers per patch of age class 1 (newly founded populations). Genotype A is annual with fecundity of 6; genotype P is perennial with fecundity of 2 and adult survival rate of 0.9; genotype I is intermediate with fecundity of 4 and adult survival rate of 0.45. All three genotypes have the same migration rate of 0.5. Landscape L2 (Table I) with $z = 11$ was assumed. In this example, the simulation was started with 100 seeds of each genotype in each patch. The evolutionarily stable state $(A + P)$ was insensitive to initial conditions.

unexpected dynamics of gene frequencies. In a finite metapopulation, it is likely that the annual genotype would go extinct before its frequency starts to increase. In this example, decreasing fecundity selects for increasing reproductive effort (annual life cycles). This is apparently because the colonization of new patches becomes the limiting factor.

When a stable polymorphism between annual and perennial types can be maintained, the frequency of the annual type is high in young populations, whereas the frequency of the perennial type increases after colonization (Fig. 9). Within populations, selection favors perennials, while between populations annuals are selected for. The two-level selection process is exactly the same as in

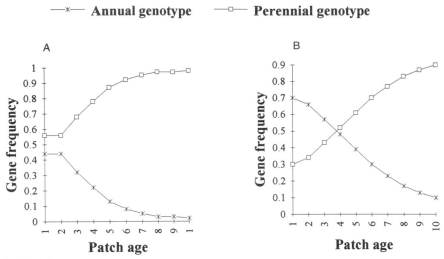

FIGURE 9 Equilibrium gene frequencies in the metapopulation as a function of population age. Landscape L2 (Table I) was assumed with $z = 11$. The perennial genotype has a fecundity of 2 and a survival rate of 0.9. (a) Net fecundity of the annual genotype is 6, as in Fig. 8. (b) Net fecundity of the annual genotype is 6.3.

the case of migration. This prediction is consistent with abundant observations about plant life histories along ecological successions, both within or among species.

In our model, we could not maintain a polymorphism in a single isolated population. Crawley and May (1987) showed that within-population polymorphisms of annuals reproducing by seeds and perennials spreading by vegetative reproduction could be maintained if ramets had a competitive advantage over seedlings. In our model, perennials have a competitive advantage over annuals (once a patch has reached its carrying capacity, seeds may germinate only after death of adult plants), but both annuals and perennials may reproduce through seeds only.

Our results are consistent with interspecific studies of life-history changes with plant successional stage. The work by Huston and Smith (1987), Tilman (1990), McCook (1994), and Huston (1994), suggest that "facility" and "inhibition" effects of species on each other, leading to sequential replacement of species, are best explained in terms of correlations of life-history traits. In insects, Brown (1985) found that along a plant succession, in which annuals were progressively replaced by perennials, a decreasing fraction of species of insects had more than one generation per year. Bengtsson and Baur (1993), in contrast, found that pioneer species of terrestrial gastropods had the same average longevity, age at first reproduction, and clutch size as nonpioneer species.

Another result of our model (not shown) is that perennials are selected against when disturbance rate increased. Although this result is intuitive, it is not always verified experimentally. McLendon and Redente (1990), in an experiment designed to study the influence of disturbance on plant communities, found that, as expected, perennial species dominated under low disturbance regimes, whereas annuals dominated under high disturbance rates. The experimental study of Fahrig *et al.* (1994) showed, however, that at very high disturbance rates, perennials with large migration rates were favored over annuals. We will discuss these results in Section VI.

C. Coevolution

1. Interacting Species

Other situations where the metapopulation effect may be expected include those in which two replicating entities (a host and its pathogen, or two symbionts) interact. *A priori,* one could expect that a host–pathogen interaction will be highly influenced by the processes of extinction–recolonization (Frank, this volume). In particular, if the host possesses both general resistance (polygenic, "horizontal" resistance) and specific resistance (oligogenic, "vertical"), the former will be very advantageous at the time of colonization, when the host colonizes a patch with uncorrelated pathogens. The latter resistance type will become progressively favored as a given patch grows older. Within each patch, the local pathogens create a selection pressure for the specific resistance genes, while colonization selects for general resistance. The same kinds of predictions may be made about predators and parasites, namely that generalist predators and parasites should be found in young host populations, whereas old host populations should harbor more specialized natural enemies. In support of this prediction, Brown (1985) showed that phytophagous Heteroptera of the young stages of the succession were more generalists than those of late successional stages.

Metapopulation structure appears to promote the coexistence of hosts and pathogens, as well as genetic variability in host–pathogen systems with gene-for-gene interactions, as suggested by Burdon and Jarosz (1992), Frank (1991b, this volume), Antonovics (1994), Antonovics and Thrall (1994), and Antonovics *et al.* (1994).

2. Interacting Genomes and Reproductive Systems

"Interactive genomes" can be different compartments of the same genome. We do not know any demonstration of the dynamics of transposons or other repeated sequences in a metapopulation, but one case of intragenomic conflict highly influenced by extinction–recolonization processes is the case of nucleo-cytoplasmic conflict in the determination of male sterility in plants. The most extensive study of this process has been made in thyme *Thymus vulgaris* and summarized in Gouyon and Couvet (1985, 1987) and Atlan *et al.* (1990). In this perennial species, female (male-sterile) individuals coexist with hermaphroditic (male-fertile) ones. The females produce more seeds than hermaphrodites, as demonstrated by various authors. Darwin (1877) found a 70% increase, Assouad *et al.* (1978) found a 200% increase. Consequently, as the fitness of cytoplasmic genes (maternally inherited) causing male sterility is directly related to seed production, the frequency of these genes increases. When there are many females, nuclear genes of hermaphrodites reproduce much better than those of females (the former reproduce through their own ovules, plus through pollen in seeds produced by hermaphrodites and by females). In thyme, the proportion of females can reach very high values locally, more than 90%. In such situations, nuclear genes able to produce hermaphrodites are highly selected for. In thyme, there exists a number of mitochondrial strains which produce male-sterility, and a number of nuclear genes which specifically restore male fertility (Belhassen *et al.*, 1991, 1993). Considered at this level, the interaction is formally equivalent to a host–pathogen interaction (see also Frank, this volume). However, in the case of male sterility, there is probably no generalist restorer: each cytoplasmic type can be restored by specific restorers only.

A metapopulation of thyme undergoes a series of events that can be described as follows. At the establishment of a new population, only few individuals are present in a patch. They mate mainly with the nearest neighbor (Couvet *et al.*, 1986). Consequently two situations may arise. (i) If one of the individuals is a female and its nearest hermaphrodite does not possess the specific nuclear restorer genes, this female will produce purely female progeny. This will create a "colony" (spatial unit) composed of females exclusively. Such "colonies" were observed in natural populations (Dommée and Jacquard, 1985). Mated with the same hermaphrodite, all these females will keep producing pure female progenies. Due to the high seed production of such females, these "colonies" will be very efficient colonizers. (ii) If, in contrast, the few cytoplasmic types locally find their specific restorer genes, hermaphroditic progeny will be produced. Since hermaphrodites produce fewer seeds than females, "colonies" of this kind colonize less efficiently the initially empty patch. Molecular markers have allowed a demonstration of this process.

After disturbance in a patch, recolonization thus occurs from large colonies of females and small colonies of hermaphrodites. Very high proportions of females are indeed found in most recently colonized patches (Belhassen *et al.*, 1987). Subsequent evolution of the population is much slower, because once the patch has been invaded, young individuals may establish only after the old ones have died. As females become more numerous in a female "colony," the probability that some of them are fertilized by pollen grains bearing the restorer gene corresponding to their cytoplasmic type increases. As the population grows older, these nuclear restorer genes invade it, and the proportion of hermaphrodites increases (Belhassen *et al.*, 1990). This metapopulation effect in this case results

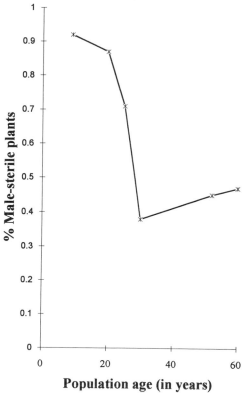

FIGURE 10 Proportion of male-sterile plants as a function of population age in the gynodioecious species *Thymus vulgaris* L. (modified from Dommée *et al.*, 1983).

in the evolution of female proportion as shown in Fig. 10. This process explains why, in disturbed habitats, the proportion of females is higher than in stable habitats (Gouyon *et al.*, 1983) and why the average proportion of females can be as high as 60% in the region of Montpellier, southern France (Belhassen *et al.*, 1991).

This example provides us with a relatively complete picture of the complexity of the metapopulation effect. There are actually processes at three levels (Couvet *et al.*, 1985): selection within a colony (favoring nuclear restorer genes), among colonies within populations (favoring cytoplasmic genes determining male sterility as well as nuclear nonrestorer genes; Denis Couvet, personal communication), and among populations, as extinction and recolonization processes generate local founder effects. In this sense, this process is related to the shifting balance process (Barton and Whitlock, this volume; Michalakis and Olivieri, 1993).

VI. INFLUENCE OF MIGRATION ON THE EVOLUTION OF LIFE-HISTORY TRAITS

In Section IV.A, we showed that in some cases, increasing the fecundity of each genotype by a constant factor had consequences for selection on adult survival. We also found that the ES migration rate increased with fecundity and adult survival. This agrees with classical predictions and observations. One could ask further whether migration might influence selection on life histories. To answer this question, we ran pairwise simulation contests between annuals and perennials as described in Section V.B, for various values of the migration rate. Results are shown Table III.

It is clear that the larger the migration rate, the more strongly the perennials are selected for. This is contrary to the classical "colonizer syndrome" (Baker and Stebbins, 1965). Recently, Fahrig *et al.* (1994) also showed, using field observations and simulations, that in highly disturbed habitats, perennials tended to dominate over annuals. Using an ecological (optimization, not ESS) model, in

TABLE III   Winners of Pairwise Contests between an Annual and a Perennial, as a Function of Population Lifespan and Migration Rate[a]

Population lifespan z	Migration rate d	Winners of pairwise contests between annuals and perennials
<9	0.20	Annual
	0.50	Annual
	0.70	Annual
9	0.20	Annual
	0.50	Annual
	0.70	Polymorphism
10	0.20	Annual
	0.50	Annual
	0.70	Perennial
11	0.20	Annual
	0.50	Polymorphism
	0.70	Perennial
12–13	0.20	Annual
	0.50	Perennial
	0.70	Perennial
14–22	0.20	Polymorphism
	0.50	Perennial
	0.70	Perennial
>22	0.20	Perennial
	0.50	Perennial
	0.50	Perennial

[a] An annual with fecundity of 6 and a perennial with constant annual fecundity of 2 and survival rate of 0.9 were assumed in landscape L2 ($A_0 = 0.4$; $A_1 = 0.9$, $A_2 = 0.95$).

which they compared the equilibrium density of monomorphic populations of either annuals or perennials, they suggested that "long-distance clonal spreading" of herbaceous perennials might explain their finding, especially if seeds were more sensitive to disturbances than stems. Our own simulation results point to the same direction, though we provide a different explanation. In highly disturbed habitats, high migration rates are expected to evolve. In such habitats, only species with high fecundity can sustain a viable metapopulation. We showed in Section III that perennial life histories are favored when fecundity and migration rates are high. For certain landscapes, migration rates and fecundities may be such that perennials are favored over annuals when disturbance rates increase, as found by Fahrig *et al.* (1994).

Ben-Shlomo *et al.* (1991) conducted a selection experiment on the migration ability of the flour beetle *Tribolium castaneum,* and they found that correlated responses to selection included shorter generation times. Roff (1986) has shown that there are fitness costs associated with the ability to disperse. These correlations are likely to affect the evolution of migration behavior, as formally shown by Cohen and Motro (1989). Den Boer (1990), in contrast, found that dispersing individuals of some arthropod species had a higher egg production than nondispersing individuals. A review of studies of genetic correlations among migration characters and other fitness components in insects may be found in Roderick and Caldwell (1992; see also Roff, 1990). The idea that life-history characters do not evolve independently is of course not new.

VII. CONCLUSION

We have shown in this chapter that the processes determining the migration rate in a metapopulation are specific to the very functioning of the metapopulation (the metapopulation effect) and interacting with the processes determining the evolution of most significant life-history traits. These processes result in a partial adaptation of the metapopulation to its landscape. This adaptation is incomplete because the processes involved act between genes at both the population and the metapopulation levels. They thus necessarily involve frequency dependence and therefore Fisher's fundamental theorem of natural selection does not apply at the metapopulation level.

One of the difficulties involved in the study of metapopulation evolutionary processes is the general confusion that characterizes the questions of levels of selection. The controversy between group selectionists (*sensu* Wynne-Edwards, 1971) and individual selectionists (e.g., Williams, 1971) has led to radical positions and made the whole point difficult to sort out. Roughly, one point of view is the one developed during the 1970s, which states that group selection acts necessarily against individual selection. A more recent point of view defends the idea that all levels can be taken into account, whether they act in different or similar directions. This latter point of view seems much more promising and tractable than any other (Lloyd, 1994). It can be formalized in diverse ways, the best known being the one developed by Price (1970), extended by Wade (1985), and applied by Frank (1986, 1987, 1994b) and Goodnight *et al.* (1992) to, for instance, migration, sex-ratio, and altruism [see also Heisler and Damuth (1987) and Damuth and Heisler (1988), for extensions to multiple characters evolution and further discussion]. It consists of a nested decomposition of forces acting on allelic frequencies, with one component being assigned to each level at which selection is supposed to act. The levels may be within individual (e.g., repeated sequences), within colony (or deme), within population, among populations, etc.

If a DNA sequence that can be repeated in the genome of an individual acts positively or negatively on its fitness, nobody will deny that the study of the evolution of this sequence has to take into account both the within- and the among-individuals component of selection to understand its fate. Oddly enough, it seems much more difficult for evolutionary biologists to jump one more level. If a gene can be selected within a population *and* influence the probability that a new patch is colonized, then, its fate can be analyzed in terms of two-level selection. Whether this is done explicitly or not in the models is just a question of words.

Unfortunately, the confusion between this hierarchical approach to levels of selection and the naive Panglossian group selection postulates still survives. In their very clear review about the evolution of migration, Johnson and Gaines (1990) still mix in their "group selection models" section a paper by Wynne-Edwards (1962) with a model by Van Valen (1971). This last model explicitly distinguished the individual and demic levels of selection, while others (e.g., Comins *et al.,* 1980; Levin *et al.,* 1984), who have treated these levels implicitly, are classified in the "individual selection" section. The fact that these models do not differ in their assumptions but in their wording is made clear by the fact that they produce the same results! (Assuming survival rate during migration is very low, all these models find that the ES migration rate is equal to the local extinction rate.) The same confusion can be found in a recent review about metapopulation dynamics and genetics (Hastings and Harrison, 1994), where consequences of the metapopulation effect described by Rice and Jain (1985), Olivieri *et al.* (1990), and Manicacci *et al.* (1992) are treated as interdemic selection, leading the authors to conclude that "however, this is open to explanation in terms of individual selection." The confusion, once again, comes from the lack of distinction between the entity that is selected (the replicator, Dawkins, 1976; or the information, Gouyon and Gliddon, 1988; Gliddon and Gouyon, 1989) and the level of integration at which differences in reproduction exist (the interactor, Dawkins, 1976; or avatar, Gouyon and Gliddon, 1988; Gliddon and Gouyon, 1989). All formalized models of migration assume that the selected entity is the genetic information. Sometimes the two levels (within and among populations) are explicitly treated and sometimes they are not. Both levels are nonetheless always included in the analyses, leading to convergent analytical and simulation results.

ACKNOWLEDGMENTS

Sandrine Maurice, Yvain Dubois, Stuart Baird, Bob Holt, Bernard Godelle, and Stéphanie Brachet made useful comments on a final version of this chapter. Susan Mazer, Yannis Michalakis, Ilkka Hanski, Simon Levin, and an anonymous reviewer raised very interesting points on the submitted version. Anne-Marie Duffour helped with the documentation, and Stéphanie Brachet was very helpful in drawing 3-D figures and Fig. 1 was drawn by Jean-Yves Pontallies. The present version benefited from the considerable help of Ilkka Hanski, who made extensive (useful!) changes on the paper. This is publication ISEM95-097 of the Institut des Sciences de l'Evolution, Montpellier.

14 Spatial Processes in Host–Parasite Genetics

Steven A. Frank

I. INTRODUCTION

Host–parasite diversity can be described in two different ways. The first is simply the observed variability among the hosts and parasites in a particular population. For example, Burdon and Jarosz (1991) classified 67 wild flax plants into 10 distinct resistance genotypes when tested against six races of flax rust. One host genotype was completely resistant to all six pathogen races, whereas another genotype was susceptible to five of six races.

The second type of variability is the range of potential genotypes that can occur over space and time. For example, Parker (1985) used field transplant experiments to study the legume *Amphicarpaea bracteata* and its fungal pathogen *Synchytrium decipiens*. Fungal infection was heavy in each of three locations. However, a plant moved to a new location developed little or no infection, suggesting that the pathogen populations differ among sites. In a second experiment, host lines derived from different locations varied in their ability to resist a single pathogen isolate, indicating spatial differentiation among the host populations.

Parker's study suggests that the potential range of diversity over space and time is often greater than the variability observed in a single location. The potential diversity is limited by the biochemistry and morphology of host–parasite traits, whereas the observed diversity is controlled by the local dynamics of disease and the global processes of extinction and colonization in the metapopulation.

The first goal of this paper is to suggest that increasing potential diversity causes a qualitative shift in metapopulation dynamics. Local processes dominate when potential diversity is low. Colonization–extinction dynamics of alleles in the metapopulation become more important with an increase in the potential number of distinct genotypes. In the next section I present a simple model to illustrate the importance of potential diversity.

After briefly discussing the model, I review evidence that many host–parasite systems do in fact have high potential diversity. Examples include plant–pathogen genetics and bacterial defense systems against viral parasites and conspecific competitors. I also discuss the antagonistic interaction between cytoplasmic and nuclear genes in cytoplasmic male sterility.

Data from these studies suggest that spatial variation and colonization–extinction dynamics are important in the observed patterns of diversity. However, the data are difficult to interpret because of limited sampling over space and time. This difficulty leads to my second goal: the emphasis of space–time scaling when interpreting host–parasite diversity. Spatial scales that are small relative to migration distance have well-mixed populations dominated by local interactions. Local processes also dominate on temporal scales that are short relative to the expected times to extinction and recolonization of genotypes. By contrast, observations aggregated over long spatial and temporal scales may obscure colonizations, extinctions, and rapid changes in genetic composition that occur on finer scales. Thus the patterns of observed variability are strongly influenced by the space–time scaling of colonizations and extinctions in the metapopulation.

II. DIMENSIONALITY AND COLONIZATION–EXTINCTION DYNAMICS

In this section I describe more precisely the relationship between potential variation and observed diversity. I define the potential number of genotypes as the *dimensionality* of the system. I begin with a verbal illustration of the link between dimensionality and colonization–extinction dynamics. I then turn to a simple model.

A. Verbal Description

The observed diversity of host–parasite genetics depends on the range of possible variants and the processes that govern local extinction or success of each genotype. For example, suppose that the host has just two alternative genotypes, h_1 and h_2, and the parasite has two genotypes, p_1 and p_2. The host h_1 can recognize and resist the matching parasite, p_1, but h_1 is susceptible to p_2. Likewise, h_2 can resist p_2 but is susceptible to p_1. In this case strong frequency dependence will favor rare genotypes, and genotype frequencies will fluctuate around 0.5. Thus diversity is controlled by the local dynamics of frequency dependence.

Now consider the same pattern of host–parasite interaction but with more genotypes. In particular, each of the n host genotypes $h_1 \ldots h_n$ matches the single corresponding parasite genotype from the set of $p_1 \ldots p_n$. Thus h_1 resists p_1 but is susceptible to all other parasite genotypes, h_2 resists p_2, so on. The same frequency dependence occurs, favoring equal abundance of all genotypes. However, the frequencies now fluctuate about $1/n$. As the dimensionality n increases, the average frequency declines, and small fluctuations are more likely to cause local extinction of a genotype.

An extinction leads to a sequence of events that changes the local dynamics. For example, suppose that host genotype h_i is locally extinct. Then the matching parasite p_i has an advantage over other parasite types because it can attack all hosts in the local population. The other parasites are resisted by their matching host genotypes. Thus p_i increases and the other parasites decline toward local extinction. The patch is now ripe for recolonization and rapid increase by h_i, which would drive p_i and the other host types toward local extinction. The point is that dynamics are now controlled by the times to extinction and recolonization. The observed variation in a particular population at a particular time will be much lower than the potential diversity.

B. The Model

I now turn to the formal model. The ideas are the same as in the verbal model just given, but the points are made more precisely. Some readers may prefer to skip ahead to the sections on natural history and return later to the details of the model.

I focus on a single-patch model with extrinsic colonizations rather than an explicit, multipatch metapopulation analysis. In the next section I discuss single-patch and multipatch models.

The model has a single haploid locus. Each of the n host alleles causes recognition and resistance to only one of the n parasite alleles. Thus each host is resistant to $1/n$ of the parasite genotypes, and each parasite can attack $(n-1)/n$ of the host genotypes (Frank, 1991a, 1993a). I call this the "matching-allele" model. In a population-genetic context the different alleles constitute a polymorphic locus of a single species. In an ecological context each allele is associated with a different species. I will use the population-genetics language of allelic polymorphism, but an ecological interpretation of species diversity is equivalent for these assumptions.

I use Lotka–Volterra equations to describe the system. These equations show the dynamics of genotype abundances rather than just the relative genotype frequencies. Thus the model tracks epidemic fluctuations in population sizes and disease intensity in addition to changes in genotype frequency. The model is

$$\Delta h_i = h_i[r(1 - H/K) - m(P - p_i)] \ \Delta t$$

$$\Delta p_j = p_j[-s + b(H - h_j)] \ \Delta t. \tag{1}$$

The values of h_i and p_j are the abundances of hosts of genotype i and parasites of genotype j. The total abundance of hosts is $H = \sum_{k=1}^n h_k$, and the total abundance of parasites is $P = \sum_{k=1}^n p_k$.

The term r is the host's intrinsic rate of increase, H/K is the strength of density dependent competition among hosts with carrying capacity of K, m is the morbidity and mortality per parasite attack, s is the parasite death rate, and b is the parasite's intrinsic birth rate per host–parasite contact. The Δt term is the size of the time step over which the interactions occur. For example, Δt may be the length of one host generation or one season in a discrete-time model. When birth, death and disease cause continuous change of the abundances of hosts and parasites, $\Delta t \rightarrow 0$.

The system in Eq. (1) is easier to analyze when rewritten in nondimensional form (Segel, 1972; Murray, 1989). Nondimensional analysis focuses attention on a minimal set of parameters and highlights relative magnitudes (scaling relations) among the processes that drive the dynamics. This is accomplished without altering the dynamics or interpretation because one can translate freely between the biologically motivated formulation and the nondimensional quantities.

The system can be rewritten with the following substitutions

$$\hat{h}_i = h_i/K, \quad \hat{p}_j = mp_j/r, \quad \tau = r \ \Delta t$$

$$\hat{s} = s/r, \quad \hat{b} = Kb/r. \tag{2}$$

Dropping the hats yields the nondimensional system

$$\Delta h_i = h_i[1 - H - (P - p_i)]\tau$$

$$\Delta p_j = p_j[-s + b(H - h_j)]\tau. \tag{3}$$

The dynamics of the system are controlled by the equilibrium with all hosts and parasites present, which occurs at $h^* = s/[b(n - 1)]$ and $p^* = (1 - H^*)/(n - 1)$, where $H^* = nh^*$ and, by the symmetry of the system, $h_i^* = h^*$ and $p_j^* = p^*$ for all i and j. This equilibrium point is unstable when there are discrete time lags in the competitive effects among hosts and in the interactions between host and parasite. This equilibrium is neutrally stable when interactions occur in continuous time ($\tau \rightarrow 0$). A detailed analysis is given in the Appendix of Frank (1993a).

Figure 1 shows the dynamics of this system with two hosts and two parasites ($n = 2$). Each panel shows how one of the two host–parasite pairs changes from an initial condition. In each case the abundances follow a stable limit cycle that

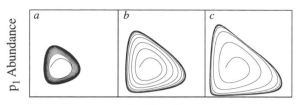

h_1 Abundance

FIGURE 1 Dynamics for the matching-allele model with two hosts and two parasites. (*a,b*) Limit cycles in which abundances fluctuate in a periodic and stable way. (*c*) Spiral from an initial condition out to a limit cycle, where parasite abundances repeatedly drop very close to zero. In this case the parasite is likely to become locally extinct, leading to colonization–extinction dynamics. The panels show the changes in abundance for one of the two host–parasite pairs in Eq. (3), with $b = 1.2$, $s = 0.4$, and $\tau = 0.125, 0.375, 0.625$ for the three panels, with increasing τ moving from left to right.

repeats at regular intervals. These cycles are stable because trajectories away from the cycle spiral toward and then remain on the cycle.

All three panels of Fig. 1 share the same parameters, equilibrium point, and initial conditions except for the size of the time step, τ. Larger time steps destabilize the system. As τ increases from the left to the right panel, the oscillations increase in magnitude. The very low parasite abundances that occur in the right panel suggest that the parasites in that system would be prone to extinction, which would change the subsequent course of the dynamics.

The difference between a repeating cycle and cyclic dynamics prone to extinctions can be seen in the next two figures. Figure 2 shows time-series plots for a model with two hosts and two parasites ($n = 2$). Extinction is simulated by setting to zero any abundance less than 0.01. In this figure abundances never drop that low and extinction never occurs. Colonization is simulated by adding 0.01 to the abundance of each host and parasite if a random number is less than the colonization rate (see figure legend). These colonizations have little effect on the dynamics because the system follows a stable limit cycle.

Figure 3 shows the same system with $n = 4$. An increase in the number of hosts and parasites has two effects on the dynamics. First, larger n lowers the equilibrium abundance of each host and parasite type. A lower equilibrium shifts the entire cycle down and to the left (see Fig. 1). Thus an increase in n shifts the cycle closer to the $p = 0$ and $h = 0$ boundaries.

The shift in the location of the cycle leads to the second effect, a tendency for genotypes to become locally extinct. When a host genotype is lost from the local population, the matching parasite genotype has a fitness advantage because it can attack all local host genotypes. Eventually the locally extinct host is reintroduced and spreads rapidly because it can resist attack by the locally dominant parasite. The spread of the resistant host causes a decline among the host's competitors and an increase among all nonmatching parasite genotypes. These extinctions followed by random immigration into the system cause unpredictable fluctuations in the composition of the four host and parasite genotypes (Fig. 3).

These theoretical examples show the qualitative shift in dynamics caused by colonization–extinction processes. Systems are more prone to extinctions of genotypes when local population sizes are small, the number of genotypes (dimensionality) is high, or nonlinear dynamics cause large, deterministic fluctuations. Colonization by locally novel alleles depends on the frequency of immigration and on the spatial variation in genotypes among populations.

Scale is clearly important. Frequent migration on a particular distance scale leads to high immigration but little differentiation among populations. Very rare migration enhances differentiation but increases the waiting time before locally extinct alleles are reintroduced by immigration. To complete the picture these spatial scalings must be tied to the temporal scales of local dynamics and extinctions (Frank, 1991b).

C. Summary

It may seem rather disappointing to have only the simplest, single-patch Lotka–Volterra model for a metapopulation theory of host–parasite genetics. However, I believe this is the right way to seek theories that apply broadly. A brief justification may be useful before turning to the observations from natural systems.

What would it take to produce a full model of host–parasite genetics within the context of metapopulation dynamics? Since genetics is the question, we need several loci, and several alleles per locus. Natural systems often have this genetic complexity, which may play an important role in determining spatial and temporal dynamics. We must also consider sex and recombination and the interaction be-

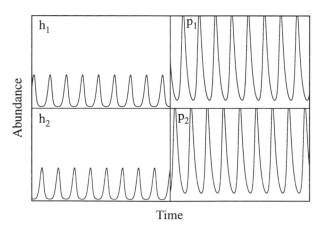

FIGURE 2 Time series for the matching-allele model with two hosts and two parasites, from Eq. (3) with $n = 2$, $b = 2.4$, $s = 0.4$, and $\tau = 0.25$. The dynamics are shown over a time period of 500 steps of length τ. Extinction is simulated by setting to zero any abundance less than 0.01. Colonization is simulated by adding 0.01 to the abundance of each host and parasite in each time step τ if a random number between zero and one is less than 0.01. Thus the average time between colonization events for each type is 100τ.

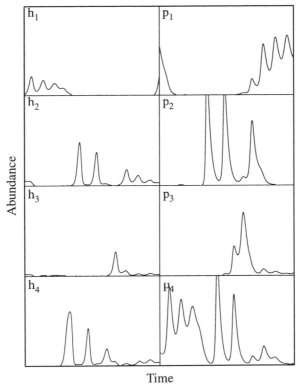

FIGURE 3 Time series for the matching-allele model with four hosts and four parasites. The parameters and methods are the same as in Fig. 2 except that $n = 4$.

tween host and parasite. Mutation is important because rare events can have a large impact on diversity. We now have many parameters, but have not yet specified ecological processes. So we must add in birth and death rates, and explicit descriptions of spatial movement in the metapopulation.

We are ready to see that host–parasite genetics is like the weather. A epidemic arises seemingly without warning in the northwest, caused by a rare migrant parasite genotype that sweeps through the local host population. The patch is ravaged, perhaps extinct or left with only a depauperate set of genotypes and a few individuals. Colonizations occur over time. The new composition is very different from the original composition. And so on over space and time. Dial the migration parameter, and a different but equally beautiful map appears on the computer screen. We have many parameters, each with an effect over some range of the parameter space.

Of course, what we would really like to know about is invariance, regions where changes in a parameter do not matter, and "bifurcation" in the generic sense, parameter changes that cause a qualitative shift in the dynamics. We want to know about qualitative properties of invariance and change over this vast and immensely complex parameter space.

Here is my conjecture. Previous population genetic models missed the most interesting point because they always studied one or two loci with two alternative alleles per locus. At that dimensionality, one finds the usual nonlinear dynamics of cycles and chaos. Local dynamics dominate because, in each patch, all possible genotypes are usually present. However, if one increases the number of loci and alleles, the system bifurcates, changes generically. Colonization–extinction dynamics matter, times to extinction and recolonization dominate. Local dynamics are much less important.

Having discovered one major axis along which qualitative aspects are controlled, one can now pursue other interesting questions. Scale always comes up, but one has to put the problem in the context of the biology and the first major axis.

In summary:

1. The goal is to search for invariance over an interesting domain and bifurcation between domains because that is only way to learn something general about a complex problem.

2. Host–parasite systems bifurcate as they move from low to high dimension. At low dimension, local dynamics are probably more important for understanding genetic diversity. At high dimension, spatial processes dominate. This shift appears inevitable. The purpose of the simple one-patch model is to illustrate this point.

3. What about real systems? There is good evidence that many systems have surprisingly high dimension. The evidence is presented in the following sections. Data about spatial dynamics is sketchy, but where available, suggest the importance of colonization–extinction dynamics at the level of genotypes.

4. When analyzing these systems one is inevitably measuring diversity. One has to be aware of scale. Observed diversity can be understood only within the context of potential diversity and the spatial and temporal dynamics.

In the following sections I turn to data and theory for natural systems. As expected, dimension and scale are important. In addition, the details of extinctions, migration, and the genetic system determine the particular attributes of each case. When one can measure these details, it may pay to consider a complex metapopulation model tuned to that system, although the size of the parameter space will make the analysis difficult. I summarize the general conclusions that can be drawn from current empirical and theoretical studies, which are still in an early stage of development.

III. INTRODUCTION TO THE EXAMPLES

Researchers working on two different host–parasite systems have recently turned their attention to spatial variation in allele frequencies. In plant–pathogen interactions the hosts often have numerous resistance genotypes and the pathogens have correspondingly diverse host-range genotypes. The limited data from natural populations suggest spatial variation both in the frequency of successful infections and in allele frequencies. Several authors propose metapopulation dynamics as the cause of spatial variation (e.g., Burdon *et al.,* 1989; Thompson and Burdon, 1992; Frank, 1992, 1993b; Antonovics *et al.,* 1994).

The second system is cytoplasmic male sterility in hermaphroditic plants. I will describe the details of this system later. The important feature is conflict between cytoplasmic genes and nuclear genes over the production of pollen. There are different cytoplasmic genotypes, each of which is "resisted" by specific, matching nuclear genes. The interaction is similar to a system with several matching host (nuclear) and parasite (cytoplasmic) genotypes. Preliminary studies show spatial variation in the frequencies of nuclear and cytoplasmic genotypes. Several authors have analyzed this variation in terms of metapopulation dynamics (e.g., Gouyon and Couvet, 1985; Van Damme, 1986; Frank, 1989; Olivieri *et al.,* 1990).

In the following sections I summarize the natural history and observations for plant–pathogen genetics and cytoplasmic male sterility. I then list other host–parasite interactions of high dimension that are candidates for metapopulation dynamics. These later examples include bacterial defense against viral pathogens and polymorphism of plant–herbivore systems. Finally, I consider how to test different explanations for the observed patterns of variation.

IV. PLANT—PATHOGEN INTERACTIONS

Genetic specificity is common in plant–pathogen systems. Each host genotype resists only specific pathogen genotypes; each pathogen genotype attacks only specific host genotypes. In this section I describe the details of genetic specificity, the dimensionality of the interaction, and spatial variation in natural populations.

Flor (1956, 1971) conducted the first detailed study of genetic polymorphisms for resistance in plants and the complementary polymorphisms for host range in pathogens. The interaction between plant and pathogen genotypes turned out to have simple properties that Flor referred to as a "gene-for-gene" system. In an idealized gene-for-gene system, each pair of resistance and susceptibility alleles in the host has a matching pair of host-range alleles in the pathogen.

Recent biochemical models suggest that resistance occurs only when a pathogen allele produces a particular gene product (elicitor) that can be recognized by a matching host receptor (Gabriel and Rolfe, 1990). If an elicitor–receptor match occurs, then the host induces a defensive response and resists attack. If the same pathogen elicitor is present, but the host produces a nonmatching receptor, then disease develops. Infection also occurs when a pathogen lacks an elicitor that matches the specific host receptor.

In multilocus interactions each host polymorphism is matched to a unique, complementary locus in the pathogen. The host resists attack when at least one of the matching pairs of host–pathogen loci leads to recognition and resistance. The pathogen succeeds only when it escapes recognition at all the complementary loci.

The relation between plant and pathogen factors is simple in a gene-for-gene system, but the total interaction is complex because many loci are involved. Flor and others have identified 29 separate host resistance factors in flax, each with a complementary host-range factor in flax rust (Flor, 1971; Lawrence *et al.*, 1981). Similar gene-for-gene interactions are now known or suspected for over 25 different host–pathogen pairs (Burdon, 1987). These systems do not conform exactly to the idealized gene-for-gene assumptions (Christ *et al.*, 1987), but these systems do have complementary major-gene interactions between hosts and pathogens.

These genetic analyses have been conducted in agricultural systems. They establish the possibility that plant–pathogen interactions in natural populations have genetic specificities of very high dimension. According to the theory described earlier, high dimensionality suggests that observed polymorphisms and the dynamics of disease are strongly influenced by colonization–extinction dynamics in a metapopulation.

That story of dimensionality and metapopulation dynamics is intriguing, but is it true? Data from natural populations are suggestive of metapopulation dynamics, but there is not enough information to draw firm conclusions. I briefly summarize the available data in the remainder of this section.

A. Dimensionality

The few studies on wild populations suggest widespread genetic polymorphisms for host resistance (Burdon, 1987; Alexander, 1992; Parker, 1992). For example, the matrix in Fig. 4 shows the frequencies of different host phenotypes of wild flax when tested against seven races of flax rust. This matrix implies

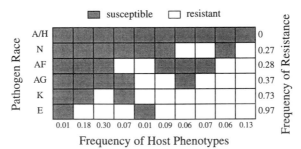

FIGURE 4   Qualitative resistance in a wild population of flax. The matrix shows the frequency distribution of resistant patterns from 67 different host plants collected from a single population when tested against seven pathogen races of flax rust (races A and H are grouped together). Redrawn from Burdon and Jarosz (1991).

complementary major-gene effects at multiple loci with extensive polymorphism in the host. Similar studies of pathogen isolates in both natural and agricultural systems show that pathogen populations are often highly polymorphic (Wolfe and Caten, 1987; Burdon and Leather, 1990).

The most detailed study of a natural plant–pathogen system in natural populations has been on an annual weed, *Senecio vulgaris* (groundsel), and its fungal pathogen, *Erysiphe fischeri* (Clarke et al., 1990). In a recent study the authors obtained 5 pathogen isolates from each of two locations. These 10 isolates were known to have different genotypes for the ability to attack specific host genotypes. The same two locations were used to obtain 360 host plants (Bevan *et al.*, 1993b).

Progeny from 215 plants were tested against 5 of the pathogen isolates, and progeny from the other 145 plants were tested against all 10 isolates. These two tests yielded large matrices of susceptible or resistant interactions. In both cases 70% of the hosts were susceptible to all pathogens tested. The case with five test races of pathogen yielded 12 different resistance phenotypes among the hosts. Each phenotype has a unique resistance/susceptibility classification against the pathogen test races. The case with 10 test races yielded 14 different host phenotypes.

Variation in natural isolates of the pathogen was measured in a second study (Bevan *et al.*, 1993a). Twelve isolates were obtained from each of the two locations used for the host study described above. These 24 pathogen isolates were tested against 50 inbred lines of host plants. Pathogen growth on each host was scored on a scale ranging from 0 (complete resistance) to 4 (vigorous fungal growth and sporulation). For the purposes of classifying genotype, each host–pathogen pair was labeled as either "resistant" or "susceptible," by splitting the continuous scale of fungal growth.

Table I shows that the majority of host and pathogen isolates have unique genotypes. The extensive variability in a limited sample suggests that natural populations are tremendously diverse for this particular plant–pathogen system. Put another way, the community matrix that describes the interactions between plant and pathogen genotypes has very high dimension.

Do other plant–pathogen systems have high dimensionality, or is the groundsel system unusual? The data are too limited to draw firm conclusions. There are several hints that diversity is high, but also some apparent exceptions.

Multilocus genetic diversity for resistance to fungal, viral, and bacterial pathogens is typical in agricultural varieties and wild relatives of crops (Burdon, 1987). The pathogens of cultivated plants evolve quickly in response to changing host genotypes, suggesting complementary genetic complexity (Vanderplank, 1984).

Studies of natural plant–pathogen populations have often revealed high diversity. Examples include the groundsel study summarized here and Burdon and Jarosz's (1991) study of wild flax and flax rust (Fig. 4). All analyses do not find variability of both host and pathogen in every sample. A study of a perennial herb, *Silene alba* and anther-smut fungus, *Ustilago violacea,* found genetic vari-

TABLE I   Numbers of Resistance and Host-Range Phenotypes Inferred from a Test Matrix of 25 Pathogen Isolates by 50 Inbred Host Lines[a]

Infection type category used to define resist-ance/host-range	No. of differ-ent groundsel resistance phenotypes discriminated	No. of different fungus host-range pheno-types discrimi-nated	Minimum number of hypo-thetical resistance/host-range gene pairs required to explain the observed variation[b]
0	12	12	10
1	21	18	14
2	34	23	23
3	43	24	24

[a] Each row shows results when a particular infection intensity is used to define the binomial split between resistance and susceptibility. For example, if category 2 is used for the split, then categories 0–2 are defined as resistant, and categories 3–4 are defined as susceptible (from Bevan *et al.*, 1993a).

[b] The calculation assumes that resistance occurs when the host has a resistance allele that matches a particular host-range allele at a complementary locus in the pathogen. One match at any of the complementary host–pathogen loci is sufficient to cause resistance. See Bevan *et al.* (1993a).

ability among hosts for resistance to the pathogen (Alexander, 1989). However, no variability of the pathogen was detected when six isolates from a single location were tested against 15 host lines (Alexander *et al.*, 1993).

B. Colonization–Extinction Dynamics of Alleles

Given high dimensionality it seems inevitable that there will be occasional extinctions of alleles from a local population and subsequent recolonizations by immigration. The problems now concern pattern, process, and inference. What are the temporal and spatial patterns of allele frequencies, disease intensity, and population sizes? What is the relative influence (scaling) of colonization–extinction processes compared with other ecological and genetic processes? What measurable properties can be used to infer process?

Only a few studies of natural systems have measured spatial variation. I briefly summarize two projects that have focused on genetic variation.

Parker (1985) used field transplant experiments to study variability in the legume *Amphicarpaea bracteata* and its fungal pathogen *Synchytrium decipiens*. I describe the details of his work because transplant experiments are a relatively simple method of measuring the scale of spatial variation in host–parasite interactions. The first experiment analyzed three sites: the focal population, 1 km away from the focal population, and 100 km away from the focal population. Seeds were collected from two self-fertilized plants in each of the three populations. For each of the six groups of selfed progeny, 15–20 seedlings were transplanted into the focal population.

All of the seedlings derived from the focal population developed severe infection when transplanted back into their natal location. Progeny from one of the plant lines derived from 1 km away was free of disease when transplanted and grown in the focal population. The other line from 1 km away had 88% of the progeny infected, but the average intensity of infection was about one-fifth that of the native plants. Infection intensity was measured as number of sori per plant (a sorus is the initial fungal lesion). All of the progeny derived from 100 km away were completely free of infection when transplanted into the focal population.

This transplant experiment suggests spatial variation in the genotypes of hosts and pathogens over distances of 1 km or greater. Fungal infection was heavy in each of the three locations. When a plant was moved to a new location, it developed little or no infection, suggesting that the pathogen populations differ between the focal site and the other two sites. The variation in infection among the host lines derived from different locations and transplanted into the focal site suggests spatial differentiation among the host populations.

In a second experiment Parker (1985) obtained stronger evidence for spatial variation over 1 km. He tested one pathogen isolate from the focal population against 13 plant families from the focal population and 11 families from 1 km away. All 13 local families developed infection, but 10 of the 11 families from 1 km away were completely resistant to this pathogen isolate.

The final experiment analyzed variation on a smaller spatial scale within the 100-km population. Plant lines were established by collecting along a linear transect from six sites separated by 30 m. The sites were labeled in order from one end of the transect to the other. A pathogen isolate from site 5 was tested against each plant line. I describe the details to show the difficulties that often arise when measuring variability in the interactions between host and parasite. Three different measures of resistance provide information about genetic variation.

First, when resistance or susceptibility was measured as the presence or absence of initial infection, there was no significant variation among sites, with a mean infection frequency of 74%. Second, if resistance was measured by percentage of sori that abort before fungal reproduction, then all plants from site 6 were 100% resistant. The other five sites aborted 0–20% of sori. Third, the sites varied significantly when the number of sori per plant was used to measure response. For example, site 1 was the least resistant, with a mean $\pm$ SE of 11.8 $\pm$ 3.2. Site 6 was the most resistant, with 2.3 $\pm$ 0.4, but neighboring site 5, where the pathogen was derived, was the second highest, with 8.0 $\pm$ 2.5. These results suggest that quantitative components of resistance may be race specific, as in the groundsel study discussed above. In Parker's study, details about race-specific quantitative variation would require tests of the plant lines with different pathogen isolates.

Parker's work shows that genetic variation can occur over short distances. In this case, pathogens are highly successful on plants near the location at which they were found, but had poor success on plants from other locations. It appears that immigrant host genotypes, with resistance to local pathogens, could increase in frequency and change the spatial patterns of differentiation. Experimental

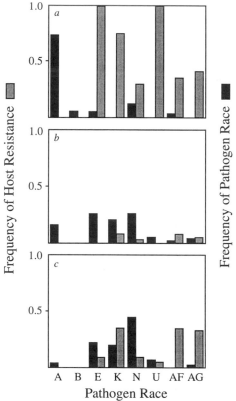

FIGURE 5   Spatial variation in pathogen genotypes and host resistance among wild populations of flax *(Linum marginale)* and flax rust *(Melampsora lini)*. Both host and pathogen isolates were obtained from several different sites. Each panel shows the racial composition of the pathogen population and the frequency of host resistance to each pathogen race when summarized over a different geographic scale. (a) Data from a 1-ha plot for 67 host lines and 94 pathogen isolates. (b) Combined data for 40 host lines and 37 pathogen isolates from two populations 300 m and 2.7 km away from the plot summarized in the first panel. (c) Combined data for 108 host lines and 80 pathogen isolates from six populations 13.8–75 km away from the plot summarized in the first panel. Redrawn from Jarosz and Burdon (1991).

movement of genotypes followed by time-series monitoring of consequences may provide a method for inferring the joint roles of selection and colonization–extinction dynamics.

The second major study of spatial variation in natural populations was conducted on flax *(Linum marginale)* and its pathogen, flax rust *(Melampsora lini)* (Jarosz and Burdon, 1991; Burdon and Jarosz, 1992). A summary of spatial variation in genotype is shown in Fig. 5. To study the role of metapopulation dynamics, the authors measured the composition of nine pathogen populations over 2 to 4 consecutive years. This is the most extensive study of temporal and spatial variation in natural populations, but limitations of the data must be considered before drawing any conclusions. First, the host plant is perennial, so the time span of the study does not cover genetic changes in the host populations. Second, it is not clear how far the wind-borne spores can move each year, in other words, the scaling of spatial distance relative to migration distance is not known. In spite of these constraints, a few tentative conclusions are interesting.

(i) Four pathogen races dominated the metapopulation over all 4 years of the study. (ii) The majority of host populations contained little or no resistance to any of the four dominant pathogen races. Thus host resistance alone cannot explain temporal and spatial variation in the pathogen. (iii) Pathogen races occasionally became locally extinct in a particular population but were often reintroduced within a year or two. (iv) Fluctuations in the genetic composition of local pathogen populations may be strongly influenced by the dynamics of population size. Twenty-two host populations were sampled for the presence or absence of infection in 2 consecutive years. One population had no pathogen infections in the first year but was infected in the next year. Another population had infections in the first year but was free of disease in the next year. Finally, two populations were free of infection in both years.

Burdon and Jarosz (1992) suggest that, over the temporal and spatial scale of their study, the observed fluctuations in the pathogen's genetic structure were driven by colonization–extinction events and drift. Thus the populations in that region may act as a cohesive unit linked by frequent migration, with selection playing a limited role in the dynamics of allele frequency. Put another way, the time scale of pathogen movement among these populations may be on the order of host generation time and thus too short for colonization–extinction dynamics of alleles among these populations to exert strong coevolutionary pressures. Perhaps at a larger spatial scale the migration rate is small relative to the length of host generations—the time scale over which selection is effective. At that scaling between migration and selection, the colonization–extinction dynamics of alleles may cause occasional major shifts in genotypic composition.

To summarize these two studies on plant–pathogen systems, it is easy to imagine how metapopulation dynamics can influence genetics, but very difficult to measure space–time variation over the proper scales. How can convincing data be obtained? One way is to observe colonizations of locally absent alleles and the subsequent local dynamics. It may be difficult to observe such rare events, but there is one suggestive study of cytoplasmic male sterility that I describe in the next section.

V. CYTOPLASMIC MALE STERILITY

Most organisms inherit mitochondrial DNA from their mother, with no input from their father. By contrast, most other genetic material is obtained equally from the mother and father. Typically these different modes of transmission, matrilineal versus biparental, have no consequences for the direction of evolutionary change favored by selection. For example, efficient respiration increases both matrilineal and biparental transmission.

The allocation of resources to sons and daughters affects matrilineal and biparental transmission differently. Traits that enhance the production of daughters at the expense of sons always increase the transmission of matrilineally inherited genes. For example, in some hermaphroditic plants the mitochondrial genes may inhibit pollen development and simultaneously enhance the production of seeds (Edwardson, 1970; Hanson, 1991). Selection of genetic variants in the mitochondria would favor complete loss of pollen production in exchange for a small increase in seed production because the mitochondrial genes are transmitted only through seeds (Lewis, 1941).

Reallocation of resources from pollen to seeds can greatly reduce the transmission of nuclear genes because biparental transmission depends on the sum of the success through seeds and pollen. Thus there is a conflict of interest between the mitochondrial (cytoplasmic) and nuclear genes over the allocation of resources to male (pollen) and female (ovule) reproduction (Gouyon and Couvet, 1985; Frank, 1989). Consistent with this idea of conflict, nuclear genes often restore male fertility by overcoming the male-sterility effects of the cytoplasm.

The nuclear–cytoplasmic conflict is very similar to a host–parasite system: there is antagonism over resources for reproduction, cytoplasmic (parasite) genes determine the host-range for exploitation, and cytoplasmic genes interact with nuclear (host) resistance genes to determine the specificity of the interaction. Cytoplasmic inheritance influences the patterns of "parasite" transmission but, on the whole, the genetics and population dynamics are typical of host–parasite interactions (Gouyon and Couvet, 1985; Frank, 1989; Gouyon et al., 1991).

The reduction of pollen caused by cytoplasmic genes is called cytoplasmic male sterility (CMS). Laser and Lersten (1972) list reports of CMS in 140 species from 47 genera across 20 families. More than one-half of these cases occurred naturally, about 20% were uncovered by intraspecific crosses, and the rest were observed in interspecific crosses. Moreover, this listing is an underestimate of the true extent of CMS because detecting a cytoplasmic component to a male sterile phenotype requires genetic analysis of polymorphism (Frank, 1994a).

Wild populations of CMS maintain several distinct cytoplasmic genotypes (cytotypes). Each cytotype is capable of causing male sterility by an apparently different mechanism because each is susceptible to a particular subset of nuclear restorer alleles. Nuclear restorer alleles are typically polymorphic at several loci, with each allele specialized for restoring pollen fertility when associated with particular cytotypes. The observations are summarized in Frank (1989), Couvet et al. (1990), and Koelewijn and Van Damme (1995a,b).

CMS has reciprocal genetic specificity of nucleus and cytoplasm and widespread polymorphism. The basic questions of dimensionality and colonization–extinction dynamics are similar to those of other host–parasite systems. What are the temporal and spatial patterns of female and hermaphrodite (phenotype)

TABLE II Summary of Available Evidence on Number of Cytoplasmic Genotypes and Nuclear Loci That Have Been Detected in Various Agricultural and Wild Species[a]

Species	Cross type	Molecular evidence	Cytoplasmic genotypes	Nuclear genes
Agricultural species[b]				
Beta vulgaris	Within	+	2	2–7
Daucus carota	Within	+	2	3
Helianthus spp.	Between	−	Many	?
Nicotiana spp.	Between	+	8	?
Oryza spp.	Between	+	2	3
Solanum spp.	Both	+	4	Many
Triticum spp.	Between	+	2	2
Zea mays	Both	+	3–4	5
Wild species				
Beta maritima[c]	—	+	2	?
Origanum vulgare[d]	Within	−	2	2–7
Nemophila menziesii[e]	Within	−	2	2
Thymus vulgaris[f]	Within	+	2–Many	?
Plantago lanceolata[g]	Within	+	2	3–5
Plantago coronopus[h]	Within	−	2	3–5

[a] Copied from Koelewijn and Van Damme (1995a).
[b] Compiled from Hanson and Conde (1985), Kaul (1988).
[c] Boutin et al. (1987).
[d] Kheyr-Pour (1980, 1981).
[e] Ganders (1978).
[f] Belhassen et al. (1991).
[g] Van Damme and Van Delden (1982), Van Damme (1983).
[h] Koelewijn and van Damme (1995a,b).

frequencies, allele frequencies, and population sizes? What is the relative influence of colonization–extinction processes compared with other ecological and genetic processes? What measurable properties can be used to infer process? As before, the data are not sufficient to answer all these questions, but the literature provides intriguing hints about dimensionality and colonization–extinction dynamics.

A. Dimensionality and Spatial Variation

Two or more different cytoplasmic genotypes may cause CMS within a particular species. The cytoplasms are recognized as distinct because they react differently to particular nuclear restorer genotypes. The dimensionality of the system increases with the number of different cytoplasmic types that cause male sterility, each with its own associated set of specific nuclear restorer loci.

Table II summarizes data on the dimensionality of agricultural and wild species. The "cross type" describes whether variability was discovered with intraspecific crosses or with hybridizations between species. Molecular evidence matches different mitochondrial markers to genetic and phenotypic properties observed in crosses. The "nuclear genes" column lists the total number of loci involved in male sterility. Although the existence of nuclear–cytoplasmic specificity is clear, the details are very difficult to work out. The numbers must be considered minimum estimates because a cytoplasmic polymorphism can be detected only when present in a study that also has matching nuclear polymorphism for restoration. Similarly, nuclear polymorphism requires matching cytoplasmic polymorphism. Each study requires tedious crosses and nurturing of many progeny to draw unambiguous conclusions. As mentioned above, CMS is widespread. The table shows only those studies in which attempts have been made to analyze the number of genotypes.

The data in Table III show that the frequency of females varies widely among populations of the same species. The column for "genetics" describes how information was obtained on the spatial variation of cytoplasmic and nuclear genes. Evidence is "direct" for *Plantago lanceolata* because the two cytoplasmic genotypes are associated with different morphological abnormalities of failed pollen production and anther development. In addition, crosses were performed to measure the frequency of the cytoplasms and associated restorer alleles. Spatial variation in *P. lanceolata* will be discussed below. For *Beta maritima*, crosses were performed to infer the frequency of cytoplasmic types and restorer alleles for each population. It appeared that cytoplasmic frequencies did not vary between the two populations. The large difference in female frequency was the result of variation in the frequency of restorers between the two locations.

Spatial variation was inferred from crosses between different populations in *Thymus vulgaris* and *P. coronopus*. These long-distance crosses yielded higher frequencies of females than were observed within each population. High frequencies of females in the crosses imply that, within each population, restorers are common for the locally common cytoplasm but relatively rare for other cytoplasms. If different populations are dominated by different cytoplasms, then the

TABLE III   Spatial Variation in Wild Populations with Cytoplasmic Male Sterility[a]

Species	Range	Median	N	Study	Genetics
Origanum vulgare	1–62	?	100	Kheyr-Pour (1980)	
Thymus vulgaris	5–95	>50	110	Gouyon and Couvet (1985)	Inferred
Plantago lanceolata	1–23	8	27	Van Damme and Van Delden (1982)	Direct
With IN	1–34	15	27		
Plantago coronopus	0–35	13	8	Koelewijn (1993)	Inferred
With IN	13–61	31	8		
Beta maritima	19–62		2	Boutin-Stadler *et al.* (1989)	Direct

[a] The second and third columns show the range and median in percentage of females per population for samples from N populations.

crosses will expose cytoplasmic genotypes from the female parent to nuclear backgrounds of the male parent that have a low frequency of matching restorers.

For the *Plantago* species, the "IN" rows show the frequency of partially male-sterile (IN) plants. Partial male sterility also depends on an interaction of cytoplasmic and nuclear genes. As noted by Koelewijn (1993), partial male sterility is a common phenomenon in CMS, but phenotypes are often reported with dichotomous classification. This is similar to the partial resistance that is common in plant–pathogen interactions, as noted in the previous section. In both CMS and plant–pathogen interactions, the intermediate phenotypes often depend on specific interactions between genetic polymorphisms of the host and parasite.

B. Colonization–Extinction Dynamics

The frequency of females in a population is the frequency of unrestored male-sterile cytoplasms. The data suggest that the frequency of females varies among populations. Phenotypic variation appears to be associated with widespread genotypic variation in cytoplasmic types and restorer frequencies.

Two related metapopulation scenarios have been proposed to explain phenotypic and genetic variation. The first theory concerns the colonization–extinction dynamics of alleles among existing populations (Gouyon and Couvet, 1985; Frank, 1989). The second theory focuses on the colonization–extinction dynamics of populations (Gouyon and Couvet, 1985). I will briefly outline the allelic theory, along with a field study that hints at how natural populations may be influenced by these processes. At the end of this section I mention the population colonization–extinction theory.

To understand the colonization–extinction dynamics of alleles one must imagine a sequence of events.

(i) Initially, one of the cytoplasmic type is lost from a local population. Loss may occur by drift or because the alternative types have higher fitness. Increasing dimensionality (more types) raises the probability that one or more cytoplasms will be absent locally.

(ii) When a cytoplasmic type is absent, the associated nuclear restorer alleles do not have any beneficial effects. These specific restorer alleles may be lost from the local population by a variety of processes. If there are no fitness differences between restorer and alternative nonrestorer alleles, then the restorers may be lost by drift. If the restorers, which must in some way influence pollen development, reduce efficiency when their matching cytoplasm is absent, then the specific restorers will be lost by selection.

(iii) After steps (i) and (ii), a cytoplasmic genotype and its specific restorers are absent locally. If an unrestored cytoplasm arrives by immigration, it will have a fitness advantage and spread quickly in the population. The fitness advantage occurs because an unrestored cytoplasm causes a male-sterile phenotype. Male-sterile plants typically produce more seeds than hermaphrodites (Lloyd, 1976; Van Damme, 1984; Van Damme and Van Delden, 1984). Because cytoplasmic fitness depends only on success through the maternal line (seeds) and not on pollen success, the male-sterile plants have greater cytoplasmic fitness than hermaphrodite plants. Thus the cytoplasms that cause male sterility spread in the local population, causing an increase in the frequency of females.

(iv) Cytoplasmic genotypes are essentially alternative alleles at a haploid locus. When one genotype increases in frequency, then the other genotypes necessarily decline in frequency. In the case of mitochondria, an increase in the frequency of one mitochondrial type will cause a decline in other mitochondrial types. Thus the selective spread of an unrestored cytoplasm may cause the local extinction of alternative cytoplasmic genotypes. Loss of cytoplasmic genotypes may be associated with loss of matching restorers, as in step (ii).

(v) The population now has a high frequency of females and a dominant cytoplasmic genotype. The restorers matching the dominant cytoplasm are locally extinct. If a matching restorer arrives by immigration, it will combine with the dominant cytoplasmic type to produce hermaphrodites. The restorer allele spreads rapidly because pollen is rare locally, thus the few hermaphrodites are the source of paternal alleles for all members of the population. As the restorer spreads, the frequency of females declines. The frequency of cytoplasmic genotypes may be unaffected by the initial spread of restorers.

(vi) A locally absent cytoplasmic genotype can invade and spread if its specific restorers are absent. The cycle then repeats, with a genotypic turnover in the local population. The greater the dimensionality – the number of cytoplasmic genotypes and matching specific restorers – the more likely an immigrant cytoplasmic type will be locally absent and can start a new round of genotypic turnover.

Van Damme's study of *P. lanceolata* provides just enough detail to show how parts of the above scenario may work in a natural population. Van Damme and Van Delden (1982) distinguished two cytoplasmic genotypes in *P. lanceolata* each with its own set of nuclear restorers. Table IV shows phenotypic frequencies in 12 populations in two habitat groups; the original paper lists data for 27 populations in five categories. The labels for each population are abbreviations for locations.

The cytoplasmic genotype *R* causes the male-sterile phenotype MS1 when unrestored and IN1 when partially restored. The cytoplasm *P* causes MS2 when unrestored and IN2 when partially restored. All four types are morphologically distinct and can be scored by direct examination. Restored cytoplasms of either type are hermaphroditic, *H*. The cytoplasmic type of a hermaphrodite can be determined only by crossing until the cytoplasm is exposed in an unrestored nuclear background.

The two population groups shown in Table IV are the most differentiated of the five groups listed in the original paper. Five of the hayfield populations either lacked the *R* cytoplasm or were fixed for the *R* restorers. In the pasture popula-

TABLE IV   Phenotype Percentages in Natural Populations of *Plantago lanceolata*[a]

Population	MS1	IN1	MS2	IN2	H	Sample size
Hayfield						
Dr	0	0	0.2	0.2	99.6	811
Ze	0	0	5.0	0.8	94.1	742
An	0	0	8.2	1.3	90.5	754
Re	0	0	5.0	3.6	91.4	695
Me1	0	0	3.9	5.5	90.6	688
Ve	12.2	2.2	0	0.6	85.0	623
Br	23.0	7.0	0.3	0.2	69.5	601
Pasture						
Wd	4.6	0.6	0.5	0.9	93.4	6902
Bm2	7.3	3.9	0	1.3	87.5	386
Pa	7.6	7.8	0.5	0.9	83.2	437
Ac2	11.8	10.8	0	1.0	76.4	305
Ju	21.5	7.0	0	0.5	71.0	414

[a] From Van Damme and Van Delden (1982).

tions, either the *P* cytoplasm was very rare or the *P*-specific restorers were common. The other three population groups were relatively more mixed for MS1 and MS2 phenotypes.

Van Damme (1986) made an intensive study of spatial variation within the Westduinen (*Wd*) population listed in Table IV. A picture of the field at Westduinen is shown in Fig. 6, with some of the data listed in Table V. Females were rare over the whole population, with MS1 more common than MS2. However, in a few locations the frequency of MS1 was high (Fig. 6). Within the larger clusters of MS1, *p1–p4*, the frequency of MS1 phenotypes was close to zero at the borders and rose to 60% near the center.

The field as a whole was dominated by the *P* (MS2) cytoplasm, with an overall frequency of 0.94. The frequencies of the *P*-specific restorer alleles were also high. Thus most plants were hermaphrodites with a *P* cytoplasm and *P* restorers. The overall frequency of the *R* cytoplasm was 0.06, and the *R*-specific restorers at the two restorer loci had frequencies of 0.02 and 0.08.

Genotypic composition was very different in those few areas that had high frequencies of the MS1 phenotype (Fig. 6 and Table V). The *R* (MS1) cytoplasm, rare in the population as a whole, had frequencies ranging between 26 and 39% in populations *p1–p4*. The *R*-specific restorers, also rare in the whole field, were more frequent in the MS1 clusters, although the exact frequencies were difficult to estimate.

Van Damme's interpretation agrees with the scenario outlined above. Initially most of the field was dominated by *P* cytoplasms and *P*-specific restorers. *R*-bearing colonists founded the MS1 spots and, since the *R*-specific restorers were initially rare, the MS1 females spread from a central focus. MS1 plants produce

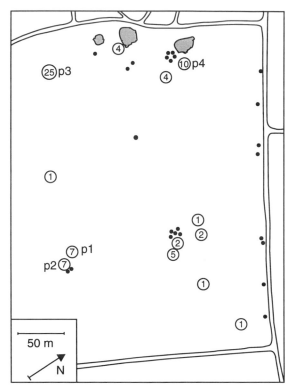

FIGURE 6   Distribution in a pasture of areas with MS1 and IN1 plants of *P. lanceolata*. Single plants are represented by dots and groups of plants by circles. The number within each circle is the area in square meters covered by the local group of plants. The four largest groups are labeled *p1–p4*. The shaded areas are pools that cattle use for water. Redrawn from Van Damme (1986).

more seeds that are larger and survive better than seeds from hermaphrodites (Van Damme and Van Delden, 1984), so the females have a competitive advantage locally. Seeds disperse at a slow rate (8 cm/year; Bos *et al.,* 1986), thus well-defined patches can form. As the frequency of unrestored *R* cytoplasms rises in an area, selection favors an increase in *R*-specific restorers. In an area with a high concentration of *R* cytoplasms, the main pollen donors will be *R*-restored hermaphrodites.

The low frequency of the *R*-specific restorers in the overall population suggests that these alleles are at a selective disadvantage when the *R* cytoplasm is absent. If so, then a population dominated by the *P* cytoplasm is likely to lack the *R* restorers, as at Westduinen. That genotypic composition is susceptible to invasion by *R* cytoplasms, followed by a subsequent change in genotypic composition.

Many details about *P. lanceolata* require further study. However, this first glimpse does suggest that colonization–extinction dynamics and the strong se-

TABLE V Percentage of Phenotypes of *Plantago lanceolata* at a Westduinen Field[a]

Description	MS1	IN1	MS2	IN2	MS3	H	Sample size
Total field	4.6	0.6	0.5	0.9	0.5	92.9	6902
p1	28.6	5.4	0	0	0	66.1	112
p2	25.6	1.6	0	0	0	72.8	188
p3	21.3	3.0	0.3	1.2	0	74.3	695
p4	22.7	1.6	0	4.7	0	71.1	688
Remainder	1.9	0.4	0.6	0.9	0.5	95.7	6140

[a] MS3 is a rare phenotype controlled by variation at autosomal loci. The locations of populations p1–p4 are shown in Fig. 6. Data from Van Damme (1986).

lective pressures on cytoplasm and nucleus may be responsible for the observed spatial variability.

The *P. lanceolata* example emphasizes the colonization and spread of a locally novel genotype into an existing population. Cytoplasms may also "escape" their restorers when a empty patch is founded by one or a few colonists (Gouyon and Couvet, 1985). Species that are subject to local population extinctions and colonizations of empty habitat patches may be particularly variable in the frequency of females and the spatial variation in genotypes. Studies of *Thymus vulgaris* in southern France suggest that disturbed populations and recently colonized patches are likely to have higher frequencies of females than undisturbed, older populations (Gouyon and Couvet, 1985; Belhassen *et al.,* 1989; Olivieri and Gouyon, this volume). However, it is difficult to obtain convincing data on processes that cover large temporal and geographic scales.

All studies do not find evidence of a dynamic process. Koelewijn (1993) summarized his work on *Plantago coronopus* by noting that the frequency of phenotypes in each of four locations was "remarkably constant" over 10 years. The evidence also suggested that both cytoplasmic male-sterile genotypes occurred at intermediate frequency in all four locations. Koelewijn's (1993) comments serve as a reminder that we have only the haziest picture of a few cases, with no empirical guidelines about appropriate temporal and spatial scales at which nonequilibrium fluctuations may be important.

VI. OTHER SYSTEMS OF HIGH DIMENSION

I have discussed dimensionality and spatial variation for plant diseases and cytoplasmic male sterility. Are these systems unusual, or are other host–parasite interactions similarly diverse?

There are very few systems with good data available on both the host and parasite. I briefly describe a few cases to illustrate the kind of information that has been collected. My interpretation is that high dimensionality occurs often, although there will certainly be many exceptions.

Bacteria have a simple recognition-based immunity system that protects them from invasion by foreign DNA (Wilson and Murray, 1991). There are two components to the system. Restriction enzymes cut DNA molecules that carry a particular sequence of nucleotides. Modification enzymes recognize the same nucleotide sequence but, instead of cutting the DNA, these enzymes modify the recognition site in a way that protects that molecule from restriction. A bacterial cell's own DNA is modified, otherwise the restriction enzymes would cut the DNA and kill the cell.

Restriction-modification (RM) enzymes are known for over 200 different recognition sites (Kessler and Manta, 1990; Roberts, 1990). Circumstantial evidence suggests that defense against bacteriophage viruses has been a powerful force promoting diversity. (1) RM can protect host cells from invading phage (Luria and Human, 1952; Arber, 1965). (2) Phage that develop in bacteria with a particular RM type are modified for the associated recognition sequence. These modified phage can attack other bacteria of the same RM type, but are sensitive to restriction by different RM systems. Rare RM types are favored because few phage will be modified for their recognition sequence. This frequency-dependent selection promotes diversity of RM as a defense against phage (Levin, 1986, 1988). (3) Phage carry a variety of antirestriction mechanisms (Kruger and Bickle, 1983; Sharp, 1986; Korona *et al.*, 1993). For example, many phage lack particular RM recognition sequences. The probability of having these recognition sequences is very high if no selective pressure were acting on sequence composition.

These details suggest that the interaction between RM and phage is of high dimension. Not enough sampling has been done to draw any conclusions about spatial variation, but it seems likely that the genotypic composition of communities varies widely among different locations.

A second bacterial system acts in a very different way from RM but is also highly diverse. This allelopathic system affects competition between bacterial strains rather than what is usually thought of as a host–parasite interaction. However, the genetic specificity of attack and defense promotes widespread polymorphism in much the same way as in host–parasite dynamics.

In this system of bacterial allelopathy, cells often carry plasmids that encode a bacterial toxin (bacteriocin) and immunity to that toxin (Reeves, 1972; Lewin, 1977; Hardy, 1975). Immunity works by neutralizing the toxin after it has entered the cell. Bacteria may also be resistant to bacteriocins because they lack a compatible receptor through which the toxin can enter the cell.

Many distinct bacteriocin types are found within a population. A type is defined by its susceptibility to a set of toxin-producing test strains. With n test strains, there are 2^n possible types. Epidemiological studies frequently use bacteriocin typing to identify and follow pathogenic strains of bacteria. These studies provide information about the diversity of bacteriocin production and susceptibility in populations. For example, Chhibber *et al.* (1988) summarize data on the number of isolates, test strains, and bacteriocin susceptibilities for 10 studies of

Klebsiella pneumoniae. The fewest number of observed types occurred in a study with 200 isolates, four test strains, and 11 types of a possible $2^4 = 16$; the most occurred in a study with 553 isolates, seven test strains, and 64 types of a possible $2^7 = 128$. Similar levels of diversity have been reported for a variety of species (Gaston *et al.*, 1989; Senior and Vörös, 1989; Rocha and de Uzeda, 1990; Traub, 1991; Riley and Gordon, 1992).

The next example is disease resistance in vertebrates. There is great diversity at the loci that encode specific recognition of parasites, the major histocompatibility complex (MHC) genes. However, this host diversity has rarely been matched to specific polymorphisms of parasites.

The molecules that bind intracellular protein fragments and bring them to the surface are coded by genes that reside within the MHC region. Each antigen-presenting molecule from the MHC has a groove that accommodates nine amino acids. Each particular MHC molecule can recognize and present on the cell surface only a subset of protein fragments. An individual has several different MHC types that, taken together, determine the set of protein fragments that can be recognized and carried to the cell surface for presentation.

The MHC loci are highly polymorphic, with between 10 and 80 different alleles known for each locus. Two lines of evidence suggest that resistance to particular diseases can strongly affect the frequency of MHC alleles. First, most of the variation among alleles occurs in the groove that binds protein fragments— the specific recognition area. Second, a few cases are known in which there is a strong spatial correlation between endemic diseases and MHC alleles that are associated with resistance to those diseases. For example, the allele HLA-B53 is associated with resistance to a severe strain of malaria that occurs in children in The Gambia. HLA-B53 occurs at a frequency of 25% in this west African nation; by contrast, the frequency of this allele in Europe is 1% (McMichael, 1993). Other MHC alleles are implicated in resistance to HIV, the cause of AIDS, and to Epstein—Barr virus, the cause of various cancers. Disease correlations with MHC alleles suggest that selective pressures influence the evolution of the immune system polymorphisms (Thomson, 1991; Mitchison, 1993).

The final example concerns genetic variation in plant resistance to herbivores. The resistance may be biochemical or structural. Karban (1992) lists 37 studies that show evidence of genetic variability in resistance to herbivore attack. These studies usually demonstrate genetic variability by growing different plant genotypes in a common environment and measuring variation in herbivore damage. The details of variable resistance and the number of independent traits involved (dimensionality) are typically unknown.

Insect herbivores are often genetically variable in their ability to attack different plant varieties (Gould, 1983). Edmunds and Alstad (1978) suggested that insect species often differentiate into populations that are locally adapted for the host genotypes in their area. Karban (1992) summarizes studies that examine geographic specialization of insect herbivores. He concludes that the data are not convincing because of limited sampling, but the hypothesis that herbivores are geographically specialized remains an important idea that deserves further study.

VII. THEORIES AND TESTS

The evidence summarized in the previous sections suggests that host–parasite interactions can be very diverse. The few careful studies of natural populations indicate a spatial component of diversity when the system is viewed on the appropriate spatial scale. In this final section I review the processes that can explain spatial variation in host–parasite allele frequencies. I then summarize the plant–pathogen and cytoplasmic male sterility studies in light of the alternative explanations for spatial variation. Five factors may influence spatial variation in host–parasite genetics.

(i) Migration–drift dynamics occur when selection is a relatively weak force and allele frequencies fluctuate stochastically. Locally extinct alleles can return to a population by immigration if populations are connected in a metapopulation. Drift is relatively more important than selection in causing fluctuations when local populations are small or have frequent bottlenecks. Migration can overcome selection when the movement of alleles occurs more quickly than selection can change local allele frequencies.

(ii) Local, nonlinear dynamics cause spatial variation when populations fluctuate in an uncoupled manner. Selection causes changes in allele frequencies, and migration does not cause major perturbations of local dynamics. Migration must be sufficiently rare to prevent synchronization of population fluctuations.

(iii) Environmental heterogeneity can favor different allelic combinations in particular locations. This will be particularly important in inbreeding or asexual species, where chance linkage will occur between alleles involved in host–parasite interactions and alleles affecting success in different habitats.

(iv) The sexual system will, in general, determine the role of linkage in changing allele frequencies. With low recombination, selection at one host–parasite locus can change allele frequencies at many other loci.

(v) Local extinctions caused by selection coupled with global migration lead to colonization–selection–extinction dynamics. These are the processes that I emphasized in my descriptions of plant–pathogen and CMS systems. In this case immigration of locally extinct alleles will sometimes cause a major perturbation of local dynamics. Spatial variation may be dominated by the timing of local extinctions of alleles and the waiting time until those alleles are reintroduced by immigration. The complicated details of nonlinear dynamics (limit cycles, chaos, etc.) may be relatively unimportant in systems of high dimension because the timing of extinctions and colonizations determines local and regional variation.

These five processes can all occur in a single system when measured over different spatial and temporal scales. For example, in the Burdon and Jarosz (1992) study of a plant–pathogen system, the evidence suggested that migration–drift dynamics dominated the spatial distribution of pathogen genotypes over approximately 75 km. Local pathogen populations apparently experience frequent bottlenecks, with immigration or recolonization from neighboring populations. The movement of pathogen alleles occurs on a time scale that is shorter than host generation time, suggesting that migration is more powerful than coevolutionary selection pressures in determining the spatial dynamics of pathogen allele frequencies.

The rate of pathogen migration will be low at some sufficiently long spatial scale. The colonization of a region by a long-distance migrant allele could cause a major local perturbation. For example, once in 100 years a locally novel resistance allele may land in a region, changing the selective pressures on the pathogens and favoring the immigration of new host-range alleles. These perturbations may be rare compared to the usual scale of study, but could be a major cause of regional variation. Other systems, such as CMS in *P. lanceolata,* may have relatively low rates of migration over short distances. Thus colonization–selection–extinction dynamics may occur over smaller scales that are easier to study (Van Damme, 1986).

Problems of inference can be severe. On the measurement side, polymorphism in coevolutionary systems can be difficult to detect (Frank, 1994a). For example, two different male sterile cytoplasms both yield the same hermaphroditic phenotype when the study sample contains matching nuclear loci that are fixed for restorer alleles. Thus the potential diversity (dimensionality) of systems is difficult to measure. On the statistical side, very different processes may yield the same patterns of host–parasite polymorphism when the observer uses a particular sampling scheme. For example, drift models and strong selective, coevolutionary models often have similar patterns when sampled without long time-series data (Frank, 1996). The only remedy is thorough understanding of both the patterns expected under alternative processes *and* the consequences of different sampling schemes and methods of data analysis.

Two standards of empirical progress will help. First, manipulation experiments in the field are an easy way to discover spatial variation. Parker's (1985) study of a plant–pathogen system is an excellent example of how transplant experiments can be used to document the extent and scale of genetic variability. Van Damme (1986) did not manipulate his populations of cytoplasmically male sterile plants. However, it is easy to imagine an experiment in which locally absent cytoplasms or restorers are introduced into fields in which those alleles are extinct. Then, over several years, the natural spread of the alleles could be monitored.

The second avenue of progress will come from molecular methods of sampling. At present, host and parasite genotypes are identified by laborious methods of genetic crossing experiments and phenotypic testing. The time required severely restricts the scope of data collection. Molecular probes will eventually allow widespread sampling of host and parasite genotypes over different spatial and temporal scales. The preliminary work on plant diseases and male sterility suggests that host–parasite systems will be highly variable and strongly influenced by metapopulation dynamics.

ACKNOWLEDGMENTS

I thank P. Amarasekare, R. M. Bush, and D. R. Campbell for helpful comments. My research is supported by NSF Grant DEB-9057331 and NIH Grants GM42403 and GRSG-S07-RR07008.

IV

CASE STUDIES

Empirical metapopulation studies were scarce in 1991, at the time of publication of our previous volume on metapopulation dynamics. Today, the situation is changing rapidly, with valuable field studies appearing monthly if not weekly. We distinguish between two kinds of empirical studies, though there is also substantial overlap: studies driven by a desire to test theories and concepts and studies motivated by conservation management questions. The former metapopulations are more likely to be at equilibrium and to fit the classical metapopulation ideas. The latter metapopulations, most often created by human fragmentation of the landscape, may not be at equilibrium and often have other features that make them problematic from the perspective of testing theory. The studies we have chosen to feature in this section are entirely of the former type, that is, directly oriented toward testing theory and elucidating concepts. This is not to say that metapopulation studies in conservation biology would not ultimately prove equally rewarding scientifically, if for no other reason than that they are often backed by significantly better funding than purely academic research. One obstacle at present is that much of the conservation-oriented research on metapopulations is reported in the gray literature.

The four case studies in this section represent four taxa: butterflies (Thomas and Hanski), mammals (Smith and Gilpin), plants (Giles and Goudet), and plant–herbivore–parasitoid metapopulations (van der Meijden and van der Veen-van Wijk). The focus varies from single-species studies to interspecific interactions and from population dynamics to population genetics. The mechanisms of extinction and colonization in these metapopulations are diverse. Especially in Europe, the taxon which is presently receiving the greatest attention is butterflies, which now play a role in the development of metapopulation theory similar to that played by birds in the development of the dynamic theory of island biogeography. Butterflies are attractive study animals not only because they fly in nice places on sunny days, but also because many species in many landscapes are structured into proper metapopulations of many but small local populations and because butterflies present many advantages for field studies. Furthermore, many butterflies especially in northern Europe have greatly declined over the past decades, apparently because the amount of suitable habitat has diminished, and there is therefore much conservation interest in butterflies. Other taxa that have been prominent in metapopulation studies include birds and small mammals on true and habitat islands, frogs in their naturally patchy environments, and many other insect species apart from butterflies. The patch-oriented metapopulation approach is apparently suitable for only a small number of plant species, mostly short-lived species in sparse vegetation, as exemplified by the chapter of van der Meijden and van der Veen-van Wijk.

Returning to butterflies, they have produced much empirical data because it is relatively easy to define, for many though not all species, what is suitable habitat, often based on the presence of one or two relatively scarce and patchily distributed host plants. It is easy to delineate habitat patches for many butterflies, it is possible to map large areas for suitable habitat and the presence of local populations, and it is convenient to study many aspects of population biology of butterflies, including individual movement behavior. So far, the work on butterflies has been largely restricted to ecology and metapopulation dynamics (Thomas and Hanski), but very soon we should have related studies on metapopulation genetics and life-history evolution. The quality of the empirical data has been sufficient to parameterize metapopulation models, and model predictions have been successfully tested in the field.

Nonetheless, even with all these virtues of butterflies, not all is as clear as might first appear. The causes and mechanisms of population extinction is a case in point. Extinctions are often simply attributed to environmental stochasticity, but exactly what is "environmental stochasticity"? The standard answer is: Changing environmental conditions affecting many individuals in a population in similar manner. A drought causing mass mortality of caterpillars is environmental stochasticity, but if all plants happen to die, then we have a "deterministic" extinction. Yet, if conditions improve by next year, was it not, after all, a minor catastrophe, extreme environmental stochasticity? Or take small populations, which go extinct because small populations in small habitat patches have a high risk of stochastic extinction. Butterflies may, however, be very likely to fly away from very small patches, with large perimeter to surface ratio, and hence populations in small patches may go deterministically extinct due to emigration losses. Thomas and Hanski discuss such ecological issues with the large data base that butterflies now provide. As with spatial population structures, the processes of population extinction and establishment are more diverse than the models might imply. Though at some level the details may not matter very much, the reward to an ecologist is a mechanistic understanding of what is actually happening.

The mammalian case study by Smith and Gilpin is on the American pika, a metapopulation that Smith has followed for more than 20 years. Habitat patches consist of rock tailings from previous mining operations. These patches are of the same quality and they are located in the midst of a practically uniform sagebrush vegetation, which is entirely unsuitable for pikas. Local populations are small, mostly less than 12 animals, with extinction largely attributable to demographic stochasticity. Movements among the patches are infrequent, as shown by both behavioral and genetic studies. This fragmented landscape and the pika metapopulation living in it satisfy closely the assumptions of the classical metapopulation concept, and the pika metapopulation does indeed exhibit classical metapopulation dynamics with population turnover, effect of patch size on extinctions and effect of isolation on colonizations. A spatially correlated pattern of patch occupancies has evolved over the past 20 years, possibly driven by regional stochasticity, in other words by year-to-year variation in the extinction and colonization rates.

In the pika metaapopulation, mustelid predators may significantly increase the risk of local extinction. Van der Meijden and van der Veen-van Wijk describe another 20-year field study, in which interspecific interactions are evidently critically important: the ragwort plant on a Dune area in the Netherlands, its specialist herbivore the cinnabar moth, and a braconid parasitoid attacking the latter. In contrast to the pika metapopulation, among-patch movements are common in the ragwort–cinnabar moth–parasitoid system, with the moth apparently being the most dispersive species of all. Movement patterns go a long way in explaining the dramatic spatial and temporal patterns in plant biomass and herbivore numbers that van der Meijden and van der Veen-van Wijk have observed. What they describe is not a classical metapopulation scenario, but spatial structuring of populations, which nonetheless has critical consequences for dynamics. The tempting (but possibly difficult) experiment would be to create one very large patch of ragwort in this system, and observe the dynamics of the three species in the absence of habitat subdivision. As van der Meijden and van der Veen-van Wijk point out, the outcome is difficult to predict, partly because, in the absence of habitat fragmentation, the poorly dispersing parasitoid might interact much more strongly with the moth and fundamentally change the now so dramatic plant–herbivore oscillations.

The fourth case study by Giles and Goudet is entirely genetical. Their study is focused on a structured metapopulation in which patch age is known and in which the history of local population size can be estimated. They study a plant species on a system of approximately 50 islands off the coast of Sweden. The plant population undergoes colonization and growth, followed by a slow decline due to successional factors; otherwise there is no significant turnover in this system. The key question which they ask is whether extinction–colonization dynamics in a metapopulation tend to increase genetic differentiation among populations relative to the situation where there is no population turnover. Using electrophoretic data, Giles and Goudet estimate F_{ST} for various age and size classes of islands and populations, and they find significant genetic structuring both within and between islands. They also find that the F_{ST} structure depends on patch size and patch age in ways that accord with theoretical predictions. Their results underscore the conclusion of Hedrick and Gilpin that a pattern of F_{ST} variation can be produced by different combinations of metapopulation parameters, in this case propagule size, colonization frequency, gene flow and effective population sizes. This chapter points to the next generation of metapopulation field studies, where the ecology, genetics, and ultimately evolution of species are investigated in a natural landscape.

15   Butterfly Metapopulations

Chris D. Thomas   Ilkka Hanski

I. INTRODUCTION

Some taxa have played a disproportionate role in the development of ecological concepts and theories. The dynamic theory of island biogeography (Mac-
Arthur and Wilson, 1967) was illuminated by numerous examples on birds, which
are sufficiently well known even on remote oceanic islands. At present, we feel
that butterflies have gained a somewhat similar status in the study of metapopulation dynamics, especially in Europe (Hanski, 1996b). There are several reasons
for this. First, butterfly populations are often structured in space in a manner that
is broadly consistent with the metapopulation concept. Second, the ecology of
butterflies is well known in most countries in Europe and elsewhere. Third, an
exceptionally large fractions of butterflies, in northern Europe in particular, have
declined, become endangered, or gone extinct already. Athough the examples
used in this chapter exclusively concern butterfly metapopulations, we expect
many of the patterns and processes described to be applicable to a much wider
range of organisms.

Despite a great deal of effort and some notable conservation successes, attempts to conserve rare butterflies at the scale of entire countries have largely
failed (New, 1991; New *et al.*, 1995; Pullin, 1995). In Britain, despite extensive
knowledge of the ecological requirements of individual species, rates of popu-

Metapopulation Biology
Copyright © 1997 by Academic Press, Inc. All rights of reproduction in any form reserved.

lation extinction have been nearly as high on nature reserves as in the rest of the landscape (Heath, *et al.,* 1984; J. A. Thomas, 1984, 1991; Emmet and Heath, 1990; New, 1991; New, *et al.,* 1995; Warren, 1992, 1993). The traditional conservation approach has been to protect and manage habitat in isolated reserves as effectively as possible, but it now appears that the fate of the rest of the landscape may play an important role in the long-term maintenance of butterfly populations within protected areas. As a result, the metapopulation approach has been widely adopted by butterfly ecologists and conservationists, and has spawned studies on metapopulations of a number of butterfly species. In this chapter, we review the ecological insights that have been gained from these studies. The conservation implications have been reviewed elsewhere (C. D. Thomas, 1994a,b,c; New, *et al.,* 1995).

We commence by describing the general features of spatial population structure in butterflies, and the effects of landscape structure on butterfly populations. The two critical elements in the metapopulation framework are the effects of habitat patch area and isolation on the distribution of species. We then turn to the two key metapopulation processes, extinction and colonization, which are better studied for butterflies than perhaps for any other comparable taxon. The next section discusses empirical tests of theoretical predictions, another area where butterfly studies have played a critical role in recent years. With rapidly accumulating knowledge, certain model assumptions appear increasingly suspicious, and we have a long section on various particulars that are now missing from metapopulation models but should probably be included in the next generation of models. This leads to the final section of conclusions and what the future may look like.

II. CONSEQUENCES OF LANDSCAPE STRUCTURE FOR POPULATION STRUCTURE

A. Population Structure

More than 50 years ago, E. B. Ford and colleagues (Ford, 1945; Ford and Ford, 1930; Dowdeswell *et al.,* 1940) recognized that most individuals of many species of butterfly remain within their natal habitat (patch), but that a few individuals may stray some kilometers from the breeding populations, and could be influential in the establishment of new local populations. Ehrlich and colleagues developed this understanding in a long-term study of three local populations of the checkerspot *Euphydryas editha bayensis* on a serpentine outcrop at Jasper Ridge, in California (Ehrlich, 1984; Ehrlich *et al.,* 1975; Ehrlich and Murphy, 1987; Singer and Ehrlich, 1979). Mark–release–recapture experiments showed that 97% of individuals remained within the local population in which they were first marked and where most of them must have emerged. The smallest local population became extinct, but the level of exchange among populations was adequate for this site to be recolonized rapidly from the surviving local populations. Subsequent work on *E. editha bayensis* on other serpentine outcrops confirmed that small local populations are most susceptible to extinction and revealed that potential breeding areas of serpentine grassland have a good chance of becoming recolonized if they are within 5 km of an existing large local population (Harrison *et al.*, 1988; Harrison, 1989; Weiss *et al.*, 1988).

The foregoing research programs were conducted or initiated before Levins (1970) coined the term "metapopulation," and before metapopulation ideas became popular in the late 1980s (Hanski and Gilpin, 1991). Nonetheless, the authors of these early studies recognized the same components of population structure which are now part of the metapopulation concept: distinct breeding areas (habitat patches) within which large numbers of adult butterflies spend their entire life, with limited migration between local populations. About three-quarters of British butterflies and 60% of Finnish butterflies appear to have a population structure of this general type, and most endangered species that have been studied in detail conform to this pattern (Arnold, 1983; New *et al.*, 1995; Hanski and Kuussaari, 1995; C. D. Thomas, 1996). Local extinctions and colonizations have been observed in most studies of butterfly metapopulations, but we do not regard this as a prerequisite before using the term metapopulation to describe a particular system (Hanski and Kuussaari, 1995); nor do we wish to imply that local extinctions and colonizations are necessarily in balance, as this is probably rare in many modern landscapes (Harrison, 1991, 1994b; Hanski, 1994b, 1996b, this volume; Hanski *et al.*, 1996b; C. D. Thomas, 1994a,b,c). Butterfly metapopulations come in all sorts of shapes and sizes, but the term metapopulation is nonetheless helpful in that it focuses the attention of researchers and conservation managers on processes at regional scales. Despite a wide range of specific differences among metapopulations, we are encouraged that a rather small number of patterns and processes appear to be important in many cases that have been intensively studied. This is illustrated by the empirical results reviewed in the following sections.

B. Effects of Habitat Patch Area and Isolation on Distribution

A serious concern in any empirical metapopulation study is to recognize suitable habitat independently of the presence of the focal species. If this cannot be done, the appropriate habitat network cannot be identified. The skipper butterfly *Hesperia comma* is mostly restricted to southerly facing dry calcarous grasslands in southern England, lays its eggs on one species of grass *(Festuca ovina)*, and only on plants of certain size growing in a particular microhabitat (J. A. Thomas *et al.*, 1986). Depending on their aspect, slope, local grazing, and disturbance, only a subset of calcareous grassland fragments represent suitable habitat patches for this butterfly. It is the distribution of these special habitat patches, not the overall distribution of calcarous grassland, which forms the habitat patch network for this species. Using a similar approach, Harrison *et al.* (1988) were able to define which patches of serpentine grassland could be regarded as potential habitat for *E. editha bayensis*.

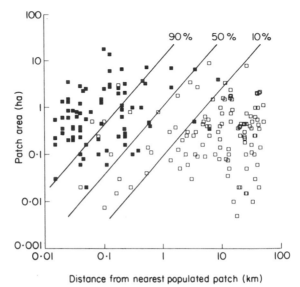

Distance from nearest populated patch (km)

FIGURE 1   Occupied (solid) and vacant (open) habitat patches for *Hesperia comma* in calcareous grasslands in England in relation to patch area and isolation. Lines show combinations of area and isolation which give 90, 50, and 10% probability of patch occupancy, from logistic equations. Reprinted with permission from C. D. Thomas and Jones, 1993, Blackwell Science Ltd., [Osney Mead, Oxford OX2 OEL, UK].

Once potential habitat has been identified, it is possible to map the distribution of occupied and empty habitat. Studies of many butterfly metapopulations reveal that large habitat patches are usually occupied, especially if they are close to other patches, whereas relatively small and/or isolated patches are the ones most likely to be vacant (Fig. 1; Harrison *et al.,* 1988; C. D. Thomas and Harrison, 1992; C. D. Thomas *et al.,* 1992; C. D. Thomas and Jones, 1993; Hanski, 1994a,b, this volume; cf. Arnold, 1983). These patterns are consistent with the idea that small local populations are prone to extinction and that isolated habitat patches are least likely to be (re)colonized. The same pattern extends to networks of habitat patches; the fraction of suitable habitat occupied by Glanville fritillaries (checkerspots) *Melitaea cinxia* is greatest in patch networks in which patches are large and close together (Table I) (Hanski *et al.,* 1995a, this volume). Both patterns are consistent with model predictions (Hanski, 1991, 1994a,b, this volume; Hanski and Thomas, 1994).

The distribution of patch sizes in a fragmented landscape is thus of great importance (Harrison and Taylor, this volume). Metapopulations can be placed on a continuum from a "mainland–island" structure in which one habitat patch is much larger than all the others, and long-term persistence is dominated by the persistence of the largest (mainland) population (Boorman and Levitt, 1973), through to systems in which all local populations are equally important (Levins-type metapopulations). At one extreme, the Morgan Hill metapopulation of *E.*

TABLE I   Effects of Average Patch Area and Regional Density (Number of Patches in Squares of 4 km²) on the Fraction of Patches Occupied (P) by the Glanville Fritillary *Melitaea cinxia* on the Åland Islands in SW Finland[a]

Average patch area (ha)	Occupancy		No. of patches per 4 km²	Occupancy	
	n	P		n	P
<0.01	23	0.24	1	61	0.21
0.01–0.1	138	0.24	2–3	70	0.32
0.10–1.0	88	0.40	4–7	58	0.25
>1.0	6	0.56	>7	66	0.41

[a] Effects of both average patch area and regional density on occupancy are highly significant (from Hanski *et al.*, 1995).

editha bayensis is dominated by one very large and apparently persistent "mainland" population with transient "island" populations found nearby in smaller habitat patches (Fig. 2; Harrison *et al.*, 1988). At the other extreme, the metapopulation of *M. cinxia* on the Åland islands in the Baltic is an extensive system of hundreds of small local populations, each of which is potentially susceptible to extinction (Fig. 1 in Hanski, this volume; Hanski *et al.*, 1994, 1995a,b). Important differences between these two metapopulations appear to be due to differences in the distribution of habitat patch areas in the landscape, rather than to a fundamental difference in the biology of these closely related species (Hanski *et al.*, 1994). Indeed, different metapopulations of the blue butterfly *Plebejus argus* show nearly as much variation in their spatial structure as do all comparisons of metapopulation structure across all species studied to date. Differences in the rates and patterns of local extinction and colonization in different metapopulations of *P. argus* are apparently largely due to differences in the distributions of patch sizes and vegetation dynamics, not to differences in the butterfly (C. D. Thomas and Harrison, 1992). Most metapopulations occupy an intermediate position along this continuum, with some relatively large, but not necessarily permanently populated patches, and other small and/or isolated patches with higher turnover and lower probabilities of being occupied.

One difficulty in making deductions about the persistence and dynamics of metapopulations from a "snapshot" distribution of "occupied" and "empty" habitat is that empty habitat might not be suitable after all, because of some subtle unrecognised attribute of particular habitat patches. Although this potential problem should be considered carefully in every empirical study, we believe that it is a relatively minor issue in most butterfly studies. A few misclassified patches are unlikely to change the overall conclusions in any study of which we are aware. Introductions of butterflies to empty habitats have succeeded in establishing new local populations on numerous occasions (Oates and Warren, 1990; C. D. Thomas, 1992; C. D. Thomas and Harrison, 1992; Nève *et al.*, 1995), and other studies have reported natural colonization of patches which had previously been

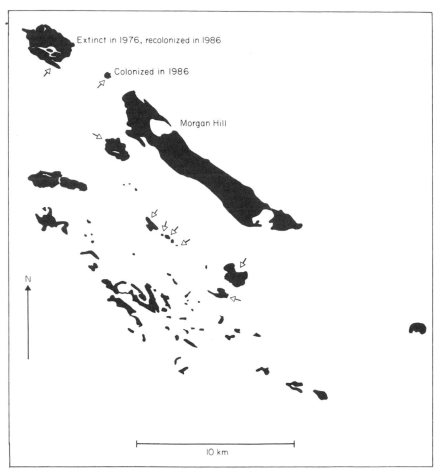

FIGURE 2   The Morgan Hill metapopulation of *Euphydryas editha bayensis*. Black areas are serpentine outcrops. "Mainland" Morgan Hill supported $\approx 10^6$ butterflies, with "island" outcrops supporting 10^1–10^2 butterflies each (arrowed). Eighteen of the more isolated outcrops were suitable but unoccupied. Reprinted with permission from Harrison *et al.*, 1988, University of Chicago Press.

identified as unoccupied but suitable habitat (Fig. 3; C. D. Thomas and Jones, 1993). Introductions and natural colonizations both demonstrate that at least some empty habitat was indeed suitable.

III. POPULATION TURNOVER IN BUTTERFLIES

A . Local Extinction

Empirical evidence on local extinctions and colonizations is still incomplete, given the need for long-term study. Excluding extinctions caused by complete destruction of habitat, such as deforestation, cultivation, and urbanization, relatively high rates of butterfly extinction are generally observed in small populations in small habitat patches; this pattern has been found in *E. editha bayensis* (Ehrlich *et al.*, 1975; Ehrlich and Murphy, 1987; Harrison *et al.*, 1988), *M. cinxia* (Fig. 3, Hanski *et al.*, 1994, 1995b), *P. argus* (C. D. Thomas and Harrison, 1992; C. D. Thomas, 1994b, 1996), and *H. comma* (C. D. Thomas and Jones, 1993). Local populations which contain hundreds of adult butterflies are frequently seen to become extinct during studies lasting 3 to 30 years (though some such populations survive for decades), whereas populations of thousands or tens of thousands rarely become extinct during studies of the same duration, provided the environment remains relatively stable. Historical records indicate that extremely large populations can become extinct over longer periods, usually associated with declines in habitat quality (J. A. Thomas, 1984, 1991; J. A. Thomas and Morris, 1994; see also *E. editha* below).

Population extinction appears to be affected by many factors. (i) Demographic stochasticity, the chance variation inevitably associated with death and birth, may cause the extinction of a small population even in a constant environment. This form of stochasticity is most likely to be significant where habitat patches are very small, and especially in species with gregarious caterpillars, because the fates of caterpillars in one group are to a large extent correlated. Small local populations of the Glanville fritillary *M. cinxia* often have just one or a few groups of caterpillars (50–100 individuals per group in late summer, but fewer in spring; Hanski *et al.*, 1995a; Fig. 3). Pure chance must be a major

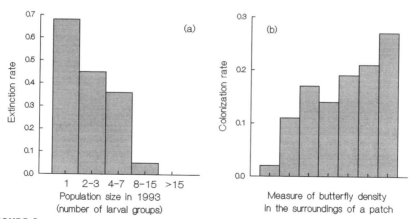

FIGURE 3   Rates of population extinction and colonization of empty meadows in the large metapopulation of *Melitaea cinxia* (Fig. 1 in Hanski, this volume). (a) Fraction of populations that went extinct from 1993 until 1994 as a function of the size of the population, measured by the number of larval groups detected in 1993. (b) Rate of recolonization as a function of isolation. Isolation was measured by the sum of the sizes of the surrounding populations, weighted by their distance to the focal meadow (see Eq. (4) in Hanski, 1994a).

factor in the dynamics of such small populations, which have a high probability of extinction even if there is nothing wrong with the habitat patch. (ii) Environmental stochasticity; drought, aseasonal cold, and other extreme weather events, is known to have generated extinctions in populations of the blue *Glaucopsyche lygdamus*, *E. editha*, *M. cinxia*, the ringlet *Aphantopus hyperantus*, and the fritillary *Boloria aquilonaris* (Ehrlich *et al.*, 1972, 1980; Harrison *et al.*, 1988; Hanski *et al.*, 1995a, unpublished; Pollard and Yates, 1993; T. Ebenhard, personal communication). (iii) The interaction between environmental stochasticity and habitat heterogeneity can cause extinctions. Large patches usually contain several microhabitats, not all of which will become equally inhospitable as a result of a single environmental event. Variation in local topography and turf height appear to buffer populations of *E. editha bayensis* and *H. comma*, respectively, from local extinction in large habitat patches (Singer, 1972; Weiss *et al.*, 1988; Murphy *et al.*, 1990; C. D. Thomas, 1994a; see Hanski, this volume, and below). (iv) When a metapopulation inhabits more than one type of habitat, rare environmental events may cause widespread extinctions in some habitat types but not in others. Again, *E. editha* provides an example (Ehrlich *et al.*, 1980; C. D. Thomas *et al.*, 1996). Rare events may be particularly likely to cause extinction from suboptimal habitat. One local population of the blue butterfly, *Lysandra bellargus*, became extinct from relatively poor habitat patch, during a drought in 1976, while two nearby populations survived in better quality habitat (J. A. Thomas, 1983a). Meanwhile, *A. hyperantus* populations appeared to survive the same drought better in areas of intermediate shade than in the open or in denser shade (Sutcliffe *et al.*, 1996c). (v) Vegetation dynamics/grazing patterns and other habitat changes render particular patches unsuitable, and the respective populations decline deterministically to local extinction. This appears to be the most common cause of local extinction for medium-sized and large local populations and may therefore be of particular importance to the persistence of metapopulations. There are examples of this mechanism for all species covered in this chapter (e.g., Harrison, 1991, 1994b; J. A. Thomas, 1991; Warren, 1992, 1993; C. D. Thomas, 1994a,b,c, 1996; see Section V.B below). Because vegetation dynamics can occur at a variety of scales, loss of all suitable habitat from a small patch (population) is more likely than complete loss of suitable habitat from an initially large patch. (vi) Isolated local populations may suffer increased rates of extinction (C. D. Thomas and Jones, 1993; Hanski *et al.*, 1995b) as a result of a reduction in the number of immigrants. When such rescue effects (Brown and Kodric-Brown, 1977) become weak, populations in poor (sink) habitat may become isolated from source populations and become extinct (cf. Warren, 1994; Rodríguez *et al.*, 1994), and populations in small habitat patches may disappear because immigration no longer replaces individuals lost to emigration (we discuss this possibility in detail below).

As more evidence becomes available, it is becoming apparent that environmental stochasticity, isolation, vegetation dynamics, and both subtle and extreme human activities interact to generate patterns of extinction and that we cannot expect a single factor to explain all extinctions even within a single metapopulation.

B. Colonization

Empirical observations of colonization present compelling evidence that migration occasionally takes place over distances considerably greater than those normally revealed in mark–release–recapture studies conducted within existing metapopulations (C. D. Thomas *et al.,* 1992; Hanski and Kuussaari, 1995). Patterns of colonization have been particularly clear when entire networks of empty habitat patches have been available for colonization (C. D. Thomas *et al.,* 1992; C. D. Thomas, 1994a,b,c). *Hesperia comma* invaded habitat patches which had recently become suitable for this species in part of southern England. Over a 9-year period, population establishment occurred with a high probability in large patches which were close to existing populations; beyond 1 km, the probability of colonization declined, and the maximum observed colonization distance was 8.6 km (C. D. Thomas and Jones, 1993). The black hairstreak *Srymonidia pruni,* the skipper *Thymelicus acteon,* the heath fritillary (checkerspot) *Mellicta athalia, M. cinxia,* and *P. argus* have similarly showed decreased probabilities of colonizing relatively isolated habitat patches (Fig. 3; J. A. Thomas, 1983b; C. D. Thomas *et al.,* 1992; C. D. Thomas and Harrison, 1992). These results indicate declining rates of migration with increasing distance. Over short distances, immigration rates are sufficiently high that successful colonization is practically assured (for *H. comma,* this was attained within 1 km, after 9 years). At somewhat greater distances, colonization is observed in empirical studies but with decreasing probability. At longer distances still, colonization is not observed on the time-scale of empirical studies.

Several species have been deliberately released and have become established in a new region (see C. D. Thomas and Harrison, 1992; C. D. Thomas *et al.,* 1992; C. D. Thomas, 1992, 1994a, 1996, for examples). Nève *et al.* (1995) report the sequential colonization of 24 *Polygonum bistorta* (host plant) meadows by the bog fritillary *Proclossiana eunomia,* following the butterfly's introduction to the Morvan region in France. Within a network of habitat patches, there was a noisy, but approximately linear, relationship between distance achieved and time since introduction. Diffusion models of colonization predict a linear relationship between distance and time if there are no major barriers to colonization (Andow *et al.,* 1990; Nash *et al.,* 1995), suggesting that the patch structure may not have been a major constraint on colonization in this metapopulation, though the scatter of data points leaves the question open. Migration distances recorded in mark–release–recapture studies on *P. eunomia* revealed substantial movements of individuals among patches (Baguette and Nève, 1994; Nève *et al.,* 1995), again indicating that patch isolation may not be great within metapopulations. In other species, migration rates between local populations vary between about 1.4% in a metapopulation of the particularly sedentary *P. argus* in North Wales (O. T. Lewis *et al.*, unpublished) and 3% in Jasper Ridge *E. editha* to >20% in several species (e.g., *A. hyperantus,* Sutcliffe *et al.,* 1996c; *H. comma,* Hill *et al.,* 1996; see Warren, 1987a; C. D. Thomas, 1994a; Hanski and Kuussaari, 1995, for other exchange rates). Overall, high exchange rates among patches and rapid colonization suggest that the rescue effect is common in butterfly meta-populations and that isolation may not be a major constraint on colonization of empty habitat *within* many existing metapopulations. Often only a small fraction of habitat patches is empty within a connected patch network, unless patch sizes are extremely small (C. D. Thomas and Harrison, 1992). High migration rates and consequent strong rescue effects may generate alternative stable equilibria in metapopulation dynamics; i.e., almost all patches occupied or metapopulation extinction. A putative example is described for *M. cinxia* by Hanski *et al.* (1995b; see Fig. 4 in Gyllenberg *et al., this volume*).

In contrast to the situation within patch networks, isolation is an extremely important reason why empty habitat does not become colonized beyond the rec-ognized boundaries of existing metapopulations (C. D. Thomas and Harrison, 1992). Groups of habitat patches are often separated by distances which will prevent colonization from another patch network (1 to >20 km separation may provide an effective barrier, depending on the species). Once such metapopula-tions become extinct, reestablishment can be very slow, which highlights the importance of obtaining an empirical and theoretical understanding of the factors that contribute to the persistence of entire metapopulations.

IV. THEORETICAL PREDICTIONS TESTED

Despite many complications in specific butterfly metapopulations, empirical data lend considerable support to the central tenets of metapopulation theory, namely that extinction is related to patch area/population size and colonization is distance-dependent and that these relationships contribute to the observation that populations are most likely to be present in habitat patches which are large and close together (Hanski, 1994b). These general results give us confidence that the approach is valuable, even if more specific and complex models may be required to predict the dynamics of a particular system.

The potential ability to test model assumptions and predictions is one of the attractions of using specific, spatially explicit models, of the type outlined in Hanski (1994a, this volume; Hanski and Thomas, 1994; Sjögren Gulve and Ray, 1996). In these models, the location and area of each habitat patch is specified, with rules governing the probability of local extinction (mainly dependent on population size) and the probability of colonization (mainly dependent on isola-tion and source population sizes). In this section, we deal with the predictions of existing models and consider some of the model assumptions in the following section.

We have been able to test the predictions of a spatially explicit metapopulation model with data for *H. comma* (Hanski and Thomas, 1994). This model explicitly iterates local dynamics in each occupied patch, which are connected to each other via distance-dependent migration (modeled at the population rather than individual level). Colonization is a mechanistic consequence of immigration to an empty patch. Altogether, the model has nine parameters. Where possible, parameter values should be estimated independently, but some parameters are very difficult or extremely time-consuming to estimate reliably. In our case, parameter values which could not be measured independently were estimated by simulation from a metapopulation which was assumed to be at equilibrium; this was done by choosing parameter values which generated model predictions that matched empirical patterns of patch occupancy. The model was then applied, using these parameter values, to a *different* metapopulation, which was clearly not at equilibrium, to test whether the model could successfully predict dynamics in a nonequilibrium metapopulation.

In England, *H. comma* occupies short, sparse, dry grasslands. Myxomatosis removed rabbit grazing in the mid-1950s and the grassland habitat became overgrown, with the result that the skipper butterfly became very localized between 1960 and 1975 (J. A. Thomas *et al.,* 1986). In East Sussex, *H. comma* became restricted to one large population. In 1982, the skipper was still thriving in this large refuge population and had colonized two small nearby habitat patches. By 1982, it was clear that substantial areas of habitat were again suitable for breeding by *H. comma,* as rabbits had partially recovered from myxomatosis and conservation organizations had begun to undertake active grazing management on some of the previously overgrown grasslands. This represented an ideal situation in which to test the predictions of the model; a network of empty patches which could be mapped and a known distribution of the skipper in 1982. The predictions of the models could then be compared with the observed skipper distribution after 9 years of colonization (C. D. Thomas and Jones, 1993). The results were a qualified success. The model predicted that the skipper would spread in this region and that it would not spread in other regions where it failed, in reality, to expand its distribution, but the model underestimated the real rate of expansion (Hanski and Thomas, 1994). *Hesperia comma* actually occupied 21 patches after 9 years (Fig. 7 in Hanski and Thomas, 1994), which was more than the model predicted (mean 8.6 patches occupied after 9 years, maximum 11, in 100 replicate simulations). The quantitative mismatch between prediction and observation prompts a series of new questions. Was the region used to parameterize the model really at equilibrium (not quite; C. D. Thomas and Jones, 1993)? Was the negative exponential distribution used for migration/colonization appropriate (new mark–release–recapture data were collected to resolve this issue, and migration distances were found to fit a negative power function better, implying more long-distance migrants than in the original simulations; Hill *et al.,* 1996)?

Hanski *et al.* (1996c) have attempted to predict the distribution of *M. cinxia* over some 1000 km² on Åland in the Baltic, using an incidence function model

(Hanski, 1994a). Again the results are encouraging but mixed. The model was parameterized using data collected from a small part of Åland in 1991 (Hanski *et al.*, 1994) and model predictions were tested with independent data collected in 1993. Over large parts of Åland, the model-predicted and observed fractions of occupied habitat were in good agreement (Fig. 8 in Hanski, this volume), but this began to break down in drier areas of the island and in areas where the level of grazing was an important determinant of the butterfly's distribution. As with *H. comma*, testing the model revealed that there were further aspects of the biology of the species that needed to be understood.

Another kind of problem for predicting species' distribution in fragmented landscapes is posed by the possibility of multiple equilibria, for which there is empirical evidence in the case of *M. cinxia* (Fig. 4 in Gyllenberg *et al.*, this volume; Hanski *et al.*, 1995b). Multiple equilibria impose inherent uncertainties on our ability to predict, accurately, the distribution of species in particular networks. History can also play a crucial role when habitat is itself dynamic. Although a habitat network may be extensive enough to support a persistent metapopulation now, the whole network may be too isolated to have been colonized. In some areas where *H. comma* used to occur, and where it became extinct when the habitat was overgrown, the habitat has now recovered. Simulations suggest that a substantial *H. comma* metapopulation could be reestablished in at least one such area, but the butterfly has failed to recolonize this area naturally and simulations indicate that it is likely to take more than 100 years to do so (C. D. Thomas and Jones, 1993; Hanski and Thomas, 1994).

A. Metapopulation Persistence and Establishment

One of the major potential uses of models is to predict whether metapopulations are likely to persist in specific networks of habitat patches. The models referred to above (Hanski, 1994a; Hanski and Thomas, 1994) can be used for this purpose, assuming of course that one has been able to parameterize them. One may use the models to explore how the size and persistence of a particular metapopulation is affected by further loss of suitable habitat, which is likely to increase the risk of local extinction and to decrease the probability of recolonization. Hanski (1994a,b) and Hanski and Thomas (1994) give examples. More generally, one may ask how the expected metapopulation lifetime depends on the number of extinction-prone local populations that are connected to each other. Hanski *et al.* (1996c) have explored this question with the incidence function model using the observed patch networks for *M. cinxia* on the Åland islands. The expected time to metapopulation extinction is closely related to the product $p\sqrt{H}$, where p is the fraction of occupied habitat patches at stochastic steady state and H is the number of suitable patches (Hanski, this volume). A rough rule of thumb is that if this product exceeds 3 the metapopulation is likely to persist much longer than the expected lifetime of a local population. Assuming that the species occupies most patches in the patch network in which it is present, a minimum of some 15–20 well-connected patches are required for long-term persistence. With spatial correlation in local dynamics and much environmental stochasticity, even this would not be enough (Hanski *et al.*, 1996b).

Empirical tests of the persistence of entire metapopulations are very limited and, understandably, difficult to accomplish. The extinction of an entire metapopulation is much rarer and takes place on a longer time scale than does the extinction of individual local populations. As Harrison (1991, 1994b) and C. D. Thomas (1994a,b) have noted, metapopulation extinction is generally related to an overall decline in the amount of suitable habitat due to widespread human-caused changes in the landscape. Here, we are concerned whether a metapopulation is likely to persist in a network of habitat patches which is not becoming further degraded. In studies of *P. argus* and *H. comma,* C. D. Thomas (1994b) found that regions with over 15 to 20 habitat patches tended to be populated, but that regions with < 10 patches were rarely populated, suggesting that the latter may be inadequate for long-term metapopulation persistence. Figure 4 shows just the same pattern for *M. cinxia*. T. Ebenhard (personal communication) was able to examine metapopulation extinctions directly, when working on the cranberry fritillary *B. aquilonaris* in an extremely cloudy and wet year. *Boloria aquilonaris* occurs in bogs within remnant forest patches in southern–central Sweden, with the number of cranberry-containing bogs per forest fragment representing the

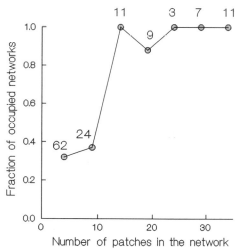

FIGURE 4 The frequency of occupancy of habitat patch networks by *M. cinxia* on Åland as a fraction of the number of patches in the network. For this analysis, the habitat patches (shown in Fig. 1 in Hanski, this volume) were divided into 127 semi-isolated patch networks, isolated by a physical barrier to dispersal or by *ca.* 1 km from the nearest other patches. Numbers of networks in each class is given by the number in the figure.

number of habitat patches per network. Ebenhard found that many of the meta-populations with fewer than 20 patches became extinct while those with > 20 patches survived a massive population collapse.

At this stage, theory and empirical data appear to be in approximate accord. A metapopulation rarely persists for very long at < 10 local populations and usually does so for extended periods at > 20 local populations. There is obviously a trade-off between the number of patches and area per patch, with fewer patches needed for some fixed length of metapopulation survival when each patch is large; one patch alone may ensure persistence if it is vast and heterogeneous (the theory described above predicts metapopulation lifetime in relation to the lifetime of local populations, and the results should be thus interpreted). When the habitat consists of transient or successional vegetation, much larger areas may be required to ensure that appropriate habitat is continuously available, the amount of extra vegetation required depending on the natural dynamics of the vegetation and on patterns of human management. Finally, if there is much regional stochasticity (spatially correlated environmental stochasticity), even very large metapopulations may go extinct, the extreme case being a catastrophe of some kind sweeping away the species from a large area (see below and Hanski, this volume). Given these and other uncertainties, we must stress that the patch numbers mentioned above should not be taken as rules or guidelines in conservation planning. Much larger numbers of patches will be required in some circumstances, but one vast patch may suffice in some cases. In the context of conservation, each case must be considered individually.

One of the most valuable uses of specific models is in predicting the consequences of possible future management options. Predicting the actual probability of long-term persistence is bound to be fraught with uncertainties, but predicting the likely direction of response to changes in the distribution of habitat over periods of a few decades is potentially of immense value to managers, and may be possible to achieve. We have already seen that a spatially explicit metapopulation model predicted correctly, in qualitative terms, where *H. comma* would expand its distribution in response to increased habitat quality and where it would not. Even without any further refinements, such models could be of use to conservation managers. It has been possible to draw maps of existing habitat around surviving *H. comma* populations and to assess the potential for further spread. In East Sussex, managers can relax in the sense that continuing the present management is likely to permit continued expansion of *H. comma*. In three other regions in southeast England, further expansion is not predicted (C. D. Thomas and Jones, 1993; Hanski, 1994a; Hanski and Thomas, 1994). When the species is unable to spread within the existing patch network, the consequences of many different management options can be explored by changing the distribution and sizes of specific habitat patches in the model. For example, would enhanced management to increase existing population sizes facilitate further spread, and would management of the surrounding areas to increase target patch areas or to decrease distances between patches facilitate spread? In a model, many different options may be compared quickly, before embarking on time-consuming and expensive management work. Even if the predictions are not quantitatively quite correct, the *relative* merits of different practical options may be robustly assessed.

An additional important application is likely to be in predicting the potential success of species translocation projects. Hanski and Thomas (1994) found that one set of parameter values correctly predicted the existing distribution of *P. argus* on limestone grassland, and its invasion of two new networks of habitat patches to which it had been successfully introduced in the past. For *H. comma,* the model has been used to identify at least one network which could potentially support a viable metapopulation of the skipper, but which is too isolated for colonization to occur naturally. This approach may help to eliminate some of the effort wasted on releasing rare butterflies in single or very small groups of habitat patches where long-term persistence is improbable (Oates and Warren, 1990; C. D. Thomas, 1992). Using spatially explicit models, the target species can be introduced to the same patch network repeatedly, and the value of releasing a fixed number of individuals at one versus many sites can be explored. There are several lessons to learn. For example, managers should not necessarily give up even if the first attempt fails; the introduced population may have to exceed some threshold before establishment is likely. This is especially true if there are multiple equilibria (Hanski *et al.,* 1995b; Gyllenberg *et al., this volume*). In most cases, the largest, high quality and least isolated habitat patches in the network should be targeted for releases (Hanski, 1994b), even if these patches fall outside existing nature reserves in the network.

V. ADDING REALISM TO THE METAPOPULATION CONCEPT

At this relatively early stage in the development of specific and predictive metapopulation models, field studies are needed to test model assumptions and to identify additional behavioral and ecological factors which may need to be incorporated into the next generation of models. Having identified existing short-comings, it is important to find out whether the predictions of refined models differ substantially from the predictions of simpler models, because adding many more parameters to models creates new problems. In this section we address model assumptions and complications that we believe will have important implications for metapopulation persistence, patterns of distribution, and rates of colonization.

A. Migration

In Hanski and Thomas (1994, p. 170), we highlighted *"the need for good empirical data on emigration and immigration rates in butterflies."* In the model, we assumed that the distribution of dispersing individuals could be fitted to a negative exponential function, which is typical of metapopulation models (Harrison *et al.,* 1988; Hanski, 1994a; Hanski and Thomas, 1994; Akçakaya, 1994; Sjögren Gulve and Ray, 1996). This distribution clearly gives the right general pattern, of many short-distance movements and a few long-distance movements, but needs to be tested more rigorously with empirical data.

When adult *P. argus* were released into an extensive area of empty habitat, it was found that the distribution of migration distances was a close fit to a negative exponential (O. T. Lewis, C. D. Thomas and J. K. Hill, unpublished). However, individuals that moved the longest distances may have left the whole study area, and hence the tail of the distribution is likely to have been underestimated. Mark–release–recapture work has shown that the between-patch distribution of distances moved by *H. comma* (Hill *et al.,* 1996) fits a negative power function well,

$$M = zD^{-k}, \tag{1}$$

where M is the fraction of individuals reaching distance D, and z and k are constants. For *H. comma,* the negative exponential distribution gives a slightly poorer fit to the data than does Eq. (1); in particular, the negative exponential underestimates the proportion of butterflies that fly relatively long distances. We surmise that this tail of the distribution is generated either by individuals that change behavior during migration, after they have initially failed to locate new habitat, or by individuals that are inherently dispersive. When appropriate, this alternative assumption about migration could easily be incorporated into spatially explicit models, and would presumably generate more long-distance colonizations than the negative exponential distribution.

In the absence of good field data, Hanski and Thomas (1994) assumed, with misgivings, that a constant proportion of individuals emigrates from each local population. Empirical data now show that the fraction of individuals emigrating are relatively high when patches are small and have high perimeter-to-area ratios (Hill *et al.,* 1996; Sutcliffe *et al.,* 1996c; Kuussaari *et al.,* 1996; M. Baguette and G. Nève, personal communication; see Kareiva, 1985). Therefore, it would be desirable to incorporate area-dependent emigration rates in spatially explicit metapopulation models. Unfortunately, it is very difficult to calculate the *actual* (as opposed to relative) rate of emigration in relation to patch area, because emigrants are rarely detected unless they immigrate into another habitat patch; substantial numbers of emigrants may fail to arrive in any patch. Since the spread of migrating individuals is neither entirely random (migrants may be attracted to new patches from some distance, and may then stay there) nor entirely directed, it may not be feasible to estimate the fraction lost accurately.

C. D. Thomas, O. T. Lewis, and J. K. Hill (unpublished) have used a simulation approach in order to estimate emigration rates (see also Buechner, 1987; Stamps *et al.,* 1987). We took imaginary habitat patches with a range of areas and placed butterflies in random locations within those patches. Butterflies were then allowed to migrate, by making each move away from its origin at a random angle, and for a distance chosen at random from the empirical ($\approx$ per generation) distribution of migration distances that we had already recorded for *P. argus* and

H. comma. The fraction of individuals leaving the patch was then recorded. These estimates of emigration rate have two main biases: (i) In reality, individuals may perceive patch boundaries and be reluctant to leave (we assumed that boundaries were fully permeable), leading to an overestimate of the emigration rate, and (ii) the proportion of individuals moving long distances may be under-represented in the empirical distributions of migration distances, leading to an underestimate of the emigration fraction. These two biases act in opposing directions. In any case, this exercise produced a strong and negative relationship between patch area and the fraction of individuals emigrating, with most individuals emigrating from small patches with high perimeter-to-area ratios. For a given area, the emigration rate differed considerably between the two species, reflecting differences in dispersiveness.

The implications of such emigration–area relationships have not yet been explored in the context of metapopulation models, but they are likely to be important. Within central parts of metapopulations, high perimeter-to-area ratios also result in high per-unit-area rates of immigration into small patches and potentially to high local densities (Hanski and Thomas, 1994; *H. comma*, Hill *et al.*, 1996; *M. cinxia*, Hanski *et al.*, 1994). However, the consequences are much more significant when habitat patches are relatively isolated. Changes in population size in isolated patches with no immigration can be described by (C. D. Thomas, O. T. Lewis, and J. K. Hill (unpublished)

$$N_{t+1} = RN_t e^{r(1 - RN_t/K)}, \tag{2}$$

where N_t is the number of individuals in generation t, r is the intrinsic rate of population increase, R is the proportion of individuals which are resident in the local population (proportion $1 - R$ emigrates), and K is the local carrying capacity. The equilibrium population size in this model is $(K/R)(1 + ln R/r)$ and extinction takes place when $ln R/r < -1$, that is, when local reproduction fails to replace losses due to emigration. Furthermore, even when isolated populations can sustain emigration losses, depression of local population size by emigration would make the populations more susceptible to other causes of extinction.

Incorporating the emigration–area relationship described above in the model, and using independent estimates of r and K, allows us to predict the expected population size of *H. comma* in habitat patches which are isolated enough for immigration to have little effect on local population size, but not so isolated that colonization is unlikely (Fig. 5). In Fig. 5, we assume that emigrants for that generation have already left at the time of census, but this assumption makes little difference to the overall pattern. The good match between the model predictions and empirical data suggests that many isolated patches of apparently suitable habitat are not populated because the losses due to emigration are too high for local reproduction to match (see Kareiva, 1985). The match is equally good for *P. argus*, though much smaller patches can be populated in isolation by this less dispersive insect. These results are encouraging and suggest that the immigration–emigration balance may determine patch occupancy patterns to a previously unexpected extent. At the same time, as these patterns can also be explained by

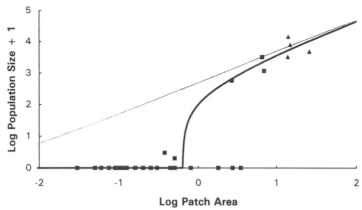

FIGURE 5   Predicted and actual population sizes in relation to $\log_{10}$ patch area (ha) for *Hesperia comma*. Thin line, predicted population size in the absence of emigration and stochastic extinction. Thick line, predicted patch occupancy and population sizes, with carrying capacity reduced by area-dependent emigration, according to Eq. (2) (adult butterflies were assumed to have emigrated at the time of census). One was added to all measured and predicted population sizes to allow zero values to be plotted. Measured population sizes in 1991 (■) and 1982 (▲). Patches plotted were >0.6 to 5 km from the nearest (other) population, which are regarded as sufficiently isolated that immigration is likely to have negligible effects on local density, but not so isolated that sites could not be colonized by occasional migrants (from C. D. Thomas, O. T. Lewis, and J. K. Hill, unpublished).

alternative hypotheses (area-dependent extinction rate for reasons other than emigration), critical field experiments would be welcome.

Area-dependent emigration rates have important dynamic implications. For example, successful invasion of empty patch networks is expected to proceed by the colonization of large patches first (as observed in *H. comma;* C. D. Thomas and Jones, 1993), or by stepwise colonization of small patches, gradually eroding isolation before population establishment is possible. Loss of a part of an isolated patch may result in rapid population extinction. Finally, the area of habitat needed to support a single isolated population is much larger than the minimum area of habitat that can be populated within a metapopulation where immigration roughly equals emigration in each patch.

B. Deterministic Population Responses to Habitat Change

Perhaps the most serious shortcoming of metapopulation theory has been the general assumption that the distribution of suitable habitat remains constant through time. Many butterfly species occupy successional vegetation. Even when habitats are potentially permanent, landscape changes brought about by human activities may drive significant changes in species distributions (Arnold, 1983; J. A. Thomas, 1991; C. D. Thomas, 1994a,b,c; New *et al.,* 1995). For example, many local populations of fritillary butterflies inhabiting woodland clearings in the United Kingdom have gone extinct because of successional changes in the vegetation, not because of stochastic fluctuations in local population size (Fig. 6; Warren, 1991; Warren and Thomas, 1992). Reviews of local and regional extinctions have argued that successional changes in vegetation, changes in human management of surviving habitat fragments, and outright habitat loss are principally responsible for most extinctions of substantial butterfly populations in modern landscapes; extinctions are frequently a deterministic consequence of the deterioration of local breeding conditions (Harrison, 1991, 1994b; Warren, 1993; J. A. Thomas, 1991; J. A. Thomas and Morris, 1994; C. D. Thomas, 1994a,b,c, 1996). Similarly, most colonizations appear to take place when environmental conditions improve locally (C. D. Thomas, 1994a,b,c, 1996). The spatial dynamics of many species appear to be driven by the changing distribution of their habitats. These insects are tracking a shifting habitat mosaic.

Examples described in this chapter clearly illustrate the importance of stochastic extinctions too, but these "traditional" metapopulation dynamics are superimposed on a dynamic habitat mosaic. When the dynamics of the butterfly are fast relative to vegetation dynamics, ignoring the latter may still leave a good fit between observed and predicted species distributions in the short term. This may be one reason why insects, with their fast dynamics, have become popular subjects for metapopulation studies. However, these models may not be particularly useful for predicting long-term trends or long-term probabilities of persistence if long-term changes are determined by underlying vegetation dynamics. The task of superimposing stochastic dynamics of a butterfly on top of vegetation dynamics has hardly begun.

We give two examples to show how butterfly metapopulation dynamics may lag behind changes in the spatial distribution of suitable habitat/vegetation. After earlier declines, *H. comma* has recently enjoyed an increase in the amount and extent of suitable habitat in southern England, particularly in the late 1970s and early 1980s (described above). After 9 years of documented colonizations, this butterfly had still not colonized all of the "new" habitat available to it, and it is likely that the distribution will take several more decades to reach an equilibrium (C. D. Thomas and Jones, 1993; Hanski and Thomas, 1994). Over the past 100+ years, there has been no period of 30 years or more when the distribution of *H. comma* habitat has remained even approximately stable, and it is hard to imagine that it will remain stable over the next 30 years. At least in modern landscapes, specialized species may be continuously chasing after their habitats.

Another common scenario may be continuing loss of habitat. In parts of Åland, *M. cinxia* habitat has been lost in recent decades, and predictions of spatially explicit models suggest that the dynamics may not be at equilibrium in all areas. Figure 10 in Hanski (this volume) illustrates such nonequilibrium dynamics with an example from *M. cinxia* on Åland. Within an area of *ca* 25 km², the total area of suitable habitat has declined to one-third and the number of habitat patches has decreased from 55 to 42 over the past 15–20 years. The metapopulation of *M. cinxia* is predicted to have followed this decline rather rapidly, apparently

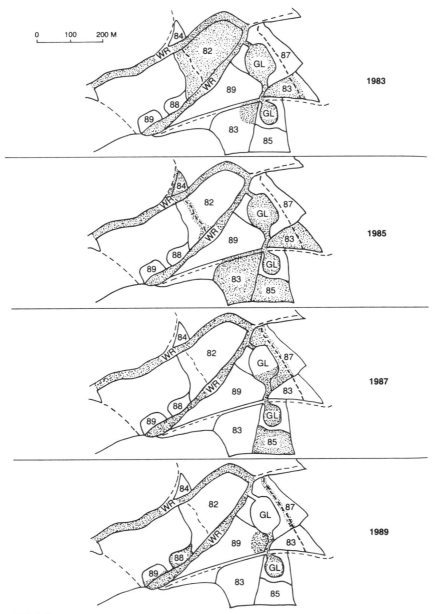

FIGURE 6   Changes in the distribution of *Mellicta athalia* in response to rotational cutting (coppicing) of deciduous woodland in Blean Woods, Kent, England. Numbers show year of cutting; shaded areas indicate the distribution of adult butterflies. Glades (GL) and wide rides (WR) were only sporadically used for breeding. Reprinted from Warren, 1991, with kind permission from Elsevier Science Ltd, The Boulevard, Langford Lane, Kidlington OX5 1GB, UK.

because it occupied most of the habitat before patch reduction, and most of the dynamics has occurred in small patches with fast turnover. However, an entirely different picture is predicted if the area of each patch is halved again over the next 20 years. In this case, further loss of habitat is predicted to lead to a network smaller than that required for long-term persistence (Fig. 10 in Hanski, this volume). However, the actual extinction of the metapopulation is predicted to take a long time, and for tens of years we would see a metapopulation slowly but inevitably oscillating to extinction. The final decline to extinction is slow because the last populations to go are typically the largest ones with the smallest risk of local extinction.

This latter finding is especially worrying, in that the status of many or even most species in landscapes which are gradually being degraded may presently be "better" than the expected status at equilibrium (Hanski, 1996b; Hanski *et al.*, 1996b,c). Nonequilibrium systems of this type may lead us to conclude that potential colonization distances are greater than they really are and to overestimate local population persistence in small patches. These biases result in overestimates of metapopulation lifetimes. Some currently surviving metapopulations may be doomed even if all further habitat loss is prevented.

C. Habitat Heterogeneity

Habitat heterogeneity is probably crucial to the persistence of many local butterfly populations and metapopulations. For convenience, there is a strong tendency in metapopulation ecology to define some parts of the landscape as "habitat" and the remainder as "nonhabitat" or "matrix"—and to ignore the latter. There are several problems with this (see also Wiens, this volume). (i) The environment may exist as a series of habitats which vary in suitability, and it may be difficult to distinguish between habitat and nonhabitat (Rodríguez *et al.*, 1994). Typically, not all suitable habitat will be of equivalent quality (Hochberg *et al.*, 1992, 1994; C. D. Thomas, 1996). (ii) Vegetation dynamics and human activities can change the suitability of a given habitat patch; some of these changes will be predictable (succession in a woodland clearing resulting in a gradual decline in patch quality for clearing butterflies), but other changes may be unpredictable and reversible. (iii) Temporal environmental variability may affect an entire environmental gradient. For example, the warmest and driest parts of a habitat patch may represent the environmental optimum in normal years, but be inhospitable in a drought year, requiring the drought-affected species to move into taller vegetation (*H. comma*, C. D. Thomas, 1994a) or to more mesic slopes (*E. editha bayensis*, Weiss *et al.*, 1988; the large blue *Maculinea arion*, J. A. Thomas, personal communication). Singer (1972) and Weiss *et al.* (1988) found that complex spatial variation in the microdistribution of disturbance and of two host plant species, and the aspect of the slope, interacted with climatic variation to affect population persistence and changes in population size in *E. editha bayensis* populations on serpentine grassland. Since a greater range of microhabitats is more likely to be present in large than in small areas, this may be an extremely important reason why populations are often most persistent in large habitat fragments. (iv) Some occupied habitat patches may not be suitable for population persistence; populations may be present in these habitats (sinks) only because of immigration from source populations (Pulliam, 1988; Rodríguez *et al.*, 1994; Warren, 1994).

Adding realistic variation in habitat quality is likely to be one of the key issues facing empirical and theoretical metapopulation biologists in coming years. The following example (C. D. Thomas *et al.*, 1996) is almost certainly unusual, but nonetheless highlights the potential complexity of metapopulation dynamics when more than one habitat type is present. A metapopulation of the checkerspot *E. editha* (an unnamed race which differs from *bayensis*) occurs at 2000 to 3000 m elevation in openings in coniferous forest in Sequoia National Forest and Sequoia National Park in California. Before 1967, the butterflies were restricted to natural rocky outcrops and laid most of their eggs on *Pedicularis semibarbata* (Scrophulariaceae). Around 1967, clearings in the forest were made by logging. *Pedicularis semibarbata* disappeared from clear-cut areas, but the butterfly invaded this habitat and colonized *Collinsia torreyi,* which is also in the Scrophulariaceae, but which is not used on outcrops (Singer, 1983; Singer *et al.*, 1993, 1994). By 1985, a patchwork of host use had been established over 100+ km², with *P. semibarbata* as the principal host on unlogged outcrops and *C. torreyi* as the principal host in clear-cuts.

Clear-cuts acted as population sources during the 1980s. Although the clear-cut habitat received fewer eggs, it generated more adults due to higher survival there than on outcrops (Singer, 1983; Moore, 1989). The butterflies moved from clear-cut to outcrop about twice as frequently as they moved in the opposite direction. Biased movement generated a gradient in insect density, such that emigration from clear-cuts raised insect densities on nearby outcrops (C. D. Thomas *et al.*, 1996).

Then, a severe summer frost killed virtually all of the *C. torreyi* in the clear-cut habitat in 1992 (Singer *et al.*, 1994). Although *E. editha* eggs and larvae were not damaged by the cold, the larvae starved. The populations in this habitat declined from $\sim 10^4$ egg batches in 1992 to two in 1993, even though *C. torreyi* regenerated in abundance in all clear-cuts in 1993. *Pedicularis semibarbata* was unaffected by the 1992 frost, *E. editha* survival was apparently normal on outcrops, and there was no mass extinction in this habitat.

Extinction of the source populations set up a fascinating natural experiment. In the source–sink theory, sources are areas of habitat which generate individuals and sinks consume them; sinks are areas which are populated because there is a net influx of migrants into the habitat, and they are predicted to become extinct in the absence of immigration (Pulliam, 1988). "Pseudosinks" are areas which can support a population without immigration, but where immigration increases population density above the local equilibrium; removal of immigration should result in a decline in density to the local carrying capacity rather than in extinction

(Holt, 1985; Watkinson and Sutherland, 1995). Following the sudden extinction of the population sources, overall egg densities fell on *P. semibarbata,* the decline was greatest close to former population sources, and the 1993 densities on outcrops were no longer correlated with isolation from former sources. In this case, we know that natural outcrops were pseudosinks, and not true sinks, because *E. editha* populations occurred on outcrops before the clear-cut habitat was created, they survived on outcrops after the frost, and they persisted throughout the study period at moderate abundance on undisturbed outcrops in Sequoia National Park, to the south of our disturbed study sites (C. D. Thomas *et al.,* 1996).

Source populations are often considered especially important for metapopulation persistence—they clearly are when long-term survival is impossible in sink habitats, as has been shown for the blue butterfly *Cyanaris semiargus* (Rodríguez *et al.,* 1994). However, it is resistance to extinction that really matters to persistence, not the balance of birth and death in "normal" years. In this particular example, pseudosinks were more resilient to a particular type of extreme environmental event, although it may well be that in most metapopulations the sources are usually the more resilient to environmental extremes. Empirical evidence is lacking. Both source and pseudosink populations may be prone to extinction for all of the reasons given in Section III.A. Unfortunately, there is almost no empirical information with which to assess the relationship between local population productivity in a "normal" year, and ability to survive environmental extremes.

D. Spatial Synchrony in Population Dynamics

When populations fluctuate in synchrony, and particularly when they become extinct in synchrony, the probability of metapopulation persistence may be much lower than predicted by standard models (Hanski, 1991). When the chance of extinction is completely independent in each local population, the probability of metapopulation extinction rapidly declines with increasing number of local populations (Fig. 4; Hanski, 1991, this volume), but if extinction probabilities are correlated, for instance because local populations are responding in a similar way to climatic variability, metapopulations with even large numbers of local populations may be susceptible to extinction (Hanski, 1991, this volume). There are examples of synchronous butterfly extinctions in response to single climatic events in *E. editha* (Ehrlich *et al.,* 1980; C. D. Thomas *et al.,* 1996; above) and *Aphantopus hyperantus* (Pollard and Yates, 1993; Sutcliffe *et al.,* 1996b), and evidence that extinction probability varies between years in *M. cinxia* (Hanski, this volume). As yet, these events have rarely been shown to cause metapopulationwide extinction, but the above examples concern large metapopulations or ones which contain either some very large patches or some relatively safe habitat type. The example in which *B. aquilonaris* became extinct from a number of forest fragments that contained fewer than 20 habitat patches, after a wet and cloudy summer (T. Ebenhard, personal communication, above), shows that extreme environmental events can cause entire metapopulations to become extinct.

Over longer time periods, landscape and habitat changes that cause deterministic extinctions have often been relatively synchronized over large areas. *Hesperia comma* and *Lysandra bellargus* both showed a period of rapid decline in England when myxomatosis killed rabbits, and their short-grass habitats disappeared throughout the UK (J. A. Thomas, 1983a; C. D. Thomas and Jones, 1993). Similarly, economic pressures and technological innovations that cause changes in farming or forestry practices usually do so over very large areas in a relatively short space of time, causing widespread changes in the fortunes of associated species. The information that exists at the moment suggests that extinctions are normally at least partially synchronized.

In the absence of better data on the spatial synchrony of extinctions and colonizations (see also Hanski, this volume), we must rely on analyses of extant populations, and presume that different levels of synchrony in their dynamics provide some insight into the extent to which local extinctions might be synchronized over wide areas. Analyses of butterfly, moth, and aphid population dynamics over wide areas (Britain) suggest that populations fluctuate in synchrony over areas of at least 10^5 km^2 (Pollard and Yates, 1993; Hanski and Woiwod, 1993), which is orders of magnitude greater than anyone's metapopulation study areas. For many of these species, correlated population fluctuations occur over areas that are so much larger than their potential migration distances that climate must be a major determinant of year-to-year population variability (Pollard and Yates, 1993, and references therein). These conclusions are based on counts of insects (transect counts for butterflies, traps for the others) from widely scattered locations across the landscape. However, each sampling point may to some extent lump together insects from more than one local population, and local populations may be fluctuating partly out of synchrony. In the case of the ≈ 1- to 3-km butterfly transect walks, several habitats are sampled. For aphids and moths, traps may attract insects from more than one habitat. If each sampling location counts insects from more than one habitat patch, local population variability may have been averaged out, leaving only residual large-scale variability caused by the climate to be detected.

Studies of population fluctuations at a smaller scale provide a rather different picture. Small-scale analyses are possible for butterflies because the British butterfly transects are usually divided into about 10 sections, and separate counts are made for each section. Populations in individual sections often fluctuate in parallel with regional fluctuations, apparently because of weather effects, but changes in local habitat management can cause deviations from regional trends (Pollard and Yates, 1993). When local habitats improve in quality, the change in local population size is upward relative to the overall regional trend, and downward when local habitats deteriorate (Pollard and Yates, 1993). Such deviations are particularly clear in species which are associated with successional vegetation. *Plebejus argus* fluctuates out of synchrony in areas which are only 500 m apart on successional habitats in heathland (C. D. Thomas, 1991), and *M. athalia,* which inhabits freshly made clearings in a forest in southeast England, increases in new clearings, but simultaneously declines in others as they become overgrown (Warren, 1987b, 1991). There are good and bad years for these species over wide areas, associated with climatic fluctuations, but local populations behave idiosyncratically in response to local habitat conditions.

Even in nonsuccessional species, analyses of local population fluctuations show that there is a great deal of heterogeneity in local dynamics, often but not always associated with different responses of local populations in different (micro)habitats to annual variation in the climate (Ehrlich *et al.,* 1975; Sutcliffe *et al.,* 1996a,b; see Fig. 2 in Hanski, this volume). Such heterogeneity in local dynamics may be crucial to long-term persistence. As found with *E. editha* above, what appears to be relatively poor habitat in most years may be crucial to persistence after some extreme environmental event

The scale over which systemwide extinction is likely to take place as a result of extreme events is one of the most crucial, but least well understood, aspects of metapopulation biology and needs to be addressed by long-term and large-scale field studies.

VI. Conclusions

Many of the predicted patterns and processes are widely observed in studies of butterfly metapopulations. However, we have also come to realize that a key empirical challenge is to identify the critical habitat requirements of different species and the factors causing changes in the distribution of habitats. Species have individualistic habitat and host plant requirements, hence the habitat patch networks available to each species are specific and not generally congruous with human definitions of general vegetation type. The specific habitat mosaic for each species is likely to be dynamic, especially in modern, human-dominated landscapes, and changes in the distributions of species are often driven by spatial changes in the distribution of suitable habitats.

Populations in large habitat patches have low rates of extinction for several reasons, including large initial population size, high habitat heterogeneity, and low risk of extinction from habitat dynamics. There is also some evidence to show that isolated local populations are relatively prone to extinction. Colonization probability is determined by isolation, by the sizes of source populations, and also by the size of the patch to be colonized (large patches are more likely to become colonized). The dynamic processes of extinction and colonization can thus generate the widely observed pattern in which large patches that are located close to each other are likely to be populated but in which small and isolated patches are usually empty. It appears that the flow of migrants in and out of habitat patches is also an important determinant of patch occupancy and local population sizes, and this needs to be addressed more specifically in the next generation of models.

Considerable progress has been made with spatially realistic simulation models, which have successfully predicted the observed patterns of patch occupancy based on the dynamic processes of extinction and colonization. These models have predicted persistence where metapopulations do survive and extinction where they do not, and qualitatively, the models have predicted which empty networks of habitat patches butterflies will invade and which networks they will not invade. Quantitative differences between model predictions and field data have been useful in revealing where further biological information is required, for example on migration and the effects of habitat quality. Use of the models in specific systems has also helped us to identify some general problems, including the need for butterfly population dynamics to be superimposed on shifting habitat mosaics.

We return now to two serious issues which have not yet been settled and where much more information is required. The first relates to migration and population structure, and the second to the importance of specific habitats to persistence. Metapopulation research has stimulated a considerable reappraisal of the migration capacities of butterflies. Results that are presently available suggest that the notion of a local population, which is at the core of metapopulation ideas, may be under threat. Exchange rates of individuals among adjacent but distinct habitat patches may be so high (sometimes $> 20\%$) that local populations have only limited demographic independence. If the population in an individual patch includes many immigrants, stochastic breeding failure will not result in extinction of that "local" population unless (i) the habitat changes in such a way that immigrants no longer enter or remain in the patch or (ii) regional stochasticity produces simultaneous breeding failures in a group of patches, thus interrupting immigration. Observed local extinction rates may therefore be much lower than the underlying rate (Hanski *et al.*, 1995b; Hanski, this volume). Yet, even when a metapopulation consists of an assemblage of such populations with a high degree of connectance, direct interactions (migration) between local populations at the opposite ends of the same network may never occur. In such systems, immigration and emigration are important determinants of local dynamics, but the whole network is certainly not one panmictic population. Migration is vital to local dynamics as well as to metapopulation-wide processes. Most current metapopulation thinking (if not modeling) limits the role of migration to one of seeding empty habitat patches (with little effect of immigration on abundance after colonization), and to a lesser extent as propping up small local populations which are under threat from stochastic extinction (rescue effect). The role of migration in local dynamics needs to be explored more fully in structured metapopulation models (Gyllenberg and Hanski, 1992; Hanski and Gyllenberg, 1993; Gyllenberg *et al.*, this volume) and through analyses of spatial patterns in population variability, colonization, and extinction (Hanski, this volume).

Because emigration and immigration rates vary with patch size and isolation, real metapopulations do not fall easily into the various categories of metapopulation types (Harrison, 1991; Harrison and Taylor, this volume). In some parts of a patch network, persistence may largely depend on the existence of one or a few large blocks of habitat (mainlands), but other parts of the same system may persist because there is a high density of small patches. Parts of metapopulations with a high density of small patches, each with a high emigration and immigration rate, resemble scaled-down versions of the population structure of highly mobile species (patchy populations, *sensu* Harrison, 1991). In mobile species, such as the nettle-feeding nymphalids in Europe, practically no single patch could support a local population for more than a few generations in isolation, and each individual will enter and leave several such patches. The distinction between relatively sedentary species which are regarded as existing as metapopulations and the more mobile species with "patchy populations" is becoming increasingly vague. In some cases, it is just a matter of scaling. The relative contribution of local versus regional population processes in different local "populations" within metapopulations and in different metapopulations is a much more important issue than trying to pigeonhole each system and give it an approved name. General models and field studies exploring these notions would be very useful.

The contribution of specific habitats to persistence is also becoming an important question. Most metapopulation models and field studies examine probabilities of extinction in relation to patch area, local population size and isolation, but pay limited attention to variation in habitat quality. Nonetheless, there is already enough evidence to suggest that the type of habitat can be just as important. If populations respond differently to environmental stochasticity in different habitats or microhabitats, habitat heterogeneity can buffer populations against large fluctuations and extinction. We should ask whether large populations in large habitat patches survive best because they are large, or because large patches usually contain several microhabitats; and whether large metapopulations persist because they have much habitat or because they have more kinds of habitats and microhabitats than small patch networks? A small metapopulation in 10 patches, each of a slightly different habitat type, might possibly persist for longer than a metapopulation in 50 identical patches.

Allied to this is the question of whether some habitats always hold the key to persistence. J. A. Thomas (1983a) suggested that *L. bellargus* may spread in good years, but is confined to population refuges in bad years. The same argument has been put forward for several mobile species which may breed over large areas at favorable times of year, but retract to specific habitats at other times (Shapiro, 1979; Jordano *et al.,* 1991). If this phenomenon is widespread, the existence of specific habitats within patches or patch networks may be more important to persistence than patch size or number. An entire program of empirical research is required to evaluate to what extent populations in different habitats vary in their responses to environmental stochasticity, whether habitat heterogeneity buffers populations against extinction, and whether specific habitats hold the key to persistence.

Finally, a metapopulation approach is becoming important in several other areas of butterfly population biology. A metapopulation approach provides the potential to bridge the gap between studies of local population dynamics and species distributions. Densities, sizes, and average suitabilities of habitat patches may vary geographically, but this aspect of metapopulation biology has barely been considered for butterflies. Comparing central and marginal parts of species ranges is almost bound to reveal interesting results (J. A. Thomas, 1993; J. A. Thomas *et al.*, 1994). Another area of interest is the extent to which metapopulation structure and migration affect local adaptations to different habitats (C. D. Thomas and Singer, 1987; Thompson, 1993, 1994; Singer and Thomas, 1996; Barton and Whitlock, this volume) and levels of genetic variation (Descimon and Napolitano, 1993a,b; Hedrick and Gilpin, this volume), and whether habitat geometry itself affects the evolution of migration (Dempster *et al.*, 1976; Dempster, 1991; Olivieri and Gouyon, this volume). In an ever changing landscape, evolutionary changes may play an increasingly important role in population persistence and extinction.

16 Tritrophic Metapopulation Dynamics
A Case Study of Ragwort, the Cinnabar Moth, and the Parasitoid Cotesia popularis

Ed van der Meijden

Catharina A. M. van der Veen-van Wijk

I. INTRODUCTION

Many short-lived monocarpic plant species have a markedly patchy distribution. These species reproduce only once in their lifetime, they typically exploit disturbed habitats, and they often have a high rate of local extinction (Harper, 1977; Gross and Werner, 1978; van Baalen, 1982; Reinartz, 1984; Grubb, 1977; de Jong and Klinkhamer, 1988; see van der Meijden et al., 1992, for a review). Well-known examples are biennial plant species colonizing windfalls in woodlands, species exploiting locally grazed or otherwise disturbed vegetation on sand dunes and chalk grasslands, and species of "old fields." To survive over long periods of time, such extinction-prone biennials depend on regional regeneration through interacting local populations (metapopulations). Critical processes for long-term persistence include seed dispersal and dormancy in variable and patchily distributed environments (Kuno, 1981; Klinkhamer et al., 1987). Also of potential importance are biotic interactions with species at higher trophic levels. Often the herbivores and their parasites are monophagous, and their populations too may function as metapopulations. Dynamics of species at the higher trophic levels are necessarily affected by dynamics of the host plant, but the herbivores and predators can also play a more active role by modifying the extinction probabilities at other trophic levels (Nee et al., this volume; Holt, this volume).

In this chapter, we analyze whether and to what extent interactions within a tritrophic system are affected by the spatial distribution of habitat patches. In doing so, we follow the suggestion of Harrison *et al.* (1995): "many populations appear patchy to the human eye, but [that] critical examination is required to deduce the dynamic consequences of this patchiness." Specifically, we review our long-term data (two decades) on the relationships between the plant ragwort *(Senecio jacobaea),* its most important herbivore, the monophagous cinnabar moth *(Tyria jacobaeae),* and the specialist parasitoid of the herbivore, *Cotesia popularis.*

Ragwort has been the subject of intensive studies in several countries over a long time, starting in 1935 when Cameron summarized his early biocontrol project in a study entitled "Natural Control of Ragwort." Ragwort seeds had been accidentally introduced into New Zealand, and the plant had grown into a major pest by colonizing the entire country in a few decades, between 1874 and 1900. The powerful colonizing and weedy behavior of ragwort is undoubtedly based on its capacity to efficiently exploit scattered disturbed habitat patches. Ragwort is a "pest" thanks to the alkaloids that it produces, which are toxic to cattle but not to the cinnabar moth (Harper and Wood, 1957), which was used as a biocontrol agent. Subsequently, two additional long-term population studies of this plant–moth system have been carried out in the United Kingdom (Dempster, 1982; Crawley and Gillman, 1989). Dempster concluded that "the moth's population is buffered against extinction by the heterogeneity within the habitat," indicating that some sort of spatial effects are important.

In the early 1970s, we commenced our studies of ragwort in The Netherlands on three small local dune populations. Within 3 years, two of the three populations had become extinct. As ragwort density in the dune area as a whole did not continue to decrease, we became convinced that population dynamics of this species should be studied on a much broader scale and that spatial aspects are crucial for understanding the mechanism of persistence, which is the focus of the present chapter. Apart from the patchy distribution of the plant and, consequently, of the herbivore and its parasitoid, typical features of this tritrophic system on sand dunes include frequent complete defoliation of plants by the cinnabar moth (not only leaves are consumed, but also buds and flowers, thus reducing ragwort seed production to zero). The lifetimes of local plant populations are restricted, with the cinnabar moth and its parasitoid continuously tracking these ephemeral populations (van der Meijden, 1979a).

In this chapter, we will first give an outline of the population dynamics of the three organisms (van der Meijden *et al.,* 1991, 1992) and describe how they interact with each other. Next, we will calculate parameters describing the degree of synchrony between local populations and the metapopulation to reveal to what extent the dynamics of local populations differ from each other. We also pay attention to spatial correlations within and between the three species. Finally, we will discuss the mechanisms that appear to play a role in the persistence of the plant, the herbivore, and the parasitoid in their patchy environment. This information will be used to infer the type of metapopulation (Hanski and Gilpin, 1991, this volume) that best describes the organisms of this tritrophic interaction.

II. MATERIALS AND METHODS

A. Study System

Ragwort is a facultative biennial plant. It is native to Europe and has invaded overgrazed areas throughout the world (Harper and Wood, 1957; Dempster, 1982; van der Meijden, 1979b). Its weedy character is largely due to its extremely powerful reproductive potentials. Individual plants may produce up to 20,000 seeds with pappus that enable wind dispersal. Poole and Cairns (1940) refer to seed numbers per plant ranging from 50,000 to 150,000 in New Zealand. Mowing, plowing, and other such conditions that reduce the opportunities of generative reproduction stimulate vegetative reproduction (Poole and Cairns, 1940; Harris *et al.*, 1978). Even *in vivo*, small root fragments may develop into mature plants (van der Meijden, 1979b). Seeds may remain viable for more than 8 years (Poole and Cairns, 1940).

Ragwort is a common weed of sand dunes, roadsides, and waste lands. Local disturbances create suitable circumstances for establishment, whereas vegetation succession may lead to loss of suitable growing sites. Often, however, populations disappear without any changes in the vegetation. Such sites may become recolonized at a later time.

The cinnabar moth is a univoltine insect and monophagous on ragwort (elsewhere it has been reported to occasionally use the closely related *Senecio vulgaris;* Aplin and Rothschild, 1972). It lays its eggs in small batches of ca. 30 eggs. Fecundity varies from 100 to 400 eggs (Dempster, 1982). First or second instar larvae may become parasitized by the specialist braconid parasitoid *Cotesia popularis.* Up to 15 parasitoid larvae may develop per host larva. These larvae leave their host shortly before it would otherwise have pupated, and the host then dies. From the third instar onward larvae show a tendency to disperse. This is especially so in the fourth and fifth instar when many larvae leave their original food plant before it is fully defoliated. This dispersal tendency is related to the numbers of larvae on the plant and the plant size (van der Meijden, 1976; Sjerps and Haccou, 1996).

Population data on the three species were collected in a coastal sand dune area near The Hague in The Netherlands. The patchy distribution pattern of local ragwort populations is brought about by the geomorphology of the dunes in combination with grazing activity of rabbits. The landscape is a mosaic of north-facing slopes with a closed vegetation of trees, shrubs, and grasses, poorly vegetated south-facing slopes, and valleys with a vegetation depending on the groundwater level. Ragwort can potentially grow anywhere in this landscape provided that neither the grass layer nor the shrub/tree canopy is closed.

An area of about 6 km^2 of a much larger system was searched for local ragwort populations in 1973. Plants are considered to belong to one local population if they are not spatially separated from each other by ragwort-free distances of more than 5 m. Of 150 local populations, 102 were selected, based on differences in population density and habitat characteristics, such as the amount of shade by woody perennials and the presence of the predatory ant *Formica polyctena*. Local populations covered areas ranging from 8 to 3000 m^2 in 1974 (mean 900 m^2). Plant numbers (from small vegetative rosettes to large flowering plants) per population varied from 1.5 to 62 per m^2. The distance between populations ranged from 10 to over 200 m, with at least 5 m without any ragworts. Local populations were usually separated from each other by barriers like scrubs, forested areas, dune lakes, or blowouts.

B. Census Data

Census data were collected from 1974 to 1994. Relevant data for this paper are:

1. *The amount of ragwort biomass per local population sample.* During the period of cinnabar moth egg laying, in May–June in each year, the same permanent squares (4 m^2) in each local population were visited three to four times, and ragwort cover in dm^2 was estimated (a measure of biomass; van der Meijden, 1979a) just prior to oviposition by the cinnabar moth. The highest value per local population in each year was used as the final estimate. Herbivory by cinnabar moth larvae may reduce biomass (and seed production) from June onward. Defoliated plants often produce regrowth foliage shortly after herbivory. A population of ragwort was supposed to have gone extinct if no living plants (either seedling, rosette plant, or mature plant) were present on any census dates 1 year after the last herbivory episode by the cinnabar moth. Based on observations made in the vicinity of the sampling squares, we concluded that disappearance from a sample typically meant extinction of that particular local population.

2. *The number of cinnabar moth egg batches per local population.* Egg batches were counted in the above-mentioned 4 m^2 squares at four to six visits with weekly intervals to each local population in May–July each year. A population of the cinnabar moth was supposed to have gone extinct if no eggs were found on the plants following a year in which eggs were present in that population.

3. *Percentage parasitism by Cotesia.* Percentage parasitism by *Cotesia* was determined (from 1988 onward) by collecting fourth or fifth instar cinnabar moth larvae in five census populations at three moments during the larval season (from the end of May until the beginning of August) because of a seasonal trend in percentage parasitism (Soldaat, 1991). At every date, 50 larvae were collected per site, yielding 750 larvae which were reared to pupation every year.

III. POPULATION DYNAMICS OF RAGWORT

Local extinction of ragwort and recolonization of empty patches are frequent events on sand dunes. Figure 1 (A) shows the number of extant populations over time. In two extreme seasons (1975–1976 and 1981–1982), 40 populations disappeared in a single year (Fig. 1, middle). The cumulative extinction curve over time demonstrates that not all local populations are equally vulnerable. Eighteen of 102 populations never became extinct during the period of 20 years, whereas 56 populations disappeared (and were recolonized) once or twice and 26 even three to five times. Apparently ragwort has refuge populations with a low extinction probability. A habitat analysis (van der Meijden *et al.*, 1992) revealed that this group of populations is located in areas with a mixed vegetation of trees and shrubs and a not fully closed ground vegetation with grasses and/or mosses and lichens. The populations with the highest extinction risk are situated in open sandy areas.

To test whether the probability of population extinction was related to their size, we analyzed data for the period from 1974 until 1976 (Table I). Contrary to the general expectation (Hanski, this volume), small populations had a higher probability of survival than large populations. The reason for this is that populations in areas with trees and/or shrubs, with a low extinction risk, tend to be small (Table I). Apparently vegetation structure determines the size of suitable habitat patches. In the open dune areas suitable habitat patches are considerably larger than in areas with trees and/or shrubs.

Even on a regional (metapopulation) scale, ragwort shows huge fluctuations in total ground cover (Fig. 1, A). The difference in ground cover between 1980

TABLE I Extinction, Survival, and Habitat Type of Ragwort Populations between 1974 and 1976 in Relation to Patch Size in 1974

| | No. of populations | | | |
| | Fate | | Habitat | |
Patch size (m²)	Extant	Extinct (%)	Trees or shrubs	Open sandy habitat
<10	9	2(18)	10	1
10–100	13	4(24)	10	7
101–1000	17	17(50)	23	11
1001–2000	13	10(43)	9	14
>2000	3	12(80)	8	7
χ^2	11.03		15.68	
Significance level	0.026		0.003	

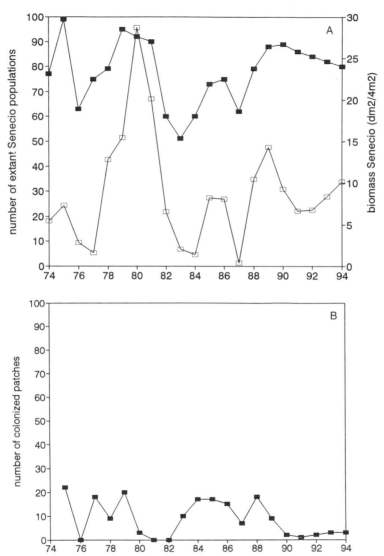

FIGURE 1   (A) Time course of the number of extant ragwort populations (black squares) and the metapopulation fluctuation in ragwort ground cover (summed over all local populations; open squares). (B) Time course of ragwort colonization of empty patches. (C) Time course of extinct ragwort populations (= sites without ragwort in the current year) and the cumulative function of populations that disappeared at least once.

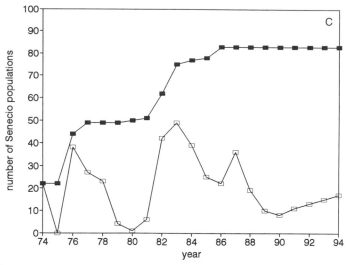

FIGURE 1   (Continued)

and 1987 was almost 100-fold. There is a fair correlation between total ground cover of the metapopulation and the number of extant populations, which suggests that it is unlikely that there is much asynchrony in fluctuations among local populations. As a measure of synchrony we used the correlation coefficient calculated between the time series of ragwort ground cover (dm^2/m^2) of every local population and the metapopulation (Fig. 2, top). No significant negative correlations were found. Fifty-eight local populations were significantly positively correlated with the overall fluctuation, but ground cover in 44 local populations was not significantly correlated with metapopulation fluctuations. Sixteen populations of this latter group are located in sites that were completely overgrown by dense vegetation (the percentage cover of trees, shrubs, and grasses increased from 1973 to 1994 from 71 to 100%). Mechanical reduction of immigrating seeds, reduction of "safe sites" for germination, and reduction of penetrating light by the closed vegetation, may have rendered these sites unsuited for ragwort germination, growth, or survival.

To test whether sites where ragwort had disappeared had indeed become unsuitable for plant growth, seeds were added experimentally (Table II). Seeds germinated successfully in all sites and a number of the rosettes survived in most sites, indicating that ragwort is seed limited. However, twin plots that were cleared of the grass-herb vegetation showed considerably higher germination and survival, demonstrating that the suitability of growing sites was affected by vegetation development.

Two main factors are responsible for the fluctuations in biomass and the shortage of seeds contributing to local extinction: herbivory by cinnabar moth

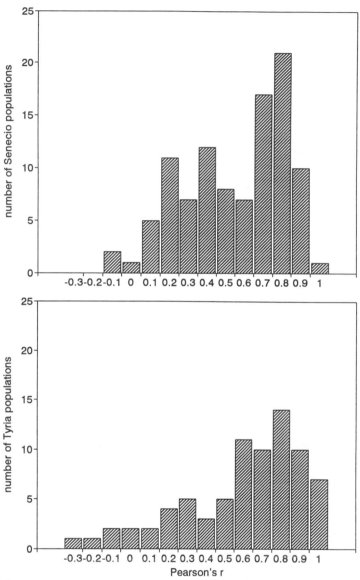

FIGURE 2   (Top) Pearson's correlation coefficient (r) between the amount of ground cover of ragwort in each local population and the metapopulation (the sum of all local populations), reflecting the level of synchrony. (Bottom) Pearson's correlation coefficient (r) between the numbers of cinnabar moth eggs per dm^2 of ragwort in local populations and in the metapopulation (the sum of all local populations).

TABLE II Experimental Seed Addition to 25 Sites in Different Habitat Types without Ragwort[a]

Habitat	Seedlings		Rosette plants	
	A	B	A	B
Open woodland	193	624	12	21
Grass–herbs	123	531	4	37
Moss–herbs	175	348	12	19

[a] Three thousand seeds were sown per a plot of 30 by 30 cm. The table gives the mean numbers of seedlings and rosette plants per plot after 1 year. Plots were either undisturbed (A) or cleared of the grass–herb vegetation (B).

larvae (the quantitative effect of herbivory will be described below) and drought (van der Meijden *et al.*, 1985). Defoliation may not only mean complete loss of above-ground biomass, but also lack of seed production over periods of 2 or even more years. The effects of drought on plant performance were also mentioned by Dempster (1982) and Crawley and Gillman (1989).

In Figure 1 (C) colonization by ragwort of empty patches is plotted over time. Years in which total defoliation in the entire metapopulation was observed are 1975 and 1976, 1981 and 1982, and 1986 and 1987. During such periods seed production is reduced to zero. This figure immediately reveals that colonization of empty patches may take place in years following total defoliation and destruction of the seed crop. Consequently, colonization probably occurs through germination from the seed bank. However, relatively high rates of colonization were found also in the second and third year following defoliation, when plants in the extant populations produced again seeds, suggesting that seed dispersal is also important to colonization.

IV. MECHANISMS OF PERSISTENCE IN RAGWORT

Ragwort has three mechanisms that enhance the probability of survival on local scale:

1. *Regrowth capacity.* Ragwort has spectacular powers of regeneration, which make it such a powerful weed. Even completely defoliated plants may regenerate (van der Meijden *et al.*, 1988). Plants that were artificially damaged by removing the whole shoot in a field experiment suffered only 5% more mortality than control plants. Biomass was reduced by 35%. However, the negative effect of herbivory may be much greater in combination with adverse weather

conditions (drought) during or after defoliation (van der Meijden *et al.*, 1985; Prins and Nell, 1990). Repeated defoliation reduces regrowth as well (McEvoy, 1985). Despite the fact that ragwort is a biennial plant, and normally dies after a single production of seeds, the life cycle may be prolonged by herbivory. Analogous to a seed bank we can talk about a rosette bank in this species.

2. *Dormancy*. Ragwort has only a small seed bank, but it seems to be effective, judging from germination in many open sites after at least 2 years in which seed production was reduced to zero by cinnabar moth herbivory (van der Meijden *et al.*, 1985). Fifty populations were sampled in 1983, after two successive seasons of total defoliation, by taking eight circular (diameter 10 cm) soil cores per population. Samples of 31 populations did not contain any viable seeds. In 9 populations one seed was found, in 4 two seeds, in 1 three seeds, and in 5 more than five seeds (which means 55 viable seeds in 3.14 m^2 soil samples). These numbers are small compared with a production of 20,000 seeds per individual flowering plant. Many local populations do not seem to have a seed bank at all, though one cannot be sure based on small samples of soil. Experiments on the longevity of buried seeds showed that viability was positively related to the depth at which they were buried. Viability of seeds buried at a depth of 10 cm was hardly reduced after 5 years (E. van der Meijden, unpublished results).

3. *Seed dispersal*. Experiments by Poole and Cairns (1940) showed that the majority of ragwort seeds landed within a few meters from the parent plant. Yet a considerable number of seeds was found to have been dispersed up to 20 m, and especially in the direction of prevailing winds the dispersal curve seemed to have a long tail. We (van der Meijden *et al.*, 1985) have demonstrated that both germination of dormant seeds and colonization through seed dispersal were important mechanisms in the reestablishment of local populations. Because no seeds were formed in the 2 earlier years, the occurrence of seedlings in empty patches after a period of defoliation indicates germination from a seed bank. However, the level of germination in empty patches in subsequent years was much too high to be explained by germination of dormant seeds alone, and much of this germination must have been the result of seed dispersal. Unfortunately, our data do not allow us to judge whether the small number of refuge populations played an essential role as sources of dispersing seeds.

V. POPULATION DYNAMICS OF THE CINNABAR MOTH AND THE PARASITOID *COTESIA POPULARIS*

The cinnabar moth is conspicuous not only because of its appearance (a bright red and black colored moth and even brighter yellow and black colored larva), but also because of its behavior. Periodically, larvae completely defoliate their food plants over large areas (in the Dutch coastal dunes defoliation seems to be highly synchronized). During such years, thousands of larvae can be observed to disperse in search of food. Fourth and especially fifth instar larvae can cover hundreds of meters and even leave their habitat by crossing beaches and motor-ways. Eventually they may defoliate all local populations of ragwort. During such episodes thousands of larvae die from starvation. Dempster (1982) and Crawley and Gillman (1989) also found food shortage to be the key factor in population dynamics in the United Kingdom, determining the magnitude of population fluc-tuations of the cinnabar moth. At first sight, the frequent total defoliation and the enormous fluctuations in plant and insect populations might give the impression that this insect–plant relationship is one of the best examples for refuting coad-aptation and regulation. Nevertheless, both ragwort and the cinnabar moth are still common species in sand dunes and some other habitats of western Europe. The pest characteristics of ragwort, especially its regenerative powers, might well be the result of selection by the insect. The insect, on the other hand, sequesters the alkaloids of its food plant, probably as defense against potential natural en-emies (van Zoelen and van der Meijden, 1991; Ehmke *et al.,* 1990).

Figure 3 shows fluctuations in the metapopulation size of the cinnabar moth expressed as numbers of eggs per plant biomass. Egg numbers fluctuated more than 200-fold. For comparison biomass fluctuations of ragwort are plotted in the same graph. Above a level of 10 eggs per dm^2 of ragwort, total defoliation will take place (van der Meijden *et al.,* 1991). The moth lays the largest numbers of eggs per unit of plant biomass in open areas, without ants (*F. polyctena;* Table III). These are also the areas that receive most eggs in absolute terms, because there are many more plant populations in open than fully shaded sites. Experi-ments on oviposition demonstrated an almost absolute preference for open areas

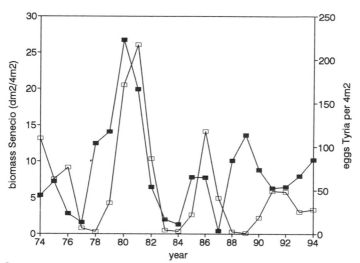

FIGURE 3   Metapopulation dynamics of the number of cinnabar moth eggs (mean number of eggs per population sample of 4 m^2; open squares) and ragwort ground cover (black squares).

TABLE III   Mean Yearly Egg Load of the Cinnabar Moth per dm² of Ragwort in Local Populations of Ragwort in Different Habitat Types from 1974 to 1994[a]

Habitat type	No. of populations	Egg numbers/ dm² ragwort	Standard error
−S−F	65	9.08	0.83
S−F	14	3.88	1.03
−S F	16	4.79	1.57
S F	7	0.99	0.41

[a] −S−F, unshaded, *Formica*-free; S−F, shaded, *Formica*-free; −SF, unshaded, *Formica*; SF, shaded, *Formica* habitat. Effects of shade and *Formica* are both significant ($P < 0.001$ in two-way ANOVA); the interaction between shade and *Formica* is not significant.

over fully shaded areas (van der Meijden *et al.*, 1991). A similar result was found with respect to daily light intensity: the moth does not oviposit in cloudy days.

The amount of ragwort biomass in open habitat is probably the most important parameter determining the population change of the herbivore. The correlation coefficient between the amount of ragwort in the open populations and the number of cinnabar moth eggs in the next year is highly significant (Table IV). This correlation was not significant in the shaded plant populations.

Cinnabar moth population growth is followed by an inevitable reduction of plant biomass as explained above. The correlation coefficient between the number of eggs in year *t* and ragwort biomass in year *t* + 1 is negative and significant for the open populations (Table IV), indicating the effect of herbivory on plant production. The correlation is negative but not significant in the shaded plant populations.

The interaction between the plant and the moth produces the striking cycles

TABLE IV   Correlations between the Amount of Food (Expressed as Ragwort Ground Cover) in Year *t* and the Number of Cinnabar Moth Eggs in Year *t* + 1 (A) and between the Amount of Herbivory (Expressed as the Number of Cinnabar Moth Eggs) in Year *t* and Ragwort Ground Cover in Year *t* + 1

	A		B	
Ragwort population	*r*	*P*	*r*	*P*
Open populations	0.74	0.00	−0.74	0.02
Shaded populations	0.34	NS	−0.46	NS

in their populations shown in Fig. 3, with plant biomass followed by insect (egg) numbers. After overshooting the egg/biomass ratio of 10 eggs/dm² ragwort in the open populations, a complete defoliation follows. Next larvae disperse to the still undefoliated plants in other populations. Finally this leads to regional defoliation and total loss of seed production. A second season of total defoliation usually follows, resulting in a strong reduction of egg numbers in the following year. Reduced plant biomass is now driving insect numbers. The final increase in insect numbers in Fig. 3 was unexpectedly ended in 1991, without the moth overshooting the carrying capacity and without further reduction in plant biomass. It was only in 1995 that the ragwort population became again completely defoliated.

There is no effective regulatory mechanism for the insect on the scale of a local plant population. With the numerical advantage for an individual to lay more eggs than its competitors, this makes food shortages among larvae inevitable. Food shortages in turn result in mass migration and starvation. The reduction in plant biomass leads to local disappearance of the herbivore. Local ragwort populations with relatively more food are selected by the adult moth for oviposition. Figure 4 (top) illustrates the dynamics of the numbers in local cinnabar moth populations over time and (bottom) local moth extinctions. Within 6 years from the beginning of this study, in 1980, all local insect populations had become extinct at least once. Contrary to its food plant, the cinnabar moth has no refuge populations that never became extinct.

Figure 3 shows that after the crash in herbivore (egg) numbers due to food shortage and starvation it took another year before egg numbers started to increase again (1978, 1984 and 1989). This is unexpected because plant biomass did not appear to be limiting in these years, especially not so in 1978 and 1989. This lag period implies that during these periods insect numbers were not driven by available plant biomass. It also implies (Fig. 3) that as herbivory is relaxed for 1 year after a period of heavy herbivory, local ragwort populations can recover and biomass can increase.

The absence of moth population growth in the year following the crash in egg numbers is probably due to two factors. In the first place adult females are much smaller than in a normal year (Dempster, 1982), because they suffered food shortage as a larva in the previous year. Mean pupal size is reduced from 0.52 to 0.45 cm, which corresponds to a reduction in fecundity of more than 60%. Second, the effect of the parasitoid *C. popularis* is inversely related to the density of the cinnabar moth (Dempster, 1982). The highest percentage parasitism is found after a population crash of the cinnabar moth. Similarly Cameron (1935) found the highest percentage parasitism after a population crash of the cinnabar moth in the United Kingdom in 1932. We have hypothesized (van der Meijden *et al.*, 1991) that the mechanism underlying this inverse relationship is that *Cotesia* has a lower fecundity and a lower dispersal capacity than its host. Especially in the years of crash in cinnabar moth numbers (1977, 1983, and 1988), few new patches are colonized by the moth (Fig. 4, bottom). This allows a buildup of parasite numbers and consequently an increase in parasitism. In the following years the host colonizes other local ragwort populations and temporally escapes from its parasitoid. If the rate of increase of the host is higher than that of the parasitoid, percentage parasitism should decrease with the rapidly increasing host numbers. Dempster's (1982) data suggest such a lower rate of increase in the parasitoid. Table V gives further information on parasitism, colonization of new sites by the cinnabar moth, and population change of the moth. The 1988 crash resulted in a

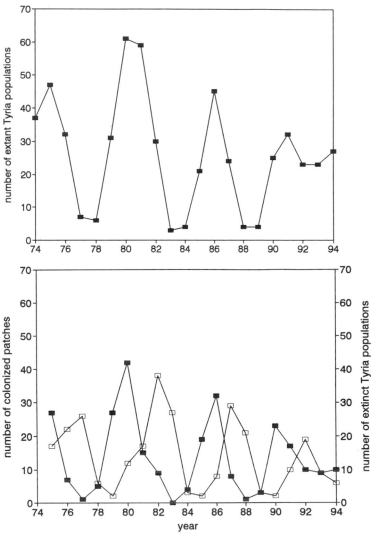

FIGURE 4   (Top) Time course of the number of extant populations of the cinnabar moth (= ragwort-growing sites with cinnabar moth eggs in the current year). (Middle) Time course of the number of extinct populations of the cinnabar moth (populations that went extinct in the current year; open squares) and the number of ragwort-growing sites colonized by the cinnabar moth in the current year (black squares).

TABLE V Parasitism of the Cinnabar Moth by *Cotesia popularis*[a]

Year	Percentage parasitism	Moth population turnover rate	Change in moth population
1988	37.9	25	0.08
1989	3.2	75	0.33
1990	0.0	92	17.12
1991	0.4	53	2.55
1992	1.0	43	1.05
1993	5.1	39	0.52
1994	10.5	37	1.10

[a] Population turnover in the cinnabar moth was calculated as $100 \times$ number newly colonized sites/(number of persistent populations + newly colonized sites). Change in host population size was calculated as the number of cinnabar moth eggs in year $t - 1$/number of eggs in year t.

high percentage of parasitism, apparently due to the small number of new sites colonized by the moth and the reduction in its population size. 1989 and 1990 show a fast decrease in percentage of parasitism together with an increase in population turnover and general increase of the moth. The increase in parasitism in 1993 and 1994 occurs concurrently with a reduction in host population turnover rate and a reduction in population growth of the host.

The level of synchrony between egg numbers of the cinnabar moth per plant biomass in local plant populations and the pooled egg numbers in the metapopulation was calculated as the correlation coefficient between these variables (Fig. 2, bottom). As eggs will only be found on ragwort plants we skipped those sites that had no ragwort or had a ragwort cover of $< 1\%$. Half of the correlation coefficients in Fig. 2 are positive and significant, indicating considerable synchrony in the local fluctuations of the numbers of cinnabar moth eggs. The local populations with a low correlation with metapopulation fluctuations invariably had many zero values due to low ragwort density or less suitable habitat (shaded ragwort populations or populations with the predatory ant *F. polyctena*). These results confirm the notion that the cinnabar moth is able to track efficiently suitable local populations of its host plant.

VI. MECHANISMS OF PERSISTENCE IN THE CINNABAR MOTH

There are at least five mechanisms that enhance the probability of persistence in the cinnabar moth:

1. *Dispersal of adult moths and the capacity to locate isolated spots of ragwort.* The female moths do not seem to be strong flyers, yet the distribution of eggs in relation to plant biomass across local populations of ragwort shows good synchrony, which can only be explained by an effective dispersal/searching ca-

pacity of the female moths. The information in Fig. 4 on newly colonized local sites is a clear illustration of the dispersal capacity. Harrison *et al.* (1995) came to a similar conclusion.

2. *Dispersal of larvae.* During periods of food shortage, larvae are observed to move long distances to find food. Especially fourth and fifth instar larvae can cover hundreds of meters and reach yet undefoliated sites. A remarkable feature of larval behavior is that some of them leave their food plants before defoliation has caused any food shortage. Using a gametheoretical model Sjerps and Haccou (1994) demonstrated that such behavior, especially when sibs are involved, may have an evolutionary advantage.

3. *The temporal distribution of oviposition.* Oviposition is extended from early May until early July. There are marked differences in the beginning of the oviposition season between years which undoubtedly relate to effects of temperature on the development of pupae. The wide distribution of oviposition guarantees that some egg batches have a considerable head start over others. During years of food shortage larvae from the first egg batches will have the highest chance of reaching the threshold weight for pupation.

4. *Flexible size for pupation.* During periods of food shortage pupal size can be considerably reduced (Dempster, 1982). Successful pupation is possible only when the larva has reached a threshold weight of 140–150 mg (van der Meijden, 1976), which lies far below the weight that a larva may reach in a year of abundant food supply (300–500 mg).

5. *Distribution of eggs over different habitats.* The egg load expressed as the number of eggs per unit of plant biomass varies significantly and consistently between different habitat types (Table III). Open habitat without the predatory ant is preferred, shaded habitat with the ant is avoided. This suggest that the less favored habitats may act as short-term refuges in periods of food shortage. As mentioned earlier, the cinnabar moth has no refuge populations over long periods of time.

VII. DISCUSSION

A. Population Dynamics and Persistence

Populations of ragwort show tremendous fluctuations in biomass at both local and regional scales, with local fluctuations frequently resulting in extinction. Herbivory by the cinnabar moth, especially in open habitats, contributes to the local extinction risk. The metapopulation of ragwort, however, does not seem to be in a great danger of extinction. Figure 1 shows that we never observed more than half of the local populations to have gone extinct in 1 year. The lowest percentage ground cover was observed in 1987, when the plant was still found in 62 of 102 patches. On the other hand, the extinction risk of ragwort is not independent of location: the plant has refuge populations that never went extinct during 20 years, and it has local populations with a very high risk of extinction.

On a local scale the cinnabar moth has an ephemeral existence of one or only a few years. Its survival is closely linked with the presence of ragwort. By numerically overshooting the local carrying capacity, larvae are subject to scramble competition for food, which frequently leads to mass starvation and local extinction. Even on the regional scale, the cinnabar metapopulation seems to be in great danger of extinction during such events. In 1984 and 1989, we found only six and five egg batches, respectively, which could have been produced by a single female, in all samples (408 m^2). Dempster (1982) who studied a similar system in the United Kingdom found only one egg batch in 1969 in 150 m^2 samples. He observed immigration of adult moths from outside his study system. In a smaller, more isolated population, Dempster (1971) observed an extinction of the moth in 1968, and it was only 10 years later that the site was reoccupied. Apart from the risk of extinction one would expect extreme loss of genetic diversity to result from such severe population bottlenecks (Hedrick and Gilpin, this volume).

B. Do Spatially Extended Dynamics Affect the Probability of Survival?

Do these organisms survive because assemblages of local populations persist in a balance between local extinctions and colonizations, as in the classical metapopulation scenario (Hanski, this volume)? In some biennial plants with a fairly constant appearance and decay of growing sites that might indeed be the case (van der Meijden et al., 1992). In species like ragwort, with refuge populations and rather synchronous fluctuations in biomass, the situation appears to be different. Although some localities became unsuitable for ragwort, the majority remained suitable and could have been recolonized from the seed bank or through seed dispersal from extant local populations. Survival over long periods of time is enhanced by these two buffer mechanisms, by the low vulnerability to extinction in the refuges and by the capacity of individual plants to recover from extreme damage. The metapopulation type (Hanski and Gilpin, this volume) that fits ragwort best would be the source–sink metapopulation consisting of patches with mostly negative population growth rate in the absence of seed dispersal and patches with a positive growth rate. Even so, seeds from open populations may occasionally disperse to the refuge populations and contribute to their stability.

Regional survival of the cinnabar moth is brought about, first of all, by the capacity of its food plant to recover soon after complete defoliation. Heterogeneity of the ragwort habitat, which leads to differences in egg load per unit of plant biomass, and the temporally extended oviposition period also increase the chance of at least a few larvae to reach the threshold weight for pupation even when food becomes completely exhausted on the regional scale (van der Meijden, 1979a, van der Meijden, et al., 1991; Dempster, 1982). The positive, and often high, correlations between temporal fluctuations in local populations and the metapopulation indicates that the searching capacity of the female moth is high and apparently not much hampered by interpatch distances. In this respect, the cinnabar moth may be considered to live in only one, but patchy, population and not in a classical metapopulation (Harrison, 1991; Harrison and Taylor, this volume).

An important question in metapopulation theory is whether an interacting system of local populations is more stable or has a higher probability of persistence than the separate local populations (Hanski, 1991). This case study has demonstrated that many sites where ragwort went extinct are recolonized through seed dispersal from existing local populations. Such dynamics must lead to higher regional densities and biomass production and, consequently, to a higher probability of persistence. This study has also demonstrated that habitat heterogeneity in local ragwort populations adds to the probability of cinnabar moth survival. Finally, it seems very plausible that the parasitoid *C. popularis,* with its limited powers of dispersal between local populations, causes a delay in the recovery of the cinnabar moth and consequently enables ragwort populations to grow for one season without herbivory. In conclusion, spatial effects related to habitat heterogeneity and parasitoid dispersal probably contribute greatly to cinnabar moth survival.

We would expect that even if ragwort were not patchily distributed, but all patches were combined to one large population, it would still survive given its specific characteristics and the present level of spatial heterogeneity. Because dispersal of cinnabar moth larvae would not be affected by distances in such a uniform environment, food would become exhausted sooner. This would lead to greater fluctuations in insect numbers and probably in plant biomass as well (Sabelis *et al.,* 1991). As its survival in a network of connected patches seems to be risky, the survival of the cinnabar moth would be even more questionable. However, in such a system without any interpatch distances the parasitoid *C. popularis* might become much more effective, because dispersal, its weak point, would be less critical. We thus conclucde that every possible outcome, from cinnabar moth extinction to regulation by the parasitoid, remains a possibility in a large uniform habitat.

Harrison *et al.* (1995) studied ragwort and a guild of its herbivorous insects during 3 years in the United Kingdom to test whether coexistence of these competing species could be explained by any spatial effects. They experimentally demonstrated that insect dispersal across local patches was not seriously limited by interpatch distances. From these experiments they concluded that it was unlikely that patches were acting as separate dynamic entities with respect to competition. In other words, they refuted the idea that spatial effects were essential for coexistence. Our study demonstrates that spatial effects are not limited to those caused by distances between local populations: differences in habitat quality among local patches may be critical. Although we prefer simple models to test ecological principles, we agree with Hanski (1991) that "there is an urgent need to develop metapopulation models that include variation in habitat quality." This study has also demonstrated that many metapopulation-level processes cannot be easily detected in short-term studies.

ACKNOWLEDGMENTS

We enjoyed discussing these ideas with Tom de Jong and Peter Klinkhamer, Rinny Kooi is acknowledged for technical assistance, and we especially thank Ilkka Hanski for his extensive comments on the first version of this chapter.

17

Spatially Correlated Dynamics in a Pika Metapopulaton

Andrew T. Smith   *Michael Gilpin*

I. INTRODUCTION

Because interest in modeling fragmented landscapes and associated meta-population dynamics is relatively recent (Hanski and Simberloff, this volume; Wiens, this volume), few long-term empirical studies of metapopulations are available to guide theoretical analysis and exploration (but see Thomas and Hanski, this volume). In this chapter we present the results of a long-term study of a metapopulation that appears ideal with regard to the measurement of parameters of a metapopulation. In this study the patches are of roughly the same size, and interpatch spacing is fairly regular. The metapopulation is large relative to lifetime movements of the animals. Not all patches have been occupied in any census period. Both numerous extinctions and recolonizations have been recorded over the more than 20 years of observation of the metapopulation. The study organism is the American pika (*Ochotona princeps*), a small (132 g) alpine lagomorph; the study site is the abandoned gold-mining area of Bodie, Mono County, California. One of us (A.T.S.) has been studying pikas at Bodie since 1969 (Smith, 1974a,b, 1978, 1980), and here we present data from four complete population censuses made at varying intervals since that time.

Results from the first two censuses (1972 and 1977) were interpreted in terms of area-dependent extinction rates and distance-dependent colonization rates bor-

Metapopulation Biology
Copyright © 1997 by Academic Press, Inc. All rights of reproduction in any form reserved.

rowed directly from island biogeographic theory (MacArthur and Wilson, 1967; Smith, 1974a, 1980). The populations on patches of habitat at Bodie apparently represented a dynamic equilibrium between extinction (which was inversely related to patch size) and recolonization (which was inversely related to interpatch distance; Smith, 1974a, 1980). These results suggested that a two-dimensional "stepping-stone" metapopulation model, i.e. one for which interactions occur between neighboring patches, could be parameterized to explain these patterns. We introduce in this chapter such a spatially explicit model that successfully integrates patch-specific population growth with these size and distance effects for the Bodie pika metapopulation.

In addition, qualitative inspection of the map-based pattern of patch extinctions and recolonizations that occurred between the first and the second censuses suggested that these events did not occur randomly across the Bodie landscape. Instead, there appeared to be a clustering of extinction and recolonization events; in some neighborhoods patch occupancy declined markedly, while in other neighborhoods patch occupancy increased (Smith, 1980). Our stepping-stone model inherently incorporates such observations by accounting for certain correlations in population growth on patches within neighborhoods, particularly as a result of recolonization from neighboring patches. Thus, the model takes into consideration that immigration onto vacant patches and recurrent colonization of occupied patches (the "rescue effect;" Brown and Kodric-Brown, 1977; Smith, 1980) in regions of high average patch occupancy can lower the probability of extinction on neighboring patches and result in the increased persistence of clusters of patches.

We renewed the empirical investigation by conducting a third and fourth census (1989 and 1991) with the intention of examining closer the spatial structure of local population dynamics in the system. These recent censuses paint a picture of the Bodie pika metapopulation that rather is different from that presented earlier by Smith (1974a, 1980), with the scale of correlated turnover events dramatically widened. Almost all populations in one subregion at Bodie (covering almost half of the area of the metapopulation) had gone extinct, while the remaining subregion hardly suffered any change in distribution or abundance. The "collapse" of populations in the one subregion represents a departure from the supposed dynamic equilibrium and clustering effects seen in the earlier censuses.

Thus, the full data set (four censuses and three census intervals) challenges our ability to construct a metapopulation model built exclusively around the ideas of island biogeography: area-dependent extinctions and distance-dependent recolonizations. The inability of the stepping-stone model to explain completely the regional effects observed in our map-based observations of pika populations at Bodie led us to consider more fully elements of "regional stochasticity" (Hanski and Gilpin, 1991) that have been explored theoretically by Gilpin (1988, 1990) and Hanski (1989). We explored these properties of the Bodie metapopulation quantitatively with an examination of the autocorrelational properties of our spatially referenced data. The results demonstrate the need to understand the landscape properties of metapopulations in nature to understand fully their behavior.

II. RELEVANT PIKA NATURAL HISTORY

Characteristics of American pikas allow measurement of the most important variables needed for metapopulation analysis (Smith, 1974a,b, 1978, 1980, 1987; Smith and Ivins, 1984; summary in Smith and Weston, 1990). Pikas are diurnally active, thus easy to observe. Pikas do not hibernate, are active all year, and spend much of their summer gathering vegetation that is stored in haypiles to serve as food overwinter. Haypiles form the figurative center of activity for each individual. Pikas, being lagomorphs, have small round feces that can be distinguished readily from those of all other mammals. They also deposit soft feces (black and elongate) unlike those of any other animal. Unlike most lagomorphs, pikas are highly vocal and utter both short and long calls (males only) that can be heard over long distances. Vocalizations, fresh haypiles, feces, and urine stains can each be used to assess current occupancy of pika habitat. Because alpine climates are dry, both haypiles and the round fecal pellets do not decompose readily; at Bodie they may persist for many years. Old haypiles and old feces may be used to determine sites that have been occupied previously by pikas, even if they no longer occur there.

American pikas are habitat-specific to talus or piles of broken rock adjoining suitable vegetation for grazing and gathering forage for their haypiles. As talus is a characteristic habitat type which is easily distinguished from surrounding vegetative habitat, it is possible to define precisely the habitat area available for pika occupancy. Additionally, talus is usually distributed patchily, and at the Bodie site is highly fragmented (see below).

American pikas are individually territorial. Males and females maintain separate territories, and there is no significant difference in territory size by gender. Population density is low and averages about four to eight animals per hectare on suitable habitat throughout the geographic range of the American pika. Within-patch nearest-neighbor distances average approximately 20 m (Smith and Weston, 1990). As a result of these factors it is possible to obtain a good estimate of carrying capacity, or percentage saturation by pikas, of each pika habitat patch. It is likely that carrying capacity can be determined more accurately for pikas than any other animal. Spacing among territories does not vary among years, and most haypile localities are traditional and built on the same site year after year (Smith and Weston, 1990).

The sex-ratio of adult pikas is near unity, and animals tend to reside on territories adjoining an animal of the opposite gender (Smith and Ivins, 1984). Male and female pikas are remarkably similar; body mass is not significantly different and even their external reproductive morphology is barely distinguishable (Smith and Weston, 1990). American pikas are relatively long-lived for small mammals. Survivorship normally exceeds 50% per year, and almost 10% of the Bodie population was 5 or 6 years old (Smith, 1978).

Throughout the range of the American pika, all females, including all yearlings, initiate two litters per summer breeding season and successfully wean only one of these litters (Smith, 1978; Smith and Ivins, 1983a). Litter size (determined from embryo counts of pregnant females) is relatively small for a lagomorph and averages 3 throughout the range of the American pika (Smith and Weston, 1990). The mean litter size of 3.7 at Bodie is the highest of all populations yet studied (Smith, 1978). Litter size does not vary significantly with age of mother, habitat productivity, or between first and second litters. Litter size at weaning is normally less than litter size determined by embryo counts (Smith and Weston, 1990).

Weaned young grow rapidly and reach adult size in their summer of birth. Juveniles must successfully colonize a vacant territory to survive to become adults. As adults tend to be long-lived, vacancies are rare and occur only sporadically, and one would expect pikas to exhibit high vagility in their search for available territories. However, there are two severe constraints on the movements of juvenile pikas. First, it is difficult for juveniles to move freely in saturated pika habitat; adults apprehend and chase unfamiliar juveniles from their territories (Smith and Ivins, 1983b, 1984). Second, pikas are cold-adapted and cannot tolerate warm temperatures. At Bodie, which is near the lower distributional boundary of pikas in the Sierra Nevada, high daytime temperatures severely restrict the movements of pikas (Smith, 1974b).

Dispersal by marked animals at Bodie has been observed in two investigations. In one study 58 animals were individually marked (J. D. Nagy, personal communication). Of 34 adults, 25 were resighted the following year. Only one adult dispersed, and this movement was 18 m to a neighboring patch. Of 24 juveniles, only 5 were resighted the following year. Two of these 5 juveniles dispersed from their natal patch, in each case to the next closest patch (60 m and 150 m, respectively). In a second study, only one of 105 marked adults dispersed; it moved to a neighboring patch (Peacock, 1995). In addition, Peacock (1995) observed dispersal by 15 marked juveniles that occurred during the summer months. Three juveniles dispersed from their natal patches after they had been tagged; the other 12 were trapped postdispersal, and their natal patches were identified by genetic paternity exclusion analysis. Four of these animals dispersed within a large patch (moving an average of 45 m from their natal home range), the other 11 dispersed between patches. Nine of the 11 originated from saturated patches, the other 2 came from unsaturated patches. Nine colonized the next closest unsaturated patch, while 2 "passed up" available neighboring patches to settle on patches farther away. The average distance moved by these 11 juveniles that dispersed between patches was 132.5 m (range 70–396 m; Peacock, 1995). These results indicate that juvenile dispersal among patches occurs at Bodie, but that these movements are usually limited to nearby and/or neighboring patches.

This restricted ability of pikas to disperse between patches has profound implications for the metapopulation dynamics at this site.

III. THE BODIE SITE

American pikas live in a patchy distribution throughout their geographic range in the mountains of western North America. Most expanses of talus are disjunct and of a size to harbor pika populations in the neighborhood of tens of animals. It is rare for continuous populations of pikas to contain more than 100 animals.

The habitat at Bodie presents an ideal situation for assessing the metapopulation dynamics of pikas (Fig. 1). Bodie is an old mining area, and most of the habitat occupied by pikas there are the tailings and scree left by prior mining activity. Average size of rock in each tailing patch is similar, hence potential aspects of habitat selection by substrate are minimal. In general, these tailing patches are smaller and contain fewer pikas than natural habitat—for example in the Sierra Nevada mountains 35 km away. Mining at Bodie began in 1859 and has continued, with bursts of activity, until the present day. Roughly 100 mine tailing patches dot the landscape at Bodie, and they vary in size and distance from one another. These patches are separated by a sea of sagebrush and other typical Great Basin sage community vegetation. Additionally, pikas have been found in several of the higher natural bluffs surrounding Bodie, habitat that was most likely colonized during the Pleistocene when it was cooler in the area (Smith, 1974a). We have no indication of any recent dispersal from these natural habitats to the mine tailing patches at Bodie; in fact, the nearest natural population on Sugarloaf Hill (estimated $K = 10$) went extinct between the second and third censuses. In summary, the Bodie pika metapopulation system fits the patch–matrix paradigm discussed by Wiens (this volume). The intervening sagebrush habitat is quite homogeneous, and animals dispersing through it would not face a series of choices between subhabitats.

The pikas at Bodie were "discovered" by the biologist Joye Harrold Severaid in the mid-1940s. We know from Severaid's (1955) 4 years of intensive observations, historical accounts, and more recent censuses (Smith, 1974a, 1980) that pikas at Bodie have at one time occupied every patch of available habitat at Bodie. Severaid (1955) reported that three long-time residents of Bodie knew about the pikas, going back to their boyhood around the turn of the century. We also know from inspection of maps, photographs, and descriptions of the mine tailing patches (Severaid, 1955; Bodie California State Historical Park records), as well as direct observation over the past 25 years, that these patches are permanent and have not changed quantitatively or qualitatively since the mid-1940s.

Severaid (1955) was the first to observe that the pika population at Bodie "was never equal to the carrying capacity of the habitat." He trapped out some

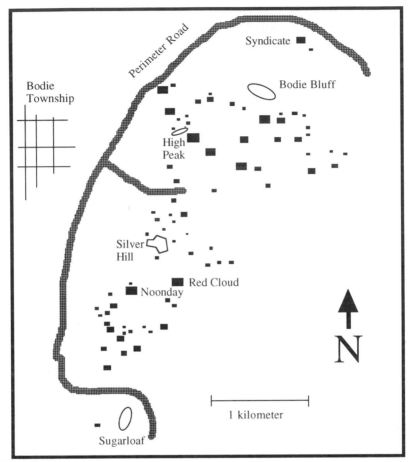

FIGURE 1   The configuration of mine tailing patches occupied by pikas at Bodie, California. The filled rectangles represent habitable patches, with the size of the rectangle proportional to the carrying capacity of the patch. Major dirt roads are portrayed with the stippled lines. Principal natural topographic features and mine tailings are named.

of the tailing patches and observed that they were colonized slowly over the course of his studies. Thus, there were early indications that pika populations on patches at Bodie were linked by occasional dispersal.

IV. METHODS

Censuses of the pika population at Bodie were made in 1972 (Smith, 1974a), 1977 (Smith, 1980), 1989 and 1991. Each census was conducted in late summer, for two reasons. First, at this time those juveniles with a chance of surviving to become adults have reached adult size and become established on available territories. Thus, at this time the resident population is at its highest level. Second, in late summer there is more available sign (haypiles, feces, etc.) to facilitate an accurate population census.

The four censuses included almost all of the available pika habitat in the Bodie region. More than 70 isolated small patches (effectively islands) of talus were included in each census. In the area of High Peak (Fig. 1), where mining activity was extreme, we sampled three large expanses of tailings, but did not take a complete census. We estimate that this large area harbors a population of approximately 50 pikas (see also Peacock, 1995). Regrettably, one area at Bodie was not censused: the expanse near the top of Bodie Bluff (several small patches) and the downslope from the bluff in the direction of the Syndicate Mine (a small number of medium-sized patches; Fig. 1). Access was denied to this area for the first two censuses.

Patch size (measured by perimeter in meters), degree of isolation (measured by distance to the nearest patch inhabited by three or more pikas in meters), and the number of resident pikas were determined for each patch (following Smith, 1974a, 1980). Patch perimeter provided the most meaningful assessment of patch size, because territories on the mine tailings are spaced linearly and adjacent to the rock–vegetation interface. None of the patches are so tall as to contain a "second story" of pikas. Interpatch distances were measured between patches inhabited by three or more pikas because those with fewer animals were unlikely to be sources of colonizing individuals (Smith, 1974a, 1980).

Percent saturation of pikas on patches was defined as N_i/K_i, where N_i is the number of pikas found on the ith patch, and K_i is the carrying capacity of each patch, based on patch size and the number of potential territories that would fit into this area (following Smith, 1974a).

A spatially explicit structured metapopulation model that incorporates our data on patch locations, population change on patches between census intervals, and observations of extinction of populations on patches and recolonization of vacant patches was created. Conceptually the model assumes independent stochastic growth, λ, on each of the habitat patches and the possibility of rescue (*sensu* Brown and Kodric-Brown, 1977) or recolonization from nearby patches. Each patch has a population ceiling, K, based on its size (perimeter in meters), and patch populations that rise above K in a single time step are truncated back to K at the end of the time step. For the one area of High Peak that was sampled, rather than completely censused, we extrapolated our population samples to incorporate the entire patch area (new $K = 50$). We calibrated our stochastic growth on the turnover seen between the 1972 and the 1977 censuses (when the metapopulation was assumed to be in equilibrium, given that the occupancy remained near 60%), and accordingly our time step was set at 5 years. Immigration from nearby patches is based on the number of individuals living within an "effects radius" of the target patch. We utilized a rectangular dispersal function such that dispersal probabilities drop to zero beyond the maximal dispersal (effects) radius.

The model was run for 20 time steps (100 years) into the future, and this was repeated 20 times for each λ and effects radius. A sensitivity investigation was performed by varying both λ and the effects radius over a range of biologically plausible values, based on the ecology of pikas at Bodie. While there are a number of directions that such a model may take, we limited ourselves to an investigation of the following question: what is the probability that, within 100 years, all populations in a subregion will go extinct? In other words, our model is designed to allow us to determine, for a variety of combinations of model parameters, the extent to which the landscape properties of the patches at Bodie may contribute to regional persistence or collapse of pika populations. Copies of the compiled model are available by contacting M.E.G. at mgilpin@ucsd.edu.

We have used two approaches to quantify spatial autocorrelation of population-level events on patches and, most important, integrated population changes in neighborhoods or subregions centered on each patch. First, we analyzed net population growth or decline in neighborhoods surrounding each focal patch, excluding the change on the focal patch itself. Positive spatial autocorrelations were said to have occurred when population size increased on both the focal patch and in the surrounding neighborhood, or if both the focal patch and surrounding patches decreased in population size. Negative spatial autocorrelations occurred when population growth or decline experienced on a focal patch was the opposite of the net population trend in the surrounding neighborhood. We ran this analysis for three "effects radii" (0.5, 1.0, and 2.0 km) surrounding each focal patch. Results of these analyses include only those focal patches that exhibited a change in population size during the respective interval.

Second, we quantified spatial autocorrelation of discrete growth rates among neighboring patches more precisely using Moran's I statistic (Haining, 1990). Spatial autocorrelation occurs when occupancy patterns of pikas on patches within specific distances are significantly associated. For this analysis, we examined the spatial autocorrelation of the discrete growth rate per census interval. This growth rate, λ, is defined as $(n_{t+1} - n_t)/n_t$ for each of the i patches. The general approach is to look at a kind of a spatially weighted cross product term of the form

$$I = \sum_i \sum_j W_{ij} \Delta_{ij},$$

where Δ_{ij} is a measure of the proximity of the variate, the discrete growth rate, between the ith and jth spatial positions, and where w_{ij} is an arbitrary spatial weighting function. On a regular grid, w_{ij} is sometimes taken as 1 for nearest neighbors and 0 otherwise. Because our patch structure was irregular, we approached the problem differently. We set w_{ij} equal to 1 if the Euclidean distance between points i and j was less than d, an "effects radius," and 0 otherwise. For the Moran statistic, Δ_{ij} is taken as $(\lambda_i - \langle \lambda \rangle)(\lambda_j - \langle \lambda \rangle)$, where $\langle \lambda \rangle$ is the mean value of the discrete growth rate for all patches during the time period. The Moran coefficient is then calculated as this weighted covariance term divided by the variance of the growth rates.

We computed the Moran statistic for each of the census period intervals and for a range of different effects radii, from 0.1 to 1.5 km, in 0.1-km increments. The 95% confidence limit against which our data were contrasted was produced by creating 200 independent realizations of uncorrelated discrete growth over our grid of habitat patches.

V. RESULTS

A. Patch Occupancy

As was initially observed by Severaid (1955), not all habitable mine tailing patches at Bodie contained pikas during any of our four censuses. Approximately 60% of the patches were occupied during the first two censuses (Smith, 1974a, 1980), but this level of occupancy fell to about 45% for the latter two censuses (Table I). Correspondingly, more pikas were located on the study area during the two earlier censuses than the latter two (Table I).

Typical of a metapopulation system, there was frequent turnover (extinctions of populations on patches and subsequent recolonization of vacant patches from existing local populations) between censuses (Table II). In addition, roughly 50% of patches varied in population size during each of the three census intervals (Table II).

The most striking difference in the pattern of occupancy of patches at Bodie among the censuses was the decline and near collapse of pika populations on patches in the southern half of the study area. The distribution of patches at Bodie is roughly in the shape of an hourglass; a saddle in the area of a dirt road extending up from the Bodie township figuratively divides the northern and southern half of the study area (Fig. 1). In the first two censuses, there was a mixture of occupied and unoccupied patches in both the northern and the southern parts (Fig. 2). By 1989 we noted a significant drop in the percentage of occupied patches in the southern half (Fig. 2). This decline included the extinction of the pika population on the Red Cloud tailing, a site that harbored nine pikas in 1972 and 1977. Only 2 years later pikas were absent from nearly all patches in the southern half. In the extreme southern part of the study area, only one animal was found (on the relatively large Noonday tailing; Fig. 2).

B. Area Effects

Size of habitat patch appeared to be the most important factor governing the occurrence of pikas on the habitat patches at Bodie (Table I), an effect apparently due to the relatively low probability of extinction of populations on large patches (Smith, 1980). In all censuses, average size of occupied patches was greater than the average size of vacant patches. There were, however, significant differences in the apparent effect of patch size among the four censuses. In the first two

TABLE I Descriptive Measurements and Correlational Statistics of Pikas on Mine Tailing Patches at Bodie

	1972[a]	1977[b]	1989	1991
Average size of patches[c] (perimeter in m)				
Occupied patches	96.0	90.5	96.9	85.9
Vacant patches	29.1	41.3	48.2	55.2
Average inter-patch distances[c] (m)				
Occupied patches	101.5	119.1	102.6	193.4
Vacant patches	184.8	238.8	267.7	1065.2
Correlation of percentage saturation with patch size				
All patches	$r_s{}^d = 0.65{*}{*}{*}$	$r_s = 0.53{*}{*}{*}$	$r_s = 0.37{*}{*}{*}$	$r_s = 0.18$
Occupied patches	$r_s = 0.47{*}{*}$	$r_s = 0.43{*}{*}$	$r_s = -0.07$	$r_s = -0.14$
Correlation of percentage saturation with interpatch distance				
All patches	$r_s = -0.30{*}{*}$	$r_s = -0.47{*}{*}{*}$	$r_s = -0.57{*}{*}{*}$	$r_s = -0.69{*}{*}{*}$
Occupied patches	$r_s = 0.02$	$r_s = 0.03$	$r_s = -0.48{*}{*}$	$r_s = -0.01$
Number of patches censused	78	78	77	78
Percentage occupied of mine tailing patches	60.3	57.7	44.2	43.6
Total number of pikas censused	164	140	118	129

[a] Data from Smith (1974a).
[b] Data from Smith (1980).
[c] Excluding three High Peak samples.
[d] Spearman rank correlation coefficient corrected for tied observations.
$**P < 0.01$.
$***P < 0.001$.

censuses (1972 and 1977) there were significant correlations between patch size and percentage saturation for both all patches and just those that were occupied. This correlation was weaker (although still highly significant) for all patches in 1989, but there was no correlation for occupied patches only in 1989. In 1991, neither all patches nor occupied patches showed a correlation to area. In 1989, when the "collapse" of the southern half of the Bodie pika metapopulation was underway, a number of large patches harbored very few pikas. The resulting low percentage saturation values on large patches apparently resulted in the lack of a correlation between these variables in that year. In 1991, there was no correlation between size and percentage saturation for all patches because nearly all of the patches in the southern half of the Bodie metapopulation were unoccupied, independent of their size. Interestingly, there remained no correlation

TABLE II   Between-Census Extinctions and Recolonizations and Changes in Percentage Saturation on Mine Tailing Patches at Bodie

	1972–1977	1977–1989	1989–1991
Patch extinctions	10	14	8
Patch recolonizations	8	4	7
Percentage saturation			
No. patches increased	17	20	15
No. patches decreased	30	21	21
No change (including unoccupied patches)	30	35	41
Percentage of patches that experienced a change	61.0	53.9	46.8

between size and percentage saturation for the occupied patches in the northern half of the study area. While the southern half was nearly void of pikas, the absolute number of pikas increased in 1991 over 1989 (Table I). Thus, while there were fewer occupied sites at Bodie in 1991 (Table I), those that were occupied all had relatively high percentage saturation values independent of size of patch.

C. Isolation Effects

There was a strong negative correlation for all patches of percentage saturation with interpatch distance across all years (Table I), apparently because the percentage of patches that were unoccupied increased with interpatch distance (see Smith, 1980). This correlation was more pronounced in each subsequent census year (Table I). The strongest correlation, in 1991, resulted from the long inter-patch distances from each unoccupied patch in the southern half of the study area to the closest patch with three pikas in the northern half (Table I).

At the same time, in 3 of 4 years there was no relationship between interpatch distance and percentage saturation for occupied patches (Table I). Apparently, once occupied, factors other than isolation play a major role in the determination of percentage saturation on tailing patches. The one exception was in 1989 when there was a significantly negative correlation between these two variables (Table I). At this time the collapse of the southern half of the Bodie pika metapopulation had started, and percentage saturation among occupied patches apparently was not being determined by patch size as in other years (see above). Instead, the effect of isolation was beginning to show among occupied patches; those that were occupied yet declining in percentage saturation were not receiving new propagules (i.e. no "rescue effect").

Thus the data on correlation of percentage saturation with patch size and interpatch distance portrayed in Table I show a "wave" effect from 1972 and

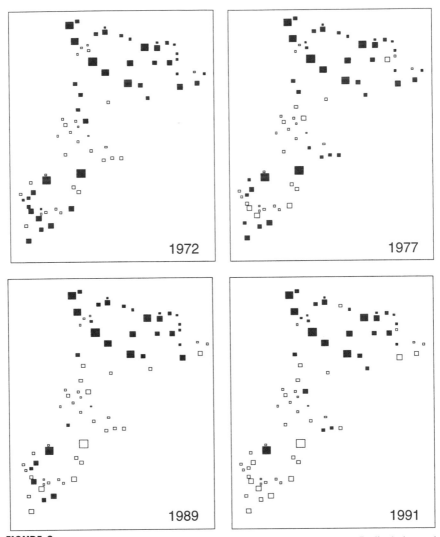

FIGURE 2 Configuration of occupied and unoccupied mine tailing patches at Bodie during each of the four census periods (1972, 1977, 1989, 1991). Patches occupied by pikas are represented by filled rectangles; open rectangles represent unoccupied patches. The rectangles represent habitable patches, with the size of each rectangle being proportional to its carrying capacity (size).

1977 (when population extinctions and subsequent recolonization on patches appeared to be in a dynamic equilibrium; Smith, 1974a, 1980), to 1989 (when the effect of isolation rather than patch size determined percentage saturation on occupied patches), to 1991 (when the effect of near extirpation of the southern half of the Bodie pika metapopulation was the predominant factor).

D. The Spatially Explicit Structured Metapopulation Model

There are many ways of constructing and parameterizing a metapopulation model (Hanski and Gilpin, 1991). We developed a spatially explicit structured metapopulation model because we were interested in exploring a specific question: is the southern region of the Bodie metapopulation more extinction prone than the northern region? To approach this we have examined λ, which governs rate of extinction, and the effects radii, which govern "rescue" and recolonization, over wide, but biologically reasonable, ranges of values (Fig. 3). Our response variable is the ratio of pikas in the northern region to the number of pikas in the southern region at the end of 20 time steps, or 100 years (Fig. 3). The outcome

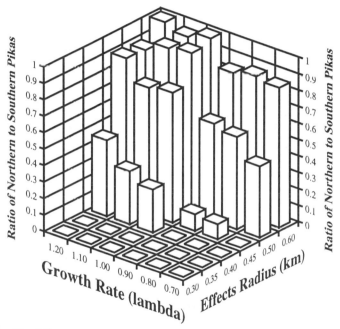

FIGURE 3   The difference between projected population behavior of pikas in the north and south subregions at Bodie versus the two parameters of the stepping-stone model. One independent axis shows a set of values for the "effects radius" (see text for explanation), while the second independent axis shows values for the expected discrete growth rate (which is absent any immigration from nearby patches). Over the full set of parameter values, the full metapopulation behavior goes from full occupancy and full saturation to complete regional extirpation. The dependent axis, plotted vertically as histogram cells, shows the ratio of the number of animals in the northmost half of the metapopulation to the number of animals in the southmost half of the metapopulation after 100 years (20 time steps) based on 50 replications. At full saturation (high effects radius and high expected growth rate), there are about 20% more animals in the northern subregion. As the parameters are each reduced in value, so does this north-to-south ratio, which indicates that the northern half of the metapopulation is predicted to be the more vulnerable subregion to subregional extirpation.

shows that the north-to-south ratio declines with declining parameter values. The model appears most sensitive to variation in the effects radii. The north and south subregions at Bodie appear to behave similarly only when parameter values are jointly high (leading to full occupancy of patches) or jointly low (leading to regional extinction). At intermediate values the north-to-south ratio falls below unity, indicating that the northern half of the metapopulation is the more vulnerable subregion to subregional extirpation (Fig. 3). This finding is the exact opposite of our empirical result in 1991.

One potential reason for the discrepancy between the model and empirical results is that average nearest-neighbor distance between each pair of patches is shorter in the southern region (0.169 km) than in the north (0.204 km). Perhaps more important is Fig. 4, a demonstration of the degree of connectivity (number of potential patches within sweeps of each of four effects radii) for each patch. Clearly, there are significantly fewer patches available to influence spatially autocorrelated changes at the shortest distance (0.25 km) than at each of the longer distances, 0.5, 1.0, and 2.0 km, respectively (Fig. 4). At an effects radius of 0.25 km, very few of the patches show connectivity, and extinction of populations on patches regionally or throughout the Bodie metapopulation is not a surprising result. At 0.5 km many of the patches in the southern region are connected, and there are two distinct clusters with high connectivity. At this distance the north is still very loosely connected. The 2.0-km effects radius figuratively "joins" each patch with a very large subset of patches within the Bodie metapopulation system (Fig. 4).

E. Regional Effects (Spatially Autocorrelated Patterns)

The spatial pattern of patch occupancy, and area and isolation effects (Fig. 2; Tables I and II; see also Smith, 1980) indicate that the dynamics of extinction and recolonization of patches at Bodie has not occurred uniformly across the landscape. In addition, our model based on stepping-stone metapopulation dynamics failed to explain our observation that the southern half of the Bodie metapopulation collapsed in the 1991 census, while the northern half remained close to fully saturated. Thus, we examined this putative spatially nonrandom pattern by determining the extent to which there was spatial autocorrelation of population-level events among patches.

Between-census correlated changes in percentage saturation were determined between focal patches and neighboring patches at three effects radii for each of the three census intervals (Table III). For the 1972–1977 and 1989–1991 intervals, there were no significant differences between the number of patches that exhibited positive spatial autocorrelation than negative spatial autocorrelation at the 2.0-km effects radii (Table III). Both intervals showed significant differences ($P < 0.1$) at the 0.5-km radius; and for the 1989–1991 interval, there was a highly significant difference at the 1.0-km radius. The opposite result obtained for the 1977–1989 census interval: there were significant differences at effects

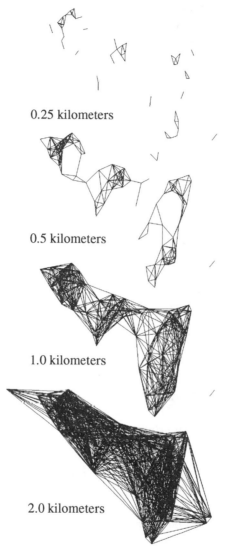

0.25 kilometers

0.5 kilometers

1.0 kilometers

2.0 kilometers

FIGURE 4   Connectivity of the mine tailing patches at Bodie for different "effects radii" (0.25 km, 0.50 km, 1.00 km, 2.00 km). In each portrayal, the effects radius is swept across the landscape from each patch, and lines join those patches within the circumscribed area.

TABLE III   Between-Census Correlated Changes in Percentage Saturation between Focal Mine Tailing Patches and Neighboring Patches at Various Fixed Distances ("Effects Radii")[a]

Effects radius (km)	Census interval	Positive spatial autocorrelations	Negative spatial autocorrelations	G-test value
2.00	1972–1977	18	21	0.101
	1977–1989	27	13	4.302**
	1989–1991	18	16	0.028
1.00	1972–1977	25	14	2.591
	1977–1989	26	14	3.062*
	1989–1991	26	8	8.895***
0.50	1972–1977	26	13	3.751*
	1977–1989	23	17	0.627
	1989–1991	25	9	6.849***

[a] Positive spatial autocorrelations occurred when the focal patches increased or decreased in percentage saturation and the average change in percentage saturation in neighboring patches (within a respective radius) correspondingly increased or decreased. Negative spatial autocorrelations occurred when percentage saturation between a focal patch and its neighboring patches were of different sign. Only those patches that exhibited a change in percentage saturation during the respective census interval were analyzed.
* $P < 0.10$.
** $P < 0.05$.
*** $P < 0.01$.

radii of 2.0 and 1.0 km, whereas there was no difference between the number of positive and negative spatially autocorrelated patches at the 0.5-km radius (Table III).

The sensitivity of spatial autocorrelation at varying radii was examined also using Moran's I statistic (Fig. 5). During the 1972–1977 census interval, there was positive spatial autocorrelation only for effects radii up to 0.3 km (Fig. 5), which is a distance that includes on average very few nearest neighbors (Figs. 1, 4). During this time overall saturation was high, there was an interspersion of occupied and unoccupied patches (see Fig. 2), and population extinctions and subsequent recolonizations on patches appeared to be in a dynamic equilibrium (Smith, 1974a, 1980). This result is consistent with the reported low vagility of pikas at Bodie (Smith, 1974b)—population sizes on focal patches were likely to be influenced only by nearby patches (either growing in part due to immigration of surplus animals from nearby patches, or declining and concomitantly not receiving immigrants from nearby patches that were also declining).

The 1977–1989 census interval shows a more interesting pattern. During this interval there was a net loss of animals (140 to 118; Table I), and this loss was confined mostly to the southern end of the study area (Fig. 2). This regional decline accounts for much of the positive spatial autocorrelation (largely negative patch growth associated with negative regional growth in the south) over distances from 0.3 to 1.5 km (Fig. 5). It is also possible that the longer time interval between censuses influenced the autocorrelation statistic by allowing more "averaging"

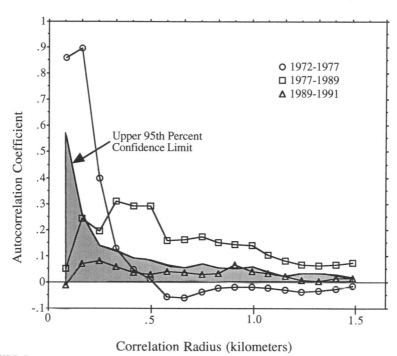

FIGURE 5   Between-census spatial correlation and 95% confidence limit of discrete growth rate per time interval analyzed at varying effects radii among mine tailing patches at Bodie using Moran's *I* statistic. See text for explanation.

over regions of the cumulative small dispersal movements of pikas at Bodie. In this way, many patches in the north increased in population size during the census interval (Fig. 4).

A lack of significant positive autocorrelation occurred over the shortest time interval (1989–1991; Fig. 5). The few remaining animals in the southern portion died out, and the northern region showed stability at a relatively high overall saturation.

VI. DISCUSSION

Our long-term multigenerational study of the Bodie pika metapopulation has provided insight into the dynamics of a real world metapopulation. We have gained understanding of the underlying mechanisms of the metapopulation dynamics, and we have learned something about the different spatial scales that are relevant to these mechanisms and to their associated dynamics.

The regional scale at which metapopulation dynamics are likely to occur has been defined as that "at which individuals infrequently move from one place (population) to another, typically across habitat types which are not suitable for their feeding and breeding activities, and often with substantial risk of failing to locate another suitable habitat patch in which to settle" (Hanski and Gilpin, 1991). The American pika population at Bodie fits this definition closely. Pikas cannot and do not occupy the Great Basin sagebrush habitat that separates their obligatory talus habitat—the mine tailing patches at Bodie. Physiologically, pikas are at risk due to high temperatures and inability to find ameliorating microclimates (such as the cool interstices deep in talus) while dispersing (Smith, 1974b). Pikas are more vulnerable to predation when off of talus (Smith, 1974a; Ivins and Smith, 1983), and a high diversity of potential predators of pikas occupy the study area (Smith, 1979). Juveniles, the primary dispersers in this system, normally settle on the closest unsaturated patch (J. A. Nagy, personal communication; Peacock, 1995).

Metapopulations are characterized also by their balance between extinction of local populations and establishment of new populations in vacant habitat patches (Hanski and Gilpin, 1991). Such extinction-colonization dynamics, as well as a pattern of incomplete patch occupancy (in the range of 60–40%), have been a characteristic of the Bodie pika population for several decades (Tables I, II; Severaid, 1955; Smith, 1974a, b, 1980). During the 1970s, extinction of populations on patches was a function of patch area, largely due to demographic stochasticity, and immigration, which was a function of the distance to the nearest' possible source patch (Table I; Smith, 1974a, 1980). Indeed, the average small patch size at Bodie, thus the small carrying capacity of pikas on these patches, appears to be a central feature of this metapopulation.

The strong effect of isolation of patches on occupancy rates of pikas in the Bodie system indicates that a basic assumption of the spatially implicit Levins (1970) metapopulation model—that dispersal occurs with equal probability between any pair of patches, independent of their location—is incorrect. Smith (1974a, 1980) formulated his early analysis on a spatially explicit framework based on island biogeographic theory (MacArthur and Wilson, 1967). In the present analysis, we have extended this with the parameterization of a spatially explicit "stepping-stone" metapopulation model.

There are many forms, varying considerably in complexity, that such a steppingstone metapopulation model can take. We chose to work with the simplest form of model that could include the most important features of the Bodie pika metapopulation system. The model includes the size and location of all patches. We have provided for distance-dependent rescue and recolonization through an "effects radius" that limits single time-step dispersal to a maximal distance, which, in our sensitivity analyses, we have kept to below 1 km. We modeled occupancy and extinction in a structured manner (see Gyllenberg *et al.*, this volume, for a review) in which we distinguish the integer population size on a patch, most of which fall in the 0 to 10 range. Transitions between states on a patch are governed by demographic stochasticity, which is independent between individuals.

Based on data from the 1970s, our analysis shows the metapopulation system to be stable, with continuing turnover and with patch occupancy in the range of

60%. If these results were to be taken as a model prediction, then the actual behavior of the system—the near extirpation of populations in the southern sub-region—would be quite surprising. Of interest, then, is whether the model was misparameterized or incompletely structured.

We have explored the possibility of misparamterization. We have performed a wide sensitivity analysis by varying the effects radius and the expected growth rate per patch. We varied these parameters over a range that takes the metapopulation from full saturation and occupancy to extirpation. What we found is that it is highly improbable (although not impossible) that the southern subregion should go extinct, while the northern subregion should remain highly saturated. Thus, we conclude that the structure of the model is incomplete.

Because of the apparent inadequacy of our stepping-stone model for predicting the future of the Bodie metapopulation, we analyzed the system for spatially correlated patterns among neighborhoods (or clusters) of patches. We did this in two ways: with analyses of correlated changes in population growth within neighborhoods, and with the development of Moran's I statistic for our map-based data in which multiple effects radii were examined.

The autocorrelation/neighborhood analysis indicated that most patches were strongly influenced by the average level of occupancy in surrounding patches—thus that entire neighborhoods rather than distance to a single potential source patch was important in determining the probability of patch occupancy (Table III). These neighborhoods were well-defined for the shortest effects radius, 0.5 km, for the two censuses taken close together (1972–1977, 1989–1991). The radius of 1.0 km yielded the highest overall difference between positive and negative autocorrelations of patches for all censuses. These results highlight the influence of neighboring patches at greater distances from target patches than is apparent from the spatially explicit metapopulation model based upon stepping-stone (or nearest-neighbor) dispersal distances. The longest census interval was between 1977 and 1989, and the strongest effect (difference between number of positive and negative autocorrelated patches) was found at 2.0 km. Apparently, given more time and the likelihood of more cumulative nearest-neighbor dispersal movements among patches, the radius affected by such movements increases.

The results of Moran's I statistic also confirm the spatially correlated pattern of occupancy of pikas at Bodie and indicate that there may be distinct differences in the nature of autocorrelation of patch occupancy among census intervals (Fig. 5). Unfortunately, because of the different intervals between censuses, we cannot discriminate between whether these different responses were caused by time of census interval or the actual spacing of occupied and unoccupied patches during the census interval. Parsimoniously, the 1972–1978 interval showed high levels of autocorrelation at relatively short effects radii distances, indicative of the extinction-colonization dynamics that characterized this time frame. At the other extreme, no autocorrelation was evident for the 1989–1991 interval, and this result could be due to either (or both) the short time between censuses or the total collapse of the southern half of the study population. The 1977–1989 interval yielded significant autocorrelations across the full range of effects radii, again indicating that there may have been more "averaging" of neighborhoods due to the cumulative effects of stepping-stone dispersal over this period.

We do not know the actual mechanism(s) that may have produced the spatial nonrandomness of extinction as demonstrated in this investigation. In his book on spatial statistics, Haining (1990) shows that very different dynamics, and different underlying mechanisms, can cause similar patterns of spatial autocorrelation. We pose two general hypotheses about what occurred in this system. The first is that the results could have been due to "position" effects. The second is that extinction events leading to the collapse of the southern half of the study population were driven by spatial autocorrelation effects.

The position effects hypothesis holds that the southern half of the study area is qualitatively different from the northern half or that external forces (such as climate, predation, competition) could affect pikas in the southern and northern halves differentially. Although we have no direct measurements, available data that we outline below indicate that there is no demonstrable difference in important parameters that may affect pika viability in the southern and northern halves of the study area. The Great Basin sage plant community is roughly similar throughout the study area, and each of the habitat patches on which pikas live was "thrown up" in the middle of this homogeneous habitat. There is very little difference in size of rock in each patch, and the structure of the patches are remarkably similar throughout the study area. Similarly, the area is too small to have been affected differentially by climatic patterns.

All available observations on potential predators and competitors (Severaid, 1955; Smith, 1979) indicate that they occur throughout the study area. The most likely predators of pikas on their preferred talus habitat are weasels (*Mustela frenata* and *M. erminea*—both of which are common at Bodie; Smith, 1979), as they can gain entrance to the dens of pikas (Ivins and Smith, 1983). Weasels at Bodie have the tendency to hunt repeatedly on the same patch (A. T. Smith, unpublished data), and it is likely that weasel predation is the most common cause of extinction of a population of pikas on a patch. If a family of weasels then moved to the next closest occupied patch (as they do when hunting *Microtus;* Fitzgerald, 1977), a cluster of patches could show "correlated extinctions." However, as mentioned above, there is no reason that such predator-caused extinctions would occur only in one region of the study area. Instead, such extinction events could be said to operate as a form of environmental stochasticity that would impact neighborhoods of patches (see below).

A final potential position effect is that the largest single patch (High Peak; Fig. 1) was found on the northern half of the study area (although there were large patches—the Noonday and the Red Cloud mine tailings [Fig.1]—found in the south). It could be that the northern area operates more like a mainland–island population, in which persistence depends on the existence of one or more extinction-resistant populations (Harrison, 1991). Two lines of evidence argue against this effect. First, the large patch ($K = 50$) was included in our spatially explicit model, and in spite of its size the northern population was more extinction-prone than the southern population. Second, from 1988 to 1991 all pikas in a portion of the High Peak patch were individually marked (Peacock, 1995). Although several patches are found in the neighborhood of this site, none of them were colonized by any of the marked animals (Peacock, 1995). Thus, within the time scale of our censuses, the continued occupancy of a large number of patches in the northern half does not appear to be explained by mainland–island colonization dynamics.

The alternative to the position effects hypothesis is that extinction events leading to the collapse of the southern half of the study population were driven by spatial autocorrelation effects. In this case the southern and northern portions of the study area are considered equivalent, and a stochastic cause of extinction may have impacted patch populations in the south rather than in the north. As fewer and fewer southern patches were occupied, then there were fewer source patches to produce propagules for the increasing number of unoccupied patches. Ultimately, we believe a threshold may have been reached that the system itself was incapable of remaining in extinction–colonization equilibrium, and the result was its total collapse. The 1989 census, which showed a greatly reduced number of occupied islands in the south, was a harbinger of the collapse observed in 1991. The 1989 census is important in that it shows that the southern half did not collapse over a short period (such as might be expected should there have been an epidemic), but rather that there was a gradual decline in the number of occupied islands. Effectively, what we observed in the southern half of the study area during the 1991 census was a regional extension of the spatial autocorrelation among patches seen throughout the Bodie landscape leading up to the 1991 census (Table III; Fig. 5).

We have no direct evidence of the stochastic event(s) that may have initiated this downward spiral of patch occupancy in the southern half of the Bodie study area. Demographically, pikas are long-lived. It is possible that the age structure on some patches could have been skewed to older animals, which in turn prohibited local settlement of their offspring (see Smith and Ivins, 1983a). A few key patches could have been structured in this way, followed by the death of all adults from "old age." Also possible is that populations on individual patches became genetically inbred, leading to inbreeding depression and local patch extinctions (Gilpin, 1991). However, a detailed study of the genetics of pikas in a subset of the Bodie population indicated that enough movement took place among patches (with juveniles primarily colonizing the closest available territory as well as occasionally dispersing greater distances) that habitat fragmentation resulted in only limited genetic subdivision within this population (Peacock, 1995). Another possibility, as mentioned above, is that weasels cleared out a few key patches, beginning the decline. Naturally, there could also have been multiple causes of extinction of patches in the south.

The results of this investigation show clearly the need for metapopulation models to incorporate explicit spatial dimensions and to examine a range of spatial and temporal scales in their analyses. Accordingly, efforts to understand and to model further the Bodie pika metapopulation continue. First, we desire to understand dispersal as a function of source patch density, degree of patch isolation, intervening habitat structure (Wiens, this volume), and size and occupancy of target patches. Second, we want to see to what degree demographic stochasticity plus rescue may explain local patch fluctuations in population size. Third, we want to explore alternative hypotheses for spatial autocorrelation of extinction. For us, the metapopulation of American pikas at the ghost town of Bodie remains a living laboratory.

ACKNOWLEDGMENTS

We thank the staff of the Bodie California State Historical Park for their support and assistance over the years. Permission for access to our field site was kindly granted by the California State Parks Association, Bureau of Land Management, and various mining companies holding patent rights at Bodie. We appreciate the contributions of Chris Ray and Mary Peacock (1989 and 1991), and David Quammen (1991), who helped us conduct the censuses. Lyle Nichols assisted in the statistical analyses, for which we are grateful. We thank Jan Bengtsson, Steve Dobson, Ilkka Hanski, John Nagy, Harriet Smith, and Chris Thomas for their conscientious reviews of the manuscript.

A Case Study of Genetic Structure in a Plant Metapopulation

Barbara E. Giles   *Jérôme Goudet*

I. INTRODUCTION

Habitats that are suitable for the establishment and maintenance of most species of plants and animals are distributed patchily across landscapes. Environmental patchiness forces species to be structured into systems of local populations within which conspecifics are more likely to interact with each other than with conspecifics from other populations (McCauley, 1995). Isolation, however, is not complete and since most real organisms have some power of dispersal, members of a local population have a low but positive probability of interacting with individuals from other localities. Depending on the rate of migration, demographic and genetic dynamics will be influenced by this migration as well as by local birth and death rates.

Local populations are seldom immortal. Demographic, environmental, and genetic stochasticities (Shaffer, 1981; Lande, 1988b), and deterministic processes such as succession (Olivieri *et al.,* 1990; 1995; Harrison, 1991) may cause the extinction of local populations, although some of these habitats may be colonized again by dispersing propagules. As a consequence, local populations come and go on temporal scales that do not allow demographic and/or genetic equilibria to be attained, and the age structure of the local populations reflects the time elapsed since they were formed (Whitlock and McCauley, 1990).

Patches may also differ in quality, size, and spatial arrangement (Hanski, this volume). Since (re)colonization and migration depend on the distances between habitat patches relative to the dispersal range of a species, patch dispersion is critical for the dynamics in individual patches and for the maintenance of the metapopulation as a whole (Harrison, 1991; Fahrig, 1992; Slatkin, 1993; Hastings and Harrison, 1994; McCauley, 1995; Hanski, this volume). Where local populations are connected by migration that is not extensive enough to entirely obliterate the local generation to generation dynamics (Gyllenberg and Hanski, 1992), treating the system as if it were one large homogeneous population at equilibrium is not appropriate (Olivieri *et al.,* 1990). Instead, understanding population and evolutionary processes in temporally and spatially structured populations requires work at a regional or metapopulation level (Antonovics *et al.,* 1994; Hanski, 1996a).

While the theory of metapopulations is now reasonably well developed (Hanski and Simberloff, this volume) and many ecological and genetic predictions have been made, empirical tests of theoretical predictions lag behind. The primary purpose of this chapter is to present a case study of a plant metapopulation. The plant, *Silene dioica,* is a dioecious perennial and a component of early stages of primary succession in northern Scandinavia. This study was carried out on islands in an area of the Baltic Sea subject to land uplift so that new islands are continuously, though slowly, being formed. The rate of land uplift allows the ages of the island populations to be estimated; the successional processes together with the continual creation of new islands imply that population turnover must occur. We have proceeded by constructing groups of islands differing in their spatial, temporal or demographic characteristics, and compared the observed changes in genetic differentiation among these groups with those predicted by metapopulation models. In this way, we have been able to test and, in many cases, confirm the predicted effects of spatial and temporal heterogeneity on genetic structuring resulting from metapopulation dynamics. We begin with a short review of the genetic theory of metapopulations to contrast the differences in the assumptions and questions of the ecological and genetics models. After presenting the results of our study, we briefly review other genetic studies of metapopulations.

II. THEORY

The consequences of genetic differentiation and gene flow among local populations for the rate and pattern of evolutionary change has been studied for a long time in population genetics. In large randomly mating populations, the major factor affecting allele and genotypic frequencies is selection. In assemblages of small populations, however, other factors come into play, the main ones being genetic drift (the random fluctuations of allele frequencies) and the degree to which drift is counterbalanced by migration. The evolutionary role of genetic drift is at the center of a long-standing debate in population biology, and many theories rely upon this force (Barton and Clark, 1990). For example, in the shifting balance theory of evolution (Wright, 1977; Barton and Whitlock, this volume), the general idea is that many small populations can explore the fitness landscape (where valleys and peaks represent low and high fitnesses, respectively) more extensively than larger populations. This is because the selective pressures need to be much stronger in small populations to be effective against random genetic drift. While the consequences of small population sizes are much debated in the literature in terms of inbreeding depression, Wright (1977, Chapter 13) showed that new favorable gene combinations could arise. In this process, individuals with gene combinations that selection would eliminate from large populations may survive to reproduce. As the size of these small populations increases, fine tuning of these gene combinations may allow the population to reach a higher adaptive peak. Migration could then spread these combinations to other localities (Barton and Whitlock, this volume). For shifting balance to work, however, populations need to be fairly isolated from one another. Too much gene flow would prevent genetic differentiation; too little would prevent the spread of favorable combinations. Much effort has therefore been put into measuring the extent to which populations are differentiated, and these measurements have been used to infer the level of gene flow between local populations.

The modeling of spatially structured populations was pioneered by Sewall Wright (1931, 1943). His first model, the island model (Wright, 1931), assumed a set of populations of finite and equal sizes with equal probabilities of migrant exchange. At equilibrium between the forces of migration and genetic drift, Wright's model showed that the degree of genetic differentiation among local populations, measured as F_{ST} (the variance of allele frequencies standardized by the maximum possible variance), was a simple function of the effective numbers of migrants among populations (Wright, 1940, 1943, 1951):

$$F_{ST} \simeq 1/4Nm + 1. \tag{1}$$

Further mathematical development of the model allowed the effects of spatial variance in migration rate to be incorporated and led Kimura to propose his stepping-stone model (Kimura, 1955; Kimura and Weiss, 1964; Weiss and Kimura, 1964), where only adjacent populations exchange migrants. Both the island and the stepping-stone models assume that individuals are members of discrete populations within which mating occurs at random (panmixia). For many species, however, it is possible that truly random mating units do not exist at all. To account for this type of spatial population structure, another set of models without panmictic units was developed. In the isolation by distance or neighborhood model of Wright (1943), the genes of each individual disperse as a decreasing function of distance, with a neighborhood being defined for each individual. The size of the neighborhood corresponds to the area from which the parents of the central individual could have been drawn at random (Wright, 1943). This area is defined as a circle of radius 2σ around the central individual, where σ is the standard deviation of a Normal distribution of parent–offspring migration distances. Both isolation by distance and stepping stone models of migration, in which there is spatial variance in migration rate, have been shown to increase the degree of population differentiation relative to the island model (Kimura and Weiss, 1964; Crow and Aoki, 1984; Whitlock, 1992a; Goudet, 1993).

The island, stepping-stone, and isolation by distance models, however, still assume that populations have constant sizes and are immortal. Although Wright (1940) was the first to suggest that extinction could increase population differentiation relative to an island model, the effects of extinction and (re)colonization (population turnover) on among-population variance in gene frequencies were not studied before 1977, when Slatkin introduced the ecological concept of the metapopulation into population genetics. Population turnover implies that local populations, or demes, vary in their degree of demographic maturity. Newly established populations are not demographically mature, and (re)colonization represents an additional source of genetic drift due to founder effects at the time of colonization. With the introduction of population turnover into models of spatial population structure, spatial variance in migration rate is ignored (i.e., migration patterns are those assumed in the island model) and the focus lies entirely on the temporal relationships among habitat patches.

Slatkin's (1977) models are variations of Wright's island model and consist of a collection of local populations of diploid monoecious individuals of identical and fixed size N, which exchange migrants (gametes) at a common rate m. Generations do not overlap and N is assumed to represent the carrying capacity, or the number of organisms which the resources of a habitat can support. Each generation, a proportion e of the local populations goes extinct, where the probability of extinction is equal for all age classes. The extinct sites are then recolonized immediately by a fixed number of (diploid) colonizers, k, which in turn reproduce and give rise to N offspring within one generation. This model differs from that of Levins (1970) in that: (1) patch size is a variable in the model (Levins was only concerned with the proportion of occupied patches) (Hanski, this volume), (2) extant populations exchange migrants (they do not matter in the Levins model), and (3) no site is empty at any time. Slatkin cast his model in terms of the variance in gene frequencies among populations and considered two forms of colonization: the migrant pool model, in which the k colonizers are drawn as a random sample from the metapopulation; and the propagule pool model, where the k colonizers are drawn from a single population chosen at random. His results indicated that under the propagule pool model, genetic drift resulting from sampling effects during population establishment would increase the differentiation among local populations, while under the migrant pool model, mixing prior to colonization resulted in additional gene flow reducing the degree of differentiation. Slatkin (1985, 1987) later concluded that if the average time (in generations) to extinction of local populations was less than or equal to the effective size of populations, then, even in the absence of migration, extinction and recolonization would prohibit the genetic differentiation of local populations by genetic drift.

This latter view was challenged by Wade and McCauley (1988), who recast the model in terms of the standardized variance of allele frequencies, F_{ST}, and asked: "under what conditions do extinction, colonisation, and dispersion bind an array of subdivided populations into a single evolutionary unit, and when do they permit local populations to assume more independent evolutionary trajectories?" For the migrant pool model, these authors found that F_{ST} would be increased compared to an island model if the number of colonists was less than or equal to twice the number of migrants, whereas F_{ST} was always increased in the propagule model. Whitlock and McCauley (1990) generalized Slatkin's migrant and propagule pool model by including a new parameter, ϕ, which corresponds to the proportion of colonizers with a common origin (in the migrant pool model, ϕ takes a value of 0, whereas it is equal to 1 in the propagule pool model). They reached the conclusion that, under this model, differentiation is enhanced in a metapopulation compared to an island model of population structure if

$$2k(1 - \phi) + \phi < (4Nm + 1), \tag{2}$$

where k is the number of colonizers, $(1 - \phi)$ is the probability that colonists come from different source populations, and Nm is the effective number of migrants per generation. With this formulation, the problem becomes an ecological one, and a critical consideration is whether or not colonization is a phenomenon different from migration. When migration and colonization are the same thing (e.g., $4Nm = 2k$), differentiation is always increased when extinction occurs compared to a case where it does not.

Measuring colonization and migration to test whether the conditions in Eq. (2) hold in the field has a number of problems. First, direct measurement is difficult or impossible for most organisms (see Slatkin, 1987). Second, while migration can be estimated "indirectly" from F_{ST} if a system behaves like an island model at equilibrium (Wright, 1931), two biological quantities, in this case k and Nm, cannot be estimated from the same estimate of F_{ST}. Whitlock and McCauley (1990) and Whitlock (1992b), however, showed that if the conditions specified in Eq. (2) hold, it then follows that F_{ST} for younger, more recently colonized populations will be greater than F_{ST} for older populations. This relationship between the F_{ST} values for local populations of different ages thus appears to provide a useful tool which can be applied to field data to assess the effects of turnover on levels of genetic differentiation. This relationship has already been described in the literature as a "simple assay for enhanced differentiation in a metapopulation" (Dybdal, 1994), but a word of caution is necessary. While the conditions described in Eq. (2) imply the stated relationships between the values of F_{ST} for younger and older populations, finding the stated relationship between the F_{ST} values does not imply that Eq. (2) has been satisfied.

III. THE *SILENE* METAPOPULATION ON ISLANDS IN THE BALTIC

A. The Environmental Setting

The Skeppsvik Archipelago (Fig. 1) is located in the Gulf of Bothnia at the mouth of the Sävar River in the province of Västerbotten, Sweden ($63°44'–48'N$, $20°34'–40'E$). The archipelago contains about 100 islands within 20 km². The islands are composed of morainic deposits initially left under water when the ice receded at the end of the last ice age 7700 years ago. This area has been, and still is, subject to a rapid and now relatively constant rate of land uplift of ca 0.9 cm per year (Ericson and Wallentinus, 1979). Because the glaciers did not deposit equal amounts of material in all places, land uplift naturally raises the highest points above the water level first. This creates islands, which will differ in age as the process continues. After an even longer period of land uplift, the islands may eventually fuse as has occurred in what is now the mainland in Fig. 1. The rate of land uplift also provides means for estimating the ages of the islands in the Skeppsvik Archipelago from their heights above sea level, corrected relative to the theoretical mean sea level at the nearest mareographic station.

Once the bare rocks are exposed, processes leading to the formation of soil begin, and the islands gradually become available for colonization by plants. The Skeppsvik Archipelago is relatively undisturbed and each island goes through autogenic primary succession. The age differences among the islands, however, mean that the plant populations on the islands differ in their ages and stages of demographic development. The ages of the plant populations on these islands can be estimated from

$$Age_{ij} = (h/m)j - t_i, \qquad (3)$$

where Age_{ij} is the age of population i on island j, $(h/m)j$ is the age of island j (i.e., island height, h, divided by the rate of land uplift, m), and t_i is the average time from island exposure to colonization and establishment of species i (Ericson and Wallentinus, 1979; Carlsson *et al.*, 1990). t_i has been estimated from data on new colonizations, changes in the species composition, and changes in population sizes of target species occurring between 1972, 1986, and 1990 (L. Ericson, unpublished). Since a number of islands were colonized during this period, the time of colonization could be compared with known island age. Time to colonization also increases with exposure to wave and wind action, so t_i was calculated separately for different parts of the archipelago, and the appropriate t_i value was substituted to obtain the age of each population. Since successional processes, i.e., "nonseasonal, directional, and continuous patterns of colonization and extinction on a site by species populations" (Begon *et al.*, 1986), necessarily include turnover dynamics, we chose to work with a plant species lying within the series of succes-

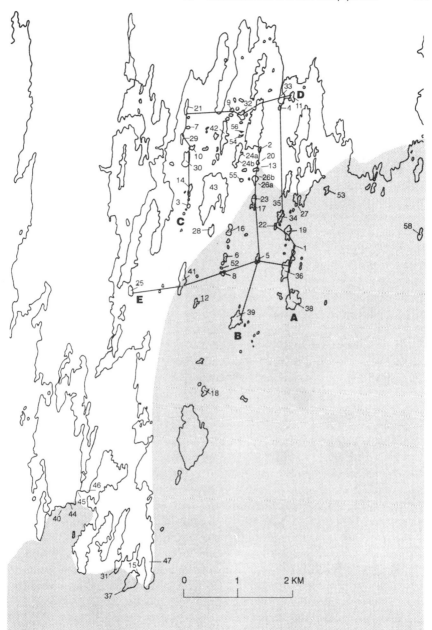

FIGURE 1   Skeppsvik Archipelago, in the Gulf of Bothnia, Sweden (63°44′–48′N, 20°34′–40′E). The shaded and unshaded areas are the exposed and protected parts of the archipelago, respectively. As discussed in the test, **A, B,** and **C** are the outer, middle, and inner chains, respectively; **D** and **E** are the north and south transects, respectively.

sional replacements occurring on these islands. In this way, we are able to work with a large number of populations at different stages of development to study the effects of colonization and extinction on genetic differentiation among local populations.

B. Life History of the Plant

The species we have chosen is the red bladder campion, *S. dioica* (L.) Clairv. (Caryophyllaceae), which requires fertile, disturbed habitats and is a member of the deciduous phase of primary succession in the study area. This species disappears as later successional species close in the habitat and it is never found in evergreen forests (Baker, 1947), which dominate the later stages of primary succession in the region. *Silene dioica* is insect pollinated and dioecious, and the dynamics of population growth are dependent on sexual reproduction. The bumblebees *Bombus hortorum, B. lucorum, B. hypnorum,* and *B. pascuorum* ssp. *sparreanus* are the primary pollinators. This herb has a perennial life cycle; the average life span of an individual surviving to adulthood is about 10 years, and plants begin flowering in their second or third year. Flowering occurs in June and July, and the seeds, which ripen in August, are dispersed by gravity and germinate the following spring. Although seeds may remain viable for 2–3 years (Baker, 1947), a number of studies have shown that most viable seeds germinate within the first year (Baker, 1947 and references therein; Elmqvist and Gardfjell, 1988), suggesting that seed bank dynamics may not be important for this species.

C. Population History

Colonization and establishment of *S. dioica* populations in the Skeppsvik Archipelago occur via individual diploid seeds. Seeds are transported between islands by water in the drift material that is moved around the archipelago with the prevailing winds, storms, and rising water levels each autumn. Colonizers are found growing singly in the decomposing shore deposits (B. E. Giles, personal observation). The advanced successional status of the vegetation along the shores of the mainland (Fig. 1) suggests that the major contributors of pollen or seeds entering island populations are other islands in the archipelago. Colonization and demographic histories of *S. dioica* populations follow a repeatable pattern on these islands (Carlsson *et al.,* 1990; Giles and Goudet, 1996a). The range of ages given below reflects the fact that colonization and successional processes are quicker in the inner archipelago since the outer islands (shaded part in Fig. 1) are subjected to stronger wind, wave, and ice action.

Seeds of *S. dioica* are not able to germinate and establish populations on islands that are less than 70–150 years old. This is the time required for sufficient soil and nutrients to accumulate and for an island to attain a height at which the risks of being washed over by storm waves are small. Successful colonization by *S. dioica* is observed when the central parts of an island have been colonized by the nitrogen-fixing *Alnus incana*. The number of individuals founding new island populations appears to be small. The first evidence of population expansion is seen as a dense circle of even-aged seedlings about 0.5 m in diameter growing under female plants. We have observed this with about 5–10 individuals on an island, as long as both sexes were present and flowering simultaneously (B. E. Giles, personal observation). The first few (2–6) individuals to arrive are also observed in locations up to 1 or 2 m apart, suggesting that it is unlikely that the seed capsule is the unit of colonization. While it is clear that new populations expand from seeds produced by the original colonists and their offspring, migrants arriving as seeds and pollen will be able to establish within these expanding populations since space and nutrients are plentifully available. Germination success is high in newly colonized habitats (Carlsson, 1995).

Silene dioica populations expand rapidly and attain large sizes and high densities on islands 120–250 years old. Succession, however, continues as the *S. dioica* populations expand. *Sorbus aucuparia* succeeds *A. incana* in the middle of the islands, and as even later successional species (*Betula pendula, Picea abies, Juniperus communis,* and *Vaccinium myrtillus,*) establish in the centers, *A. incana/ S. aucuparia* form a border which is forced toward the shores. These later arriving species change the soil and light conditions and germination success of *S. dioica* decreases dramatically in these darker habitats (B. E. Giles, personal observation). At this stage, the *S. dioica* populations thin out and disappear from the centers of the islands, but form thick rings with high densities and numbers in the *A. incana/S. aucuparia* border. Populations at this stage are probably at their largest.

When *S. dioica* populations reach this thick ring stage, they are often invaded by *Microbotryum violaceum,* which is a systemic, perennial, sterilizing anther– smut fungus and an obligate parasite of *S. dioica.* The incidence of infection varies from 5 to 60% in these middle-aged populations (Carlsson *et al.,* 1990). There are no differences in longevity between infected and healthy individuals (Carlsson and Elmqvist, 1992), and infected individuals continue to use space and other resources. Infected male and female individuals no longer contribute to the gene pool, and *M. violaceum* may therefore effectively reduce the numbers of reproducing individuals within an island population.

On islands 200–400 years old, *P. abies* populations rapidly expand outward. The deciduous border becomes increasingly restricted to the shores and the rings of *S. dioica* become thinner. Deposition of drift decreases in regularity because islands continue to rise and increase in height, so that even the deciduous habitat becomes more stable allowing the invasion of later successional competitors. As these events occur, the *Silene* populations decrease in size and the ring of *S. dioica* eventually breaks up into small discontinuous patches or single individuals. On islands 250–500 years old, the later successional species have successfully occupied the islands to their shorelines and *S. dioica* populations go extinct. Recolonization of successionally advanced patches does not occur.

D. Skeppsvik Archipelago and the Assumptions of the Metapopulation Model

The continual formation of new islands and the patterns of population development described above suggest that the Skeppsvik Archipelago is a spatially and temporally structured metapopulation. Since models seldom fit the real world exactly, but our inferences about the world are to be drawn relative to model predictions, it is important to be aware of the differences between the two and how they might affect these inferences. In our case, the deviations of the system from the assumptions of the model (Section II) relax the stringency of the conditions required for increased differentiation.

Both ecological and genetic metapopulation models assume discrete generations and growth to carrying capacity within a single generation. *Silene dioica* has overlapping generations and the 50–100 years taken by the island populations to reach maximum size and density strongly suggests that growth to carrying capacity takes much longer than the model stipulates. Recent theoretical work (Ingvarsson, 1995), however, suggests that delayed growth to carrying capacity is either without effect or it reduces the stringency of the conditions under which colonization increases the degree of differentiation. This occurs because a period of delayed population growth keeps the newly founded populations small for several generations. Since the accumulated effect of genetic drift in finite populations depends on the reciprocal of population size (Falconer, 1989), the effective population size is reduced and the levels of differentiation obtained during the prolonged growth period will always be higher than when populations grow to size N in one generation. The presence of close relatives from different generations will also increase inbreeding during this initial phase (Crow and Kimura, 1970). Thus, relative to a system with discrete generations started from the same number but expanding to fill the niche in one generation, there is a greater opportunity for differentiation among the young subpopulations in our system than in the model. The model also specifies monoecious organisms. Dioecy should have little effect on genetic differentiation if sex ratio is balanced and migration the same for both sexes (Chesser, 1991). While biased sex ratios have not been observed in established populations on the islands (Carlsson, 1995), biases toward one or another sex have been seen on five islands colonized since this study began in 1991 (B. E. Giles, personal observation). Thus biased sex ratios, which reduce effective population size (Wright, 1938), would also act to inflate differentiation arising from founder effects relative to the metapopulation model.

We also believe that the colonization and migration dynamics in Skeppsvik Archipelago satisfy Eq. (2). Our reasons are as follows. The first seedlings may be produced in new populations with fewer than 10 individuals (Section III.C), suggesting that k is small. If the number of migrants (Nm) per generation are of the same magnitude as k, differentiation will be enhanced, but other evidence suggests that migration rates may actually be higher. Only seeds can colonize, but seeds and pollen may contribute to migration. If the pollinator-borne spores of *M. violaceum* (Section III.C) are regarded as "natural" color markers for pollen, the proportion of flowers from uninfected populations bearing spores provides a rough estimate of the extent of pollen carryover between populations. Two studies (P. K. Ingvarsson, unpublished; B. E. Giles, unpublished) have shown that 3–5% of the female flowers from uninfected but well established populations bore smut spores. While the number of fertilizations will be lower than the number of transfers, the magnitude of these numbers plus continued seed flow suggests that the numbers of migrants are likely to be higher than the number of colonists. It is therefore not unreasonable to assume that our metapopulation system corresponds with the conditions stated in Eq. (2).

One final difference between our system and the model can be dealt with in the analyses. In metapopulation models, extinction is stochastic, independent of population age and a requirement for recolonization (Levins, 1970; Slatkin, 1977; Wade and McCauley, 1988; Whitlock and McCauley, 1990; Whitlock, 1992a). In Skeppsvik Archipelago, extinction is determined by succession and is therefore deterministic, dependent on population age, and not a requirement for colonization. Except in terms of the rate of local population extinction, the metapopulation models cited above make no predictions about the specific effects of extinction on levels of genetic differentiation since the immediate recolonization of extinct sites equates extinction and recolonization; increased genetic differentiation in metapopulation models arises solely from founder effects during recolonization. Since colonization and extinction are independent processes in the archipelago, we expect that the continual formation of new islands will enhance levels of genetic differentiation as predicted by the models and that the drawn out extinction process could have an additional affect (see Section IV.C for details). We have dealt with these differences between our system and the model by removing those populations going toward extinction from the data set used to test the model predictions; the effects of successively driven extinction on differentiation within the metapopulation are treated separately. Note also that the persistence of *S. dioica* in Skeppsvik Archipelago depends on the continual formation of colonisation sites, which is consistent with one of the main inferences drawn from the ecological metapopulation models (e.g., Levins, 1969a, 1970; Hanski and Simberloff, this volume).

IV. PATTERNS OF GENETIC STRUCTURE ON THE ISLANDS

A. Methods

The positions of the islands, their ages, demographic stages, population sizes and degree of exposure are given in the Appendix and Fig. 1. We have surveyed the genetic structures of our populations using allozymes. Six polymorphic loci, with 20 alleles in total, were presumed to represent neutral characters (Giles and Goudet, 1996a). Individuals were collected at random from the entire area occupied on each island and 4500 individuals were screened from 52 islands (Appendix). Genotypic and allelic compositions were analyzed using the program FSTAT V1.2 (Goudet, 1995), which calculates Weir and Cockerham (1984) estimators of Wright's F statistics. Deviations from random mating were assessed by means of F statistics (F_{it} estimates the overall heterozygote deficit, F_{ST} estimates the degree of isolation of populations and F_{is} estimates the heterozygote deficit within islands). The significance of each F value was tested using a permutation procedure. The null hypothesis, $F_{xy} = 0$ (where $xy =$ is, it, or ST), was obtained from 5000 permutations and the observed value of F_{xy} was then compared with the null distribution and tested against the alternative hypothesis, $F_{xy} > 0$ (Giles and Goudet, 1996a). For F_{is} and F_{it}, alleles were permuted within and among populations, respectively. Since the two alleles in an individual may not be independent, the permutation units for F_{ST} were the genotypes permuted among populations.

B. Genetic Structure among Islands

While it is essential that the local populations composing a metapopulation are connected by migration, migration must also be shown to be restricted. If no evidence of population structure can be obtained, then it must be assumed that the local populations belong to the same panmictic evolutionary unit. Using the data from all study sites within the archipelago, we first asked whether there was any evidence of population structure and restricted gene flow arising from habitat subdivision within and among islands. As Table Ia shows, the values $F_{it} = 0.110$, $F_{ST} = 0.038$, and $F_{is} = 0.075$, calculated over all loci, are significantly greater than zero ($P < 0.0002$), providing strong evidence of restricted gene flow among (F_{ST}) and within (F_{is}) islands in the archipelago (F_{it}). Note that no spatial or temporal dynamics have been taken into account; this analysis reveals only whether the populations are structured where the habitat is structured.

C. Genetic Differentiation among Cohorts of Plant Populations Varying in Age

The geological and successional processes occurring in the Skeppsvik Archipelago clearly indicate that our populations are characterized by colonization/

TABLE I   *F*-Statistics Averaged over All Loci and Populations (a) and over All Loci for the Three Age Classes of Populations (b)

	Class	N	F_{it}	F_{ST}	F_{is}
(a)	All	52	0.110 (.039)	0.038 (.010)	0.075 (.035)
(b)	Young	13	0.105 (.061)	0.057 (.028)	0.052 (.046)
	Intermediate	30	0.107 (.033)	0.030 (.006)	0.080 (.032)
	Old	9	0.158 (.068)	0.066 (.009)	0.098 (.068)

[a] Standard errors in parentheses. All F values are significantly greater than zero ($P < 0.0002$).

extinction dynamics and differ in terms of their demographic maturity. The question then is whether these particular dynamics increase genetic differentiation among populations relative to an island model that does not take turnover into account. The predicted increase in differentiation in the model arises solely from founder effects during recolonization since the effect of extinction is seen as recolonization (Section III.D). This also partly explains why the F_{ST} values of only two age groups have been used to show whether population turnover increases differentiation (see Whitlock, 1992b; Dybdal, 1994; McCauley *et al.,* 1995). In these comparisons, the younger group contains the newly founded populations, while all other populations are combined into a single older group since these populations are assumed to be approaching equilibrium between migration and genetic drift. Where extinction in natural systems is stochastic, dividing populations into two groups may be appropriate. In contrast, our populations decline slowly and eventually become extinct as succession proceeds (Section III.C). The decrease in size is likely to be due to a reduction in germination success in the presence of later successional species. This reduction in germination appears to have two effects on these populations; the age distribution shifts toward adult individuals (Carlsson, 1995), and since migrants can be recruited only by the growth of seeds, migration (m) cannot compensate for the reduction in N which is expected under a model which assumes that Nm is constant at equilibrium. It thus appears that not only are the oldest populations moving away from demographic and genetic equilibria in our system, the reductions in numbers and migration are likely to increase the F_{ST} among these populations. Since our declining populations do not fulfill the criterion for membership in a single group called "older" populations (Whitlock and McCauley, 1990; Whitlock, 1992b; Dybdal, 1994; McCauley *et al.,* 1995), we divided our populations into three groups, called young, intermediate, and old, where our young and intermediate populations correspond to the younger and older populations in the papers cited above. To test whether colonization increases differentiation, we compared the F_{ST} values of young and intermediate populations; to test whether the population decline associated with succession also inflates levels of differentiation at the metapopulation level, we looked to see whether F_{ST} values increased in the old populations. We proceeded as follows.

The young group contains recently founded populations. These populations still occupy the centers of the islands and are less than 30 years old. The intermediate group contains the populations which were well into the expansion phase. These had the thick ring form and were between 30 and 250 years old. The old populations were decreasing in size, had the broken ring form, and were older than 250 years. The average census sizes of populations in the young, intermediate, and old groups were 670, 13000, and 1300 individuals, respectively. The classification of each population into an age group is given as "stage class" in the Appendix. We then calculated the F statistics for each group separately and tested whether the F_{ST} of young populations was larger than that of intermediate populations, and whether the F_{ST} of intermediate populations was smaller than that of old populations. We tested our hypotheses using a Monte Carlo method, in which partitions of the total sample into three classes were generated at random. Each three-class partition was made up of 30, 12, and 9 islands, sampled without replacement. Differences in F_{ST} between groups were calculated for 5000 replicates. The distributions of these differences are our null hypotheses; (1) $F_{ST,y} = F_{ST,i}$, and (2) $F_{ST,i} = F_{ST,o}$, which we tested against the alternative hypotheses; (1) $F_{ST,y} > F_{ST,i}$, and (2) $F_{ST,i} < F_{ST,o}$ (y, i, o = young, intermediate, and old, respectively).

All F statistics estimated for the young, intermediate and old age classes are significantly greater than zero (Table Ib). F_{ST} is high among young islands (0.057), decreases to 0.030 in the intermediate age class, and reaches its highest value of 0.066 among the old islands. The F_{ST} of the young populations was significantly larger than that of the intermediate populations ($P < 0.05$), which suggests that colonization increases differentiation (Whitlock and McCauley, 1990). Continued migration among populations also appears to contribute to the decrease in F_{ST} observed in the intermediate class. This inference is supported by noting that the means and standard deviations of the numbers of alleles observed at all loci (maximum 20) in the young, intermediate, and old groups were 14.33 ± 3.39, 18.56 ± 0.73, and 15.25 ± 2.05, respectively. The F_{ST} for the intermediate class remains significantly greater than zero, suggesting that gene flow is not high enough to link these populations into a single evolutionary unit.

In the second test, which is not part of the model, the F_{ST} for the old class was significantly higher than that for the intermediate class ($P < 0.04$). We strongly suspect that this increase is associated with extinction processes, and that in our metapopulation system, both founding events and population decay may increase levels of differentiation relative to an island model at equilibrium. We know of no other study in which such an increase has been reported, and it remains to be seen whether these results will be found for other metapopulation systems with similar extinction dynamics. However, in studies where the aim is to see whether colonization increases differentiation using the F_{ST} relationships which follow from the Whitlock–McCauley model, it may be wise to check whether F_{ST} values for old or decaying populations are exceptionally high. If they are, these populations should be removed from the group of populations assumed to be approaching equilibrium since their inclusion could obscure the expected decrease between younger and older populations.

D. Genetic Structure within Islands

While we have demonstrated that extinction/colonization processes increase the genetic variance among island populations, this is not the entire story. The within-island and overall values of F_{is} obtained in this study are large and significantly greater than zero (Appendix and Table I). One of the implications of these significant F_{is} values is that the island populations themselves consist of more than one random mating unit. To test this, a pilot study was carried out on island 23 (Fig. 1, Appendix), in which all *S. dioica* individuals within six 50 cm^2 patches, separated by 1.5 to 35 m, were surveyed electrophoretically. F_{is} values for all patches were not statistically different from zero, but all F_{ST} values calculated for different combinations of patches were significantly greater than zero. Converting the distance between the two nearest patches in which F_{ST} was significant into area revealed that random mating units could be as small as 0.2– 6 m^2 (E. Lundqvist, J. Goudet, and B. E. Giles, unpublished). Some of this within-island differentiation probably results from restrictions to gene flow. Seed dispersal by gravity may lead to the establishment of groups of offspring near the mother plant which are likely to be at least half-sibs. Nearest-neighbor pollination has also been observed (Carlsson, 1995), and the compounding of these two processes over a number of generations could lead to an apparently continuous population consisting of a patchwork of differentiated family groups (see Turner *et al.,* 1982).

Spatial subdivision, however, does not appear to be the only dynamic affecting within-island population structure. Because an island is rising all the time, the middle is always older than the outer parts. With the exception of shoreline species, each successional stage invades the middle first and moves outward. Close observation of any single patch of ground on an island reveals that the plant species colonizing it go through a series of successional and demographic changes similar to those described at the island level (Section III.C). A patch is first colonized by *S. dioica* when conditions are right. Initially, we observe that a few flowering adults, few if any seedlings, and shoreline species may still share the habitat. When seedlings begin to be produced within the patch, the numbers and densities of the *S. dioica* "population" increase, plants at all life-history stages are present, and *S. dioica* is the most common plant. As *J. communis, V. myrtillus, B. pendula,* and *P. abies* invade, fewer and fewer *S. dioica* seedlings are observed, the number of adults decreases, and *S. dioica* eventually disappears from the patch (Carlsson, 1995; B. E. Giles, unpublished). These changes associated with patch colonization and extinction suggest that each island population is itself an age-structured metapopulation. In a recent study of three islands (23, 35, and 39; Fig. 1) in which we collected *S. dioica* from patches of different ages, the F_{ST} values from young, intermediate, and old patches within an island followed the same trend as those between young, intermediate, and old islands (Giles and Goudet, 1996b). Clearly a new model will be needed to investigate how the dynamics occurring at these two nested spatial and temporal scales could affect each other and the system as a whole. Nonetheless, finding that the degree of demographic maturity of mating units at two spatial and temporal scales has a consistent effect on the degree of genetic differentiation among those units suggests that this phenomenon may be general.

E. The Effects of Distance on Genetic Differentiation among Islands

If an organism's potential for migration is limited relative to the typical isolation of the habitat patches it occupies, the arrangement and distances among patches can affect population differentiation even in the absence of turnover. We next looked for evidence of isolation by distance which has been shown to increase the degree of differentiation among populations (e.g., Whitlock, 1992a; Goudet, 1993; Slatkin, 1993). Since population turnover and spatial variance in migration have not been jointly investigated in a single model, it is difficult to know how to disentangle their effects. To reduce the confounding effects of age and concentrate on the effects of spatial variance in migration as much as possible, we restricted our analyses to specific groups of neighboring islands which, on the basis of information about vectors of gene flow, we believed could (or could not) exchange genes. Age was ignored in these analyses.

From the time of seed release and onward into the autumn, the prevailing winds blow from the southwest, storms blowing from the southeast are common, and water moves into the archipelago in the course of the annual water cycle in the Baltic (Ericson, 1981). The joint action of these forces will be to move seeds from the south to the north in the archipelago. To investigate the spatial patterns of gene flow via seeds, we constructed three "chains" of islands, called the outer, middle, and inner chains, which are oriented in a south–north direction (A,B,C, respectively, Fig. 1). Isolation by distance was expected in all three chains.

We also looked for isolation by distance where wind and water are not likely agents of seed flow by setting up two east–west transects. Since bumblebee movement occurs among islands and appears to be more frequent in the inner (unshaded part in Fig. 1) than in the outer archipelago (L. Ericson, personal communication), the occurrence of isolation by distance is not unexpected in the north transect (D, Fig. 1). The second transect, south transect (E, Fig. 1), was set up in the most exposed part of the archipelago as a contrast to the other four groups. Since we suspect that south transect islands are not directly linked by wind and water, we do not expect to detect isolation by distance.

We looked for evidence of isolation by distance within these chains and transects using Mantel tests (Mantel, 1967) of the correlations between the matrices of pairwise genetic (based on F_{ST}) and physical distances. A significant correlation is taken as evidence of isolation by distance. Since there is little information as to what type of transformation of F_{ST} and distance is appropriate, we used three: (1) untransformed F_{ST} vs untransformed physical distances, (2) Slatkin's (1993) $\ln[1/4(1/F_{ST} - 1)]$ (the log of the number of migrants in an island model) vs log physical distances, and (3) ranked F_{ST} vs ranked physical distances.

Table II shows that the correlation coefficients (R) between the two matrices were high, 0.40 or greater, for the outer and inner chains and the north transect for the untransformed and ranked data. In contrast, the correlation coefficients for the middle chain and the south transect were very low. Thus, our predictions are supported by these tests for the outer chain, the southern transect, and marginally for the inner chain and the northern transect, but not for the middle chain. These results suggest that genetic differentiation among islands in the Skeppsvik Archipelago is inflated relative to an island model due to the dispersion of the islands, although it is as yet impossible to estimate the relative contributions of

TABLE II Mantel Tests of Isolation by Distance in Groups of Islands Formed on the Basis of Hypotheses about Vectors of Gene Flow[a]

	Group	N	R	P	sdr	Threshold
(a)	Inside chain	7	0.4669	0.019	0.26	0.0125
	Middle chain	9	0.0767	0.327	0.27	
	Outer chain	11	0.4296	0.008	0.16	0.01
	North transect	6	0.5629	0.054	0.29	
	South transect	5	−0.5034	0.932	0.36	
(b)	Inside chain	7	−0.3947	0.052	0.24	
	Middle chain	9	−0.2548	0.128	0.21	
	Outer chain	11	−0.3289	0.021	0.14	0.01
	North transect	6	−0.2275	0.196	0.27	
	South transect	5	0.2922	0.759	0.34	
(c)	Inside chain	7	0.3961	0.057	0.24	
	Middle chain	9	0.1681	0.268	0.25	
	Outer chain	11	0.3980	0.011	0.15	0.01
	North transect	6	0.5673	0.024	0.29	
	South transect	5	−0.3697	0.868	0.33	

[a] Three transformations of pairwise genetic and physical distances were used: (a) F_{ST} against distance, (b) ln $(1/4\ F_{ST} - 1/4)$ against ln distances, (c) ranked F_{ST} against ranked distances. Threshold specifies the sequential Bonferroni level of significance required for rejection of the null hypothesis. N, R, P, and sdr are sample size, correlation coefficient, probability, and standard deviation, respectively.

temporal and spatial heterogeneity to the total levels of differentiation in this metapopulation. The occurrence of isolation by distance in this archipelago, however, also allows us to examine another factor which is critical for the degree to which colonization and extinction dynamics will increase differentiation, namely, the parameter ϕ and the degree of propagule mixing at colonization (Whitlock and McCauley, 1990; McCauley, 1991).

F. Numbers and Sources of Colonists

The genetic consequences of founding events depend not only on the typical numbers of individuals involved in a colonization (k), but also on the number of sources (ϕ) from which the individuals entering empty patches are drawn (see Eq. (2); Slatkin, 1977; Wade and McCauley, 1988; Whitlock and McCauley, 1990; McCauley et al., 1995). The observation that groups of seedlings are found on an island when 5 to 10 individuals representing both sexes have arrived and flowered simultaneously suggests that k is small. As a first attempt to find out the probability of common origin of the colonists, we have looked for isolation by distance within age groups. Isolation by distance detected among young populations, or among both young and intermediate populations, will be taken as evidence that propagules come from one or a small number of nearby islands (high ϕ) and a propagule pool model will best describe the archipelago. If, however, isolation by distance is detected among intermediate but not among young populations, it is more likely that colonizers come from a wide range of source populations rather than neighboring ones, and a migrant pool model will be inferred. We carried out Mantel tests of the correlations between the genetic and physical distances of the young and intermediate island populations. As above, a significant correlation between the two matrices was taken as evidence of isolation by distance. No effect of distance was detected in any of the tests for the young populations, but all three Mantel tests were highly significant for islands in the intermediate age class (Table III). These results suggest that colonizers are a mixture from several locations in the archipelago and that the migrant pool model is an appropriate description of founding events.

Finding isolation by distance among islands in the intermediate age class raises some interesting problems and questions. According to existing genetic metapopulation models, postcolonization migration among populations is supposed to follow an island model. Finding isolation by distance among intermediate-age islands in Skeppsvik Archipelago means that their degree of genetic differentiation will be inflated by these distance effects. It also means that this metapopulation system combines migrant pool colonization with subsequent distance-dependent migration. Since the degree of differentiation is lower in migrant pool than in propagule pool colonization, and distance-dependent migration inflates differentiation relative to an island model, the first problem is that this combination of colonization and migration modes is a "worst-case scenario" for detecting overall increases in differentiation arising from colonization. Too few studies of metapopulation systems have been carried out to know whether this combination is a common one. In our case, the F_{ST} of intermediate populations was still significantly lower than that of the young populations (Table Ib) as predicted by the model (which also strengthens our belief that the conditions specified in Eq. (2) are met in this metapopulation). It is, however, possible that these expected changes in F_{ST} with age may not be detectable in other systems showing a similar combination of colonization and migration patterns. Further

TABLE III Mantel Tests for Isolation by Distance among Islands in Young and Intermediate Age Classes[a]

	Group	N	R	P	sdr	Threshold
(a)	Young	13	−0.0561	0.405	0.207	0.05
	Intermediate	30	0.3662	0.002	0.118	0.025
(b)	Young	13	−0.0011	0.43	0.159	0.05
	Intermediate	30	−0.3231	0.00	0.093	0.025
(c)	Young	13	−0.0032	0.461	0.161	0.05
	Intermediate	30	0.3298	0.002	0.106	0.025

[a] See Table 2 legend for description of transformations (a), (b), (c) and column entries.

development of the theoretical framework to include the effect of isolation by distance, in addition to the effect of turnover dynamics is clearly required.

A second question is whether colonization and migration are really different phenomena. In plants at least, the answer is probably yes, since migrants come in two forms, as diploid seeds and haploid pollen. Colonization can only occur through seed migration, while migration into and out of established populations may occur as seeds and pollen. While we do not know which form of migration is more responsible for the gene flow which reduces the genetic differentiation among maturing populations, we strongly suspect that gene flow via pollen becomes more prevalent as populations age. As the islands age and rise, seed deposition even on the shores becomes more difficult. In addition to the large proportion of flowers from uninfected populations which bore *M. violaceum* spore "pollen-markers" (see Section III.D), other evidence suggests that the potential for pollen carryover increases with population size (Ågren, 1996; Handel, 1983; Fritz and Nilsson, 1994). At this stage of the study, we can only speculate. The conclusive evidence will come from a study using markers which enable gene flow via pollen to be distinguished from that via seed. This can be done by comparing the patterns of gene flow inferred from maternally inherited organelle markers (e.g., mtDNA or cpDNA) with the patterns inferred from biparentally inherited nuclear markers (e.g., allozymes), since only the nuclear markers will be carried from one population to another as pollen (Ennos, 1994; McCauley, 1994). F_{ST} values of newly founded populations of the white bladder campion, *Silene alba,* have been found to be higher than those of older populations belonging to the same metapopulation (McCauley *et al.,* 1995), and a comparison of estimates from allozymes and cpDNA suggested that the rate of gene flow in nuclear genes was greater than the effective rate of organellar gene flow in this species (McCauley, 1994).

G. Environmental Heterogeneity

Our studies have also revealed that the degree of genetic differentiation is influenced by environmental factors. The values of F_{is}, which describe the degree of structuring within single islands (Appendix), were not only significantly greater than zero, they were also highly heterogeneous and ranged from -0.18 to 0.38. It is therefore of interest to know whether particular factors associated with high levels of within-island structuring can be identified. Successional changes occur more quickly in the protected than in the exposed parts of the archipelago. Exposure is likely to increase both the difficulty with which migrants (of any plant species) are deposited on islands and the degree of local patch destruction by salty waves and ice-blocks blown onto the islands during storms. We thus suspected that exposure could increase F_{is}, but we could not ignore population age because of the differences in rates of succession. We therefore used age (stage class, Appendix) and exposure as factors in a two-way analysis of variance on

TABLE IV   Two-Way Analysis of Variance of Ranked Within-Island F_{is} Values for Exposure and Stage Classes (Appendix)

Source	DF	MS	F	P
Exposure	1	1819	10.36	0.002
Age	2	284.3	1.62	0.209
Interaction	2	217.9	1.24	0.299
Residual	46	175.5		

the ranked F_{is} values for the islands. Islands were grouped into two exposure classes, protected and exposed, according to their position in the archipelago relative to the shaded and unshaded sections of Fig. 1 (see exposure, Appendix). The shading marks the boundary between those islands in the lee of other islands (unshaded) and those that are directly exposed to wind and wave actions by autumn storms (shaded) and the area which is frozen solid during the winter (unshaded) and the area subject to pack ice driven by winds (shaded) (Ericson, 1981). Table IV indicates that exposure is significant ($P < 0.05$), but age and the interaction between age and exposure are not. With few exceptions, F_{is} values for islands in the exposed area are greater than 0.09 (Appendix). We have confirmed this effect in our recent study of genetic structure within three of the islands (Giles and Goudet, 1996b); when islands were ranked according to their degree of exposure, within-island F_{ST} values increased with exposure. We strongly suspect that local patch destruction by wave and ice action followed by recolonization increases the genetic differentiation over and above that induced by the successional dynamics within individual exposed islands, but this is currently a working hypothesis.

H. Conclusion

The results of our study of isozyme markers in 52 island populations of *S. dioica* in Skeppsvik Archipelago show that migration cannot bind these populations into a single evolutionary unit and that each of the island populations are likely to have independent evolutionary trajectories. Furthermore, overall levels of genetic differentiation in this metapopulation will be inflated relative to an island model at equilibrium. In our case, the founder events dealt with in the genetic metapopulation models (Slatkin, 1977; Wade and McCauley, 1988; Whitlock and McCauley, 1990) do not appear to be the only cause of this inflation since we detected additional affects arising from other temporal, spatial and environmental factors. Our study has both confirmed model predictions and shed light on other areas which need further theoretical and empirical development. We found that newly colonized populations were more differentiated than those nearer their demographic and genetic equilibria, implying that founder events could increase differentiation at the metapopulation level as predicted by the Whitlock–McCauley model. However, we also observed that differentiation increased among the oldest populations undergoing demographic and genetic changes as they approached extinction by succession. While these "old" populations will contribute to the total increase in levels of differentiation measured at a metapopulation level, it is not yet known what kind of long-term influences these old populations could exert on the system (e.g., on the maintenance of genetic diversity) since the relative contributions of old, intermediate, and young populations to migrant and colonizer pools are unknown. It will be interesting to see whether these results will be found for other metapopulation systems with similar successionally driven extinction dynamics. A spatial factor known to increase differentiation, namely isolation by distance, a form of spatial variance in migration rate (Whitlock, 1992a; Goudet, 1993; Slatkin, 1993), was also detected in Skeppsvik Archipelago among several small groups of neighboring islands and among the islands of the intermediate age group. For the past 30 years, field population geneticists have been focusing on the consequences of spatial subdivision on population differentiation. Our study indicates that spatial and temporal factors interact in a manner which may be complex, and the few other studies which account for temporal as well as spatial effects (e.g., Whitlock, 1992a,b; McCauley *et al.*, 1995) imply that both are crucial for understanding population structure. Clearly, the interplay of these two factors needs to be more thoroughly investigated in theoretical and empirical studies. Our results showing that the degree to which habitats are exposed to storm driven wind, water, and ice influences the degree of genetic differentiation also suggests that more attention should be paid to environmental factors likely to affect the dynamics of gene flow (see Whitlock, 1992a).

V. OTHER GENETIC METAPOPULATION STUDIES

Studies of single species metapopulation dynamics involving plants are rare (Silvertown, 1991; Husband and Barrett, 1995), as are empirical studies designed to test the predictions of genetic metapopulation models regardless of organism. To our knowledge, there are three such genetic studies in addition to our own (Giles and Goudet, 1996a).

Whitlock (1992b) studied the forked-fungus beetle, *Bolitotherus cornutus,* which lives on patchily distributed fruiting bodies of several species of polypore bracket fungi growing on dead and dying trees. Whitlock was able to estimate migration and extinction rates, numbers of colonizers, and their probabilities of common origin using a combination of methods based mainly on mark–recapture. He also created new habitat patches to estimate colonization rate and numbers and resurveyed habitat patches known to be occupied 10 years prior to his study. These experiments enabled him to show that the conditions specified by Eq. (2) were met. To confirm that these conditions led to increased genetic differentiation, Whitlock divided the populations of forked fungus beetles into young and older groups using the size of the fungal resource as an indicator of demographic maturity. An electrophoretic study confirmed that the young populations were more differentiated than the older populations.

Dybdal (1994) studied populations of the bottom-dwelling marine copepod, *Tigriopus californicus,* which inhabits intertidal splash-zone tidepools on rocky shores of the Pacific coast of North America. In contrast to the studies of Whitlock (1992b), Giles and Goudet (1996a), and McCauley *et al.* (1995, see below), he found that the older populations of *T. californicus* were genetically more differentiated than the younger populations and concluded that extinction/recolonization dynamics decreased genetic differentiation in this metapopulation system. Dybdal (1994), suggested that this was due to violations of two of the assumptions of the metapopulation model, namely, not all extant populations were equally likely to be sources of colonists, and not all extant populations had equal probabilities of extinction. In the *T. californicus* system, the extinction probabilities were related to the size and location of the tidepools; there was a tendency for the older tidepools to be larger and more isolated than the younger pools, which reduce, according to Dybdal, the likelihood of older populations acting as sources of colonists. In the light of our earlier discussions about the importance of migrant exchange among local populations, we suggest an alternative interpretation. Isolated populations are less likely to exchange migrants after colonization, and therefore some of the older, more isolated populations may be more influenced by the very founder events that the genetic metapopulation model suggests are responsible for the higher F_{ST} values normally observed among young populations.

McCauley *et al.* (1995) have worked on *S. alba,* a shorter lived but close relative to *S. dioica,* which grows along roadsides in the vicinity of Mountain Lake Biological Station in Giles County, Virginia. The dynamics of this metapopulation system differ in several ways from the *S. dioica* system described above. The *S. alba* populations chosen for study are not separated by an uninhabitable matrix. As a consequence, the metapopulation was defined as an area (25 × 25 km) and the local populations as those individuals inhabiting each 40-m segment of 150 km of roadside contained in this area (Antonovics *et al.,* 1994). These populations, which have been followed since 1988, are often much smaller than ours (see Appendix and Antonovics *et al.,* 1994), and the system can be thought of as a large series of small patches, not necessarily very far apart, which show high rates of turnover (Antonovics *et al.,* 1994; D. E. McCauley, personal communication). Extinction rates have been shown to be higher in small than in large populations and the extinction–colonization dynamics of the *S. alba* system, which are influenced by human activities (e.g., mowing, road maintenance), are highly stochastic (Antonovics *et al.,* 1994; D. E. McCauley, personal communication). McCauley *et al.* (1995) showed that the F_{ST} values calculated from both isozyme and cpDNA data were higher for newly colonized populations than those for established populations. Thus, even though there are differences in the nature and rates of the turnover dynamics in the *S. dioica* and *S. alba* meta-populations, the two systems show qualitatively similar results.

The degree of genetic differentiation between 11 established and 7 newly founded populations of the perennial cowslip, *Primula veris,* was studied by Antrobus and Lack (1994). While this study is concerned with the effects of subdivision per se and whether founder effects associated with colonization may lead to a loss of genetic variation, the metapopulation concept is never mentioned. These authors concluded that there were no differences in the genetic structure of new and established populations because the genetic distances and degree of differentiation among all the populations (F_{ST} = 0.039) were small and because new populations did not possess fewer rare or uncommon alleles than established populations. However, Antrobus and Lack (1994) report F_{ST} values of 0.046 for new and 0.033 for established populations. While the F_{ST} values have not been tested for differences from zero or from each other, the relative magnitudes of these values are consistent with the predictions of the metapopulation model.

Metapopulation dynamics have been shown to account for changes in the relative proportions of male sterile and hermaphroditic plants in patchily distributed populations of the gynodioecious, perennial shrub, *Thymus vulgaris* (Couvet *et al.,* 1986; Belhassen *et al.,* 1989; Olivieri *et al.,* 1990; see also Frank, this volume). Male sterility is determined by cytoplasmic, maternally inherited genes, but specific nuclear genes may restore male fertility. *Thymus vulgaris* lives in areas which are regularly destroyed by fire. While fire leads to the extinction of local populations, the burned sites are available for recolonization. Newly colonized and young populations contain high proportions of male sterile individuals, but as the populations age, the proportions of hermaphrodites increase. New sites are colonized by seeds from different sources and there is a strong probability that the founding propagules will contain male steriles and hermaphrodites which do not contain the matching nuclear genes required to restore the fertility of the male steriles in just that patch. Since male sterile plants produce more seeds (Gouyon and Couvet, 1987), they increase more quickly than hermaphrodites, leading to young populations with high proportions of females. With time and migration among populations in the form of pollen, the restorer genes specific to the various male sterile forms eventually appear in the population. Since most pollination occurs within local populations, the matching restorer gene(s) spread(s) through the population and the proportion of hermaphrodites increases as the local populations approach equilibrium.

In a similar manner, recurrent extinction and recolonization dynamics have also been shown to affect sex ratio in local populations of the African butterfly, *Acraea encedon* (Heuch, 1978). Two kinds of females occur in these populations, normal females which produce equal numbers of sons and daughters and abnormal females (thought to contain a Y-linked gene causing meiotic drive of the Y chromosome) which only produce daughters. Since females are the heterogametic sex, daughters always have the same phenotype as their mothers. Population and eventually species extinction is expected since the numbers of females only producing daughters increase in a population with time. Species extinction is not observed in nature, even though population extinction has been recorded. Heuch (1978) showed that turnover within the metapopulation could explain the persistence of these butterfly populations by allowing recolonization by normal females.

Selection associated with high rates of population turnover may also account for the maintenance of a seed polymorphism in *Carduus pycnocephalus* and *C. tenuiflorus* (Olivieri *et al.*, 1983, 1990, 1995; Olivieri and Gouyon, this volume). These thistles produce two kinds of seeds: those produced in the center of the capitulum, which are not dormant and have a pappus allowing wide dispersal, and the outer seeds, which have no dispersal mechanism but show dormancy. The proportion of seeds producing pappuses is under partial genetic control (Olivieri and Gouyon, 1985). Older populations have been observed to produce higher proportions of nondispersed seed, and it has been proposed that there is selection within populations against interpopulation dispersal since most dispersing seeds will be lost to their source populations (Olivieri and Gouyon, 1985). However, population extinction and colonization occur frequently and because all new populations must be founded by migrants, traits aiding dispersal cannot be lost at a metapopulation level if the species is to persist. This suggests that these traits are subject to different selection pressures in populations of different ages. Olivieri *et al.* (1995) and Olivieri and Gouyon (this volume) showed that between-deme selection for high migration and within-deme selection for low migration could maintain this seed polymorphism.

As evidenced by the papers written and cited in this volume, the metapopulation concept is more than just a "rage." This concept recognizes and studies the consequences of movement among the spatially separated and smaller-than-infinite populations that typify nature. It also allows us to start to let go of our comfortable view of nature as an immortal, homogeneous, and constant entity and to examine the consequences of extinction, fluctuation, and randomness where it exists. At some level, metapopulation dynamics probably influence most species (Husband and Barrett, 1995). In our case study and several of the others that we have described, the focus has been on neutral genetic variation. These are important studies. That their results support the predictions of existing models suggest that the concept is not simply an empty fashion, but one that describes nature. For any particular species, studies involving neutral characters are an important first step in establishing the spatial and temporal scales at which turnover occurs and for coupling the intimate details of the biology of each species to the magnitude of the metapopulation effect. It is, however, too much to believe that the effects of metapopulation dynamics we have seen so well reflected in neutral characters will not effect the rates and patterns of evolution (see Barton and Whitlock, this volume). The difficult but necessary theoretical and empirical next step demands that we learn to understand and measure the consequences of this dynamic population structure for adaptation and speciation.

ACKNOWLEDGMENT

A grant (to BEG) from the Natural Sciences Research Council of Sweden (NFR) has funded this work.

APPENDIX:

Information About the Studied Islands

Island no.	Exposure	Age	Stage	Size	Sample	F_{is}
1	2	5	1	400	53	0.093
2	1	5	1	500	94	0.063
3	1	10	1	3600	91	−0.040
4	1	12	1	430	47	0.050
5	2	13	1	1500	93	0.113
6	2	31	2	4000	220	0.000
7	1	35	1	400	150	0.026
8	2	39	2	4000	354	0.152
9	1	41	2	16000	256	0.045
10	1	46	1	530	98	0.084
11	1	53	2	5000	168	0.072
12	2	53	2	3900	48	−0.060
13	1	58	2	6600	49	−0.140
14	1	61	2	10500	235	0.112
15	2	63	1	800	90	0.106
17	2	75	2	3200	49	0.168
18	2	80	2	15000	35	0.238
19	2	80	2	66000	228	0.094
20	1	84	2	2000	49	0.100
21	1	85	2	3000	50	0.096
22	2	91	2	3000	50	0.053
23	2	96	2	16000	360	0.096
24a	1	107	2	11000	48	0.100
24b	1	107	2	11000	49	0.047
25	1	111	2	4900	181	−0.050
26a	2	134	2	4700	51	0.152
26b	2	134	2	4700	26	0.120
27	2	156	2	4600	27	0.215
28	2	164	3	150	46	0.133
29	1	178	2	15000	50	0.033
30	1	180	2	12200	49	0.034
32	1	185	3	2400	45	0.021
33	1	187	2	9800	58	−0.030
34	2	191	2	25000	100	0.098
35	2	196	2	8500	46	0.040
36	2	221	2	30000	96	0.160
37	2	265	3	600	28	0.376

(continues)

Information About the Studied Islands   *(continues)*

Island no.	Exposure	Age	Stage	Size	Sample	F_{is}
38	2	265	2	43000	96	0.104
39	2	276	2	18000	60	0.226
40	2	281	3	4700	10	0.000
41	2	300	3	1500	30	0.210
42	1	318	2	12000	46	0.000
44	2	360	3	960	58	0.086
45	1	370	3	150	46	−0.130
46	1	370	3	700	66	0.063
47	2	381	3	600	29	0.103
52	2	5	1	20	12	−0.180
53	2	5	1	13	16	−0.180
54	1	15	1	300	117	0.077
55	1	5	1	18	18	0.010
56	1	10	1	200	55	−0.010
58	2	*	2	*	82	0.188

Note. Use Island No. for location in Fig 1. Exposure is the degree to which islands are subjected to wind, wave, and ice action; classes 1 and 2 are protected and exposed islands in unshaded and shaded parts of Fig 1, respectively. Age is population age or time since colonization, while Stages 1, 2, and 3 divide the islands into young, intermediate, and old age classes, respectively. Size is census size. Sample describes the numbers studied electrophoretically. F_{is} gives within-island values.

Bibliography

Aars, J., Andreassen, H. P., and Ims, R. A. (1995). Root voles: Litter sex ratio variation in fragmented habitat. *J. Anim. Ecol.* **64:**459–472.

Abrams, P. (1992). Predators that benefit prey and prey that harm predators: Unusual effects of interacting foraging adaptations. *Am. Nat.* **140:**573–600.

Addicott, J. F. (1978). The population dynamics of aphids on fireweed: A comparison of local populations and metapopulations. *Can. J. Zool.* **56:**2554–2564.

Adler, F. R., and Nuernberger. B. (1994). Persistence in patchy irregular landscapes. *Theor. Popul. Biol.* **45:**41–75.

Ågren, J. (1996). Population size, pollinator limitation and seed production in the self-incompatible herb *Lythrum salicaria. Ecology* (in press).

Akçakaya, H. R. (1994). "RAMAS/GIS. Linking Landscape Data with Population Viability Analysis." Applied Biomathematics, Setauket, NY.

Akçakaya, H. R., and Ferson, S. (1992). "RAMAS/Space User Manual: Spatially Structured Population Models for Conservation Biology." Applied Biomathematics, Setauket, NY.

Akçakaya, H. R., and Ginzburg, L. R. (1991). Ecological risk analysis for single and multiple populations. *In* "Species Conservation: A Population-Biological Approach" (A. Seitz and V. Loeschcke, eds.), pp. 73–87. Birkhäuser, Basel.

Alberts, S. C., and Altmann, J. (1995). Balancing costs and opportunities: Dispersal in male baboons. *Am. Nat.* **145:**279–306.

Alexander, H. M. (1989). An experimental field study of anther-smut disease of *Silene alba* caused by *Ustilago violacea:* Genotypic variation and disease resistance. *Evolution (Lawrence, Kans.)* **43:**835–847.

Alexander, H. M. (1992). Evolution of disease resistance in plant populations. *In* "Ecology and

Evolution of Plant Resistance" (R. S. Fritz and E. L. Simms, eds.), pp. 326–344. University of Chicago Press, Chicago.

Alexander, H. M., Antonovics, J., and Kelly, A. (1993). Genotypic variation in plant disease resistance—physiological resistance in relation to field disease transmission. *J. Ecol.* **81:** 325–333.

Alexander, K. A., and Appel, M. J. G. (1994). African wild dogs *(Lycaon pictus)* endangered by a canine distemper epizootic among domestic dogs near the Masai Mara National Reserve, Kenya. *J. Wildl. Dis.* **30:**481–485.

Allee, W. C., Emerson, A. E., Park, O., Park, T., and Schmidt, K. P. (1949). "Principles of Animal Ecology." Saunders, Philadelphia.

Allen, E. J., Harris, J. M., and Allen, L. J. S. (1992). Persistence-time models for use in viability analyses of vanishing species. *J. Theor. Biol.* **155:**33–53.

Allen, J. C., Schaffer, W. M., and Rosko, D. (1993). Chaos reduces species extinction by amplifying local population noise. *Nature (London)* **364:**229–232.

Alvarez-Buylla, E. R. (1994). Density dependence and patch dynamics in tropical rain forests: matrix models and applications to a tree species. *Am. Nat.* **143:**155–191.

Alvarez-Buylla, E. R., and Shatkin, M. (1991). Finding confidence limits on population growth rates. *Trends Ecol. Evol.* **6:**221–224.

Amarasekare, P. (1994). Spatial population structure in the banner-tailed kangaroo rat, *Dipodomys spectabilis. Oecologia* **100:**166–176.

Andersen, M. (1991). Mechanistic models for the seed shadows of wind-dispersed plants. *Am. Nat.* **137:**476–497.

Anderson, R. M., and May, R. M. (1991). "Infectious Diseases of Humans: Dynamics and Control." Oxford University Press, New York.

Andow, D. A., Kareiva, P. M., Levin, S. A., and Okulso, A. (1990). Spread of invading organisms. *Landscape Ecol.* **4:**177–188.

Andreassen, H. P., Ims, R. A., Stenseth, N. C., and Yoccoz, N. G. (1993). Investigating space use by means of radiotelemetry and other methods: A methodological guide. *In* "Biology of Lemmings" (Stenseth, N. C. and Ims, R. A., eds.), pp. 589–618. Academic Press, London.

Andreassen, H. P., Halle, S., and Ims, R. A. (1996). Optimal design of movement corridors in root voles—not too wide and not too narrow. *J. Appl. Ecol.* **33:**63–70.

Andreassen, H. P., Ims, R. A., and Steinset, O. K. (1996). Discontinuous habitat corridors: The effects on male root vole movements. *J. Appl. Ecol.* (in press).

Andrén, H. (1992). Corvid density and nest predation in relation to forest fragmentation: A landscape perspective. *Ecology* **73:**794–804.

Andrén, H. (1994). Effects of habitat fragmentation on birds and mammals in landscapes with different proportions of suitable habitat: A review. *Oikos* **71:**355–366.

Andrén, H. (1995). Effects of landscape composition on predation rates at habitat edges. *In* "Mosaic Landscapes and Ecological Processes" (L. Hansson, L. Fahrig, and G. Merriam, eds.), pp. 225–255. Chapman & Hall, London.

Andrewartha, H. G., and Birch, L. C., eds. (1954). "The Distribution and Abundance of Animals." University of Chicago Press, Chicago.

Angelstam, P. (1992). Conservation of communities—the importance of edges, surroundings and landscape mosaic structure. *In* "Ecological Principles of Nature Conservation" (L. Hansson, ed.), pp. 9–70. Elsevier, London.

Antonovics, J. (1994). The interplay of numerical and gene-frequency dynamics in host-pathogen systems. *In* "Ecological Genetics" (L. A. Real, ed.), pp. 129–145. Princeton University Press, Princeton, NJ.

Antonovics, J., and Thrall, P. H. (1994). Cost of resistance and the maintenance of genetic polymorphism in host-pathogen systems. *Proc. R. Soc. London, Ser. B* **257:**105–110.

Antonovics, J., Thrall, P., Jarosz, A. and Stratton, D. (1994). Ecological genetics of metapopulations: The *Silene-Ustilago* plant-pathogen system. *In* "Ecological Genetics" (L. A. Real, ed.), pp. 146–170. Princeton University Press, Princeton, NJ.

Antrobus, S., and Lack, A. J. (1994). Genetics of colonizing and established populations of *Primula veris*. *Heredity* **71**:252–258.

Aplin, R. T., and Rothschild, M. (1972). Poisonous alkaloids in the tissues of the garden tiger moth (*Arctia caja* L.) and the cinnebar moth (*Tyria* (= *Callimorpha*) *jacobaeae* L.) (Lepidoptera). *In* "Toxins of Animal and Plant Origin" (A. de Vries and K. Kochva, eds.), pp. 579–595. Gordon & Breach, London.

Arber, W. (1965). Host controlled modification of bacteriophage. *Annu. Rev. Microbiol.* **19**:365–368.

Arditi, R., and Ginzburg, L. R. (1989). Coupling in predator-prey dynamics: Ratio dependence. *J. Theor. Biol.* **139**:311–326.

Armstrong, R. (1987). A patch model of mutualism. *J. Theor. Biol.* **125**:243–246.

Arnason, A. N. (1973). The estimation of population size, migration rates and survival in a stratified population. *Res. Popul. Ecol.* **15**:1–8.

Arnold, L. (1974). "Stochastic Differential Equations." Wiley, New York.

Arnold, M. L., and Hodges, S. A. (1995). Are natural hybrids fit or unfit relative to their parents? *Trends Ecol. Evol.* **10**:67–71.

Arnold, R. A. (1983). Ecological studies of six endangered butterflies *(Lepidoptera, Lycaenidae):* Island biogeography, patch dynamics and the design of habitat preserves. *Univ. Calif. Publ. Entomology,* Vol. **99**.

Askew, R. R. (1971). "Parasitic Insects." Heinemann, London.

Assouad, M. W., Dommée, B., Lumaret, R., and Valdeyron, G. (1978). Reproductive capacities in the sexual forms of the gynodioecious species *Thymus vulgaris*. *Biol. J. Linn. Soc* **77**: 29–39.

Atlan, A., Gouyon, P.-H., and Couvet, D. (1990). Sex allocation in a hermaphroditic plant: *Thymus vulgaris* and the case of gynodioecy. *Oxford Surv. Evol. Biol.* **7**:225–249.

Atkinson, T. C., Briffa, K. R., Coope, G. R. (1987). Seasonal temperatures in Britain during the past 22,000 years, reconstructed using beetle remains. *Nature (London)* **325**:587–593.

Atkinson, W. D., and Shorrocks, B. (1981). Competition on a divided and ephemeral resource: A simulation model. *J. Anim. Ecol.* **50**:461–471.

Augspurger, C. K., and Franson, S. E. (1987). Wind dispersal of artificial fruits varying in mass, area, and morphology. *Ecology* **68**:27–42.

Augspurger, C. K., and Kitajima, K. (1992). Experimental studies of seedling recruitment from contrasting seed distributions. *Ecology* **73**:1270–1284.

Babbitt, B. (1995). Science: Opening the next chapter on conservation history. *Science* **267**:1954–1955.

Back, C. E. (1988a). Effects of host plant size on herbivore density: Patterns. *Ecology* **69**:1090–1102.

Back, C. E. (1988b). Effects of host plant patch size on herbivore density: Mechanisms. *Ecology* **69**: 1103–1117.

Baguette, M., and Nève, G. (1994). Adult movements between populations in the specialist butterfly *Proclossiana eunomia* (Lepidoptera, Nymphalidae). *Ecol. Entomol.* **19**:1–5.

Baker, H. G. (1947). Biological flora of the British Isles. Melandrium (Roehling em,) Fries. *J. Ecol.* **35**:271–292.

Baker, H. G., and Stebbins, G. L. (1965). "The Genetics of Colonizing Species." Academic Press, New York.

Baker, R. R. (1969). The evolution of the migratory habit in butterflies. *J. Anim. Ecol.* **38**:703–746.

Barrett, G. W., and Bohlen, P. J. (1991). Landscape ecology. *In* "Landscape Linkages and Biodiversity" (W. E. Hudson, ed.), pp. 149–161. Island Press, Washington, DC.

Barrett, G. W., Ford, H. A., and Recher, H. F. (1994). Conservation of woodland birds in a fragmented rural landscape. *Pacif. Conserv. Biol.* **1**:245–256.

Barton, N. H. (1979). The dynamics of hybrid zones. *Heredity* **43**:341–359.

Barton, N. H. (1986). The maintenance of polygenic variation through a balance between mutation and stabilising selection. *Genet. Res.* **47**:209–216.

Barton, N. H. (1987). The probability of establishment of an advantageous mutation in a subdivided population. *Genet. Res.* **50:**35–40.

Barton, N. H. (1989). Founder effect speciation. *In* "Speciation and its Consequences" (D. Otte and J. A. Endler, eds.), Chapter 10. Sinauer Assoc., Sunderland, MA.

Barton, N. H. (1992). On the spread of new gene combinations in the third phase of Wright's shifting balance. *Evolution (Lawrence, Kans.)* **46:**551–557.

Barton, N. H. (1993). The probability of fixation of a favoured allele in a subdivided population. *Genet. Res.* **62:**149–158.

Barton, N. H. (1994). The reduction in fixation probability caused by substitutions at linked loci. *Genet. Res.* **64:**199–208.

Barton, N. H. (1995). Linkage and the limits to natural selection. *Genetics* **140:**821–841.

Barton, N. H., and Charlesworth, B. (1984). Genetic revolutions, founder effects, and speciation. *Annu. Rev. Ecol. Syst.* **15:**133–164.

Barton, N. H., and Clark, A. (1990). Population structure and processes in evolution. *In* "Population Biology" (K. Wohrmann and S. K. Jain, eds.), pp. 115–173. Springer-Verlag, New York.

Barton, N. H., and Hewitt, G. M. (1985). Analysis of hybrid zones. *Annu. Rev. Ecol. Syst.* **16:**113–148.

Barton, N. H., and Hewitt, G. M. (1989). Adaptation, speciation and hybrid zones. *Nature (London)* **341:**497–503.

Barton, N. H., and Rouhani, S. (1987). The frequency of shifts between alternative equilibria. *J. Theor. Biol.* **125:**397–418.

Barton, N. H., and Rouhani, S. (1991). The probability of fixation of a new karyotype in a continuous population. *Evolution (Lawrence, Kans.)* **45:**499–517.

Barton, N. H., and Rouhani, S. (1993). Adaptation and the 'shifting balance.' *Genet. Res.* **61:**57–74.

Bascompte. J., and Solé, R. V. (1994). Spatially induced bifurcations in single-species population dynamics. *J. Anim. Ecol.* **63:**256–264.

Bascompte. J., and Solé, R. V. (1995). Rethinking complexity: Modelling spatiotemporal dynamics in ecology. *Trends Ecol. Evol.* **10:**361–366.

Bazzaz, F. (1986). Life history of colonizing plants: Some demographic, genetic and physiological features. *In* "Ecology of Biological Invasions of North America and Hawaii" (H. A. Mooney and J. A. Drake, eds.), pp. 96–110. Springer-Verlag, Berlin.

Beddington, J. R., and Hammond, P. S. (1977). On the dynamics of host-parasite-hyperparasite interactions. *J. Anim. Ecol.* **46:**811–821.

Beddington, J. R., Free, C. A., and Lawton, J. H. (1975). Dynamic complexity in predator-prey models framed in difference equations. *Nature (London)* **255:**58–60.

Begon, M., Harper, J. L., and Townsend, C. R. (1986). "Ecology: Individuals, Populations and Communities." Blackwell, Oxford.

Begon, M., Harper, J. L., and Townsend, C. R. (1990). "Ecology: Individuals, Populations and Communities, 2nd ed." Blackwell, Boston.

Beier, P. (1993). Determining minimum habitat areas and habitat corridors for cougars. *Conserv. Biol.* **7:**94–108.

Belhassen, E., Dockes, A.-C., Gliddon, C., and Gouyon, P.-H. (1987). Dissémination et voisinage chez une espèce gynodioïque: Le cas de *Thymus vulgaris* (L.). *Génét., Sél., Evol.* **19:**307–320.

Belhassen, E., Trabaud, L., Couvet, D., and Gouyon, P.-H. (1989). An example of nonequilibrium processes: Gynodiecy of *Thymus vulgaris* L. in burned habitats. *Evolution (Lawrence, Kans.)* **43:**662–667.

Belhassen, E., Beltran, M., Couvet, D., Dommée, B., Gouyon, P.-H., and Olivieri, I. (1990). Evolution des taux de femelles dans les populations naturelles de thym, *Thymus vulgaris* L. Deux hypothèses alternatives confirmées. *C. R. Sciences Acad. Sci., Ser. 3* **310:**371–375.

Belhassen, E., Dommée, B., Atlan, A., Gouyon, P.-H., Pomete, D., Assouad, M. W., and Couvet, D. (1991). Complex determination of male sterility in *Thymus vulgaris* L.: Genetic and molecular analysis. *Theor. Appl. Genet.* **82:**137–143.

Belhassen, E., Atlan, A., Couvet, D., and Gouyon, P.-H. (1993). Mitochondrial genome of *Thymus vulgaris.* L. (Labiatae) is highly polymorphic between and among natural populations. *Heredity* **71:**462–472.

Bell, W. J. (1991). "Searching Behaviour." Chapman & Hall, London.

Bengtsson, B. O., and Christiansen, F. B. (1983). A two-locus mutation-selection model and some of its evolutionary implications. *Theor. Popul. Biol.* **24:**59–77.

Bengtsson, J. (1989). Interspecific competition increases local extinction rate in a metapopulation system. *Nature (London)* **340:**713–715.

Bengtsson, J. (1991). Interspecific competition in metapopulations. *In* "Metapopulation Dynamics: Empirical and Theoretical Investigations" (M. E. Gilpin and I. Hanski, eds.), pp. 219–237. Academic Press, London.

Bengtsson, J. (1993). Interspecific competition and determinants of extinction in experimental populations of three rockpool Daphnia species. *Oikos* **67:**451–464.

Bengtsson, J., and Baur, B. (1993). Do pioneers have r-selected traits? Life history patterns among colonizing terrestrial gastropods. *Oecologia* **94:**17–22.

Bengtsson, J., and Milbrink, G. (1995). Predicting extinctions: Interspecific competition, predation and population variability in experimental Daphnia populations. *Oecologia* **101:**397–406.

Bennett, A. F. (1990). "Habitat Corridors: Their Role in Wildlife Management and Conservation." Department of Conservation and Environment, Melbourne, Australia.

Bennett, A. F. (1991). Roads, roadsides and wildlife conservation: A review. *In* "Nature Conservation 2: The Role of Corridors" (D. A. Saunders and R. J. Hobbs, eds.), pp. 99–118. Surrey Beatty & Sons, Chipping Norton, NSW, Australia.

Ben-Shlomo, R., Motro, U., and Ritte, U. (1991). The influence of the ability to disperse on generation length and population size in the flour beetle, *Tribolium castaneum. Ecol. Entomol.* **16:**279–282.

Berthold, P., and Pulido, F. (1994). Heritability of migratory activity in a natural bird population. *Proc. R. Soc. London, Ser. B* **257:**311–315.

Bevan, J. R., Crute, I. R., and Clarke, D. D. (1993a). Variation for virulence in *Erysiphe fischeri* from *Senecio vulgaris. Plant Pathol.* **42:**622–635.

Bevan, J. R., Clarke, D. D., and Crute, I. R. (1993b). Resistance to *Erysiphe fischeri* in two populations of *Senecio vulgaris. Plant Pathol.* **42:**636–646.

Bickhan, R. J., and Baker, J. W. (1986). Speciation by monobrachial centric fusions. *Proc. Natl. Acad. Sci. U.S.A.* **83:**8245–8248.

Boag, D. A., and Murie, J. O. (1981). Population ecology of Columbian ground squirrels in southwestern Alberta. *Can. J. Zool.* **59:**2230–2240.

Boecklen, W. J., and Simberloff, D. (1986). Area-based extinction models in conservation. *In* "Dynamics of Extinction" (D. K. Elliott, ed.), pp. 247–276. Wiley, New York.

Boerlijst, M. C., Lamers, M. E., and Hogeweg, P. (1993). Evolutionary consequences of spiral waves in a host-parasitoid system. *Proc. R. Soc. London, Ser. B* **253:**15–18.

Bondrup-Nielsen, S. (1992). Emigration of meadow voles, *Microtus pennsylvanicus. Oikos* **65:**358–360.

Bonham, C. D., and Lerwick, A. (1976). Vegetation changes induced by prairie dogs on shortgrass range. *J. Range Manage.* **29:**221–225.

Boorman, S. A., and Levitt, P. R. (1973). Group selection on the boundary of a stable population. *Theor. Popul. Biol.* **4:**85–128.

Bos, M., Harmens, H., and Vrieling, K. (1986). Gene flow in *Plantago* I. Gene flow and neighbourhood size in *P. lanceolata. Heredity* **56:**43–54.

Boutin, V., Pannenbecker, G., Ecke, W., Schewe, G., Samitou-Lapgrade, P., Jean, R., Verne, P., and Michaelis, G. (1987). Cytoplasmic male sterility and nuclear restorer genes in a natural population of *Beta maritima:* Genetical and molecular aspects. *Theor. Appl. Genet.* **73:** 625–629.

Boutin-Stadler, V., Saumitou-Laprade, P., Valero, M., Jean, R., and Vernet, P. (1989). Spatio-temporal variation of male sterile frequencies in two natural populations of *Beta maritima. Heredity* **63:**395–400.

Boycott, A. E. (1930). A re-survey of the fresh-water mollusca of the parish of Aldenham after ten years with special reference to the effect of drought. *Trans. Herts. Nat. Hist. Soc.* **19**:1–25.

Bright, P. W., and Morris, P. A. (1994). Animal translocation for conservation: Performance of dormice in relation to release methods, origin and season. *J. Appl. Ecol.* **31**:699–708.

Brown, J. H. (1971). Mammals on mountaintops: Nonequilibrium insular biogeography. *Am. Nat.* **105**:467–478.

Brown, J. H. (1978). The theory of insular biogeography and the distribution of boreal birds and mammals. *Great Basin Nat. Mem.* **2**:209–227.

Brown, J. H. (1995). "Macroecology." University of Chicago Press, Chicago.

Brown, J. H., and Gibson, A. C. (1983). "Biogeography." Mosby, St. Louis, MO.

Brown, J. H., and Kodric-Brown, A. (1977). Turnover rates in insular biogeography: Effect of immigration on extinction. *Ecology* **58**:445–449.

Brown, V. K. (1985). Insect herbivores and plant succession. *Oikos* **44**:17–22.

Brownie, C., Hines, J. E., Nichols, J. D., Pollock, K. H., and Hestbeck, J. B. (1992). Capture-recapture studies for multiple strata including non-Markovian transitions. *Biometrics* **49**:1173–1187.

Bruening, G. (1977). Plant covirus systems: Two component systems. *In* "Comprehensive Virology" (H. Fraenkel-Conrat and R. R. Wagner, eds.), Vol. 11, pp. 55–142. Plenum, New York.

Brussard, P. F., and Vawter, A. T. (1975). Population structure, gene flow, and natural selection in populations of *Euphydryas phaeton*. *Heredity* **34**:407–415.

Brussard, P. F., Murphy, D. D., and Noss, R. F. (1992). Strategy and tactics for conserving biological diversity in the United States. *Conserv. Biol.* **6**:157–159.

Buechner, M. (1987). A geometric model of vertebrate dispersal: Tests and implications. *Ecology* **68**: 310–318.

Bulmer, M. (1991). The selection-mutation-drift theory of synonymous codon usage. *Genetics* **129**: 897–907.

Burdon, J. J. (1987). "Diseases and Plant Population Biology." Cambridge University Press, Cambridge, UK.

Burdon, J. J., and Jarosz, A. M. (1991). Host-pathogen interactions in natural populations of *Linum marginale* and *Melampsora lini:* I. Patterns of resistance and racial variation in a large host population. *Evolution (Lawrence, Kans.)* **45**:205–217.

Burdon, J. J., and Jarosz, A. M. (1992). Temporal variation in the racial structure of flax rust *(Melampsora lini)* populations growing on natural stands of wild flax *(Linum marginale):* Local versus metapopulation dynamics. *Plant Pathol.* **41**:165–179.

Burdon, J. J., and Leather, S. R., eds. (1990). "Pests, Pathogens and Plant Communities." Blackwell, Oxford.

Burdon, J. J., and Thompson, J. N. (1995). Changed patterns of resistance in a population of *Linum marginale* attacked by rust pathogen *Melampsora lini. J. Ecol.* **83**:199–206.

Burdon, J. J., Jarosz, A. M., and Kirby, G. C. (1989). Pattern and patchiness in plant-pathogen interactions—causes and consequences. *Annu. Rev. Ecol. Syst.* **20**:119–136.

Burgman, M. A., Ferson, S., and Akçakaya, H. R. (1993). "Risk Assessment in Conservation Biology." Chapman & Hall, London.

Burnham, K. P., Anderson, D. R., White, G. C., Brownie, C., and Pollock, K. H. (1987). Design and analysis methods for fish survival experiments based on release-recapture. *Monogr.—Am. Fish. Soc.* **5**.

Burt, A. (1995). The evolution of fitness. *Evolution (Lawrence, Kans.)* **49**:1–8.

Caballero, A. (1994). Review article: Developments in the prediction of effective population size. *Heredity* **73**:657–679.

Caley, M. J. (1991). A null model for testing distributions of dispersal distances. *Am. Nat.* **138**:524–532.

Cameron, E. (1935). A study of the natural control of ragwort (*Senecio jacobaea* L.). *J. Ecol.* **23**: 265–322.

Carlsson, U. (1995). Anther-smut disease in *Silene dioica*. Ph.D. Thesis, Umeå University, Sweden.

Carlsson, U., and Elmqvist, T. (1992). Epidemiology of the anther-smut disease *Microbotryum violaceum* and numeric regulation of populations of *Silene dioica*. *Oecologia* **90**:509–517.

Carlsson, U., Elmqvist, T., Wennström, A., and Ericson, L. (1990). Infection by pathogens and population age of host plants. *J. Ecol.* **78**:1094–1105.

Case, T. J. (1991). Invasion resistance, species build-up and community collapse in metapopulations models with interspecies competition. *In* "Metapopulation Dynamics: Empirical and Theoretical Investigations" (M. Gilpin and I. Hanski, eds.), pp. 239–266. Academic Press, London.

Caswell, H. (1982). Life history theory and the equilibrium status of populations. *Am Nat.* **120**:317–339.

Caswell, H. (1989). "Matrix Population Models." Sinauer Assoc., Sunderland, MA.

Caswell, H., and Cohen, J. E. (1993). Local and regional regulation of species-area relations: A patch-occupancy model. *In* "Species Diversity in Ecological Communities" (R. E. Ricklefs and D. Schluter, eds.), pp. 99–107. University of Chicago Press, Chicago.

Caswell, H., and Etter, R. J. (1993). Ecological interactions in patchy environments: From patch-occupancy models to cellular automata. *In* "Patch Dynamics" (S. A. Levin, T. M. Powell, and J. H. Steele, eds.), pp. 93–109. Springer-Verlag, Berlin.

Caughley, G. (1976). Plant-herbivore systems. *In* "Theoretical Ecology" (R. M. May, ed.), pp. 94–113. Blackwell, Oxford.

Caughley, G. (1994). Directions in conservation biology. *J. Anim. Ecol.* **63**:215–244.

Caughley, G., and Sinclair, A. R. E. (1994). "Wildlife Ecology and Management." Blackwell, Boston.

Charlesworth, B. (1994a). The effect of background selection against deleterious mutations on weakly selected, linked variants. *Genet. Res.* **63**:213–228.

Charlesworth, B. (1994b). "Evolution in Age-Structured Populations," 2nd ed. Cambridge University Press, Cambridge, UK.

Chatfield, C. (1989). "The Analysis of Time Series." Chapman & Hall, London.

Chepko-Sade, B. D., and Halpin, Z. T., eds. (1987). "Mammalian Dispersal Patterns: The Effects of Social Structure on Population Genetics." University of Chicago Press, Chicago.

Chesser, R. K. (1983). Genetic variability within and among populations of the black-tailed prairie dog. *Evolution (Lawrence, Kans.)* **37**:320–331.

Chesser, R. K. (1991). Influence of gene flow and breeding tactics on gene diversity within populations. *Genetics* **129**:573–583.

Chesser, R. K., Rhodes, O. E., Sugg, D. W., and Schnabel, A. (1993). Effective sizes for subdivided populations. *Genetics* **135**:1221–1232.

Chesson, P. L. (1981). Models for spatially distributed populations: The effect of within-patch variability. *Theor. Popul. Biol.* **19**:288–325.

Chesson, P. L. (1986). Environmental variation and the coexistence of species. *In* "Community Ecology" (J. Diamond and T. J. Case, eds.), pp. 240–256. Harper & Row, New York.

Chesson, P. L. (1990). Geometry, heterogeneity and competition in variable environments. *Philos. Trans. R. Soc. London, Ser. B* **330**:165–173.

Chesson, P. L., and Case, T. J. (1986). Overview: Nonequilibrium community theories: Change, variability, history, and coexistence. *In* "Community Ecology" (J. M. Diamond and T. J. Case, eds.), pp. 229–239. Harper & Row, New York.

Chesson, P. L., and Huntly, N. (1989). Short-term instabilities and longterm community dynamics. *Trends Ecol. Evol.* **4**:293–298.

Chesson, P. L., and Murdoch, W. W. (1986). Aggregation of risk: Relationships among host-parasitoid models. *Am. Nat.* **127**:696–715.

Chhibber, S., Goel, A. Kapoor, N., Saxena, M., and Vadehra, D. V. (1988). Bacteriocin (klebocin) typing of clinical isolates of *Klebsiella pneumoniae*. *Eur. J. Epidemiol.* **4**:115–118.

Christ, B. J., Person, C. O., and Pope, D. D. (1987). The genetic determination of variation in pathogenicity. *In* "Populations of Plant Pathogens: Their Dynamics and Genetics" (M. S. Wolfe and C. E. Caten, eds.), pp. 7–19. Blackwell, Oxford.

Cipollini, M. L., Wallacesenft, D. A., and Whigham, D. F. (1994). A model of patch dynamics, seed dispersal, and sex ratio in the dioecious shrub *Lindera benzoin* (Lauraceae). *J. Ecol.* **82:**621–633.

Clark, C. W., and Rosenweig, M. L. (1994). Extinction and colonization processes: Parameter estimates from sporadic surveys. *Am. Nat.* **143:**583–596.

Clarke, D. D., Campbell, F. S., and Bevan, J. R. (1990). Genetic interactions between *Senecio vulgaris* and the powdery mildew fungus *Erysiphe fischeri. In* "Pests, Pathogens and Plant Communities" (J. J. Burdon and S. R. Leather, eds.), pp. 189–201. Blackwell, Oxford.

Clarke, P. J. (1993). Dispersal of grey mangrove *(Avicennia marina)* propagules in southeastern Australia. *Aquat. Bot.* **45:**195–204.

Clay, K. (1982). Environmental and genetic determinants of cleistogamy in a natural population of the grass *Danthonia spicata. Evolution (Lawrence, Kans.)* **36:**734–741.

Clutton-Brock, T. H. (1989). Female transfer and inbreeding in social mammals. *Nature (London)* **337:**70–72.

Cohen, D., and Levin, S. A. (1991). Dispersal in patchy environments: The effects of temporal and spatial structure. *Theor. Popul. Biol.* **39:**63–99.

Cohen, D., and Motro, U. (1989). More on optimal rates of dispersal: Taking into account the cost of the dispersal mechanism. *Am. Nat.* **134:**659–663.

Cohen, J. E. (1970). A Markov contingency-table model for replicated Lotka-Volterra systems near equilibrium. *Am. Nat.* **104:**547–560.

Comins, H. N. (1982). Evolutionarily stable strategies for localized dispersal in two dimensions. *J. Theor. Biol.* **94:**579–606.

Comins, H. N., Hamilton, W. D., and May, R. M. (1980). Evolutionary stable dispersal strategies. *J. Theor. Biol.* **82:**205–230.

Comins, H. N., Hassell, M. P., and May, R. M. (1992). The spatial dynamics of host-parasitoid systems. *J. Anim. Ecol.* **61:**735–748.

Conroy, M. J., Cohen, Y., James, F. C., Matsinos, Y. G., and Maurer, B. A. (1995). Parameter estimation, reliability, and model improvement for spatially explicit models of animal populations. *Ecol. Appl.* **5:**17–19.

Cook, R. R., and Hanski, I. (1995). On expected lifetimes of small bodied and large-bodied species of birds on islands. *Am. Nat.* **145:**307–315.

Cormack, R. M. (1964). Estimates of survival from the sighting of marked animals. *Biometrika* **51:** 429–438.

Couvet, D., Gouyon, P., Kjellberg, F., Olivieri, I., Pomente, D., and Valdeyron, G. (1985). De la métapopulation au voisinage: La génétique des populatons en déséquilibre. *Génét., Sél., Evol.* **17:**407–414.

Couvet, D., Bonnemaison, F., and Gouyon, P.-H. (1986). The maintenance of females among hermaphrodites: The importance of nuclear-cytoplasmic interactions. *Heredity* **57:**325–330.

Couvet, D. Atlan, A., Belhassen, E., Gliddon, C., Gouyon, P.-H., and Kjellberg, F. (1990). Coevolution between two symbionts: The case of cytoplasmic male-sterility in higher plants. *Oxford Surv. Evol. Biol.* **7:**225–249.

Cox, D. R., and Isham, V. (1980). "Point Processes." Chapman & Hall, London.

Crawley, M. J., and Gillman, M. P. (1989). Population dynamics of cinnabar moth and ragwort in grassland. *J. Ecol.* **58:**1035–1050.

Crawley, M. J., and May, W. D. (1987). Population dynamics and plant community structure: Competition between annuals and perenials. *J. Theor. Biol.* **125:**475–489.

Crawley, M.J. and Pattrasudhi, R. (1988). Interspecific competition between insect herbivores: Asymmetric competition between cinnabar moth and the ragwort seed-head fly. *Ecol. Entomol.* **13:** 243–249.

Cressie, N. A. C. (1993). "Statistics for Spatial Data." Wiley, New York.

Crist, T. O., Guertin, D. S., Wiens, J. A., and Milne, B. T. (1992). Animal movement in heterogeneous landscapes: An experiment with *Eleodes* beetles in shortgrass prairie. *Funct. Ecol.* **6:** 536–544.

Crome, F. H. J. (1993). Tropical forest fragmentation: Some conceptual and methodological issues.

In "Conservation Biology in Australia and Oceania" (C. Moritz and J. Kikkawa, eds.), pp. 61–76. Surrey Beatty & Sons, Chipping Norton, NSW, Australia.

Crow, J. F., and Aoki, K. (1984). Group selection for a polygenic behavioural trait: Estimating the degree of population subdivisions. *Proc. Nat. Acad. Sci. U.S.A.* **81:**6073–6077.

Crow, J. F., and Kimura, M. (1970). "An Introduction to Population Genetics Theory." Harper & Row, London.

Crow, J. F., Engels, W. R., and Denniston, C. (1990). Phase three of Wright's shifting-balance theory. *Evolution (Lawrence, Kans.)* **44:**233–247.

Crowell, K. L. (1973). Experimental zoogeography: Introduction of mice to small islands. *Am. Nat.* **107:**535–558.

Crowley, P. H. (1981). Dispersal and the stability of predator-prey interactions. *Am Nat.* **118:**673–701.

Czaran, T., and Bartha, S. (1992). Spatiotemporal dynamic models of plant populations and communities. *Trends Ecol. Evol.* **7:**38–42.

Dale, V. H., Pearson, S. M., Offerman, H. L., and O'Neill, R. V. (1994). Relating patterns of land-use change to faunal biodiversity in the central Amazon. *Conserv. Biol.* **8:**1027–1036.

Damuth, J., and Heisler, I. L. (1988). Alternative formulations of multilevel selection. *Biol. Philos.* **3:**407–430.

Danielson, B. J. (1992). Habitat selection, interspecific interactions and landscape composition. *Evol. Ecol.* **6:**399–411.

Danielson, B. J., and Gaines, M. S. (1987). The influences of conspecific and heterospecific residents on colonization. *Ecology* 68:1778–1784.

Danthanarayana, W. (1986). "Insect Flight. Dispersal and Migration." Springer-Verlag, Berlin.

Darwin, C. R. (1859). "On the Origin of Species by Means of Natural Selection." Murray, London.

Darwin, C. R. (1877). Polygamous, dioecious, and gyno-dioecious plants. *In* "The Different Forms of Flower on Plants of the Same Species," pp. 278–309. Murray, London.

Davidson, W. R., Nettles, V. F., Hayes, L. E., Howerth, E. W., and Couvillion, C. E. (1992). Diseases diagnosed in grey foxes *(Urocyon cinereoargenteus)* from the south-eastern United States. *J. Wildl. Dis.* **28:**28–33.

Davis, G. J., and Howe, R. W. (1992). Juvenile dispersal, limited breeding sites, and the dynamics of metapopulations. *Theor. Popul. Biol.* **41:**184–207.

Dawkins, R. (1976). "The Selfish Gene." Oxford University Press, Oxford.

DeAngelis, D. L. (1992). "Dynamics of Nutrient Cycling and Food Webs." Chapman & Hall, London.

DeAngelis, D. L., and Gross, L. J., eds. (1992). "Individual-based Models and Approaches in Ecology: Populations, Communities and Ecosystems." Chapman & Hall, New York.

Debinski, D. M. (1994). Genetic diversity assessment in a metapopulation of the butterfly *Euphydryas gillettii. Biol. Conserv.* **70:**25–31.

de Jong, G. (1979). The influence of the distribution of juveniles over patches of food on the dynamics of a population. *Neth. J. Zool.* **29:**33–51.

de Jong, G. (1981). The influence of dispersal pattern on the evolution of fecundity. *Neth. J. Zool.* **32:**1–30.

de Jong, T. J., and Klinkhamer, P. G. L. (1988). Population ecology of the biennials *Cirsium vulgare* and *Cynoglossum officinale* in a coastal sand dune area. *J. Ecol.* **76:**366–382.

Dempster, J. P. (1971). The population ecology of the cinnabar moth, *Tyria jacobaeae* L. (Lepidoptera, Arctiidae). *Oecologia* **7:**26–67.

Dempster, J. P. (1982). The ecology of the cinnabar moth, *Tyria jacobaeae* L. (Lepidoptera, Arctiidae). *Adv. Ecol. Res.* **12:**1–36.

Dempster, J. P. (1991). Fragmentation, isolation and mobility of insect populations. *In* "Conservation of Insects and their Habitats" (N. M. Collins and J. A. Thomas, eds.), pp. 143–154. Academic Press, London.

Dempster, J. P., King, M. L., and Lakhani, K. H. (1976). The status of the swallowtail butterfly in Britain. *Ecol. Entomol.* **1:**51–56.

den Boer, P. J. (1968). Spreading of risk and stabilization of animal numbers. *Acta Biotheor.* **18:** 165–194.

den Boer, P. J. (1981). On the survival of populations in a heterogeneous and variable environment. *Oecologia* **50:**39–53.

den Boer, P. J. (1987). Detecting density dependence. *Trends Ecol. Evol.* **2:**77–78.

den Boer, P. J. (1990). The survival value of migration in terrestrial arthropods. *Biol. Conserv.* **54:** 175–192.

den Boer, P. J. (1991). Seeing the tree for the wood: Random walks or bounded fluctuations of population size? *Oecologia* **86:**484–491.

Dennis, B., and Taper, M. L. (1994). Density dependence in time series observations of natural populations: Estimating and testing. *Ecol. Monogr.* **64:**205–224.

Dennis B., Munholland, P. L., and Scott, J. M. (1991). Estimation of growth and extinction parameters for endangered species. *Ecol. Monogr.* **61:**115–143.

Dennis, B. *et al.* (1996). In preparation.

Denno, R. F. (1994). The evolution of dispersal polymorphisms in insects: The influence of habitat, host plants and mates. *Res. Popul. Ecol.* **36:**127–135.

Denno, R. F., Roderick, G. K., Olmstead, K. L., and Dobel, H. G. (1991). Density-related migration in planthoppers (Homoptera, Delphacidae)—The role of habitat persistence. *Am. Nat.* **138:** 1513–1541.

De Roos, A. M., McCauley, E., and Wilson, W. G. (1991). Mobility versus density-limited predator-prey dynamics on different spatial scales. *Proc. R. Soc. London, Ser. B* **246:**117–122.

Descimon, H., and Napolitano, M. (1993a). Enzyme polymorphism, wing pattern variability, and geographical isolation in an endangered butterfly species. *Biol. Conserv.* **66:**117–123.

Descimon, H., and Napolitano, M. (1993b). Les populations de *Parnassius mnemosyne* (Linné) á la Sainte Baume (Bouches-du-Rhône, France): Structure génétique, origine et histoire (Lepidoptera: Papilionidae). *Ecol. Mediterr.* **19:**15–28.

Diamond, J. M. (1972). Biogeographic kinetics: Estimation of relaxation times for avifaunas of Southwest Pacific islands. *Proc. Nat. Acad. Sci. U.S.A.* **69:**3199–3203.

Diamond, J. M. (1975a). The island dilemma: Lessons of modern biogeographic studies for the design of natural reserves. *Biol. Conserv.* **7:**129–146.

Diamond, J. M. (1975b). Assembly of species communities. *In* "Ecology and Evolution of Communities" (M. L. Cody and J. M. Diamond, eds.), pp. 342–444. Harvard University Press, Cambridge, MA.

Diamond, J. M. (1984). "Normal" extinction of isolated populations. *In* "Extinctions" (M. H. Nitecki, ed.), pp. 191–246. University of Chicago Press, Chicago.

Diamond, J. M., and May, R. M. (1976). Island biogeography and the design of nature reserves. *In* "Theoretical Ecology: Principles and Applications" Saunders, Philadelphia. (R. M. May, ed.), pp. 163–186.

Diamond, J. M., and Pimm, S. (1993). Survival times of bird populations. A reply. *Am. Nat.* **142:** 1030–1035.

Dias, P., Verheyen, G. R., and Raymond, M. (1996). Source-sink populations in mediterranean blue tits: Evidence using single-locus minisatellite probes. *J. Evol. Biol.* (in press).

Diekmann, O. (1993). An invitation to structured (meta)population models. *In* "Patch Dynamics" (S. A. Levin, T. M. Powell, and J. H. Steele, eds.), pp. 162–175. Springer, Berlin.

Diekmann, O., Metz, J. A. J., and Sabelis, M. W. (1988). Mathematical models of predator-prey-plant interactions in a patchy environment. *Exp. Appl. Acarol.* **5:**319–342.

Diekmann, O., Metz, J. A. J., and Sabelis, M. W. (1989). Reflections and calculations on a prey-predator-patch problem. *Acta Appl. Math.* **14:**23–25.

Diekmann, O., Heesterbeek, J. A. P., and Metz, J. A. J. (1990). On the definition of the basic reproduction ratio R_0 in models for infectious diseases in heterogeneous populations. *J. Math. Biol.* **28:** 365–382.

Diekmann, O., Gyllenberg, M., and Thieme, H. R. (1993a). Perturbing semi-groups by solving Stieltjes renewal equations. *Differ. Integral Equations* **6:**155–181.

Diekmann, O., Gyllenberg, M., Metz, J. A. J., and Thieme, H. R. (1993b). The "cumulative" formulation of (physiologically) structured population models. *In* "Evolution Equations, Control Theory and Biomathematics" (P. Clément and G. Lumer, eds.), pp. 145–154. Dekker, New York.

Diekmann, O., Gyllenberg, M., Thieme, H. R., and Verduyn Lunel, S. M. (1993c). A cell-cycle model revisited. *Proc. Conf. Equations Differ.*, Marrakech, Maroc, *1991*, Report AM-R9305.

Diekmann, O., Gyllenberg, M., and Thieme, H. R. (1995a). Perturbing evolutionary systems by cumulative outputs and step responses. *Differ. Integral Equations* **8:**1205–1244.

Diekmann, O., Gyllenberg, M., Metz, J. A. J., and Thieme, H. R. (1995b). On the formulation and analysis of general deterministic structured population models. I. Linear theory. Preprint.

Dingle, H., Blackley, N. R., and Miller, E. R. (1980). Variation in body size and flight performance in milkweed bugs *(Oncopeltus). Evolution (Lawrence, Kans.)* **34:**371–385.

Diver, C. (1938). Aspects of the study of variation in snails. *J. Conchol.* **21:**91–142.

Doak, D. F., and Mills, L. S. (1994). A useful role for theory in conservation. *Ecology* **75:**615–626.

Dobson, F. S. (1994). Measures of gene flow in the Columbian ground squirrel. *Oecologia* **100:**190–195.

Dodd, A. P. (1959). The biological control of prickly pear in Australia. *Monogr. Biol.* **8:**565–577.

Doebeli, M. (1995). Dispersal and dynamics. *Theor. Popul. Biol.* **47:**82–106.

Dommée, B., and Jacquard, P. (1985). Gynodioecy in thyme, *Thymus vulgaris* L.: Evidence from successional populations. *In* "Genetics Differentiation and Dispersal in Plants" (P. Jacquard *et al.*, eds.), pp. 141–164. Springer-Verlag, Berlin.

Dommée, B., Guillerm, J.-L., and Valdeyron, G. (1983). Régime de reproduction et hétérozygotie des populations de thym, *Thymus vulgaris* L., dans une succession postculturale. *C. R. Seances Acad. Sci. Sér. 3* **296**111–114.

Doncaster, C. P., Micol, T., and Jensen, S. P. (1996). Determining minimum habitat requirements in theory and practice. *Oikos* **75:**335–339.

Dowdeswell, W. H., Fisher, R. A., and Ford, E. B. (1940). The quantitative study of populations in the Lepidoptera. *Ann. Eugen.* **10:**123–136.

Drummond, D. (1966). Rats resistant to warfarin. *New Sci.* **30:**771–772.

Dunning, J. B., Stewart, D. J., Danielson, B. J., Noon, B. R., Root, T. L., Lamberson, R. H., and Stevens, E. E. (1995). Spatially explicit population models: Current forms and future uses. *Ecol. Appl.* **5:**3–11.

Durelli, P., Studer, M., Marchand, I., and Jacob, S. (1990). Population movements of arthropods between natural and cultivated areas. *Biol. Conserv.* **54:**193–207.

Durrett, R. (1989). "Lecture Notes on Particle Systems and Percolation." Wadsworth and Brooks/Cole, Pacific Grove, CA.

Durrett, R., and Levin, S. A. (1994). Stochastic spatial models: A user's guide to ecological applications. *Philos. Trans. R. Soc. London, Ser. B* **343:**329–350.

Dybdal, M. F. (1994). Extinction, recolonisation, and the genetic structure of tidepool copepod populations. *Evol. Ecol.* **8:**113–124.

Dytham, C. (1995). Competitive coexistence and empty patches in spatially explicit metapopulation models. *J. Anim. Ecol.* **64:**145–146.

Ebenhard, T. (1987). An experimental test of the island colonization survival model: Bank voles *(Clethrionomys glareolus)* populations with different demographic parameter values. *J. Biogeogr.* **14:**213–223.

Ebenhard, T. (1991). Colonization in metapopulations: A review of theory and observations. *Biol. J. Linn. Soc.* **42:**105–121.

Eber, S., and Brandl, R. (1994). Ecological and genetic spatial patterns of *Urophora cardui* (Diptera: Tephritidae) as evidence for population structure and biogeographical processes. *J. Anim. Ecol.* **63:**187–190.

Eberhardt, L. L., Blanchard, B. M., and Knight, R. R. (1994). Population trend of the Yellowstone grizzly bear as estimated from reproductive and survival rates. *Can. J. Zool.* **72:**360–363.

Edmunds, G. F., Jr., and Alstad, D. N. (1978). Coevolution in insect herbivores and conifers. *Science* **199:**941–945.

Edwardson, J. R. (1970). Cytoplasmic male sterility. *Bot. Rev.* **36:**341–420.

Edwards, S. V. (1993). Long distance gene flow in a cooperative breeder detected in genealogies of mitochondrial DNA sequences. *Proc. R. Soc. London, Ser. B* **252**:177–185.

Ehmke, A., Witte, L., Biller, A., and Hartmann, T. (1990). Sequestration, N-oxidation and transformation of plant pyrrolizidine alkaloids by the Arctiid moth *Tyria jacobaeae* L. *Z. Naturforsch., C: Biosci.* **45C**:1185–1192.

Ehrlich, P. R. (1983). Genetics and extinction of butterfly populations. *In* "Genetics and Conservation: A Reference for Managing Wild Animal Populations," (C. M. Schonewald-Cox, S. M. Chambers, B. MacBryde, and L. Thomas, eds.) pp. 152–163. Benjamin/Cummings, Menlo Park, CA.

Ehrlich, P. R. (1984). The structure and dynamics of butterfly populations. *In* "The Biology of Butterflies" (R. I. Vane-Wright and P. R. Ackery, eds.), pp. 25–40, Princeton University Press, Princeton, NJ.

Ehrlich, P. R., and Murphy, D. D. (1987). Conservation lessons from long-term studies of checkerspot butterflies. *Conserv. Biol.* **1**:122–131.

Ehrlich, P. R., and Birch, L. C. (1967). The "balance of nature" and "population control." *Am. Nat.* **101**:97–107.

Ehrlich, P. R., and Raven, P. H. (1969). Differentiation of populations. *Science* **165**: 1228–1232.

Ehrlich, P. R., Breedlove, D. E., Brussard, P. F., and Sharp, M. A. (1972). Weather and the "regulation" of subalpine butterfly populations. *Ecology* **53**:243–247.

Ehrlich, P. R., White, R. R., Singer, M. C., McKechnie, S. W., and Gilbert, L. E. (1975). Checkerspot butterflies: A historical perspective. *Science* **188**:221–228.

Ehrlich, P. R., Murphy, D. D., Singer, M. C., Sherwood, C. B., White, R. R., and Brown, I. L. (1980). Extinction, reduction, stability and increase: The responses of checkerspot butterfly *(Euphydryas)* populations to California drought. *Oecologia* **46**:101–105.

Einstein, A. (1905). Investigations on the theory of Brownian motion. *Ann. Phy. (Leipzig)* **17**:549.

Ellner, S. (1987). Competition and dormancy: A reanalysis and review. *Am. Nat.* **130**(5):798–803.

Ellner, S., and Hairston, N. G. (1994). Role of overlapping generations in maintaining genetic variation in a fluctuating environment. *Am. Nat.* **143**:403–417.

Ellner, S., and Shmida, A. (1981). Why are adaptations for long-range seed dispersal rare in desert plants? *Oecologia* **51**:133–144.

Elmqvist, T., and Gardfjell, H. (1988). Differences in response to defoliation between males and females of *Silene dioica*. *Oecologia* **77**:225–230.

Emmet, A. M., and Heath, J. (1990). "The butterflies of Great Britain and Ireland." Harley Books, Colchester.

Ennos, R. A. (1994). Estimating the relative rates of pollen and seed migration among plant populations. *Heredity* **72**:250–259.

Epperson, B. K. (1993). Recent advances in correlation studies of spatial patterns of genetic variation. *Evol. Biol.* **27**:95–155.

Ericson, L. (1981). Aspects of the shore vegetation of the Gulf of Bothnia. *Wahlenbergia* **7**:45–60.

Ericson, L., and Wallentinus, H.-G. (1979). Sea-shore vegetation around the Gulf of Bothnia. Guide for the International Society for Vegetation Science, July-August, 1977. *Wahlenbergia* **5**:1–142.

Ewald, P. W. (1994). "Evolution of Infectious Disease." Oxford University Press, Oxford.

Ewens, W. J. (1979). "Mathematical Population Genetics." Springer-Verlag, Berlin.

Ewens, W. J. (1989). The effective population size in the presence of catastrophes. *In* "Mathematical Evolutionary Theory" (M. W. Feldman, ed.), pp. 9–25. Princeton University Press, Princeton, NJ.

Ewens, W. J., Brockwell, P. J., Gani, J. M., and Resnick, S. I. (1987). Minimum viable population size in the presence of catastrophes. *In* "Viable Populations for Conservation" (M. E. Soulé, ed.), pp. 59–68. Cambridge University Press, Cambridge, UK.

Fahrig, L. (1990). Interacting effects of disturbance and dispersal on individual selection and population stability. *Comments Theor. Biol.* **1**:275–297.

Fahrig, L. (1992). Relative importance of spatial and temporal scales in a patchy environment. *Theor. Popul. Biol.* **41**:300–314.

Fahrig, L., and Freemark, K. (1993). Landscape-scale effects of toxic events for ecological risk assessment. *In* "Toxicity Testing at Different Levels of Biological Organization" (J. Cairns and B. R. Niederlehner, eds.), Lewis Publishers, Ann Arbor, MI.

Fahrig, L., and Merriam, G. (1985). Habitat patch connectivity and population survival. *Ecology* **66:** 1762–1768.

Fahrig, L., and Merriam, G. (1994). Conservation of fragmented populations. *Conserv. Biol.* **8:** 50–59.

Fahrig, L., and Paloheimo, J. (1988). Determinants of local population size in patchy habitats. *Theor. Popul. Biol.* **34:**194–212.

Fahrig, L., Coffin, D. P., Lauenroth, W. K., and Shugart, H. H. (1994). The advantage of long-distance clonal spreading in highly disturbed habitats. *Evol. Ecol.* **8:**172–187.

Falconer, D. S. (1981). "Introduction to Quantitative Genetics." Longman, New York.

Falconer, D. S. (1989). "Introduction to Quantitative Genetics," 3rd ed. Longman, Essex, UK.

Falk, D. A., and Holsinger, K. E. (1991). "Genetics and Conservation of Rare Plants." Oxford University Press, Oxford.

Feder, J. L., Chilcote, C. A., and Bush, G. L. (1990a). The geographic pattern of genetic differentiation between host-associated populations of *Rhagoletis pomonella* (Diptera: Tephritidae) in the Eastern United States and Canada. *Evolution (Lawrence, Kans.)* **44:**570–594.

Feder, J. L., Chilcote, C. A., and Bush, G. L. (1990b). Regional, local and microgeographic allele frequency variation between apple and hawthorn populations of *Rhagoletis pomonella* in Western Michigan. *Evolution (Lawrence, Kans.)* **44:**595–608.

Feldman, M. W., ed. (1989). "Mathematical Evolutionary Theory." Princeton University Press, Princeton, NJ.

Feldman, M. W., and Roughgarden, J. (1975). A population's stationary distribution and chance of extinction in a stochastic environment with remarks on the theory of species packing. *Theor. Popul. Biol.* **7:**197–207.

Feller, W. (1971). "An Introduction to Probability Theory and Its Applications." Vol. II. Wiley, New York.

Felsenstein, J. (1977). Multivariate normal genetic models with a finite number of loci. *In* "Quantitative Genetics" (E. Pollak, O. Kempthorne, and T. B. Bailey, eds.), pp. 227–246. Iowa State University Press, Ames.

Fenner, F., and Myers, K. (1978). Myxoma virus and myxomatosis in retrospect: The first quarter century of a new disease. *In* "Viruses and Environment" (E. Kurstak and K. Maramorosch, eds.), pp. 539–570. Academic Press, New York.

Fiedler, P. L., and Jain, S. K. (1992). "Conservation Biology: Theory and Practice of Nature Conservation, Preservation and Management." Chapman & Hall, London.

Fisher, R. A. (1922). On the dominance ratio. *Proc. R. Soc. Edinburgh* **42:**321–431.

Fisher, R. A. (1930). "The Genetical Theory of Natural Selection." Oxford University Press, Oxford.

Fisher, R. A. (1937). The wave of advance of advantageous genes. *Ann. Eugen.* **7:**355–369.

Fisher, R. A. (1958). "The Genetical Theory of Natural Selection." Dover, New York.

Fisher, R. A., and Ford, E. B. (1947). The spread of a gene in natural conditions in colony of the moth *Panaxia dominula* L. *Heredity* **1:**143–174.

Fitzgerald, B. M. (1977). Weasel predation on a cyclic population of the montane vole *(Microtus montanus)* in California. *J. Anim. Ecol.* **46:**367–397.

Flor, H. H. (1956). The complementary genic systems in flax and flax rust. *Adv. Genet.* **8:**29–54.

Flor, H. H. (1971). Current status of the gene-for-gene concept. *Annu. Rev. Phytopathol.* **9:**275–296.

Foley, P. (1987). Molecular clock rates at loci under stabilizing selection. *Proc. Natl. Acad. Sci. U.S.A.* **84:**7996–8000.

Foley, P. (1992). Small population genetic variation at loci under stabilizing selection. *Evolution (Lawrence, Kans.)* **46:**763–774.

Foley, P. (1994). Predicting extinction times from environmental stochasticity and carrying capacity. *Conserv. Biol.* **8:**124–137.

Foley, P., Taylor, R. J., and Johnson, B. (1993). "California Owl Population Simulator" (unpublished Smalltalk program).

Foley, J. E., Foley, P., Torres, S. (1996). Have mountain lions reached carrying capacity in California? unpublished manuscript.

Foltz, D. W., and Hoogland, J. L. (1983). Genetic evidence of outbreeding in the black-tailed prairie dog *(Cynomys ludovicianus)*. *Evolution (Lawrence, Kans.)* **37**:273–281.

Ford, E. B. (1945). "Butterflies." Collins, London.

Ford, H. D., and Ford, E. B. (1930). Fluctuations in numbers and its influence on variation in *Melitaea aurinia*. *Trans. Entomol. Soc. London* **78**:345–351.

Forman, R. T. T., and Godron, M. (1986). "Landscape Ecology." Wiley, New York.

Forney, K. A., and Gilpin, M. E. (1989). Spatial structure and population extinction: A study with *Drosophila* flies. *Conserv. Biol.* **3**:45–51.

Forsman, E. D., Meslow, E. C., and Wight, H. M. (1984). Distribution and biology of the spotted owl in Oregon. *Wildl. Monogr.* **87**:1–64.

Frank, S. A. (1986). Dispersal polymorphisms in subdivided populations. *J. Theor. Biol.* **122**:303–309.

Frank, S. A. (1987). Individual and population sex allocation patterns. *Theor. Popul. Biol.* **31**:47–74.

Frank, S. A. (1989). The evolutionary dynamics of cytoplasmic male sterility. *Am. Nat.* **133**:345–376.

Frank, S. A. (1991a). Ecological and genetic models of host-pathogen coevolution. *Heredity* **67**:73–83.

Frank, S. A. (1991b). Spatial variation in coevolutionary dynamics. *Evol. Ecol.* **5**:193:–217.

Frank, S. A. (1992). Models of plant-pathogen coevolution. *Trends Genet.* **8**:213–219.

Frank, S. A. (1993a). Specificity versus detectable polymorphism in host-parasite genetics. *Proc. R. Soc. London, Ser. B* **254**:191–197.

Frank, S. A. (1993b). Coevolutionary genetics of plants and pathogens. *Evol. Ecol.* **7**:45–75.

Frank, S. A. (1994a). Recognition and polymorphism in host-parasite genetics. *Philos. Trans. R. Soc. London, Ser. B* **346**:283–293.

Frank, S. A. (1994b). Kin selection and virulence in the evolution of protocells and parasites. *Proc. R. Soc. London, Ser. B* **258**:153–161.

Frank, S. A. (1996). Statistical properties of polymorphism in host-parasite genetics. *Evol. Ecol.* **10**:307–317.

Franklin, I. R. (1980). Evolutionary change in small populations. *In* "Conservation Biology: An Evolutionary-Ecological Perspective" (M. E. Soulé and B. A. Wilcox, eds.), pp. 135–149. Sinauer Assoc., Sunderland, MA.

Fritz, A.-L., and Nilsson, L. A. (1994). How pollinator-mediated mating varies with population size in plants. *Oecologia* **100**:451–462.

Fritz, R. S. (1979). Consequences of insular population structure: Distribution and extinction of spruce grouse populations. *Oecologia* **42**:57–65.

Gabriel, D. W., and Rolfe, B. G. (1990). Working models of specific recognition in plant-microbe interactions. *Annu. Rev. Phytopathol.* **28**:365–391.

Gadgil, M. (1971). Dispersal: Population consequences and evolution. *Ecology* **52**:253–261.

Ganders, F. R. (1978). The genetics and evolution of gynodioecy in *Nemophila menziesii* (Hydrophyllaceae). *Can. J. Bot.* **56**:1400–1408.

Garcia-Ramos, G., and Kirkpatrick, M. (1996). *Am. Nat.* Genetic models of rapid evolution and divergence in populations (submitted).

Gardiner, C. W. (1985). "Handbook of Stochastic Methods." Springer-Verlag, Berlin.

Gardner, R. H., Milne, B. T., Turner, M. G., and O'Neill, R. V. (1987). Neutral models for the analysis of broad-scale landscape pattern. *Landscape Ecol.* **1**:19–28.

Gardner, R. H., O'Neill, R. V., Turner, M. G., and Dale, V. H. (1989). Quantifying scale-dependent effects of animal movement with simple percolation models. *Landscape Ecol.* **3**:217–227.

Garrett, M. G., and Franklin, W. L. (1988). Behavioral ecology of dispersal in the black-tailed prairie dog. *J. Mammal.* **69**:236–250.

Gaston, K. J. (1994). "Rarity." Chapman & Hall, London.

Gaston, K. J., and Lawton, J. H. (1987). A test of statistical techniques for detecting density dependence in sequential censuses of animal populations. *Oecologia* **74**:404–410.

Gaston, M. A., Strickland, M. A., Ayling-Smith, B. A., and Pitt, T. L. (1989). Epidemiological typing of *Enterobacter aerogenes. J. Clin. Microbiol.* **27**:564–565.

Gates, J. E., and Gysel, L. W. (1978). Avian dispersion and fledging success in field-forest ecotones. *Ecology* **59**:871–883.

Gause, G. F. (1935). Experimental demonstration of Volterra's periodic oscillation in the numbers of animals. *J. Exp. Biol.* **12**:44–48.

Gibbs, J. P. (1993). Importance of small wetlands for the persistence of local populations of wetland-associated animals. *Wetlands* **13**:25–31.

Gilbert, F. S. (1980). The equilibrium theory of island biogeography: Fact or fiction? *J. Biogeogr.* **7**: 209–235.

Giles, B. E., and Goudet, J. (1996a). Genetic differentiation in *Silene dioica* metapopulations: Estimation of spatio-temporal effects in a successional plant species. *Am. Nat.* (in press).

Giles, B. E., and Goudet, J. (1996b). Metapopulations within metapopulations—The genetic consequences for *Silene dioica*. In preparation.

Gill, D. E. (1978a). The metapopulation ecology of the red-spotted newt, *Notophthalamus viridescens* (Rafinesque). *Ecol. Monogr.* **48**:145–166.

Gill, D. E. (1978b). Effective population size and interdemic migration rates in a metapopulation of the red-spotted newt, *Notophthalamus viridescens* (Rafinesque). *Evolution (Lawrence, Kans.)* **32**:839–849.

Gillespie, J. H. (1991). "The Causes of Molecular Evolution." Oxford University Press, Oxford.

Gillespie, J. H., and Guess, H. A. (1978). The effects of environmental autocorrelation on the progress of selection in a random environment. *Am. Nat.* **134**:638–658.

Gilpin, M. (1975). "Group Selection in Predator-prey Communities." Princeton University Press, Princeton, NJ.

Gilpin, M. E. (1987). Spatial structure and population vulnerability. *In* "Viable Populations for Conservation" (M. Soulé, ed.), pp. 125–139. Cambridge University Press, Cambridge, UK.

Gilpin, M. E. (1988). A comment on Quinn and Hastings: Extinction in subdivided habitats. *Conserv. Biol.* **2**:290–292.

Gilpin, M. E. (1990). Extinction of finite metapopulations in correlated environments. *In* "Living in a Patchy Environment" (B. Shorrocks and I. R. Swingland, eds.), pp. 177–186. Oxford University Press, Oxford.

Gilpin, M. E. (1991). The genetic effective size of a metapopulation. *In* "Metapopulation Dynamics: Empirical and Theoretical Investigations" (M. E. Gilpin and I. Hanski, eds.), pp. 165–175. Academic Press, London.

Gilpin, M. E., and Diamond, J. M. (1976). Calculation of immigration and extinction curves from the species-area-distance relation. *Proc. Natl. Acad. Sci. U.S.A.* **73**:4130–4134.

Gilpin, M. E., and Diamond, J. M. (1981). Immigration and extinction probabilities for individual species: Relation to incidence functions and species colonization curves. *Proc. Natl. Acad. Sci. U.S.A.* **78**:392–396.

Gilpin, M., and Hanski, I. eds. (1991). "Metapopulation Dynamics: Empirical and Theoretical Investigations." Academic Press, London.

Gilpin, M. E., and Soulé, M. E. (1987). Minimum viable populations: processes of species extinction. *In* "Conservation Biology: The Science of Scarcity and Diversity." (M. E. Soulé, ed.), pp. 19–34. Sinauer Assoc., Sunderland, MA.

Glasser, J. (1982). On the causes of temporal change in communities: Modification of the biotic environment. *Am. Nat.* **119**:375–390.

Gliddon, C. J., and Gouyong, P.-H. (1989). The units of selection. *Trends Ecol. Evol.* **4**:204–208.

Godfray, H. C. J. (1994). "Parasitoids: Behavioral and Evolutionary Ecology." Princeton University Press, Princeton, NJ.

Goldstein, D. B., and Holsinger, K. E. (1992). Maintenance of polygenic variation in spatially structured populations: Roles for local mating and genetic redundancy. *Evolution (Lawrence, Kans.)* **46:**412–429.

Goldwasser, L., Cook, J., and Silverman, E. D. (1994). The effects of variability on metapopulation dynamics and rates of invasion. *Ecology* **75:**40–47.

Gomulkiewicz, R., and Holt, R. D. (1995). When does evolution by natural selection prevent extinction? *Evolution (Lawrence, Kans.)* **49:**201–207.

Gonzales-Andujar, J. L., and Perry, J. N. (1995). Reversals of chaos in biological control systems. *J. Theor. Biol.* **175:**603.

Goodman, D. (1987a). The demography of chance extinction. *In* "Viable Populations for Conservation" (M. E. Soulé, ed.), pp. 11–34. Cambridge University Press, Cambridge, UK.

Goodman, D. (1987b). Consideration of stochastic demography in the design and management of biological reserves. *Nat. Resour. Model.* **1:**205–234.

Goodnight, C. J., Schwartz, J. M., and Stevens, L. (1992). Contextual analysis of models of group selection, soft selection, hard selection, and the evolution of altruism. *Am. Nat.* **140:**743–761.

Gosz, J. R. (1993). Ecotone hierarchies. *Ecol. Appl.* **3:**369–376.

Gotelli, N. J. (1991). Metapopulation models: The rescue effect, the propagule rain, and the core-satellite hypothesis. *Am. Nat.* **138:**768–776.

Gottfried, B. M. (1979). Small mammal populations in woodlot islands. *Amer. Midl. Nat.* **102:**105–112.

Goudet, J. (1993). The genetics of geographically structured populations. Ph.D. Thesis, University of Wales, Bangor.

Goudet, J. (1995). FSTAT V1.2. A computer program to calculate F-statistics. *J. Hered.* **86:**485–486.

Gould, F. (1983). Genetics of plant-herbivore systems: interactions between applied and basic study. *In* "Variable Plants and Herbivores in Natural and Managed Systems" (R. F. Denno and M. S. McClure, eds.), pp. 599–653. Academic Press, New York.

Gouyon, P.-H., and Couvet, D. (1985). Selfish cytoplasm and adaptation: Variations in the reproductive system of thyme. *In* "Structure and Functioning of Plant Populations" (J. Haeck and J. W. Woldendorp, eds.), Vol. 2, pp. 299–319. North-Holland Publ., Amsterdam.

Gouyon, P.-H., and Couvet, D. (1987). A conflict between two sexes, females and hermaphrodites. *In* "The Evolution of Sex and its Consequences" (S. C. Stearns, ed.), pp. 245–261. Birkhäuser, Basel and Boston.

Gouyon, P.-H., and Gliddon, C. (1988). The genetics of information and the evolution of avatars. *In* "Population Genetics and Evolution" (G. de Jong, ed.), pp. 120–123. Springer-Verlag, Berlin.

Gouyon, P.-H., Lumaret, R., Valdeyron, G., and Vernet, P. (1983). Reproductive strategies and disturbance by man. *In* "Ecosystems and Disturbance" (H. Mooney, ed.), pp. 214–225. Springer-Verlag, Berlin.

Gouyon, P.-H., Vichot, F., and Van Damme, J. M. M. (1991). Nuclear-cytoplasmic male sterility: Single-point equilibria versus limit cycles. *Am. Nat.* **137:**498–514.

Grachev, M. A., Kumarev, V. P., Mammaev, L. V., Zorin, V. L., Baranova, L. V., Denikina, N. N., Belikov, S. I., Petrov, E. A., Kolesnik, V. S., Kolesnik, R. S., Dorofeev, V. M., Beim, A. M., Kudelin, V. N., Nagieva, F. G., and Sidorov, V. N. (1989). Distemper virus in Baikal seals. *Nature (London)* **338:**209.

Gradshteyn, I. S., and Ryzhik, I. M. (1980). "Table of Integrals, Series and Products." Academic Press, New York.

Grant, B. R., and Grant, P. R. (1979). "Evolutionary Dynamics of a Natural Population: The Large Cactus Finch of the Galapagos." University of Chicago Press, Chicago.

Green, D. G. (1994). Connectivity and complexity in landscapes and ecosystems. *Pac. Conserv. Biol.* **1:**194–200.

Greene, D. F., and Johnson, E. A. (1989a). A model of wind dispersal of winged or plumed seeds. *Ecology* **70:**339–347.

Greene, D. F., and Johnson, E. A. (1989b). Particulate diffusion models and the dispersal of seeds by the wind. *Trends Ecol. Evol.* **4**:191–193.

Gross, K. L., and Werner, P. A. (1978). The biology of Canadian weeds. 28. *Verbascum thapsus* L. and *V. blattaria* L. *Can. J. Plant Sci.* **56**:401–413.

Grubb, P. J. (1977). The maintenance of species richness in plant communities: The importance of the regeneration niche. *Biol. Rev. Cambridge Philos. Soc.* **52**:107–145.

Gurney, W. S. C., and Nisbet, R. M. (1978). Single species population fluctuations in patchy environments. *Am. Nat.* **112**:1075–1090.

Gutierrez, R. J., and Harrison, S. (1996). Applications of metapopulation theory to spotted owl management: A history and critique. *In* "Metapopulations and Wildlife Conservation Management" (D. McCullough, ed.). Island Press, Covelo, CA (in press).

Gyllenberg, M., and Hanski, I. (1992). Single-species metapopulation dynamics: A structured model. *Theor. Popul. Biol.* **42**:35–61.

Gyllenberg, M., and Silvestrov, D. S. (1994). Quasi-stationary distributions of a stochastic metapopulation model. *J. Math. Biol.* **33**:35–70.

Gyllenberg, M., Söderbacka, G., and Ericsson, S. (1993). Does migration stabilize local population dynamics? Analysis of a discrete metapopulation model. *Math. Biosci* **118**:25–49.

Gyllenberg, M., Osipov, A. V., and Söderbacka, G. (1996). Bifurcation analysis of a metapopulation model with sources and sinks. *J. Nonlinear Science* **6**:1–38.

Hagen, J. B. (1989). Research perspectives and the anomalous state of modern ecology. *Biol. Philos.* **4**:433–455.

Haila, Y. (1988). The multiple faces of ecological theory and data. *Oikos* **53**:408–411.

Haila, Y. (1991). Implications of landscape heterogeneity for bird conservation. *Proc. Int. Ornithol. Congr., 20th, 1990*, pp. 2286–2291.

Haila, Y., and Hanski, I. K. (1993). Breeding birds on small British islands and extinction risks. *Am. Nat.* **142**:1025–1029.

Haila, Y., and Järvinen, O. (1982). The role of theoretical concepts in understanding the ecological theatre: A case study on island biogeography. *In* "Conceptual Issues in Ecology" (E. Saarinen, ed.), pp. 261–278. Reidel, Dordrecht, The Netherlands.

Haila, Y., Saunders, D. A., and Hobbs, R. J. (1993). What do we presently understand about ecosystem fragmentation? *In* "Nature Conservation 3: Reconstruction of Fragmented Ecosystems" (D. A. Saunders, R. J. Hobbs, and P. R. Ehrlich, eds.), pp. 45–55. Surrey Beatty & Sons, Chipping Norton, NSW, Australia.

Haining, R. P. (1990). "Spatial Data Analysis in the Social and Environmental Sciences." Cambridge University Press, Cambridge, UK.

Hairston, N. G. (1993). Diapause dynamics of 2 Diatomid copepod species in a large lake. *Hydrobiologia* **293**:209–218.

Haldane, J. B. S. (1927). A mathematical theory of natural and artificial selection. V. Selection and mutation. *Proc. Cambridge Philos. Soc.* **26**:220–230.

Haldane, J. B. S. (1931). A mathematical theory of natural selection. VI. Isolation. *Trans. Cambridge Philos. Soc.* **26**:220–230.

Halley, J. M., Comins, H. N., Lawton, J. H., and Hassell, M. P. (1994). Competition, succession and pattern in fungal communities: Towards a cellular automation model. *Oikos* **70**:435–442.

Hamilton, W. D., and May, R. M. (1977). Dispersal in stable habitats. *Nature (London)* **269**:578–581.

Handel, S. N. (1983). Pollination ecology, plant population structure, and gene flow. *In* "Pollination Biology" (L. Real, ed.), pp. 163–211. Academic Press, London.

Hansen, A. J., and di Castri, F., eds. (1992). "Landscape Boundaries: Consequences for Biotic Diversity and Ecological Flows." Springer-Verlag, New York.

Hansen, A. J., Garman, S. L., Marks, B., and Urban, D. L. (1993). An approach for managing vertebrate diversity across multiple-use landscapes. *Ecol. Appl.* **3**:481–496.

Hanski, I. (1981). Coexistence of competitors in patchy environment with and without predation. *Oikos* **37**:306–312.

Hanski, I. (1983). Coexistence of competitors in patchy environment. *Ecology* **64**:493–500.

Hanski, I. (1985). Single-species spatial dynamics may contribute to long-term rarity and commonness. *Ecology* **66**:335–343.

Hanski, I. (1987). Carrion fly community dynamics: Patchiness, seasonality and coexistence. *Ecol. Entomol.* **12**:257–266.

Hanski, I. (1989). Metapopulation dynamics: Does it help to have more of the same? *Trends Ecol. Evol.* **4**:113–114.

Hanski, I. (1990). Density dependence, regulation and variability on animal populations. *Philos. Trans. R. Soc. London, Ser. B* **330**:141–150.

Hanski, I. (1991). Single-species metapopulation dynamics. *In* "Metapopulation Dynamics: Empirical and Theoretical Investigations" (M. Gilpin and I. Hanski, eds.), pp. 17–38. Academic Press, London.

Hanski, I. (1992a). Inferences from ecological incidence functions. *Am. Nat.* **139**:657–662.

Hanski, I. (1992b). Insectivorous mammals. *In* "Natural Enemies" (M. Crawley, ed.), pp. 163–187. Blackwell, Oxford.

Hanski, I. (1993). Dynamics of small mammals on islands—a comment to Lomolino (1993). *Ecography* **16**:372–375.

Hanski, I. (1994a). A practical model of metapopulation dynamics. *J. Anim. Ecol.* **63**:151–162.

Hanski, I. (1994b). Patch-occupancy dynamics in fragmented landscapes. *Trends Ecol. Evol.* **9**:131–135.

Hanski, I. (1994c). Spatial scale, patchiness and population dynamics on land. *Philos. Trans. R. Soc. London, Ser. B* **343**:19–25.

Hanski, I. (1995). Effects of landscape pattern on competitive interactions. *In* "Mosaic Landscapes and Ecological Processes" (L. Hansson, L. Fahrig, and G. Merriam, eds.), pp. 203–224. Chapman & Hall, London.

Hanski, I (1996a). Metapopulation ecology. *In* "Population Dynamics in Ecological Space and Time" (O. E. Rhodes, Jr., R. K. Chesser, and M. H. Smith, eds.), pp. 13–43. University of Chicago Press, Chicago.

Hanski, I. (1996b). Habitat destruction and metapopulation dynamics. *In* "Enhancing the Ecological Basis of Conservation: Heterogeneity, Ecosystem Function and Biodiversity" (S. T. A. Pickett, R. S. Ostfeld, M. Shachak, and G. E. Likens, eds.), Chapman & Hall, New York (in press).

Hanski, I., and Gilpin, M. (1991). Metapopulation dynamics: Brief history and conceptual domain. *In* "Metapopulation Dynamics: Empirical and Theoretical Investigations" (M. Gilpin and I. Hanski, eds.), pp. 3–16. Academic Press, London.

Hanski, I., and Gyllenberg, M. (1993). Two general metapopulation models and the core-satellite species hypothesis. *Am. Nat.* **142**:17–41.

Hanski, I., and Hammond, P. (1995). Biodiversity in boreal forests. *Trends Ecol. Evol.* **10**:5–6.

Hanski, I., and Kuussaari, M. (1995). Butterfly metapopulation dynamics. *In* "Population Dynamics: New Approaches and Synthesis" (N. Cappuccino and P. W. Price, eds.), pp. 149–172. Academic Press, San Diego, CA.

Hanski, I., and Ranta, E. (1983). Coexistence in a patchy environment: Three species of *Daphnia* in rock pools. *J. Anim. Ecol.* **52**:263–279.

Hanski, I., and Thomas, C. D. (1994). Metapopulation dynamics and conservation: A spatially explicit model applied to butterflies. *Biol. Conserv.* **68**:167–180.

Hanski, I., and Woiwod, I. P. (1993). Spatial synchrony in the dynamics of moth and aphid populations. *J. Anim. Ecol.* **62**:656–668.

Hanski, I., and Zhang, D.-Y. (1993). Migration, metapopulation dynamics and fugitive coexistence. *J. Theor. Biol.* **163**:491–504.

Hanski, I., Kuussaari, M., and Nieminen, M. (1994). Metapopulation structure and migration in the butterfly *Melitaea cinxia*. *Ecology* **75**:747–762.

Hanski, I., Pakkala, T., Kuussaari, M., and Lei, G. (1995a). Metapopulation persistence of an endangered butterfly in a fragmented landscape. *Oikos* **72**:21–28.

Hanski, I., Pöyry, J., Pakkala, T., and Kuussaari, M. (1995b). Multiple equilibria in metapopulation dynamics. *Nature (London)* **377**:618–621.

Hanski, I., Foley, P., and Hassell, M. P. (1996a). Random walks in a metapopulation: How much density dependence is necessary for long-term persistence? *J. Anim. Ecol.* **65**:274–282.

Hanski, I., Moilanen, A., and Gyllenberg, M. (1996b). Minimum viable metapopulation size. *Am. Nat.* **147**:527–541.

Hanski, I., Moilanen, A., Pakkala, T., and Kuussaari, M. (1996c). The quantitative incidence function model and persistence of an endangered butterfly metapopulation. *Conserv. Biol.* **10**:578–590.

Hanson, F. B., and Tuckwell, H. C. (1978). Persistence times of populations with large random fluctuations. *Theor. Popul. Biol.* **14**:46–61.

Hanson, F. B., and Tuckwell, H. C. (1981). Logistic growth with random density-independent disasters. *Theor. Popul. Biol.* **19**:1–18.

Hanson, M. R. (1991). Plant mitochondrial mutations and male sterility. *Annu. Rev. Genet.* **25**:461–486.

Hanson, M. R., and Conde, M. F. (1985). Functioning and variation of cytoplasmic genomes: Lessons from cytoplasmic-nuclear interaction affecting male fertility in plants. *Int. Rev. Cytol.* **84**:213–267.

Hansson, L. (1991). Dispersal and connectivity in metapopulations. *In* "Metapopulation Dynamics: Empirical and Theoretical Investigations" (M. Gilpin, and I. Hanski, eds.), pp. 89–103, Academic Press, London.

Hardy, K. G. (1975). Colicinogeny and related phenomena. *Bacteriol. Rev.* **39**:464–515.

Harper, J. L. (1977). "Population Biology of Plants." Academic Press, London.

Harper, J. L., and Wood, W. A. (1957). Biological flora of the British Isles: *Senecio jacobaea* L. *J. Ecol.* **45**:617–737.

Harper, S. J., Bollinger, E. K., and Barrett, G. W. (1993). Effects of habitat patch shape on population dynamics of meadow voles *(Microtus pennsylvanicus)*. *J. Mammal.* **74**:1045–1055.

Harris, P., Thompson, L. S., Wilkinson, A. T. S., and Neary, M. E. (1978). Reproductive biology of tansy ragwort; climate and biological control by the cinnabar moth in Canada. *In* "Proceedings of the Fourth International Symposium on Biological Control of Weeds" (T. E. Freeman, ed.), pp. 163–173. University of Florida, Gainesville.

Harrison, R. G. (1980). Dispersal polymorphism in insects. *Annu. Rev. Ecol. Syst.* **11**:95–118.

Harrison, R. G., ed. (1993). "Hybrid Zones and the Evolutionary Process." Oxford University Press, Oxford.

Harrison, S. (1989). Long-distance dispersal and colonization in the Bay checkerspot butterfly. *Ecology* **70**:1236–1243.

Harrison, S. (1991). Local extinction in a metapopulation context: An empirical evaluation. *In* "Metapopulation Dynamics: Empirical and Theoretical Investigations" (M. E. Gilpin and I. Hanski, eds.), pp. 73–88. Academic Press, London.

Harrison, S. (1994a). Resources and dispersal as factors limiting a population of the tussock moth *(Orgyia vetusta)*, a flightless defoliator. *Oecologia* **99**:27–34.

Harrison, S. (1994b). Metapopulations and conservation. *In* "Large-Scale Ecology and Conservation Biology" (P. J. Edwards, N. R. Webb, and R. M. May, eds.), pp. 111–128. Blackwell, Oxford.

Harrison, S., and Hastings, A. M. (1996). Genetic and evolutionary consequences of metapopulation structure. *Trends Ecol. Evol.* (in press).

Harrison, S., and Quinn, J. F. (1989). Correlated environments and the persistence of metapopulations. *Oikos* **56**:293–298.

Harrison, S., and Quinn, J. F. (1990). Correlated environments and the persistence of metapopulations. *Oikos* **56**:293–298.

Harrison, S., Murphy, D. D., and Ehrlich, P. R. (1988). Distribution of the bay checkerspot butterfly, *Euphydryas editha bayensis:* Evidence for a metapopulation model. *Am. Nat.* **132**:360–382.

Harrison, S., Quinn, J. F., Baughman, J. F., Murphy, D. D., and Ehrlich, P. R. (1991). Estimating the effects of scientific study on two butterfly populations. *Am. Nat.* **137:**227–243.

Harrison, S., Stahl, A., and Doak, D. F. (1993). Spatial models and spotted owls: Exploring some biological issues behind recent events. *Conserv. Biol.* **7:**950–953.

Harrison, S., Thomas, C. D., and Lewinsohn, T. M. (1995). Testing a metapopulation model of coexistence in the insect community on ragwort *(Senecio jacobaea)*. *Am. Nat.* **145:**545–561.

Hartl, D. L., and Clark, A. G. (1989). "Principles of Population Genetics." Sinauer Assoc., Sunderland, MA.

Hassell, M. P. (1975). Density dependence in single-species populations. *J. Anim. Ecol.* **44:** 283–295.

Hassell, M. P. (1978). "The Dynamics of Arthropod Predator-Prey Systems." Princeton University Press, Princeton, NJ.

Hassell, M. P. (1979). The dynamics of predator-prey interactions: Polyphagous predators, competing predators and hyperparasitoids. *In* "Population Dynamics" (R. M. Anderson, B. D. Turner, and L. R. Taylor, eds.), pp. 283–306. Blackwell, Oxford.

Hassell, M. P. (1980). Foraging strategies, population models and biological control: A case study. *J. Anim. Ecol.* **49:**603–628.

Hassell, M. P., and May, R. M. (1973). Stability in insect host-parasite models. *J. Anim. Ecol.* **42:** 693–726.

Hassell, M. P., and May, R. M. (1985). From individual behaviour to population dynamics. *In* "Behavioural Ecology" (R. M. Sibly and R. Smith, eds.), pp. 3–32. Blackwell, Oxford.

Hassell, M. P., and May, R. M., ed. (1990). "Population Regulation and Dynamics." Royal Society, London.

Hassell, M. P., Lawton, J. N., and May, R. M. (1976). Patterns of dynamical behavior in single-species populations. *J. Anim. Ecol.* **45:**471–486.

Hassell, M. P., Waage, J. K., and May, R. M. (1983). Variable parasitoid sex ratios and their effect on host parasitoid dynamics. *J. Anim. Ecol.* **52:**889–904.

Hassell, M. P., Latto, J., and May, R. M. (1989). Seeing the wood for the trees: Detecting density dependence from existing life-table studies. *J. Anim. Ecol.* **58:**883–892.

Hassell, M. P., Comins, H. N., and May, R. M. (1991a). Spatial structure and chaos in insect population dynamics. *Nature (London)* **353:**255–258.

Hassell, M. P., Pacala, S., May, R. M., and Chesson, P. L. (1991b). The persistence of host-parasitoid associations in patchy environments. I. A general criterion. *Am. Nat.* **138:**568–583.

Hassell, M. P., Godfray, H. C. J., and Comins, H. N. (1993). Effects of global change on the dynamics of insect host-parasitoid interactions. *In* "Biotic Interactions and Global Change" (Anonymous), pp. 402–423. Sinauer Assoc., Sunderland, MA.

Hassell, M. P., Comins, H. N., and May, R. M. (1994). Species coexistence and self-organizing spatial dynamics. *Nature (London)* **370:**290–292.

Hassell, M. P., Miramontes, O, Rohani, P., and May, R. M. (1995). Appropriate formulations for dispersal in spatially structured models: Comments on Bascompte and Solé. *J. Anim. Ecol.* **64:** 662–664.

Hastings, A. (1977). Spatial heterogeneity and the stability of predator-prey systems. *Theor. Popul. Biol.* **12:**37–48.

Hastings, A. (1978). Spatial heterogeneity and the stability of predator-prey systems: Predator mediated coexistence. *Theor. Popul. Biol.* **14:**380–395.

Hastings, A. (1980). Disturbance, coexistence, history and the competition for space. *Theor. Popul. Biol.* **18:**363–373.

Hastings, A. (1990). Spatial heterogeneity and ecological models. *Ecology* **71:**426–428.

Hastings, A. (1991). Structured models of metapopulation dynamics. *In* "Metapopulation Dynamics: Empirical and Theoretical Investigations" (M Gilpin and I. Hanski, eds.), pp. 57–71. Academic Press, London.

Hastings, A. (1992). Age dependent dispersal is not a simple process: Density dependence, stability, and chaos. *Theor. Popul. Biol.* **41:**388–400.

Hastings, A. (1993). Complex interactions between dispersal and dynamics: Lessons from coupled logistic equations. *Ecology* **74**:1362–1372.

Hastings, A., and Harrison, S. (1994). Metapopulation dynamics and genetics. *Annu. Rev. Ecol. Syst.* **25**:167–188.

Hastings, A., and Higgins, K. (1994). Persistence of transients in spatially structured models. *Science* **263**:1133–1136.

Hastings, A., and Wolin, C. L. (1989). Within-patch dynamics in a metapopulation. *Ecology* **70**: 1261–1266.

Hauffe, H. C., and Searle, J. B. (1993). Extreme karyotypic variation in a *Mus musculus* domesticus hybrid zone: The tobacco mouse story revisited. *Evolution (Lawrence, Kans.)* **47**:1374–1395.

Heath, J., Pollard, E., and Thomas, J. A. (1984). "Atlas of Butterflies in Britain and Ireland." Viking, Harmondsworth.

Hedrick, P. W. (1981). The establishment of chromosomal variants. *Evolution (Lawrence, Kans.)* **35**: 322–332.

Hedrick, P. W. (1985). "Genetics of Populations." Jones & Bartlett, Boston.

Hedrick, P. W. (1996). Bottleneck(s) or metapopulation in cheetahs? *Conserv. Biol.* (in press).

Heesterbeek, J. A. P. (1992). R_0. Ph.D. Thesis, Rijksuniversiteit, Leiden.

Heisler, I. L., and Damuth, J. (1987). A method for analysing selection in hierarchically structured population. *Am. Nat.* **130**:582–602.

Henderson, M. T., Merriam, G., and Wegner, J. (1985). Patchy environments and species survival: Chipmunks in an agricultural mosaic. *Biol. Conserv.* **31**:95–105.

Hertzberg, K. (1996). Migration and establishment of collembollas in a patchy environment. (Submitted for publication).

Hestbeck, J. B. (1982). Population regulation of cyclic mammals: The social fence hypothesis. *Oikos* **39**:157–163.

Hestbeck, J. B., Nichols, J. D., and Malecki, R. A. (1991). Estimates of movement and site fidelity using mark-resight data of wintering Canada geese. *Ecology* **72**:523–533.

Heuch, I. (1978). Maintenance of butterfly populations with all-female broods under recurrent extinction and recolonization. *J. Theor. Biol.* **75**:115–122.

Heywood, V. H., Mace, G. M., May, R. M., and Stuart, S. N. (1994). Uncertainties in extinction rates. *Nature (London)* **368**:105.

Hill, J. K., Thomas, C. D., and Lewis, O. T. (1996). Effects of habitat patch size and isolation on dispersal by *Hesperia comma* butterflies: Implications for metapopulation structure. *J. Anim. Ecol.* (in press).

Hobbs, R. J. (1992). The role of corridors in conservation: Solution or bandwagon? *Trends Ecol. Evol.* **7**:389–392.

Hobbs, R. J. (1994). Landscape ecology and conservation: Moving from description to application. *Pac. Conserv. Biol.* **1**:170–176.

Hobbs, R. J. (1995). Landscape ecology. *Encycl. Environ. Biol.* **2**:417–428.

Hochberg, M. E., and Holt, R. D. (1995). Refuge evolution and the population dynamics of coupled host-parasitoid associations. *Evol. Ecol.* **9**:1–29.

Hochberg, M. E., Elmes, G. W., and Thomas, J. A. (1992). The population dynamics of a large blue butterfly, *Maculinea rebeli,* a parasite of red ant nests. *J. Anim. Ecol.* **61**:397–409.

Hochberg, M. E., Clarke, R. T., Elmes, G. W., and Thomas, J. A. (1994). Population dynamic consequences of direct and indirect interactions involving a large blue butterfly and its plant and red ant hosts. *J. Anim. Ecol.* **63**:375–391.

Hoffman, A. A., and Parsons, P. A. (1991). "Evolutionary Genetics and Environmental Stress." Oxford University Press, Oxford.

Holdren, C. E., and Ehrlich, P. R. (1981). Long range dispersal in checkerspot butterflies: Transplant experiments with *E. gillettii. Oecologia* **50**:125–129.

Holland, M. M., Risser, P. G., and Naiman, R. J., eds. (1991). "Ecotones. The Role of Landscape Boundaries in the Management and Restoration of Changing Environments." Chapman & Hall, New York.

Holsinger, K. E. (1986). Dispersal and plant mating systems: The evolution of self-fertilization in subdivided populations. *Evolution (Lawrence, Kans.)* **40:**405–413.

Holt, R. D. (1977). Predation, apparent competition and the structure of prey communities. *Theor. Popul. Biol.* **12:**197–229.

Holt, R. D. (1984). Spatial heterogeneity, indirect interactions, and the coexistence of prey species. *Am. Nat.* **124:**377–406.

Holt, R. D. (1985). Population dynamics in two-patch environments: Some anomalous consequences of an optimal habitat distribution. *Theor. Popul. Biol.* **28:**181–208.

Holt, R. D. (1992). A neglected facet of island biogeography: The role of internal spatial dynamics in area effects. *Theor. Popul. Biol.* **41:**354–371.

Holt, R. D. (1993). Ecology at the mesoscale: The influence of regional processes on local communities. *In* "Species Diversity in Ecological Communities" (R. E. Ricklefs and D. Schluter, eds.), pp. 77–88. University of Chicago Press, Chicago.

Holt, R. D. (1995). Food webs in space: an island biogeographic perspective. *In* "Food Webs: Integration of Patterns and Dynamics" (G. A. Polis and K. O. Winemiller, eds.), pp. 313–323. Chapman and Hall, London, UK.

Holt, R. D., and Gaines, M. S. (1992). Analysis of adaptation in heterogeneous landscapes: implications for the evolution of fundamental niches. *Evol. Ecol.* **6:**433–447.

Holt, R. D., and Lawton, J. H. (1993). Apparent competition and enemy-free space in insect host-parasitoid communities. *Am. Nat.* **142:**623–645.

Holt, R. D., and Lawton, J. H. (1994). The ecological consequences of shared natural enemies. *Ann. Rev. Ecol. Syst.* **25:**495–521.

Holyoak, M. and Lawler, S. P. (1996). Persistence of an extinction-prone predator–prey interaction through metapopulation dynamics. *Ecology* **77:**(in press).

Hoogland, J. L. (1982). Prairie dogs avoid extreme inbreeding. *Science* **215:**1639–1641.

Horn, H. S., and MacArthur, R. H. (1972). Competition among fugitive species in a harlequin environment. *Ecology* **53:**749–752.

Horvitz, C. C., and Schemske, D. W. (1986). Seed dispersal and environmental heterogeneity in a neotropical herb: A model of population and patch dynamics. *In* "Frugivores and Seed Dispersal" (F. Estrada, ed.), pp. 169–186. Junk Publishers, Dordrecht, The Netherlands.

Huffaker, C. B. (1958). Experimental studies of predation: Dispersal factors and predator-prey oscillation. *Hilgardia* **27:**343–383.

Huffaker, C. B., Kennett, C. E., and Tassan, R. L. (1986). Comparison of parasitism and densities of *Parlatoria oleae* (1952–1982) in relation to ecological theory. *Am. Nat.* **128:** 379–393.

Hughes, L., Dunlop, M., French, K., Leishman, M. R., Rice, B., Rodgerson, L., and Westoby, M. (1994). Predicting dispersal spectra: A minimal set of hypotheses based on plant attributes. *J. Ecol.* **82:**933–950.

Husband, B. C., and Barrett, S. C. H. (1996). A metapopulation perspective in plant population biology. *J. Ecol.* **84:**461–469.

Huston, M. A. (1994). "Biological Diversity: The Coexistence of Species on Changing Landscapes." Cambridge University Press, Cambridge, UK.

Huston, M. A., and Smith, T. (1987). Plant succession: Life history and competition. *Am. Nat.* **130:** 168–198.

Hutchinson, G. E. (1951). Copepodology for the ornithologist. *Ecology* **32:**571–577.

Ims, R. A. (1989). Origin and kinship effects on dispersal and space sharing in *Clethrionomys rufocanus. Ecology* **70:**607–616.

Ims, R. A. (1995). Movement patterns related to spatial structures. *In* "Mosaic Landscapes and Ecological Processes" (L. Hansson, L. Fahrig, and G. Merriam, eds.), pp. 85–109. Chapman & Hall, London.

Ims, R. A., and Stenseth, N. C. (1989). Divided the fruitflies fall. *Nature (London)* **343:** 21–22.

Ims, R. A., Rolstad, J., and Wegge, P. (1993). Predicting space use responses to habitat fragmentation:

can voles *Microtus oeconomus* serve as an experimental model system (EMS) for capercaillie grouse *Tetrao urogallus* in boreal forest? *Biol. Conserv.* **63**:261–268.

Inglis, G., and Underwood, A. J. (1992). Comments on some designs proposed for experiments on the biological importance of corridors. *Conserv. Biol.* **6**:581–586.

Ingvarsson, P. K. (1996). The effect of delayed population growth on the genetic differentiation of local populations subject to frequent extinctions and recolonizations. *Evolution (Lawrence, Kans.)* (in press).

International Union for the Conservation of Nature and Natural Resources (IUCN) (1980). "World Conservation Strategy." IUCN, Gland, Switzerland.

Ito, Y. (1980). "Comparative Ecology." Cambridge University Press, Cambridge, UK.

Ives, A. R., and May, R. M. (1985). Competition within and between species in a patchy environment: relations between microscopic and macroscopic models. *J. Theor. Biol.* **115**:65–92.

Ivins, B. L., and Smith, A. T. (1983). Responses of pikas (*Ochotona princeps,* Lagomorpha) to naturally occurring terrestrial predators. *Behav. Ecol. Sociobiol.* **13**:277–285.

Iwasa, Y., and Kubo, T. (1995). Forest gap dynamics with partially synchronized disturbances and patch age distribution. *Ecol. Modell.* **77**:257–271.

Iwasa, Y., and Roughgarden, J. (1986). Interspecific competition among metapopulations with space-limited subpopulations. *Theor. Popul. Biol.* **30**:194–214.

Jablonski, D. (1991). Extinctions: A paleobiological perspective. *Science* **253**:754–757.

Jackson, C. H. N. (1939). The analysis of an animal population. *J. Anim. Ecol.* **8**:238–246.

Jackson, K. S. (1992). The population dynamics of a hybrid zone in the alpine grasshopper *Podisma pedestris:* An ecological and genetical investigation. Ph.D. Thesis, University College, London.

Janzen, D. H. (1983). No park is an island: increase in interference from outside as park size decreases. *Oikos* **41**:402–410.

Jarosz, A. M., and Burdon, J. J. (1991). Host-pathogen interactions in natural populations of *Linum marginale* and *Melampsora lini*: II. Local and regional variation in patterns of resistance and racial structure. *Evolution (Lawrence, Kans.)* **45**:1618–1627.

Johnson, A. R., Milne, B. T., and Wiens, J. A. (1992a). Diffusion in fractal landscapes: Simulations and experimental studies of tenebrionid beetle movements. *Ecology* **73**:1968–1983.

Johnson, A. R., Wiens, J. A., Milne, B. T., and Crist, T. O. (1992b). Animal movements and population dynamics in heterogeneous landscapes. *Landscape Ecol.* **7**:63–75.

Johnson, C. G. (1969). "Insect Migration and Dispersal by Flight." Methuen, London.

Johnson, M. L., and Gaines, M. S. (1985). Selective basis for emigration of the prairie vole, *Microtus ochrogaster*: Open field experiment. *J. Anim. Ecol.* **54**:399–410.

Johnson, M. L., and Gaines, M. S. (1987). The selective basis for dispersal of the prairie vole, *Microtus ochrogaster*. *Ecology* **68**:684–694.

Johnson, M. L., and Gaines, M. S. (1990). Evolution of dispersal: Theoretical models and empirical tests using birds and mammals. *Annu. Rev. Ecol. Syst.* **21**:449–480.

Johnson, M. R., Boyd, D. K., and Pletscher, D. H. (1994). Serologic investigations of canine parvovirus and canine distemper in relation to wolf *(Canis lupus)* pup mortalities. *J. Wildl. Dis.* **30**:270–273.

Jolly, G. M. (1965). Explicit estimates from capture-recapture data with both death and immigration-stochastic model. *Biometrika* **59**:225–247.

Jordano, D., Retamosa, E. C., and Fernández Haeger, J. (1991). Factors facilitating the continued presence of *Colotis evagore* (Klug, 1829) in southern Spain. *J. Biogeogr.* **18**:637–646.

Judson, O. P. (1994). The rise of the individual-based model in ecology. *Trends Ecol. Evol.* **9**:9–14.

Kaitala, A. (1987). Dynamic life-history strategy of the waterstrider *Gerris thoracicus* as an adaptation to food and habitat variation. *Oikos* **48**: 125–131.

Kaitala, V. (1990). Evolutionary stable migration in salmon: A simulation study of homing and straying. *Ann. Zool. Fenn.* **27**:131–138.

Kaitala, V., Kaitala, A., and Getz, W. M. (1989). Evolutionary stable dispersal of a waterstrider in a temporally and spatially heterogeneous environment. *Evol. Ecol.* **3**:283–298.

Kaneko, K. (1992). Overview of coupled map lattices. *Chaos* **2:**279–282.

Kaneko, K. (1993). The coupled map lattice: introduction, phenomenology, Lyapunov analysis, thermodynamics and applications. *In* "Theory and Applications of Coupled Map Lattices (K. Kaneko, ed.), pp. 1–49. John Wiley, Chichester.

Karban, R. (1992). Plant variation: Its effects on populations of herbivorous insects. *In* "Ecology and Evolution of Plant Resistance" (R. S. Fritz and E. L. Simms, eds.), pp. 195–215. University of Chicago Press, Chicago.

Kareiva, P. M. (1984). Predator-prey dynamics in spatially structured populations: Manipulating dispersal in a coccinelid-aphid interaction. *Lect. Notes Biomathem.* **54:**368–389.

Kareiva, P. (1985). Finding and losing host plants by *Phyllotreta:* Patch size and surrounding habitat. *Ecology* **66:**1809–1816.

Kareiva, P. M. (1987). Habitat fragmentation and the stability of predator-prey interactions. *Nature (London)* **326:**388–390.

Kareiva, P. (1989). Patchiness, dispersal, and species interactions: Consequences for communities of herbivorous insects. *In* "Community Ecology" (J. Diamond and T. J. Case eds.), pp. 192–206. Harper & Row, New York.

Kareiva, P. (1990). Population dynamics in spatially complex environments: Theory and data. *In* "Population Regulation and Dynamics" (M. P. Hassell and R. M. May, eds.), pp. 53–68. Royal Society, London.

Kareiva, P., and Odell, G. M. (1987). Swarms of predators exhibit "prey-taxis" if individual predators use area restricted search. *Am. Nat.* **130:**233–270.

Kareiva, P., and Wennergren, U. (1995). Connecting landscape patterns to ecosystem and population processes. *Nature (London)* **373:**299–302.

Karlin, S., and Taylor, H. M. (1981). "A Second Course in Stochastic Processes." Academic Press, New York.

Karlson, R. H., and Taylor, H. M. (1992). Mixed dispersal strategies and clonal spreading of risk: Predictions from a branching process model. *Theor. Popul. Biol.* **42:**218–233.

Karr, J. R. (1982). Population variability and extinction in the avifauna of a tropical land-bridge island. *Ecology* **63:**1975–1978.

Katz, C. H. (1985). A nonequilibrium marine predator-prey interaction. *Ecology* **66:**1426–1438.

Kauffman, S., and Levin, S. A. (1987). Towards a general theory of adaptive walks on a rugged landscape. *J. Theor. Biol.* **128:**11–46.

Kaul, M. L. H. (1988). "Male Sterility in Higher Plants." Springer-Verlag, New York.

Kawecki, T. J. (1993). Age and size at maturity in a patchy environment: Fitness maximization versus evolutionary stability. *Oikos* **66:**309–317.

Kellner, C. J., Brawn, J. D., and Karr, J. R. (1993). What is habitat suitability and how should it be measured? *In* "Wildlife 2001: Populations" (D. R. McCullough and R. H. Barrett, eds.), pp. 476–488. Elsevier, London.

Kennedy, J. S. (1951). The migration of desert locust *(Schistocerca gregaria,* Forsk.). I. The behaviour of swarms. II. A theory of long range migration. *Philos. Trans. R. Soc. London, Ser. B* **235:**163–290.

Kenward, R. E. (1987). "Wildlife Radio Tagging." Academic Press, New York.

Kermack, W. O., and McKendrick, A. G. (1927). A contribution to the mathematical theory of epidemics. *Proc. R. Soc. London, Ser. A* **115:**700–721.

Kessler, C., and Manta, V. (1990). Specificity of restriction endonucleases and DNA modification methyltransferases—a review. *Gene* **92:**1–248.

Kheyr-Pour, A. (1980). Nucleo-cytoplasmic polymorphism for male sterility in *Origanum vulgare* L. *J. Hered.* **71:**253–260.

Kheyr-Pour, A. (1981). Wide nucleo-cytoplasmic polymorphism for male sterility in *Origanum vulgare* L. *J. Hered.* **72:**45–51.

Kimura, M. (1955). Solution of a process of random genetic drift with a continuous model. *Proc. Nat. Acad. Sci. U.S.A.* **41:**144–150.

Kimura, M. (1964). "Diffusion Models in Population Genetics." Methuen, London.

Kimura, M. (1983). "The Neutral Theory of Evolution." Cambridge University Press, Cambridge, UK.

Kimura, M. (1985). Evolution of an altruistic trait through group selection as studied by the diffusion equation method. *IMA J. Math. Appl. Med. Biol.* **1:**1–15.

Kimura, M., and Crow, J. F. (1963). On the maximum avoidance of inbreeding. *Genet. Res.* **4:**399–415.

Kimura, M., and Maruyama, T. (1971). Pattern of neutral polymorphism in a geographically structured population. *Genet. Res.* **18:**125–131.

Kimura, M., and Weiss, G. H. (1964). The stepping stone model of population structure and the decrease of genetic correlation with distance. *Genetics* **49:**561–576.

Kindvall, O. (1995). The impact of extreme weather on habitat preference and survival in a metapopulation of the bush cricket *Metrioptera bicolor* in Sweden. *Biol. Conserv.* **73:**51–58.

Kindvall, O. (1996a). Ecology of the bush cricket *Metrioptera bicolor* with implications for metapopulation theory and conservation. Ph.D. Thesis, Uppsala University, Sweden.

Kindvall, O. (1996b). Habitat heterogeneity and survival in a bush cricket metapopulation. *Ecology* **77:**207–214.

Kindvall, O., and Ahlén, I. (1992). Geometrical factors and metapopulation dynamics of the bush cricket, *Metrioptera bicolor* Philippi (Orthoptera: Tettigoniidae). *Conserv. Biol.* **6:**520–529.

King, A. W. (1991). Translating models across scales in the landscape. *In* "Quantitative Methods in Landscape Ecology" (M. G. Turner and R. H. Gardner, eds.), pp. 479–517. Springer-Verlag, New York.

King, J. A. (1955). Social behavior, social organization, and population dynamics in a black-tailed prairie dog town in the Black Hills of South Dakota. *Contrib. Lab. Vertebr. Biol. Univ. Mich.* **67:**1–126.

Kirkpatrick, M., and Barton, N. M. (1996). Evolution of a species' range. *Am. Nat.* (submitted).

Kitching, R. L. (1971). An ecological study of water-filled treeholes and their position in the woodland ecosystem. *J. Anim. Ecol.* **40:**413–434.

Klinkhamer, P. G. L., de Jong, T. J., Metz, J. A. J., and Val, J. (1987). Life history tactics of annual organisms: The joint effects of dispersal and delayed germination. *Theor. Popul. Biol.* **32:**127–156.

Kloeden, P. E., and Platen, E. (1992). "Numerical Solution of Stochastic Differential Equations." Springer-Verlag, Berlin.

Knowlton, N., and Jackson, J. B. C. (1994). New taxonomy and niche partitioning on coral reefs: Jack of all trades or master of some? *Trends Ecol. Evol.* **9:**7–9.

Koelewijn, H. P. (1993). On the genetics and ecology of sexual reproduction in *Plantago coronopus*. Ph. D. Thesis, University of Groningen, The Netherlands.

Koelewijn, H. P., and Van Damme, J. M. M. (1995a). Genetics of male sterility in gynodioecious *Plantago coronopus*. I. Cytoplasmic variation. *Genetics* **139:**1759–1775.

Koelewijn, H. P., and Van Damme, J. M. M. (1995b). Genetics of male sterility in gynodioecious *Plantago coronopus*. II. Nuclear genetic variation. *Genetics* **139:**1749–1758.

Kondrashov, A. S. (1984). Deleterious mutations as an evolutionary factor. I. The advantage of recombination. *Genet. Res.* **44:**199–218.

Kondrashov, A. S. (1988). Deleterious mutations and the evolution of sexual reproduction. *Nature (London)* **336:**435–441.

Korona, R., Korona, B., and Levin, B. R. (1993). Sensitivity of naturally occurring coliphages to Type I and Type II restriction and modification. *J. Gen. Microbiol.* **139:**1283–1290.

Kotliar, N. B., and Wiens, J. A. (1990). Multiple scales of patchiness and patch structure: A hierarchical framework for the study of heterogeneity. *Oikos* **59:**253–260.

Krebs, C. J. (1985). "Ecology. The Experimental Analysis of Distribution and Abundance," 3rd ed. Harper & Row, New York.

Krebs, C. J. (1994). "Ecology. The Experimental Analysis of Distribution and Abundance," 4th ed. Harper Collins, New York.

Kruger, D. H., and Bickle, T. A. (1983). Bacteriophage survival: Multiple mechanisms for avoiding deoxyribonucleic acid restriction systems of their hosts. *Microbiol. Rev.* **47:**345–360.

Kuhn, T. S. (1970). "The Structure of Scientific Revolutions," 2nd ed. University of Chicago Press, Chicago.

Kuno, E. (1981). Dispersal and persistence of populations in unstable habitats: A theoretical note. *Oecologia* **49:**123–126.

Kuussaari, M., Nieminen, M., and Hanski, I. (1996). An experimental study of migration in the butterfly *Melitaea cinxia. J. Anim. Ecol.* (in press).

Laan, R., and Verboom, B. (1990). Effect of pool size and isolation on amphibian communities. *Biol. Conserv.* **54:**251–262.

Lack, D. (1966). "Population Studies of Birds." Clarendon Press, Oxford.

Lacy, R. C. (1989). Analysis of founder representation in pedigrees: founder equivalents and founder genome equivalents. *Zoo Biol.* **8:**111–123.

Lacy, R. C., and Kreeger, T. (1992). "VORTEX Users Manual." Chicago Zoological Society, Chicago.

Lahaye, W. S., Gutierrez, R. J., and Akçakaya, H. R. (1994). Spotted owl metapopulation dynamics in Southern California. *J. Anim. Ecol.* **63:**775–785.

Lamb, H. H. (1985). "Climatic History and the Future." Princeton University Press, Princeton, NJ.

Lamberson, R. H., Noon, B. R., Voss, C., and McKelvez, R. J. (1994). Reserve design for territorial species: The effects of patch size and spacing on the viability of the northern spotted owl. *Cons. Biol.* **8:**185–195.

Lamotte, M. (1951). Recherches sur la structure génétique des populations naturelles de *Cepaea nemoralis* (L.). *Bull. Biol. Fr. Belg., Suppl.* **35.**

Lande, R. (1979). Effective deme sizes during long-term evolution estimated from rates of chromosomal rearrangement. *Evolution (Lawrence, Kans.)* **33:**234–251.

Lande, R. (1985). The fixation of chromosomal rearrangements in a subdivided population with local extinction and recolonisation. *Heredity* **54:**323–332.

Lande, R. (1987). Extinction thresholds in demographic models of territorial populations. *Am. Nat.* **130:**624–635.

Lande, R. (1988a). Demographic models of the northern spotted owl *(Strix occidental caurina). Oecologia* **75:**601–607.

Lande, R. (1988b). Genetics and demography in biological conservation. *Science* **241:**1455–1460.

Lande, R. (1993). Risks of population extinction from demographic and environmental stochasticity and random catastrophes. *Am. Nat.* **142:**911–927.

Lande, R., and Barrowclough, G. F. (1987). Effective population size, genetic variation, and their use in population management. *In* "Viable Populations for Conservation" (M. E. Soulé, ed.), pp. 87–123. Cambridge University Press, Cambridge, UK.

Lande, R. and Orzack, S. (1988). Extinction dynamics of age-structured populations in a fluctuating environment. *Proc. Natl. Acad. Sci. U.S.A.* **85:**7418–7421.

La Polla, V. N., and Barrett, G. W. (1993). Effects of corridor width and presence on the population dynamics of the meadow vole *(Microtus pennsylvanicus). Landscape Ecol.* **8:**25–37.

Larsen, K. W., and Boutin, S. (1994). Movements, survival, and settlement of red squirrel *(Tamiasciurus hudsonicus)* offspring. *Ecology* **75:**214–223.

Laser, K. D., and Lersten, N. R. (1972). Anatomy and cytology of microsporogenesis in cytoplasmic male sterile angiosperms. *Bot. Rev.* **38:**425–454.

Lavorel, S., and Chesson, P. (1996). How species with different regeneration niches coexist in patchy habitats with local disturbances. *Oikos* (submitted for publication).

Lavorel, S., Gardner, R. H., and O'Neill, R. V. (1993). Analysis of patterns in hierarchially structured landscapes. *Oikos* **67:**521–528.

Lavorel, S., Gardner, R. H., and O'Neill, R. V. (1995). Dispersal of annual plants in hierarchically structured landscapes. *Landscape Ecol.* **10:**277–289.

Lawrence, G. J., Mayo, G. M. E., and Shepherd, K. W. (1981). Interactions between genes controlling pathogenicity in the flax rust fungus. *Phytopathology* **71:**12–19.

Lawrence, W. S. (1987a). Dispersal: An alternative mating tactic conditional on sex ratio and body size. *Behav. Ecol. Sociobiol.* **21:**367–373.

Lawrence, W. S. (1987b). Effects of sex ratio on milkweed beetle emigration from host plant patches. *Ecology* **68:**539–546.

Lawton, J. H., and Woodroffe, G. L. (1991). Habitat and the distribution of water voles: Why are there gaps in a species' range? *J. Anim. Ecol.* **60:**79–91.

Lawton, J. H., Nee, S., Letcher, A. J., and Harvey, P. H. (1994). Animal distributions: Patterns and processes. *In* "Large-Scale Ecology and Conservation Biology" (P. J. Edwards, R. M. May, and N. R. Webb, eds.), pp. 41–58. Blackwell, Oxford.

Lebreton, J.-D., and North, P. M. (1992). "Marked Individuals in the Study of Animal Populations." Birkhäuser, Basel.

Lebreton, J.-D., Burnham, K. P., Clobert, J., and Anderson, D. R. (1992). Modeling survival and testing biological hypotheses using marked animals: A unified approach with case studies. *Ecol. Monogr.* **62:**67–118.

Lebreton, J.-D., Pradel, R., and Clobert, J. (1993). The statistical analysis of survival in animal populations. *Trends Ecol. Evol.* **8:**91–95.

Lees, D. R. (1981). Industrial melanism: Genetic adaptation of animals to air pollution. *In* "Genetic Consequences of Man-made Change" (J. A. Bishop and L. M. Cook, eds.), pp. 129–176. Academic Press, London.

Lefkovitch, L. P., and Fahrig, L. (1985). Spatial characteristics of habitat patches and population survival. *Ecol. Modell.* **30:**297–308.

Lei, G., and Hanski, I. (1997). Metapopulation structure of *Cotesia melitaearum,* a specialist parasitid of the butterfly *Melitaea cinxia. Oikos* (in press).

Leigh, E. G. (1981). The average lifetime of a population in a varying environment. *J. Theor. Biol.* **90:**213–239.

Lens, L., and Dhondt, A. A. (1994). Effects of habitat fragmentation on the timing of Crested Tit *Parus cristatus* natal dispersal. *Ibis* **136:**147–152.

Levin, B. R. (1986). Restriction-modification immunity and the maintenance of genetic diversity in bacterial populations. *In* "Evolutionary Processes and Evolutionary Theory" (S. Karlin and E. Nevo, eds.), pp. 669–688. Academic Press, New York.

Levin, B. R. (1988). Frequency-dependent selection in bacterial populations. *Philos. Trans. R. Soc. London, Ser. B* **319:**459–472.

Levin, S. A. (1974). Dispersion and population interactions. *Am. Nat.* **108:**207–228.

Levin, S. A. (1976). Population dynamic models in heterogeneous environments. *Annu. Rev. Ecol. Syst.* **7:**287–310.

Levin, S. A., and Paine, R. T. (1974). Disturbance, patch formation, and community structure. *Proc. Natl. Acad. Sci. U.S.A.* **71:**2744–2747.

Levin, S. A., and Paine, R. T. (1975). The role of disturbance in models of community structure. *In* "Ecosystem Analysis and Prediction" (S. A. Levin, ed.), pp. 56–67. SIAM, Philadelphia.

Levin, S. A., Cohen, D., and Hastings, A. (1984). Dispersal strategies in patchy environments. *Theor. Popul. Biol.* **26:**165–191.

Levins, R. (1966). The strategy of model building in population biology. *Am. Scientist* **54:**421–431.

Levins, R. (1968). "Evolution in Changing Environments." Princeton University Press, Princeton, NJ.

Levins, R. (1969a). Some demographic and genetic consequences of environmental heterogeneity for biological control. *Bull. Entomol. Soc. Am.* **15:**237–240.

Levins, R. (1969b). The effect of random variation of different types on population growth. *Proc. Natl. Acad. Sci. U.S.A.* **62:**1061–1065.

Levins, R. (1970). Extinction. *In* "Some Mathematical Problems in Biology" (M. Gerstenhaber, ed.), pp. 75–107. American Mathematical Society, Providence, RI.

Levins, R., and Culver, D. (1971). Regional coexistence of species and competition between rare species. *Proc. Nat. Acad. Sci. U.S.A.* **68:**1246–1248.

Lewin, B. (1977). "Gene Expression 3: Plasmids and Phages." Wiley, New York.

Lewis, D. (1941). Male sterility in natural populations of hermaphrodite plants: The equilibrium between females and hermaphrodites to be expected with different types of inheritance. *New Phytol.* **40**:56–63.

Lewontin, R. C., and Cohen, D. (1969). On population growth in a randomly varying environment. *Proc. Natl. Acad. Sci. U.S.A.* **62**:1056–1060.

Lidicker, W. Z., Jr., and Patton, J. L., (1987). Patterns of dispersal and genetic structure in populations of small rodents. *In* "Mammalian Dispersal Patterns: The Effects of Social Structure on Population Genetics" (B. D. Chepko-Sade and Z. T. Halpin, eds.), pp. 144–161. University of Chicago Press, Chicago.

Lindenmayer, D. B. (1995). Metapopulation viability of Leadbeater's possum, *Gymnobelideus leadbeateri*, in fragmented old-growth forests. *Ecol. Appl.* **5**:164–182.

Lindenmayer, D. B., and Nix, H. A. (1993). Ecological principles for the design of wildlife corridors. *Conserv. Biol.* **7**:627–630.

Liu, J., Dunning, J. B., Jr., and Pulliam, H. R. (1995). Potential effects of a forest management plan on Bachman's Sparrows *(Aimophila aestivalis):* Linking a spatially explicit model with GIS. *Conserv. Biol.* **9**:62–75.

Lloyd, D. G. (1976). The transmission of genes via pollen and ovules in gynodioecious angiosperms. *Theor. Popul. Biol.* **9**:299–316.

Lloyd, E. A. (1994). "Structure and Confirmation of Evolutionary Theory," 2nd ed. Princeton University Press, Princeton, NJ.

Loehle, C. (1987). Hypothesis testing in ecology: Psychological aspects and the importance of theory maturation. *Q. Rev. Biol.* **62**:397–409.

Lomnicki, A. (1988). "Population Ecology of Individuals." Princeton University Press, Princeton, NJ.

Lomolino, M. V. (1993). Winter filtering, immigrant selection and species composition of insular mammals of Lake Huron. *Ecography* **16**:24–30.

Ludwig, D. (1974). "Stochastic Population Theories." Springer-Verlag, Berlin.

Ludwig, D. (1976). A singular perturbation problem in the theory of population extinction. *SIAM-AMS Proc.* **10**:87–104.

Ludwig, D., and Levin, S. A. (1991). Evolutionary stability of plant communities and the maintenance of multiple dispersal types. *Theor. Popul. Biol.* **40**:285–307.

Luria, S. E., and Human, M. L. (1952). A non-hereditary host-induced variation in bacterial viruses. *J. Bacteriol.* **64**:557–559.

Lynch, J. F., and Johnson, N. K. (1974). Turnover and equilibria in insular avifaunas, with special reference to the California Channel Islands. *Condor* **76**:370–384.

Lynch, M., Burger, M., Butcher, D., and Gabriel, W. (1993). The mutational meltdown in asexual populations. *J. Hered.* **84**:339–344.

Lynch, M., Conery, J., and Burger, R. (1995). Mutation accumulation and the extinction of small populations. *Am. Nat.* **146**:489–518.

MacArthur, R. H. (1972). "Geographical Ecology." Princeton University Press, Princeton, NJ.

MacArthur, R. H., and Wilson, E. O. (1963). An equilibrium theory of insular zoogeography. *Evolution (Lawrence, Kans.)* **17**:373–387.

MacArthur, R. H., and Wilson, E. O. (1967). "The Theory of Island Biogeography." Princeton University Press, Princeton, NJ.

MacNair, M. R. (1987). Heavy metal tolerance in plants: A model evolutionary system. *Trends Ecol. Evol.* **2**:354–359.

MacNair, M. R., and Cumbes, Q. J. (1989). The genetic architecture of interspecific variation in *Mimulus. Genetics* **122**:211–222.

MacPhee, A., Newton, A., and McRae, K. B. (1988). Population studies on the winter moth *Opheroptera brumata* (L.) (Lepidoptrea: Geometridae) in apple orchards in Nova Scotia. *Can. Entomol.* **120**:73–83.

Malécot, G. (1948). "Les Mathématiques de l'Hérédité." Masson, Paris.

Malécot, G. (1969). "The Mathematics of Heredity" (D. M. Yermanos, transl.). Freeman, San Francisco.

Mallet, J. (1993). Speciation, raciation and color pattern evolution in *Heliconius* butterflies: Evidence from hybrid zones. *In* "Hybrid Zones and the Evolutionary Process" (R. G. Harrison, ed.), pp. 226–260. Oxford University Press, Oxford.

Mangel, M., and Tier, C. (1993a). A simple direct method for finding persistence times of populations and application to conservation problems. *Proc. Natl. Acad. Sci. U.S.A.* **90:**1083–1086.

Mangel, M., and Tier, C. (1993b). Dynamics of metapopulations with demographic stochasticity and environmental catastrophes. *Theor. Popul. Biol.* **44:**1–31.

Mangel, M., and Tier, C. (1994). Four facts every conservation biologist should know about persistence. *Ecology* **75:**607–614.

Manicacci, D., Olivieri, I., Perrot, V., Atlan, A., Gouyon, P.-H., Prosperi, J.-M., and Couvet, D. (1992). Landscape ecology: Population genetics at the metapopulation level. *Landscape Ecol.* **6:**147–159.

Mann, C. C., and Plummer, M. L. (1993). The high costs of biodiversity. *Science* **160:**1868–1871.

Mantel, N. (1967). The detection of disease clustering and a generalized regression approach. *Cancer Res.* **27:**209–220.

Martin, K., Stacey, P. B., and Braun, C. E. (1996). Demographic rescue in ptarmigan—mechanisms for long-term population stability in grouse. *Wildlife Biol.* (in press).

Maruyama, T. (1970). On the fixation probability of mutant genes in a subdivided population. *Genet. Res.* **15:**221–225.

Maruyama, T. (1971). Analysis of population structure. II. Two dimensional stepping stone models of finite length and other geographically structured populations. *Ann. Hum. Genet.* **35:**179–196.

Maruyama, T. (1972). Rate of decrease of genetic variability in a two-dimensional continuous population of finite size. *Genetics* **70:**639–651.

Maruyama, T. (1977). "Stochastic Problems in Population Genetics." Springer-Verlag, Berlin.

Maruyama, T., and Kimura, M. (1980). Genetic variability and effective population size when local extinction and recolonization of subpopulations are frequent. *Proc. Natl. Acad. Sci. U.S.A.* **77:**6710–6714.

Massot, M., Clobert, J., Lecomte, J., and Barbault, R. (1994). Incumbent advantage in common lizards and their colonizing ability. *J. Anim. Ecol.* **63:**431–440.

Matsuda, H., and Akamine, T. (1994). Simultaneous estimation of mortality and dispersal rates of an artificially released population. *Res. Popul. Ecol.* **36:**73–78.

Matthysen, E., Adriaensen, F., and Dhondt, A. A. (1995). Dispersal distances of nuthatches, *Sitta europaea,* in a highly fragmented forest habitat. *Oikos* **72:**375–381.

May, R. M. (1973a) "Stability and Complexity in Model Ecosystems." Princeton University Press, Princeton, NJ.

May, R. M. (1973b). Time-delay versus stability in population models with two and three trophic levels. *Ecology* **54:**315–325.

May, R. M. (1975). Patterns of species abundance and diversity. *In* "Ecology and Evolution of Communities" (M. L. Cody and J. M. Diamond, eds.), pp. 81–120. Harvard University Press, Cambridge, MA.

May, R. M., ed. (1976a). "Theoretical Ecology: Principles and Applications." Blackwell, Oxford.

May, R. M. (1976b). Models for two interacting populations. *In* "Theoretical Ecology: Principles and Applications" (R. M. May, ed.), pp. 49–70. Blackwell, Oxford.

May, R. M. (1977). Togetherness among schistosomes: Its effects on the dynamics of the infection. *Math. Biosci.* **35:**301–343.

May, R. M. (1978). Host-parasitoid systems in patchy environments: A phenomenological model. *J. Anim. Ecol.* **47:**833–843.

May, R. M. (1991). The role of ecological theory in planning reintroduction of endangered species. *Symp. Zool. Soc. London* **62:**145–163.

484   Bibliography

May, R. M. (1994). The effects of spatial scale on ecological questions and answers. *In* "Large-scale Ecology and Conservation Biology" (P. J. Edwards, R. May, and N. R. Webb, eds.), pp. 1–17. Blackwell, Oxford.

May, R. M. and Hassell, M. P. (1981). The dynamics of multiparasitoid-host interactions. *Am. Nat.* **117**:234–261.

May, R. M., and Southwood, T. R. E. (1990). Introduction. *In* "Living in a Patchy Environment." (B. Shorrocks and I. R. Swingland, eds.), pp. 1–22. Oxford University Press, Oxford.

May, R. M., Hassell, M. P., Anderson, R. M., and Tonkyn, D. W. (1981). Density dependence in host-parasitoid models. *J. Anim. Ecol.* **50**:855–865.

Maynard Smith, J. (1974). "Models in Ecology." Cambridge University Press, Cambridge, U.K.

Maynard Smith, J. (1982). "Evolution and the Theory of Games." Cambridge University Press, Cambridge, UK.

Mayr, E. (1942). "Systematics and the Origin of Species." Columbia University Press, New York.

Mayr, E. (1982). "The Growth of Biological Thought: Diversity, Evolution and Inheritance." Belknap Press, Cambridge, MA.

McArdle, B. H., Gaston, K. J., Lawton, J. H. (1990). Variation in the size of animal populations: Patterns, problems and artifacts. *J. Anim. Ecol.* **59**:439–454.

McCallum, H., and Dobson, A. (1995). Detecting disease and parasite threats to endangered species and ecosystems. *Trends Ecol. Evol.* **10**:190–194.

McCarthy, M. A., Ginzburg, L. R., and Akçakaya, H. R. (1995). Predator interference across trophic chains. *Ecology* **76**:1310–1319.

McCauley, D. E. (1989). Extinction, colonization, and population structure: A study of milkweed beetle. *Am. Nat.* **134**:365–376.

McCauley, D. E. (1991). Genetic consequences of local population extinction and recolonization. *Trends Ecol. Evol.* **6**:5–8.

McCauley, D. E. (1993). Evolution in metapopulations with frequent local extinction and recolonization. *Oxford Surv. Evol. Biol.* **10**:109–134.

McCauley, D. E. (1994). Contrasting the distribution of chloroplast DNA and allozyme polymorphism among local populations of *Silene alba:* Implications for studies of gene flow in plants. *Proc. Nat. Acad. Sci. U.S.A.* **91**:8127–8131.

McCauley, D. E. (1995). Effects of population dynamics on genetics in mosaic landscapes. *In* "Mosaic Landscapes and Ecological Processes" (L. Hansson, L. Fahrig, and G. Merriam, eds.), pp. 178–198. Chapman & Hall, London.

McCauley, D. E., Raveill, J., and Antonovics, J. (1995). Local founding events as determinants of genetic structure in a plant metapopulation. *Heredity* **75**:630–636.

McCook, L. J. (1994). Understanding ecological community succession—Causal models and theories, a review. *Vegetatio* **110**:115–147.

McEvoy, P. B. (1985). Depression in ragwort *(Senecio jacobaea)* abundance following introduction of *Tyria jacobaeae* and *Longitarsus jacobaeae* on the central coast of Oregon. *Proc. Int. Symp. Biol. Control Weeds, 6th, 1984,* pp. 57–64.

McEvoy, P. B., and Cox, C. S. (1987). Wind dispersal distances in dimorphic achenes of ragwort, *Senecio jacobaea. Ecology* **68**:2006–2015.

McEvoy, P. B., Rudd, N. T., Cox, C. S., and Huso, M. (1993). Disturbance, competition and herbivory effects on ragwort *(Senecio jacobaea)* populations. *Ecol. Monogr.* **63**:55–76.

McIntosh, R. P. (1991). Concept and terminology of homogeneity and heterogeneity in ecology. *In* "Ecological Heterogeneity" (J. Kolasa and S. T. A. Pickett, eds.), Ecol. Stud. No. 86. Springer-Verlag, Berlin.

McKelvey, K., Noon, B. R., and Lamberson, R. H. (1993). Conservation planning for species occupying fragmented landscapes: The case of the Northern Spotted Owl. *In* "Biotic Interactions and Global Change" (P. M. Kareiva, J. G. Kingslover, and R. B. Huey, eds.), pp. 424–450. Sinauer Assoc., Sunderland, MA.

McLendon, T., and Redente, E. F. (1990). Succession patterns following soil disturbance in a sage-brush steppe community. *Oecologia* **85**:293–300.

McMichael, A. (1993). Natural selection at work on the surface of virus-infected cells. *Science* **260:** 1771–1772.

McPeek, M. A., and Holt, R. D. (1992). The evolution of dispersal in spatially and temporally varying environments. *Am. Nat.* **140:**1010–1027.

Mech, L. D., and Goyal, S. M. (1993). Canine parvovirus effect on wolf population change and pup survival. *J. Wild. Dis.* **29:**330–333.

Menges, E. S. (1990). Population viability analysis for an endangered plant. *Conserv. Biol.* **4:**52–62.

Menotti-Raymond, M., and O'Brien, S. J. (1993). Dating the genetic bottleneck of the African cheetah. *Proc. Natl. Acad. Sci. U.S.A.* **90:**3172–3176.

Menotti-Raymond, M., and O'Brien, S. J. (1995). Evolutionary conservation of microsatellite loci in four species of felidae. *J. Hered.* **86:**319–322.

Merriam, G. (1988). Landscape dynamics in farmland. *Trends Ecol. Evol.* **3:**16–20.

Merriam, G. (1991). Corridors and connectivity: Animal populations in heterogenous environments. *In* "Nature Conservation 2: The Role of Corridors" (D. A. Saunders and R. J. Hobbs, eds.), pp. 133–142. Surrey Beatty & Sons, Chipping Norton, NSW, Australia.

Merriam, G., and Saunders, D. A. (1993). Corridors in restoration of fragmented landscapes. *In* "Nature Conservation 3: Reconstruction of Fragmented Ecosystems" (D. A. Saunders, R. J. Hobbs, and P.R. Ehrlich, eds.), pp. 71–87. Surrey Beatty & Sons, Chipping Norton, NSW, Australia.

Metz, J. A. J., and Diekmann, O. (1986). "The Dynamics of Physiologically Structured Populations," Springer, New York.

Michalakis, Y., and Olivieri, I. (1993). The influence of local extinctions on the probability of fixation of chromosomal rearrangements. *J. Evol. Biol.* **6:**153–170.

Middleton, D. A. J., Veitch, A. R., and Nisbet, R. M. (1995). The effect of an upper limit to population size on persistence time. *Theor. Popul. Biol.* **48:**277–305.

Middleton, J., and Merriam, G. (1981). Woodland mice in a farmland mosaic. *J. Appl. Ecol.* **18:**703–710.

Miller, G. L., and Carroll, B. W. (1989). Modeling vertebrate dispersal distances: Alternatives to the geometric distribution. *Ecology* **70:**977–986.

Mitchison, A. (1993). Will we survive? *Sci. Am.* **269**(3):136–144.

Moilanen, A., and Hanski, I. (1995). Habitat destruction and coexistence of competitors in a spatially realistic metapopulation model. *J. Anim. Ecol.* **64:**141–144.

Moilanen, A., and Hanski, I. (1997). A parameterized metapopulation model: the incidence function complemented with environmental factors. Manuscript.

Mollison, D. (1977). Spatial contact models for ecological and epidemic spread. *J. R. Stat. Soc. B* **39:**283–326.

Mollison, D., ed. (1995). "Epidemic Models: Their Structure and Relation to Data." Cambridge University Press, Cambridge, UK.

Monro, J. (1967). The exploitation and conservation of resources by populations of insects. *J. Anim. Ecol.* **36:**531–547.

Monro, J. (1975). Environmental variation and the efficacy of biological control—*Cactoblastis* in the Southern Hemisphere. *Proc. Ecol. Soc. Aust.* **9:**205–212.

Moore, F. B. G., and Tonsor, F. J. (1994). A simulation of Wright's shifting-balance process: Migration and the three phases. *Evolution (Lawrence, Kans.)* **48:**69–80.

Moore, S. D. (1989). Patterns of juvenile mortality within an oligophagous butterfly population. *Ecology* **70:**1726–1737.

Morell, V. (1994). Mystery ailment strikes Serengeti lions. *Science* **264:**1404.

Morin, P. A., Moore, J. J., Chakraborty, R., Jin, L., Goodall, J., and Woodruff, D. S. (1994). Kin selection, social structure, gene flow, and the evolution of Chimpanzees. *Science* **265:**1193–1201.

Morris, D. W. (1992). Scales and costs of habitat selection in heterogeneous landscapes. *Evol. Ecol.* **6:**412–432.

Morris, D. W. (1995). Earth's peeling veneer of life. *Nature (London)* **373:**25.

Morris, W. F. (1993). Predicting consequences of plant spacing and biased movement for pollen dispersal by honey bees. *Ecology* **74:**493–500.

Motro, U. (1982). Optimal rates of dispersal. I. Haploid populations. *Theor. Popul. Biol.* **21:**394–411.

Motro, U. (1983). Optimal rates of dispersal. III. Parent-offspring conflict. *Theor. Popul. Biol.* **23:** 159–168.

Motro, U. (1991). Avoiding inbreeding and sibling competition: The evolution of sexual dimorphism for dispersal. *Am. Nat.* **137:**108–115.

Murcia, C. (1995). Edge effects in fragmented forests: Implications for conservation. *Trends Ecol. Evol.* **10:**58–62.

Murdoch, W. W., Reeve, J. D., Huffaker, C. B., and Kennett, C. E. (1984). Biological control of scale insects and ecological theory. *Am. Nat.* **123:**371–392.

Murdoch, W. W., Chesson, J., and Chesson, P. (1985). Biological control in theory and practice. *Am. Nat.* **125:**344–366.

Murdoch, W. W., Swarbrick, S. L., Luck, R. F., Walde, S. J., and Yu, D. S. (1996). Refuge dynamics and metapopulation dynamics: An experimental test. *Am. Nat.* (in press).

Murphy, D. D., and White, R. R. (1984). Rainfall, resources and dispersal in southern populations of *Euphydryas editha* (Lepidoptera, Nymphalidae). *Pan-Pac. Entomol.* **60:**350–354.

Murphy, D. D., Freas, K. E., and Weiss, S. B. (1990). An "environment-metapopulation" approach to population viability analysis for a threatened invertebrate. *Conserv. Biol.* **4:**41–51.

Murray, J. D. (1989). "Mathematical Biology." Springer-Verlag, New York.

Myers, J. H., Monro, J., and Murray, N. (1981). Egg clumping, host plant selection and population regulation in *Cactoblastis cactorum* (Lepidoptera). *Oecologia* **51:**7–13.

Nachman, G. (1988). Regional persistence of locally unstable predator-prey interactions. *Exp. Appl. Acarol.* **5:**293–318.

Nachman, G. (1991). An acarine predator-prey metapopulation inhabiting greenhouse cucumbers. *Biol. J. Linn. Soc.* **42:**285–303.

Nagylaki, T. (1975). Conditions for the existence of clines. *Genetics* **80:**595–615.

Nagylaki, T. (1979). The island model with stochastic migration. *Genetics* **91:**163–176.

Nash, D. R., Agassiz, D. L. J., Godfray, H. C. J., and Lawton, J. H. (1995). The pattern of spread of invading species: Two leaf-mining moths colonizing Great Britain. *J. Anim. Ecol.* **64:**225–233.

Nee, S. (1994). How populations persist. *Nature (London)* **367:**123–124.

Nee, S., and May, R. M. (1992). Dynamics of metapopulations: Habitat destruction and competitive coexistence. *J. Anim. Ecol.* **61:**37–40.

Nee, S., and May, R. M. (1994). Habitat destruction and the dynamics of metapopulations—reply. *J. Anim. Ecol.* **63:**494.

Nei, M. (1987). "Molecular Evolutionary Genetics." Columbia University Press, New York.

Nève, G., Barascud, B., Hughes, R., Aubert, J., Descimon, H., Lebrun, P., and Baguette, M. (1996). Dispersal, colonization power and metapopulation structure in the vulnerable butterfly *Proclossiana eunomia* (Lepidoptera, Nymphalidae). *J. Appl. Ecol.* **33:**14–22.

New, T. R. (1991). "Butterfly Conservation." Oxford University Press, Mellbourne, Australia.

New, T. R., Pyle, R. M., Thomas, J. A., Thomas, C. D., and Hammond, P. C. (1995). Butterfly conservation management. *Annu. Rev. Entomol.* **40:**57–83.

Nichols, J. D., and Pollock, K. H. (1990). Estimation of recruitment from immigration versus in situ reproduction using Pollock's robust design. *Ecology* **71:**21–26.

Nichols, J. D., Brownie, C., Hines, J. E., Pollock, K. H., and Hestbeck, J. B. (1992). The estimation of exchanges among populations or subpopulations. *In* "Marked Individuals in the Study of Animal Populations" (J.-D. Lebreton and P. M. North, eds.), pp. 265–279. Birkäuser, Basel.

Nichols, R. A., and Hewitt, G. M. (1994). The genetic consequences of long distance dispersal during colonization. *Heredity* **72:**312–317.

Nicholson, A. J., and Bailey, V. A. (1935). The balance of animal populations. *Proc. Zool. Soc. London* **3**:551–598.

Nieminen, M. (1996). Risk of population extinction in moths: Effect of host plant characteristics. *Oikos* (in press).

Nisbet, R. M., and Gurney, W. S. C. (1982). "Modelling Fluctuating Populations." Wiley, New York.

Noon, B. R., and McKelvey, K. S. (1996). A common framework for conservation planning: Linking individual and metapopulation models. *In* "Metapopulations and Wildlife Conservation Management" (D. McCullough, ed.). Island Press, Covelo, CA (in press).

Noss, R. F. (1993). Wildlife corridors. *In* "Ecology of Greenways" (D. S. Smith and P. C. Hellmund, eds.) pp. 43–68. University of Minnesota Press, Minneapolis.

Noss, R. F., and Harris, L. D. (1986). Nodes, networks, and MUMs: Preserving diversity at all scales. *Environ. Manage. (N.Y.)* **10**:299–309.

Nowak, M. A., and May, R. M. (1994). Superinfection and the evolution of parasite virulence. *Proc. R. Soc. London, Ser. B* **255**:81–89.

Nunney, L. (1985). Group selection, altruism, and structured-deme models. *Am. Nat.* **126**:212–230.

Oates, M. R., and Warren, M. S. (1990). "A Review of Butterfly Introductions in Britain and Ireland." JCCBI/WWF, Godalming.

O'Brien, S. J., Wildt, D. E., Goldman, D., Merril, C., and Bush, M. (1983). The cheetah is depauperate in genetic variation. *Science* **221**:459–462.

O'Brien, S. J., Roelke, M. E., Marker, L., Newman, A., Winkler, C. A., Meltzer, D., Colley, L., Evermann, J. F., Bush, M., and Wildt, D. E. (1985). Genetic basis for species vulnerability in the cheetah. *Science* **227**:1428–1434.

Odendaal, F. J., Turchin, P., and Stermitz, F. R. (1988). An incidental-effect hypothesis explaining aggregation of males in a population of *Euphydryas anacia. Am. Nat.* **132**:735–749.

Odendaal, F. J., Turchin, P., and Stermitz, F. R. (1989). Influence of host-plant density and male harrassment of female *Euphydryas anacia* (Nymphalidae). *Oecologia* **78**:283–288.

Ohta, T. (1972). Population size and rate of molecular evolution. *J. Mol. Evol.* **1**:150–157.

Ohta, T. (1992). Theoretical study of near neutrality. II. Effect of subdivided populations structure with local extinction and recolonization. *Genetics* **130**:917–923.

Okland, F., Jonsson, B., Jensen, A. J., and Hansen, L. P. (1993). Is there a threshold size regulating seaward migration of brown trout and Atlantic salmon? *J. Fish Biol.* **42**:541–550.

Oksanen, T. (1993). Does predation prevent Norwegian lemmings from establishing permanent populations in lowland forests? *In* "The Biology of Lemmings" (N. C. Stenseth and R. A. Ims, eds.), pp. 425–437. Academic Press, San Diego, CA.

Okubo, A. (1980). "Diffusion and Ecological Problems: Mathematical Models." Springer-Verlag, New York.

Okubo, A., and Levin, S. A. (1989). A theoretical framework for data analysis of wind dispersal of seeds and pollen. *Ecology* **70**:329–338.

Olivieri, I., and Berger, A. (1985). Seed dimorphism and dispersal: Physiological, genetic and demographical aspects. *In* "Genetic Differentiation and Dispersal in Plants" (P. Jacquard *et al.*, eds.), pp. 413–429. Springer-Verlag, Berlin.

Olivieri, I., and Gouyon, P. G. (1985). Seed dimorphism for dispersal: Theory and implications. *In* "Structure and Functioning of Plant Populations" (J. Haeck and J. W. Woldendorp, eds.), Vol. 2, pp. 77–90. North-Holland Publ., Amsterdam.

Olivieri, I., Swan, M., and Gouyon, P.-H. (1983). Reproductive system and colonizing strategy of two species of *Carduus* (Compositae). *Oecologia* **60**:114–117.

Olivieri, I., Couvet, D., and Gouyon, P.-H. (1990). The genetics of transient populations: Research at the metapopulation level. *Trends Ecol. Evol.* **5**:207–210.

Olivieri, I., Michalakis, Y., and Gouyon, P.-H. (1995). Metapopulation genetics and the evolution of dispersal. *Am. Nat.* **146**:202–228.

Olver, F. W. J. (1974). "Introduction to Asymptotics and Special Functions." Academic Press, New York.

O'Neill, R. V., DeAngelis, D. L., Waide, J. B., and Allen, T. F. H. (1986). "A Hierarchical Concept of Ecosystems." Princeton University Press, Princeton, NJ.

O'Neill, R. V., Milne, B. T., Turner, M. G., and Gardner, R. H. (1988). Resource utilization scales and landscape pattern. *Landscape Ecol.* **2:**63–69.

Opdam, P. (1991). Metapopulation theory and habitat fragmentation: A review of holarctic breeding bird studies. *Landscape Ecol.* **5:**93–106.

Opdam, P., Foppen, R., Reijnen, R., and Schotman, A. (1995). The landscape ecological approach in bird conservation: Integrating the metapopulation concept into spatial planning. *Ibis* **137:**S139–S146.

Orr, H. A. (1995). The population genetics of speciation: The evolution of hybrid incompatibilities. *Genetics* **139:**1805–1813.

Osborne, P. (1984). Bird numbers and habitat characteristics in farmland hedgerows. *J. Appl. Ecol.* **21:**63–82.

Osmond, C. B., and Monro, J. (1981). Prickly pear. *In* "Plants and Man in Australia" (D. J. Carr and S. G. M. Carr, eds.), pp. 194–222. Academic Press, New York.

Ouborg, N. J. (1993). Isolation, population size and extinction: The classical and metapopulation approaches applied to vascular plants along the Dutch Rhine-system. *Oikos* **66:**298–308.

Pacala, S., Hassell, M. P., and May, R. M. (1990). Host-parasitoid associations in patchy environments. *Nature (London)* **344:**150–153.

Paine, R. T., and Levin, S. A. (1981). Intertidal landscapes: Disturbances and the dynamics of pattern. *Ecol. Monogr.* **51:**145–178.

Parker, M. A. (1985). Local population differentiation for compatibility in an annual legume and its host-specific fungal pathogen. *Evolution (Lawrence, Kans.)* **39:**713–723.

Parker, M. A. (1992). Disease and plant population genetic structure. *In* "Ecology and Evolution of Plant Resistance" (R. S. Fritz and E. L. Simms, eds.), pp. 345–362. University of Chicago Press, Chicago.

Peacock, M. M. (1995). Dispersal patterns, mating behavior, and population structure of pikas *(Ochotona princeps).* Ph.D. Thesis. Arizona State University, Tempe.

Pearson, S. M. (1993). The spatial extent and relative influence of landscape-level factors on wintering bird populations. *Landscape Ecol.* **8:**3–18.

Pearson, S. M., Turner, M. G., Gardner, R. H., and O'Neill, R. V. (1996). An organism-based perspective of habitat fragmentation. *In* "Biodiversity in Managed Landscapes: Theory and Practice" (R. C. Szaro, ed.). Oxford University Press, Oxford.

Pease, C. P., Lande, R., and Bull, J. J. (1989). A model of population growth, dispersal and evolution in a changing environment. *Ecology* **70:**1657–1664.

Peat, H. J., and Fitter, A. H. (1994). Comparative analyses of ecological characteristics of british Angiosperms. *Biol Rev. Cambridge Philos. Soc.* **69:**95–115.

Peck, J. R. (1994). A ruby in the rubbish: Beneficial mutations and the evolution of sex. *Genetics* **137:**597–606.

Peltonen, A., and Hanski, I. (1991). Patterns of island occupancy explained by colonization and extinction rates in shrews. *Ecology* **72:**1698–1708.

Peroni, P. A. (1994). Seed size and dispersal potential of *Acer rubrum* (Aceraceae) samaras produced by populations in early and late successional environments. *Am. J. Bot.* **81**(11):1428–1434.

Perrins, C. M. (1965). Population fluctuations and clutch-size in the great tit, *Parus major* L. *J. Anim. Ecol.* **34:**601–648.

Pickett, S. T. A., and White, P. S., eds. (1985). "The Ecology of Natural Disturbance and Patch Dynamics." Academic Press, New York.

Pimm, S. L. (1982). "Food Webs." Chapman & Hall, London.

Pimm, S. L. (1991). "The Balance of Nature." University of Chicago Press, Chicago.

Pimm, S. L., Jones, H. L., and Diamond, J. M. (1988). On the risk of extinction. *Am. Nat.* **132:**757–785.

Pimm, S. L., Gittleman, J. L., McCracken, G. F., and Gilpin, M. (1989). Plausible alternatives to bottlenecks to explain reduced genetic diversity. *Trends Ecol. Evol.* **4:**46–48.

Pimm, S. L., Diamond, J. M., Redds, T. M., Russel, G. J., and Verner, J. (1993). Times to extinction for small populations of large birds. *Proc. Nat. Acad. Sci. U.S.A.* **90**:10871–10875.

Pojar, J., Diaz, N., Steventon, D., Apostol, D., and Mellen, K. (1994). Biodiversity planning and forest management at the landscape scale. *In* "Expanding Horizons of Forest Ecosystem Management. Proceedings of the Third Habitat Futures Workshop" (M. H. Huff, L. K. Norris, J. B. Nyberg, and N. L. Wilkin, coords.), Gen. Tech. Rep. PNW-GTR-336, pp. 55–70. Department of Agriculture, Forest Service, Pacific Northwest Research Station, Portland, OR.

Pokki, J. (1981). Distribution, demography and dispersal of the field vole, *Microtus agrestis* (L.), in the Tvarminne archipelago, Finland. *Acta Zool. Fenn.* **164**:1–48.

Polis, G. A. (1991). Complex trophic interactions in deserts: An empirical critique of food web theory. *Am. Nat.* **138**:123–155.

Pollard, E. (1991). Synchrony of population fluctuations: The dominant influence of widespread factors on local butterfly populations. *Oikos* **60**:7–10.

Pollard, E., and Yates, T. J. (1993). "Monitoring Butterflies for Conservation." Chapman & Hall, London.

Pollard, E., Lakhani, K. H., and Rothery, P. (1987). The detection of density dependence from a series of annual censuses. *Ecology* **68**:2046–2055.

Pollock, K. H., Winterstein, S. R., Bunck, C. M., and Curtis, P. D. (1989). Survival analysis in telemetry studies: the staggered entry design. *J. Wildl. Manag.* **53**:7–15.

Pollock, K. H., Nichols, J. D., Brownie, C., and Hines, J. E. (1990). Statistical inference for capture-recapture experiments. *Wildl. Monogr.* **107**:1–97.

Poole, A. L., and Cairns, D. (1940). Botanical aspects of ragwort *(Senecio jacobaea* L.) control. *Dep. Sci. Ind. Res. Bull.* **82**.

Porter, J. H., and Dooley, J. L. (1993). Animal dispersal patterns: A reassessment of simple mathematical models. *Ecology* **74**:2436–2443.

Portnoy, S., and Willson, M. F. (1993). Seed dispersal curves: behavior of the tail of the distribution. *Evol. Ecol.* **7**:25–44.

Potter, M. A. (1990). Movement of North Island Brown Kiwi *(Apteryx australis mantelli)* between forest remnants. *N. Z. J. Ecol.* **14**:17–24.

Pradel, R. (1996). Utilization of capture mark-recapture for the study of recruitment and population growth rate. *Biometrics* **52**:703–709.

Preston, F. W. (1962). The canonical distribution of commonness and rarity. *J. Ecol.* **43**:185–215.

Price, G. R. (1970). Selection and covariance. *Nature (London)* **227**:520–521.

Price, M. V., and Endo, P. R. (1989). Estimating the distribution and abundance of a cryptic species, *Dipodomys stephensi* (Rodentia, Heteromyidae), and implications for management. *Conserv. Biol.* **3**:293–301.

Price, N., and Cappuccino, N., eds. (1995). "Population Dynamics: New Approaches and Synthesis." Academic Press, San Diego (in press).

Prins, A. H., and Nell, H. W. (1990). Positive and negative effects of herbivory on the population dynamics of *Senecio jacobaea* L. and *Cynoglossum officinale* L. *Oecologia* **83**:325–332.

Provine, W. (1986). "Sewall Wright and Evolutionary Biology." University of Chicago Press, Chicago.

Pruett-Jones, S. G., and Lewis, M. J. (1990). Sex ratio and habitat limitation promote delayed dispersal in superb fairy-wrens. *Nature (London)* **348**:541–542.

Pulliam, H. R. (1988). Sources, sinks and population regulation. *Am. Nat.* **132**:652–661.

Pulliam, H. R., and Danielson, B. J. (1991). Sources, sinks and habitat selection: A landscape perspective on population dynamics. *Am. Nat.* **137**:50–66.

Pulliam, H. R., Dunning, J. B., Jr., and Liu, J. (1992). Population dynamics in complex landscapes: A case study. *Ecol. Appl.* **2**:165–177.

Pullin, A. S., ed. (1995). "Ecology and Conservation of Butterflies." Chapman & Hall, London.

Quezada, J. R. (1969). Population biology of the cottony-cushion scale *Icerya purchasi* Maskell (Homoptera: Coccidae) and its natural enemies in southern California. Ph. D. Thesis, University of California, Riverside.

Quinn, J. F., and Hastings, A. (1988). Extinction in subdivided habitats. *Conserv. Biol.* **1**:293–296.

Quinn, J. F., Wolin, C. L., and Judge, M. L. (1989). An experimental study of patch size, habitat subdivision and extinction in a marine intertidal snail. *Conserv. Biol.* **3**:242–251.

Ralls, K., Ballou, J. D., and Templeton, A. (1988). Estimates of lethal equivalents and the cost of inbreeding in mammals. *Conserv. Biol.* **2**:185–193.

Rastetter, E. B., King, A. W., Cosby, B. J., Hornberger, G. M., O'Neill, R. V., and Hobbie, J. E. (1992). Aggregating fine-scale ecological knowledge to model coarser-scale attributes of ecosystems. *Ecol. Appl.* **2**:55–70.

Reeve, J. D. (1988). Environmental variability, migration, and persistence in host-parasitoid systems. *Am. Nat.* **132**:810–836.

Reeves, P. (1972). "The Bacteriocins." Springer-Verlag, New York.

Reinartz, J. A. (1984). Life history variation of common mullein *(Verbascum thapsus)*. I. Latitudinal differences in population dynamics and timing of reproduction. *J. Ecol.* **72**:897–912.

Renshaw, E. (1991). "Modelling Biological Populations in Space and Time." Cambridge University Press, Cambridge, UK.

Rice, K., and Jain, S. (1985). Plant population genetics and evolution in disturbed environments. *In* "The Ecology of Natural Disturbance and Patch Dynamics" (S. T. A. Pickett and P. S. White, eds.), pp. 287–303. Academic Press, New York.

Richter-Dyn, N., and Goel, N. S. (1972). On the extinction of a colonizing species. *Theor. Popul. Biol.* **3**:406–433.

Ricklefs, R. E., and Schluter, D., eds. (1993). "Species Diversity in Ecological Communities: Historical and Geographical Perspectives." University of Chicago Press, Chicago.

Riley, M. A., and Gordon, D. M. (1992). A survey of Col plasmids in natural isolates of *Escherichia coli* and an investigation into the stability of Col-plasmid lineages. *J. Gen. Microbiol.* **138**:1345–1352.

Risser, P. G., Karr, J. R., and Forman, R. T. T. (1984). Landscape ecology: Directions and approaches. *Spec. Publ.—Ill. Nat. Hist. Surv.* **2**.

Roberts, R. J. (1990). Restriction enzymes and their isoschizomers. *Nucleic Acids Res.* **18**(Suppl.):2331–2365.

Robinson, S. K., Thompson, F. R., Donovan, T. M., Whitehead, D. R., and Faaborg, J. (1995). Regional forest fragmentation and the nesting success of migratory birds. *Science* **267**:1987–1990.

Rocha, E. R., and de Uzeda, M. (1990). Antagonism among *Bacteriodes fragilis* group strains isolated from middle ear exudates from patients with chronic suppurative otitis media. *Ear, Nose Throat J.* **69**:614–618.

Roderick, G. K., and Caldwell, R. L. (1992). An entomological perspective on animal dispersal. *In* "Animal Dispersal: Small Mammals as a Model" (N. C. Stenseth and W. Z. Lidicker, eds.), pp 274–290. Chapman & Hall, London.

Rodríguez, J., Jordano, D., and Fernández Haeger, J. (1994). Spatial heterogeneity in a butterfly-host plant interaction. *J. Anim. Ecol.* **63**:31–38.

Roff, D. A. (1974). The analysis of a population model demonstrating the importance of dispersal in a heterogeneous environment. *Oecologia* **15**:259–275.

Roff, D. A. (1975). Population stability and the evolution of dispersal in a heterogeneous environment. *Oecologia* **19**:217–237.

Roff, D. A. (1986). The genetic basis of wing dimorphism in the sand cricket, *Gryllus firmus* and its relevance to the evolution of wing dimorphisms in insects. *Heredity* **57**:221–231.

Roff, D. A. (1990). The evolution of flightlessness in insects. *Ecol. Monogr.* **60**:389–421.

Roff, D. A. (1994a). Evolution of dimorphic traits—effect of directional selection on heritability. *Heredity* **72**:36–41.

Roff, D. A. (1994b). Habitat persistence and the evolution of wing dimorphism in insects. *Am. Nat.* **144**:772–798.

Rohani, P., and Miramontes, O. (1995). Immigration and the persistence of chaos in population models. *J. Theor. Biol.* **175**:203–206.

Rohani, P., Godfray, H. C. J., and Hassell, M. P. (1994). Aggregation and the dynamics of host-parasitoid systems—a discrete-generation model with within-generation redistribution. *Am. Nat.* **144**:491–509.

Rohani, P., May, R. M., and Hassell, M. P. (1996). Metapopulations and local stability: The effects of spatial structure. *J. Theor. Biol.* (in press).

Roitberg, B. D., and Mangel, M. (1993). Parent-offspring conflict and life-history consequences in herbivorous insects. *Am. Nat.* **142**:443–456.

Rosenzweig, M. L. (1981). A theory of habitat selection. *Ecology* **62**:327–335.

Rosenzweig, M. R. (1973). Exploitation in three trophic levels. *Am. Nat.* **107**:275–294.

Ross, R. (1909). "The Prevention of Malaria." Murray, London.

Roughgarden, J. (1979). Population dynamics in a stochastic environment. *Theor. Popul. Biol.* **7**:1–12.

Roughgarden, J. (1979). "Theory of Population Genetics and Evolutionary Ecology." Macmillan, New York.

Roughgarden, J., and Diamond, J. (1986). Overview: The role of species interactions in community ecology. *In* "Community Ecology" (J. Diamond and T. J. Case, eds.), pp. 333–343. Harper & Row, New York.

Rouhani, S., and Barton, N. H. (1987a). The probability of peak shifts in a founder population. *J. Theor. Biol.* **126**:51–62.

Rouhani, S., and Barton, N. H. (1987b). Speciation and the "shifting balance" in a continuous population. *Theor. Popul. Biol.* **31**:465–492.

Rouhani, S., and Barton, N. H. (1993). Group selection and the "shifting balance." *Genet. Res.* **61**:127–136.

Royama, T. (1992). "Analytical Population Dynamics." Chapman & Hall, London.

Ruxton, G. D. (1994). Local and ensemble dynamics of linked populations. *J. Anim. Ecol.* **63**:1002.

Sabelis, M. W., and Laane, W. E. M. (1986). Regional dynamics of spider-mite populations that become extinct locally from food source depletion and predation by phytoseiid mites (Acarina: Tetranychidae, Phytosciidae). *Lect. Notes Biomat.* **68**:345–376.

Sabelis, M. W., Diekmann, O., and Jansen, V. A. A. (1991). Metapopulation persistence despite local extinction—predator-prey patch models of the Lotka-Volterra type. *Biol. J. Linn. Soc.* **42**:267–283.

Saunders, D. A. (1990). Problems of survival in an extensively cultivated landscape: The case of Carnaby's Cockatoo *Calptorhynchus funereus latirostris*. *Biol. Conserv.* **54**:111–124.

Saunders, D. A., Hobbs, R. J., and Margules, C. R. (1991). Biological consequences of ecosystem fragmentation: A review. *Conserv. Biol.* **5**:18–32.

Schmitt, J., Ehrhardt, C. W., and Schwartz, D. (1985). Differential dispersal of self-fertilized and outcrossed progeny in jewelweed *(Impatiens capensis)*. *Am. Nat.* **126**:570–575.

Schoener, A. (1974): Experimental zoogeography: Colonization of marine mini-islands. *Am. Nat.* **108**:715–737.

Schoener, T. W. (1983). Rate of species turnover declines from lower to higher organisms: A review of the data. *Oikos* **41**:372–377.

Schoener, T. W. (1989). Food webs from the small to the large. *Ecology* **70**:1559–1589.

Schoener, T. W. (1991). Extinction and the nature of the metapopulation. *Acta Oecol.* **12**:53–75.

Schoener, T. W. (1993). On the relative importance of direct versus indirect effects in ecological communities. *In* "Mutualism and Community Organization" (H. Kawanabe, J. E. Cohen, and K. Iwasaki, eds.), pp. 365–415. Oxford University Press, Oxford.

Schoener, T. W., and Schoener, A. (1983). The time to extinction of a colonizing propagule of lizards increases with island area. *Nature (London)* **302**:332–334.

Schoener, T. W., and Spiller, D. A. (1987a). Effect of lizards on spider populations: Manipulative reconstruction of a natural experiment. *Science* **236**:949–952.

Schoener, T. W., and Spiller, D. A. (1987b). High population persistence in a system with high turnover. *Nature (London)* **330**:474–477.

Schoener, T. W., and Spiller, D. A. (1995). Effect of predators and area on invasion: An experiment with island spiders. *Science* **267**:1811–1813.

Scott, J. M., Csuti, B., and Caicco, S. (1991). Gap analysis: Assessing protection needs. *In* "Landscape Linkages and Biodiversity" (W. E. Hudson, ed.), pp. 15–26. Island Press, Washington, DC.

Searle, J. B. (1986). Factors responsible for a karyotypic polymorphism in the common shrew, *Sorex araneus*. *Proc. R. Soc. London, Serv. B* **229:**277–298.

Seber, G. A. F. (1965). A note on the multiple-recapture census. *Biometrika* **52:**249–259.

Segel, L. A. (1972). Simplification and scaling. *SIAM Rev.* **14:**547–571.

Senior, B. W., and Vörös, S. (1989). Discovery of new morganocin types of *Morganella morganii* in strains of diverse serotype and the apparent independence of bacteriocin type from serotype of strains. *J. Med. Microbiol.* **29:**89–93.

Severaid, J. H. (1955). The natural history of the pika (mammalian genus *Ochotona*). Ph.D. Thesis, University of California, Berkeley.

Shaffer, M. L. (1981). Minimum population sizes for species conservation. *BioScience* **31:**131–134.

Shapiro, A. M. (1979). Weather and the lability of breeding populations of the checkered white butterfly *Pieris protodice*. *J. Res. Lepid.* **17:**1–23.

Sharp, P. M. (1986). Molecular evolution of bacteriophages: Evidence of selection against the recognition sites of host restriction enzymes. *Mol. Biol. Evol.* **3:**75–83.

Shaw, M. W. (1995). Simulation of population expansion and spatial pattern when individual dispersal distributions do not decline exponentially with distance. *Proc. R. Soc. London, Ser. B* **259:**243–248.

Shmida, A., and Ellner, S. (1984). Coexistence of plant species with similar niches. *Vegetatio* **58:**129–155.

Shorrocks, B., and Swingland, I. R., eds. (1990). "Living in a Patchy Environment." Oxford University Press, Oxford.

Shorrocks, B., Atkinson, W. D., and Charlesworth, P. (1979). Competition on a divided and ephemeral resource. *J. Anim. Ecol.* **48:** 899–908.

Sih, A., Crowley, P., McPeele, M., Petranka, J., and Strohmeier, K. (1985). Predation, competition, and prey communities: A review of field experiments. *Annu. Rev. Ecol. Syst.* **16:**269–312.

Siitonen, J., and Martikainen, P. (1994). *Coleoptera* and *Aradus* (Hemiptera) collected from aspen in Finnish and Russian Karelia: Notes of rare and threatened species. *Scand. J. For. Res.* **9:** 185–191.

Silvertown, J. (1991). Dorothy's dilemma and the unification of plant population biology. *Trends Ecol. Evol.* **6:**346–348.

Silvertown, J., and Law, R. (1987). Do plants need niches? Some recent developments in plant community ecology. *Trends Ecol. Evol.* **2:**24–26.

Simberloff, D. (1969). Experimental zoogeography of islands. A model for insular colonization. *Ecology* **50:**296–314.

Simberloff, D. (1974). Equilibrium theory of island biogeography and ecology. *Annu. Rev. Ecol. Syst.* **5:**161–182.

Simberloff, D. (1976). Species turnover and equilibrium biogeography. *Science* **193:**572–578.

Simberloff, D. (1978a). Colonization of islands by insects: Immigration, extinction, and diversity. *In* "Diversity of Insects Faunas" (L. A. Mound and N. Waloff, eds.), pp. 139–153. Blackwell, Oxford.

Simberloff, D. (1978b). Using island biogeographic distributions to determine if colonization is stochastic. *Am. Nat.* **112:**713–726.

Simberloff, D. (1983). When is an island community in equilibrium? *Science* **220:**1275–1277.

Simberloff, D. (1988). The contribution of population and community biology to conservation science. *Annu. Rev. Ecol. Syst.* **5:**473–511.

Simberloff, D. (1994a). Conservation biology and unique fragility of island ecosystems. *In* "The Fourth California Islands Symposium: Update on the Status of Resources" (W. L. Halvorson and G. J. Maender, eds.), pp. 1–10. Santa Barbara Museum of Natural History, Santa Barbara, CA.

Simberloff, D. (1994b). The ecology of extinction. *Acta Palaeontol. Pol.* **38:**159–174.

Simberloff, D. (1996). Flagships, umbrellas, and keystones: Is single-species management passe in the landscape era? *Biol. Conserv.* (in press).

Simberloff, D. S., and Abele, L. G. (1982). Refuge design and island biogeographic theory: Effects of fragmentation. *Am. Nat.* **120:**41–50.

Simberloff, D., and Abele, L. G. (1984). Conservation and obfuscation: Subdivision of reserves. *Oikos* **42:**399–401.

Simberloff, D., Farr, J. A., Cox, J., and Mehlman, D. W. (1992). Movement corridors: Conservation bargains or poor investments? *Conserv. Biol.* **6:**493–504.

Singer, M. C. (1972). Complex components of habitat suitability within a butterfly colony. *Science* **176:**75–77.

Singer, M. C. (1983). Determinants of multiple host use by a phytophagous insect population. *Evolution (Lawrence, Kans.)* **37:**389–403.

Singer, M. C., and Ehrlich, P. R. (1979). Population dynamics of the checkerspot butterfly *Euphydryas editha*. *Forsch. Zool.* **25:**53–60.

Singer, M. C., and Thomas, C. D. (1996). Evolution of host preference in a checkerspot butterfly metapopulation under spatially and temporally variable selection. *Am. Nat. (in press)*.

Singer, M. C., Thomas, C. D., and Parmesan, C. (1993). Rapid human-induced evolution of insect-host associations. *Nature (London)* **366:**681–683.

Singer, M. C., Thomas, C. D., Billington, H. L., and Parmesan, C. (1994). Correlates of speed of evolution of host preference in a set of twelve populations of the butterfly *Euphydryas editha*. *Écoscience* **1:**107–114.

Sinsch, U. (1992). Structure and dynamics of a natterjack toad metapopulation *(Bufo calamita)*. *Oecologia* **90:**489–499.

Sjerps, M., and Haccou, P. (1994). A war of attrition between larvae on the same host plant: Stay and starve or leave and be eaten? *Evol. Ecol.* **8:**269–287.

Sjerps, M., and Haccou, P. (1996). Why do insect larvae risk migration before their host plant is defoliated? Some factors affecting leaving tendency of cinnabar moth larvae. Submitted for publication.

Sjögren, P. (1991). Extinction and isolation gradients in metapopulations: The case of the pool frog *(Rana lessonae)*. *Biol. J. Linn. Soc.* **42:**135–147.

Sjögren Gulve, P. (1994). Distribution and extinction patterns within a northern metapopulation case of the pool frog, *Rana lessonae*. *Ecology* **75:**1357–1367.

Sjögren Gulve, P., and Ray, C. (1996). Large-scale forestry extirpates the pool frog: Using logistic regression to model metapopulation dynamics. *In* "Metapopulations and Wildlife Conservation and Management" (D. R. McCullough, ed.) Island Press (in press).

Skalski, J. R., and Robson, D. S. (1992). "Techniques for Wildlife Investigations: Design and Analysis of Capture Data." Academic Press, San Diego.

Skellam, J. G. (1951). Random dispersal in theoretical populations. *Biometrika* **38:**196–218.

Slatkin, M. (1973). Gene flow and selection in a cline. *Genetics* **75:**733–756.

Slatkin, M. (1974). Competition and regional coexistence. *Ecology* **55:**128–134.

Slatkin, M. (1977). Gene flow and genetic drift in a species subject to frequent local extinctions. *Theor. Popul. Biol.* **12:**253–262.

Slatkin, M. (1978a). On the equilibration of fitnesses by natural selection. *Am. Nat.* **112:** 845–859.

Slatkin, M. (1978b). Spatial patterns in the distribution of polygenic characters. *J. Theor. Biol.* **70:** 213–228.

Slatkin, M. (1985). Gene flow in natural populations. *Ann. Rev. Ecol. Syst.* **16:**393–430.

Slatkin, M. (1987). Gene flow and the geographic structure of natural populations. *Science* **236:** 787–792.

Slatkin, M. (1993). Isolation by distance in equilibrium and non-equilibrium populations. *Evolution (Lawrence, Kans.)* **47:**264–279.

Slatkin, M. (1994). Gene flow and population structure. *In* "Ecological Genetics" (L. A. Real, ed.), pp. 3–17.

Slatkin, M. (1995). A measure of population subdivision based on microsatellite allele frequencies. *Genetics* **139**:457–462.

Slatkin, M., and Barton, N. H. (1989). A comparison of three indirect methods for estimating average levels of gene flow. *Evolution (Lawrence, Kans.)* **43**:1349–1368.

Slatkin, M., and Wade, M. J. (1978). Group selection on a quantitative character. *Proc. Natl. Acad. Sci. U.S.A.* **75**:3531–3534.

Small, R. J., Holzwart, J. C., and Rusch, D. H. (1993). Are ruffed grouse more vulnerable to mortality during dispersal? *Ecology* **74**:2020–2026.

Smith, A. T. (1974a). The distribution and dispersal of pikas: Consequences of insular population structure. *Ecology* **55**:1112–1119.

Smith, A. T. (1974b). The distribution and dispersal of pikas: Influences of behavior and climate. *Ecology* **55**:1368–1376.

Smith, A. T. (1978). Comparative demography of pikas *(Ochotona):* Effect of spatial and temporal age-specific mortality. *Ecology* **59**: 133–139.

Smith, A. T. (1979). High local species richness of mammals at Bodie, California. *Southwest. Nat.* **24**:553–555.

Smith, A. T. (1980). Temporal changes in insular populations of the pika *(Ochotona princeps)*. *Ecology* **60**:8–13.

Smith, A. T. (1987). Population structure of pikas: dispersal versus philopatry. *In* "Mammalian Dispersal Patterns: The Effects of Social Structure on Population Genetics" (B. D. Chepko-Sade and Z. T. Halpin, eds.), pp. 128–142. University of Chicago Press, Chicago.

Smith, A. T., and Ivins, B. L. (1983a). Reproductive tactics of pikas: Why have two litters? *Can. J. Zool.* **61**:1551–1559.

Smith, A. T., and Ivins, B. L. (1983b). Colonization in a pika population: Dispersal versus philopatry. *Behav. Ecol. Sociobiol.* **13**:37–47.

Smith, A. T., and Ivins, B. L. (1984). Spatial relationships and social organization in adult pikas: A facultatively monogamous mammal. *Z. Tierpsychol.* **66**:289–308.

Smith, A. T., and Peacock, M. M. (1990). Conspecific attraction and the determination of metapopulation colonization rates. *Conserv. Biol.* **4**:320–323.

Smith, A. T., and Weston, M. L. (1990). Ochotona princeps. *Mamm. Species* **352**:1–8.

Smith, C. E. G. (1970). Prospects for the control of infectious disease. *Proc. R. Soc. Med.* **63**:1181–1190.

Smith, F. E. (1975). Ecosystems and evolution. *Bull. Ecol. Soc. Am.* **56**:2.

Smith, R. E. (1958). Natural history of the prairie dog in Kansas. *Misc. Publ. Mus. Nat. Hist., Univ. Kans.* **49**:1–39.

Smouse, P. E., Vitzthum, V. J., and Neel, J. V. (1981). The impact of random and lineal fission on the genetic divergence of small human groups: A case study among the Yanomama. *Genetics* **98**:179–197.

Soberón, J. (1992). Island biogeography and conservation practice. *Conserv. Biol.* **6**:161.

Solbreck, C. (1991). Unusual weather and insect population dynamics: *Lygaeus equestris* during an extinction and recovery period. *Oikos* **60**:343–350.

Solbreck, C., and Sillen-Tullberg, B. (1990). Population dynamics of a seed-feeding bug, *Lygaeus equestris.* I. Habitat patch structure and spatial dynamics. *Oikos* **58**:199–209.

Soldaat, L. L. (1991). Seasonal variation in parasitoid attack of *Tyria jacobaeae* by *Apanteles popularis. Neth. J. Zool.* **41**:194–201.

Solé, R. V., and Valls, J. (1992). Spiral waves, chaos and multiple attractors in lattice models of interacting populations. *Phys. Lett. A* **166**:123–128.

Solé, R. V., Bascompte, J., and Valls, J. (1992). Stability and complexity of spatially extended two-species competition. *J. Theor. Biol.* **159**:469–480.

Solé, M. E. (1980). Thresholds for survival: Maintaining fitness and evolutionary potential. *In* "Conservation Biology: An Evolutionary-Ecological Perspective" (M. E. Soulé and B. A. Wilcox, eds.), pp. 111–124, Sinauer Assoc., Sunderland, MA.

Soulé, M. E., ed. (1987). "Viable Populations for Conservation." Cambridge University Press, New York.

Soulé, M. E., and Simberloff, D. (1986). What do genetics and ecology tell us about the design of nature refuges? *Biol. Conserv.* **35**:19–40.

Soulé, M. E., Wilcox, B. A., and Holtby, C. (1979). Benign neglect: A model of faunal collapse in the game reserves of East Africa. *Conserv. Biol.* **15**:259–272.

Southern, H. N. (1970). The natural control of a population of Tawny owls (*Strix aluco*). *J. Zool.* **162**:197–285.

Southwood, T. R. E. (1962). Migration of terrestrial arthropods in relation to habitat. *Biol. Rev. Cambridge Philos. Soc.* **37**:171–214.

Southwood, T. R. E. (1977). Habitat, the templet for ecological strategies. *J. Anim. Ecol.* **46**:337–365.

Southwood, T. R. E. (1987). Habitat and insect biology. *Bull. Entomol. Soc. Am.* **12**:211–214.

Spight, T. (1974). Sizes of populations of a marine snail. *Ecology* **55**:712–729.

Spiller, D. A., and Schoener, T. W. (1990). Lizards reduce food consumption by spiders: Mechanisms and consequences. *Oecologia* **85**:150–161.

Stacey, P. B. (1979). Habitat saturation and communal breeding in the acorn woodpecker. *Anim. Behav.* **27**:1153–1167.

Stacey, P. B. (1994). Metapopulation structure in the Mexican Spotted Owl. *Pap., 1st Annu. Wild. Soc. Meet.* Albuquerque, NM.

Stacey, P. B., and Johnson, V. A. (1996). In preparation.

Stacey, P. B., and Ligon, B. (1987). Territory quality and dispersal options in the acorn woodpecker, and a challenge to the habitat saturation model of cooperative breeding. *Am. Nat.* **130**:654–676.

Stacey, P. B., and Ligon, J. D. (1991). The benefits of philopatry hypothesis for the evolution of cooperative breeding: Habitat variance and group size effects. *Am. Nat.* **137**:831–846.

Stacey, P. B., and Martin, K. (1997). Metapopulation structure and population rescue in the white-tailed ptarmigan. Submitted.

Stacey, P. B., and Taper, M. (1992). Environmental variation and the persistence of small populations. *Ecol. Appl.* **2**:18–29.

Stamps, J. A. (1991). The effects of conspecifics on habitat selection in territorial species. *Behav. Ecol. Sociobiol.* **28**:29–36.

Stamps, J. A., Buechner, M. B., and Krishnan, V. V. (1987). The effects of edge permeability and habitat geometry on emigration from patches of habitat. *Am. Nat.* **129**:533–552.

Stangel, P. W., Lennartz, M. R., and Smith, M. H. (1992). Genetic variation and population structure of red-cockaded woodpeckers. *Conserv. Biol.* **6**:283–290.

Steen, H. (1994). Low survival of long distance dispersers of the root vole (*Microtus oeconomus*). *Ann. Zool. Fenn.* **31**:271–274.

Stenseth, N. C. (1983). Causes and consequences of dispersal in small mammals. *In* "The Ecology of Animal Movement" (I. R. Swingland and P. J. Greenwood, eds.), pp. 63–101. Clarendon Press, Oxford.

Stenseth, N. C., and Lidicker, W. Z., eds. (1992a). "Animal Dispersal. Small Mammals as a Model." Chapman & Hall, London.

Stenseth, N. C., and Lidicker, W. Z., Jr. (1992b). The study of dispersal: A conceptual guide. *In* Animal Dispersal. Small Mammals as a Model" (N. C. Stenseth and W. Z. Lidicker, Jr., eds.), pp. 5–20. Chapman & Hall, London.

Stenseth, N. C., and Maynard Smith, J. (1984). Coevolution in ecosystems: Red Queen evolution or stasis? *Evolution (Lawrence, Kans.)* **38**:870–880.

Stiling, P. D. (1987). The frequency of density dependence in insect host-parasitoid systems. *Ecology* **68**:844–856.

Stoddart, D. M. (1970). Individual range, dispersion and dispersal in a population of water voles (*Arvicola terrestris* (L.)). *J. Anim. Ecol.* **39**:403–425.

Stone (1993). Period doubling, reversals and chaos in simple ecological models. *Nature* **365**:617–620.

Strong, D. R. J. (1983). Density-vague ecology and liberal population regulation in insects. *In* "A New Ecology: Novel Approaches to Interactive Systems" (P. W. Price, C. N. Slobodchikoff, and W. S. Gaud, eds.) 4, pp. 313–327. Wiley, New York.

Strong, D. R. J. (1986). Density vagueness: Adding the variance in the demography of real populations. *In* "Community Ecology" (J. M. Diamond and T. J. Case, eds.), pp. 257–268. Harper & Row, New York.

Strong, D. R. J., Antolin, M. F., and Rathbun, S. (1990). Variance and pathiness in rates of population change: A planthopper's case history. *In* "Living in a Patchy Environment" (B. Shorrocks and I. R. Swingland, eds.), pp. 75–90. Oxford University Press, Oxford.

Sugihara, G., and May, R. M. (1990). Nonlinear forecasting as a way of distinguishing chaos from measurement error in time series. *Nature (London)* **344**:734–741.

Sutcliffe, O. L., Thomas, C. D., and Moss, D. (1996a). Spatial synchrony and asynchrony in butterfly population dynamics. *J. Anim. Ecol.* **65**:85–95.

Sutcliffe, O. L., Thomas, C. D., Yates, T. J., and Greatorex-Davies, J. N. (1996b). Correlated extinctions, colonizations and population fluctuations in a highly connected ringlet butterfly metapopulation. *Oecologia* (in press).

Sutcliffe, O. L., Thomas, C. D., and Peggie, D. (1996c). Area-dependent migration by ringlet butterflies generates a mixture of patchy population and metapopulation attributes. *Oecologia* (in press).

Sved, J. A., and Latter, B. D. H. (1977). Migration and mutation in stochastic models of gene frequency change. I. The island model. *J. Math. Biol.* **5**:61–73.

Swingland, I. R., and Greenwood, P. J. eds. (1983). "The Ecology of Animal Movement." Oxford University Press, Oxford.

Tachida, H., and Iizuka, M. (1991). Fixation probability in spatially changing environments. *Genet. Res.* **58**:243–251.

Talent, R. (1990). "Extinction Modelling with Finite-state Continuous-time Markov Chains: A Primer," Macquarie Math. Rep. No. 90-0061. Macquarie University, Sydney, Australia.

Tanner, J. E., Hughes, T. P., and Connell, J. H. (1994). Species coexistence, keystone species and succession: a sensitivity analysis. *Ecology* **75**:2204–2219.

Taper, M., and Stacey, P. B. (1996). In preparation.

Taylor, A. D. (1988). Large-scale spatial structure and population dynamics in arthropod predator-prey systems. *Ann. Zool. Fenn.* **25**:63–74.

Taylor, A. D. (1990). Metapopulations, dispersal, and predator-prey dynamics: An overview. *Ecology* **71**:429–433.

Taylor, A. D. (1991). Studying metapopulation effects in predator-prey systems. *In* "Metapopulation Dynamics: Empirical and Theoretical Investigations" (M. Gilpin, and I. Hanski, eds.), pp. 305–323. Academic Press, London.

Taylor, L. R. (1986). Synoptic dynamics, migration and the Rothamsted insect survey. *J. Anim. Ecol.* **55**:1–38.

Taylor, P. D., Fahrig, L., Henein, K., and Merriam, G. (1993). Connectivity is a vital element of landscape structure. *Oikos* **68**:571–573.

Tegelström, H., and Hansson, L. (1987). Evidence for long distance dispersal in the common shrew *(Sorex araneus). Z. Säugetierkd.* **52**:52–54.

Terborgh, J. (1974). Preservation of natural diversity: The problem of species extinction. *BioScience* **24**:715–722.

Terborgh, J. (1975). Faunal equilibria and the design of wildlife preserves. *In* "Tropical Ecological Systems: Trends in Terrestrial and Aquatic Research" (F. Golley and E. Medina, eds.), pp. 369–380. Springer, New York.

Thomas, C. D. (1991). Spatial and temporal variability in a butterfly population. *Oecologia* **87**:577–580.

Thomas, C. D. (1992). The establishment of rare insects in vacant habitats. *Antenna* **16**:89–93.

Thomas, C. D. (1994a). The ecology and conservation of butterfly metapopulations in the fragmented

British landscape. *In* "Ecology and Conservation of Butterflies" (A. S. Pullin, ed.), pp. 46–63. Chapman & Hall, London.

Thomas, C. D. (1994b). Local extinctions, colonizations and distributions: Habitat tracking by British butterflies. *In* "Individuals, Populations and Patterns in Ecology" (S. R. Leather, A. D. Watt, N. J. Mills, and K. F. A. Walters, eds.), pp. 319–336. Intercept Ltd., Andover, UK.

Thomas, C. D. (1994c). Extinction, colonization and metapopulations: Environmental tracking by rare species. *Conserv. Biol.* **8:**373–378.

Thomas, C. D., and Harrison, S. (1992). Spatial dynamics of a patchily-distributed butterfly species. *J. Anim. Ecol.* **61:**437–446.

Thomas, C. D., and Hochberg, M. E. (1996). Essential ingredients of real metapopulations, exemplified by the butterfly *Plebejus argus. In* "Aspects of the Genesis and Maintenance of Biological Diversity" (M. E. Hochberg, J. Clobert, and R. Barbault, eds.). Oxford University Press, Oxford (in press).

Thomas, C. D., and Jones, T. M. (1993). Partial recovery of a skipper butterfly *(Hesperia comma)* from population refuges: Lessons for conservation in a fragmented landscape. *J. Anim. Ecol.* **62:**472–481.

Thomas, C. D., and Singer, M. C. (1987). Variation in host preference affects movement patterns in a butterfly population. *Ecology* **68:**1262–1267.

Thomas, C. D., Thomas, J. A., and Warren, M. S. (1992). Distributions of occupied and vacant butterfly habitats in fragmented landscapes. *Oecologia* **92:**563–567.

Thomas, C. D., Singer, M. C., and Boughton, D. A. (1996). Catastrophic extinction of population sources in a complex butterfly metapopulation. *Am. Nat.* (in press).

Thomas, J. A. (1983a). The ecology and conservation of *Lysandra bellargus* (Lepidoptera: Lycaenidae) in Britain. *J. Appl. Ecol.* **20:**59–83.

Thomas, J. A. (1983b). The ecology and status of *Thymelicus acteon* (Lepidoptera: Hesperiidae) in Britain. *Ecol. Entomol.* **8:**427–435.

Thomas, J. A. (1984). The conservation of butterflies in temperate countries: Past efforts and lessons for the future. *Symp. R. Entomol. Soc.* **11:**333–353.

Thomas, J. A. (1991). Rare species conservation: Case studies of European butterflies. *Symp. Br. Ecol. Soc.* **31:**149–197.

Thomas, J. A. (1993). Holocene climate change and warm man-made refugia may explain why a sixth of British butterflies inhabit unnatural early-successional habitats. *Ecography* **16:**278–284.

Thomas, J. A., and Morris, M. G. (1994). Patterns, mechanisms and rates of extinction among invertebrates in the United Kingdom. *Philos. Trans. R. Soc. London, Ser. B* **344:**47–54.

Thomas, J. A., Thomas, C. D., Simcox, D. J., and Clarke, R. T. (1986). The ecology and declining status of the silver-spotted skipper butterfly *(Hesperia comma)* in Britain. *J. App. Ecol.* **23:**365–380.

Thomas, J. A., Moss, D., and Pollard, E. (1994). Increased fluctuations of butterfly populations towards the northern edges of species' ranges. *Ecography* **17:**215–220.

Thomas, J. W., Forsman, E. D., Lint, J. B., Meslow, E. C., Noon, B. R., and Verner, J. (1990). "A Conservation Strategy for the Northern Spotted Owl." USDA Forest Service, Portland, OR.

Thompson, J. N. (1993). Preference hierarchies and the geographic structure of host use in swallowtail butterflies. *Evolution (Lawrence, Kans.)* **47:**1585–1594.

Thompson, J. N. (1994). The geographic mosaic of evolving interactions. *In* "Individuals, Populations and Patterns in Ecology" (S. R. Leather, A. D. Watt, N. J. Mills, and K. F. A. Walters, eds.), pp. 419–431. Intercept Ltd., Andover, UK.

Thompson, J. N. (1996). Conserving interaction biodiversity. *In* "Enhancing the Ecological Basis of Conservation: Heterogeneity, Ecosystem Function and Biodiversity" (S. T. A. Pickett, R. S. Ostfeld, M. Shachak, and G. E. Likens, eds.), Chapman & Hall, New York.

Thompson, J. N., and Burdon, J. J. (1992). Gene-for-gene coevolution between plants and parasites. *Nature (London)* **360:**121–125.

Thompson, W. A. (1988). "Point Process Models with Applications to Safety and Reliability." Chapman & Hall, London.

Thomson, G. (1991). HLA population genetics. *Bailliere's Clin. Endocrinol. Metab.* **5:**247–260.

Thrall, P. H., Antonovics, J., and Hall, D. W. (1993). Host and pathogen coexistence in sexually transmitted and vector-borne diseases characterized by frequency-dependent disease transmission. *Am. Nat.* **142:**543–552.

Tilman, D. (1990). Constraints and tradeoffs: Toward a predictive theory of competition and succession. *Oikos* **58:**3–15.

Tilman, D. (1993). Species richness of experimental productivity gradients: How important is colonization limitation? *Ecology* **74:**2179–2191.

Tilman, D. (1994). Competition and biodiversity in spatially structured habitats. *Ecology* **75:**2–16.

Tilman, D., May, R. M., Lehman, C. L., and Nowak, M. A. (1994). Habitat destruction and the extinction debt. *Nature (London)* **371:**65–66.

Tracy, C. R., and George, T. L. (1992). On the determinants of extinction. *Am. Nat.* **139:**102–122.

Traub, W. H. (1991). Bacteriocin typing and biotyping of clinical isolates of *Serratia marcescens.* *Int. J. Med. Microbiol.* **275:**474–486.

Tsuji, N., Yamauchi, K., and Yamamura, N. (1994). A mathematical model for wing dimorphism in male *Cardiocondyla* ants. *J. Ethol.* **12:**19–24.

Tuljapurkar, S. (1990). "Population Dynamics in Variable Environments." Springer-Verlag, New York.

Turchin, P. (1986). Modelling the effect of past host size on Mexican bean beetle emigration. *Ecology* **67:**124–132.

Turchin, P. (1987). The role of aggregation in the response of Mexican bean beetles to host-plant density. *Oecologia* **71:**577–582.

Turchin, P. (1989). Beyond simple diffusion: models of not-so-simple movement in animals and cells. *Comments Theor. Biol.* **1:**65–83.

Turchin, P. (1991). Translating foraging movements in heterogeneous environments into the spatial distribution of foragers. *Ecology* **72:**1253–1266.

Turelli, M. (1977). Random environments and stochastic calculus. *Theor. Popul. Biol.* **12:**140–178.

Turner, J. R. G. (1981). Adaptation and evolution in Heliconius: A defence of neo-Darwinism. *Annu. Rev. Ecol. Syst.* **12:**99–122.

Turner, M. E., Stephens, J. C., and Anderson, W. W. (1982). Homozygosity and patch structure in plant populations as a result of nearest-neighbour pollination. *Proc. Natl. Acad. Sci. U. S. A.* **78:**203–207.

Turner, M. G. (1989). Landscape ecology: The effect of pattern on process. *Annu. Rev. Ecol. Syst.* **20:**171–197.

Turner, M. G., and Gardner, R. H., eds. (1991). "Quantitative Methods in Landscape Ecology." Springer-Verlag, New York.

Turner, M. G., Gardner, R. H., Dale, V. H., and O'Neill, R. V. (1989). Predicting the spread of disturbance across heterogeneous landscapes. *Oikos* **55:**121–129.

Turner, M. G., Arthaud, G. J., Engstrom, R. T., Hejl, S. J., Liu, J., Loeb, S., and McKelvey, K. (1995). Usefulness of spatially explicit population models in land management. *Ecol. Appl.* **5:**12–16.

Urban, D. L., O'Neill, R. V., and Shugart, H. H., (1987). Landscape ecology. *BioScience* **37:**119–127.

U. S. Department of Agriculture (1995). "Final Environmental Impact Statement for the Management of the Red-cockaded Woodpecker and its Habitat on National Forests in the Southern Region." Vol. II. U.S.D.A. Forest Service, Atlanta, GA.

Vail, S. (1993). Scale dependent responses to resource spatial pattern in simple models of consumer movements. *Am. Nat.* **141:**199–216.

Val, J., Verboom, J., and Metz, J. A. J. (1995). A deterministic size-structured metapopulation model. Preprint.

Valone, T. J., and Brown, J. H. (1995). Effects of competition, colonization and extinction on rodent species diversity. *Science* **267:**880–883.

van Baalen, J. (1982). Population biology of plants in woodland clearings. Ph.D. Thesis, Free University of Amsterdam.

Vance, R. R. (1984). The effect of dispersal on population stability in one-species, discrete-space population growth models. *Am. Nat.* **123**:230–254.

Van Damme, J. M. M. (1983). Gynodioecy in *Plantago lanceolata* L. II. Inheritance of three male sterility types. *Heredity* **50**:253–273.

Van Damme, J. M. M. (1984). Gynodioecy in *Plantago lanceolata* L. III. Sexual reproduction and the maintenance of male steriles. *Heredity* **52**:77–93.

Van Damme, J. M. M. (1986). Gynodioecy in *Plantago lanceolata* L. V. Frequencies and spatial distribution of nuclear and cytoplasmic genes. *Heredity* **56**:355–364.

Van Damme, J. M. M., and Van Delden, W. (1982). Gynodioecy in *Plantago lanceolata* L. I. Polymorphism for plasmon type. *Heredity* **49**:303–318.

Van Damme, J. M. M., and Van Delden, W. (1984). Gynodioecy in *Plantago lanceolata* L. IV. Fitness components of sex types in different life cycle stages. *Evolution (Lawrence, Kans.).* **38**: 1326–1336.

van de Klashorst, G., Redshaw, J. L., Sabelis, M. W., and Lingeman, R. (1992). A demonstration of asynchronous local cycles in an acarine predator-prey system. *Exp. Appl. Acarol.* **14**:185–199.

van der Meijden, E. (1976). Changes in the distribution pattern of *Tyria jacobaeae* during the larval period. *Neth. J. Zool.* **26**:136–161.

van der Meijden, E. (1979a). Herbivore exploitation of a fugitive plant species: Local survival and extinction of the cinnabar moth and ragwort in a heterogeneous environment. *Oecologia* **42**: 307–323.

van der Meijden, E. (1979b). The population ecology of *Senecio jacobaea* in a sand dune system. I. Reproductive strategy and the biennial habit. *J. Ecol.* **67**:131–153.

van der Meijden, E., de Jong, T. J., Klinkhamer, P. G. L., and Kooi, R. E. (1985). Temporal and spatial dynamics in populations of biennial plants. *In* "Structure and Functioning of Plant Populations" (J. Haeck and J. W. Woldendorp, eds.), Vol. 2, pp. 91–103. North-Holland Publ., Amsterdam.

van der Meijden, E., Wijn, M., and Verkaar, H. J. (1995). Defence and vegetative regrowth, alternative plant strategies in the struggle against herbivores. *Oikos* **51**:355–363.

van der Meijden, E., van Wijk, C. A. M., and Kooi, R. E. (1991). Population dynamics of the cinnabar moth *(Tyria jacobaeae):* Oscillations due to food limitation and local extinction risks. *Neth. J. Zool.* **41**:158–173.

van der Meijden, E., Klinkhamer, P. G. L., de Jong, T. J., and van Wijk, C. A. M. (1992). Metapopulation dynamics of biennial plants: How to exploit temporary habitats. *Acta Bot. Neerl.* **41**: 249–270.

Vanderplank, J. E. (1984). "Disease Resistance in Plants," 2nd ed. Academic Press, New York.

van Horne, B. (1983). Density as a misleading indicator of habitat quality. *J. Wild. Manage.* **47**:893–901.

Vanmarcke, E. (1983). "Random Fields." MIT Press, Cambridge, MA.

Van Valen, L. (1971). Group selection and the evolution of dispersal. *Evolution (Lawrence, Kans.)* **25**:591–598.

van Vuren, D., and Armitage, K. B. (1994). Survival of dispersing and philopatric yellow-bellied marmots: what is the cost of dispersal? *Oikos* **69**:179–181.

van Zoelen, A. M., and van der Meijden, E. (1991). Alkaloid concentration of different developmental stages of the cinnabar moth *(Tyria jacobaeae).* *Entomol. Exp. Appl.* **61**:291–294.

Venable, D. L. (1979). The demographic consequences of achene polymorphism in *Heterotheca latifolia* Buckl. (Compositae): Germination, survivorship, fecundity and dispersal. Ph.D. Thesis, University of Texas, Austin.

Venable, D. L. (1993). The population-dynamic functions of seed dispersal. *In* "Frugivory and Seed Dispersal: Ecological and Evolutionary Aspects" (T. Fleming and F. Estrada, eds.), pp. 31–55. Kluwer Academic Publishers, Dordrecht, The Netherlands.

Venable, D. L., and Brown, J. S. (1988). The selective interactions of dispersal, dormancy, and seed size as adaptations for reducing risk in variable environments. *Am. Nat.* **131**:360–384.

Venable, D. L., and Lawlor, L. (1980). Delayed germination and dispersal in desert annuals: Escape in space and time. *Oecologia* **46:**272–282.

Venable, D. L., and Levin, D. A. (1983). Morphological dispersal structures in relation to growth habit in the Compositae. *Plant Syst. Evol.* **143:**1–16.

Verboom, B., and van Apeldoorn, R. (1990). Effects of habitat fragmentation on the red squirrel, *Sciurus vulgaris* L. *Landscape Ecol.* **4:**117–176.

Verboom, J., Lankester, K., and Metz, J. A. J. (1991a). Linking local and regional dynamics in stochastic metapopulation models. *In* "Metapopulation Dynamics: Empirical and Theoretical Investigations" (M. Gilpin and I. Hanski, eds.), pp. 39–55. Academic Press, London.

Verboom, J., Schotman, A., Opdam, P., and Metz, J. A. J. (1991b). European nuthatch metapopulations in a fragmented agricultural landscape. *Oikos* **61:**149–156.

Vickery, W. L., and Nudds, T. D. (1991). Testing for density-dependent effects in sequential censuses. *Oecologia* **85:**419–423.

Villard, M.-A., Freemark, K. E., and Merriam, G. (1992). Metapopulation dynamics as a conceptual model for neotropical migrant birds: An empirical investigation. *In* "Ecology and Conservation of Neotropical Migrant Landbirds" (J. M. Hagan and D. W. Johnston, eds.), pp. 474–482. Smithsonian Institution Press, Washington, DC.

Wade, M. J. (1985). Hard selection, soft selection, kin selection, and group selection. *Am. Nat.* **125:** 61–73.

Wade, M. J., and McCauley, D. E. (1988). Extinction and recolonization: Their effects on the genetic differentiation of local populations. *Evolution (Lawrence, Kans.)* **42:**995–1005.

Wade, M. J., McKnight, M. L., and Shaffer, H. B. (1994). The effects of kin-structured colonization on nuclear and cytoplasmic diversity. *Evolution (Lawrence, Kans.)* **48:**1114–1120.

Wagner, D. L., and Liebherr, J. K. (1992). Flightlessness in insects. *Trends Ecol. Evol.* **7:**216–218.

Wahlberg, N., Moilanen, A., and Hanski, I. (1996). Predicting the occurrence of species in fragmented landscapes. *Science* (in press).

Walde, S. J. (1991). Patch dynamics of a phytophagous mite population: Effect of number of subpopulations. *Ecology* **72:**1591–1598.

Walde, W. J. (1994). Immigration and the dynamics of a predator-prey interaction in biological control. *J. Anim. Ecol.* **63:**337–346.

Walde, S. J., Nyrop, J. P., and Hardman, J. M. (1992). Dynamics of *Panonychus ulmi* and *Typhlodromus pyri:* Factors contributing to persistence. *Exp. Appl. Acarol.* **14:**261–291.

Walter, H. S. (1990). Small viable population: The red-tailed hawk of Socorro Island. *Conserv. Biol.* **4:**441–443.

Waples, R. S. (1989). A generalized method for estimating population size from temporal changes in allele frequency. *Genetics* **121:**379–391.

Waples, R. S. (1991). Genetic methods for estimating the effective size of cetacean populations. *In* "Genetic Ecology of Whales and Dolphins" (A. R. Hoelzel, ed.), pp. 279–300. Spec. Issue No. 13, International Whaling Commission.

Warren, M. S. (1987a). The ecology and conservation of the health fritillary butterfly, *Mellicta athalia.* II. Adult population structure and mobility. *J. Appl. Ecol.* **24:**483–498.

Warren, M. S. (1987b). The ecology and conservation of the health fritillary butterfly, *Mellicta athalia.* III. Population dynamics and the effect of habitat management. *J. Appl. Ecol.* **24:**499–513.

Warren, M. S. (1991). The successful conservation of an endangered species, the heath fritillary butterfly *Mellicta athalia,* in Britain. *Biol. Conserv.* **55:**37–56.

Warren, M. S. (1992). The conservation of British butterflies. *In* "The Ecology of Butterflies in Britain" (R. L. H. Dennis, ed.), pp. 246–274. Oxford University Press, Oxford.

Warren, M. S. (1993). A review of butterfly conservation in central southern Britain: I. Protection, evaluation and extinction on prime sites. *Biol. Conserv.* **64:**25–35.

Warren, M. S. (1994). The UK status and suspected metapopulation structure of a threatened European butterfly, the marsh fritillary *Eurodryas aurinia. Biol. Conserv.* **67:**239–249.

Warren, M. S. and Thomas, J. A. (1992). Butterfly responses to coppicing. *In* "The Ecological Effects of Coppicing" (G. P. Buckley, ed.), pp. 249–270. Chapman & Hall, London.

Waser, N. M., and Price, M. V. (1994). Crossing-distance effects in *Delphinium nelsonii:* Outbreeding and inbreeding depression in progeny fitness. *Evolution (Lawrence, Kans.)* **48:** 842–852.

Waser, P. M. (1985). Does competition drive dispersal? *Ecology* **66:**1170–1175.

Waser, P. M., and Elliott, L. F. (1991). Dispersal and genetic structure in kangaroo rats. *Evolution (Lawrence, Kans.)* **45:**935–943.

Waser, P. M., Creel, S. R., and Lucas, J. R. (1994). Death and disappearance: Estimating mortality risks associated with philopatry and dispersal. *Behav. Ecol.* **5:**135–141.

Washburn, J., and Cornell, H. V. (1981). Parasitoids, patches and phenology: Their possible roles in the local extinction of a cynipid gall wasp population. *Ecology* **62:**1597–1607.

Watkinson, A. R., and Sutherland, W. J. (1995). Sources, sinks and pseudo-sinks. *J. Anim. Ecol.* **64:** 126–130.

Wayne, R. K., Gilbert, D. A., Lehman, N., Hansen, K., Eisenhawer, A., Girman, D., Peterson, R. O., Mech, L. D., Gogan, P. J. P., Seal, U. S., and Krumenaker, R. J. (1991). Conservation genetics of the endangered Isle Royale gray wolf. *Conserv. Biol.* **5:**41–51.

Weddell, B. J. (1991). Distribution and movements of Columbian ground squirrels *(Spermophilus columbianus* (Ord.)): Are habitat patches like islands? *J. Biogeogr.* **18:**385–394.

Weir, B. S., and Cockerham, C. C. (1984). Estimating F statistics for the analysis of population structure. *Evolution (Lawrence, Kans.)* **38:**1358–1370.

Weiss, G. H., and Kimura, M. (1964). A mathematical analysis of the stepping-stone model of genetic correlation. *J. Appl. Probability* **2:**129–149.

Weiss, S. J., Murphy, D. D., and White, R. R. (1988). Sun, slope and butterflies: Topographic determinants of habitat quality in *Euphydryas editha. Ecology* **69:**1486–1496.

Welsh, H. (1990). Relictual amphibians and old-growth forests. *Conserv. Biol.* **3:**309–319.

Western, D., and Pearl, M. (1989). "Conservation Biology in the 21st Century." Oxford University Press, Oxford.

White, G. C., and Garrott, R. A. (1990). "Analysis of Wildlife Radio-tracking Data." Academic Press, San Diego, CA.

White, M. J. D. (1973). "Animal Cytology and Evolution." Cambridge University Press, London.

White, R. R. (1980). Inter-peak dispersal in alpine checkerspot butterflies (Nymphalidae). *J. Lepid. Soc.* **34:**353–362.

Whitlock, M. C. (1992a). Temporal fluctuations in demographic parameters and the genetic variance among populations. *Evolution (Lawrence, Kans.)* **46:**608–615.

Whitlock, M. C. (1992b). Nonequilibrium population structure in forked fungus beetles: Extinction, colonization, and the genetic variance among populations. *Am. Nat.* **139:**952–970.

Whitlock, M. C. (1994). Fission and the genetic variance among populations: The changing demography of forked fungus beetle populations. *Am. Nat.* **143:**820–829.

Whitlock, M. C. (1996). Sources and sinks: Asymmetric migration, population structure, and the effective population size. *Evolution (Lawrence Kans.)* (in press).

Whitlock, M. C., and Barton, N. H. (1996). The effective population size with migration and extinction. *Genetics* (submitted).

Whitlock, M. C., and McCauley, D. E. (1990). Some population genetic consequences of colony formation and extinction: Genetic correlations within founding groups. *Evolution (Lawrence, Kans.)* **44:**1717–1724.

Whitlock, M. C., Phillips, P. C., and Wade, M. J. (1993). Gene interaction affects the additive genetic variance in subdivided populations with migration and extinction. *Evolution (Lawrence, Kans.)* **47:**1758–1769.

Whittaker, R. H., and Levin, S. A. (1977). The role of mosaic phenomena in natural communities. *Theor. Popul. Biol.* **12:**117–139.

Wiens, J. A. (1976). Population responses to patchy environments. *Annu. Rev. Ecol. Syst.* **7:** 81–120.

Wiens, J. A. (1977). On competition and variable environments. *Am. Sci.* **65:**590–597.

Wiens, J. A. (1984). Resource systems, populations, and communities. *In* "A New Ecology: Novel

Approaches to Interactive Systems" (P. W. Price, C. N. Slobodchikoff, and W. S. Gand, eds.), pp. 397–436. Wiley, New York.

Wiens, J. A. (1985). Vertebrate responses to environmental patchiness in arid and semiarid ecosystems. *In* "The Ecology of Natural Disturbance and Patch Dynamics" (S. T. A. Pickett and P. S. White, eds.), pp 169–193. Academic Press, New York.

Wiens, J. A. (1989a). Spatial scaling in ecology. *Funct. Ecol.* **3:**385–397.

Wiens, J. A. (1989b). "The Ecology of Bird Communities" Cambridge University Press, Cambridge, UK.

Wiens, J. A. (1992a). What is landscape ecology, really? *Landscape Ecol.* **7:**149–150.

Wiens, J. A. (1992b). Ecological flows across landscape boundaries: A conceptual overview. *In* "Landscape Boundaries: Consequences for Biotic Diversity and Ecological Flows" (A. J. Hansen and F. di Castri, eds.), pp. 217–235. Springer-Verlag, New York.

Wiens, J. A. (1995a). Landscape mosaics and ecological theory. *In* "Mosaic Landscapes and Ecological Processes" (L. Hansson, L. Fahrig, and G. Merriam, eds.), pp. 1–26. Chapman & Hall, London.

Wiens, J. A. (1995b). Habitat fragmentation: island *v* landscape perspectives on bird conservation. *Ibis* **137:**S97–S104.

Wiens, J. A. (1996a). The emerging role of patchiness in conservation biology. *In* "Enhancing the Ecological Basis of Conservation: Heterogeneity, Ecosystem Function, and Biodiversity" (S. T. A. Pickett, R. S. Ostfeld, M. Shachak, and G. E. Likens, eds.), Chapman & Hall, New York (in press).

Wiens, J. A. (1996b). Wildlife in patchy environments: Metapopulations, mosaics, and management. *In* "Metapopulations and Wildlife Conservation Management" (D. McCullough, ed., pp. 53–84. Island Press, Washington, DC.

Wiens, J. A., and Milne, B. T. (1989). Scaling of 'landscapes' in landscape ecology, or, landscape ecology from a beetle's perspective. *Landscape Ecol.* **3:**87–96.

Wiens, J. A., Crawford, C. S., and Gosz, J. R. (1985). Boundary dynamics: A conceptual framework for studying landscape ecosystems. *Oikos* **45:**421–427.

Wiens, J. A., Stenseth, N. C., Van Horne, B., and Ims, R. A. (1993). Ecological mechanisms and landscape ecology. *Oikos* **66:**369–380.

Wiens, J. A., Crist, T. O., With, K. A., and Milne, B. T. (1995). Fractal patterns of insect movement in microlandscape mosaics. *Ecology* **76:**663–666.

Wiens, J. A., Schooley, R. L., and Weeks, R. D., Jr. (1996). Patchy landscapes and animal movements: Do beetles percolate? *Oikos* (in press).

Wilcove, D. S. (1985). Nest predation in forest tracts and the decline of migratory songbirds. *Ecology* **66:**1211–1214.

Wilcox, B. A., and Murphy, D. D. (1985). Conservation strategy: The effect of fragmentation on extinction. *Am. Nat.* **125:**879–887.

Williams, E. H. (1988). Habitat and range of *E. gillettii* (Nymphalidae). *J. Lepid. Soc.* **42:**37–45.

Williams, G. C., ed. (1971). "Group Selection." Aldine-Atherton, Chicago.

Williamson, M. H. (1981). "Island Populations." Oxford University Press, Oxford.

Williamson, M. H. (1989). The MacArthur-Wilson theory today: True but trivial? *J. Biogeogr.* **16:**3–4.

Willis, E. O. (1984). Conservation, subdivision of reserves, and the antidismemberment hypothesis. *Oikos* **42:**396–398.

Wilson, D. S. (1987). Altruism in Mendelian populations derived from sibling groups: The haystack model revisited. *Evolution (Lawrence, Kans.)* **41:**1059–1071.

Wilson, E. O. (1973). "Sociobiology: The New Synthesis." Harvard University Press, Cambridge, MA.

Wilson, E. O. (1992). "The Diversity of Life." Harvard University Press, Cambridge, MA.

Wilson, E. O., and Willis, E. O. (1975). Applied biogeography. *In* "Ecology and Evolution of Communities" (M. L. Cody and J. M. Diamond, eds.), pp. 523–534. Harvard University Press, Cambridge, MA.

Wilson, G. G., and Murray, N. E. (1991). Restriction and modification systems. *Annu. Rev. Genet.* **25:**585–627.

With, K. A. (1994). Using fractal analysis to assess how species perceive landscape structure. *Landscape Ecol.* **9:**25–36.

With, K. A. (1996). The application of neutral landscape models in conservation biology. *Conserv. Biol.* (in press).

With, K. A., and Crist, T. O. (1995). Critical thresholds in species' responses to landscape structure. *Ecology* **76:**2446–2459.

With, K. A., Gardner, R. H., and Turner, M. G. (1996). Landscape connectivity and population distributions in heterogeneous environments. *Oikos* (in press).

Woinarski, J. C. Z., Whitehead, P. J., Bowman, D. M. J. S., and Russell-Smith, J. (1992). Conservation of mobile species in a variable environment: The problem of reserve design in the Northern Territory, Australia. *Global Ecol. Biogeogra. Lett.* **2:**1–10.

Woiwod, I. P., and Hanski, I. (1992). Patterns of density dependence in moths and aphids. *J. Anim. Ecol.* **61:**619–629.

Wolda, H. (1978). Fluctuations in abundance of tropical insects. *Am. Nat.* **112:**1017–1045.

Wolda, H., and Dennis, B. (1993). Density dependence tests, are they? *Oecologia* **95:**581–591.

Wolfe, M. S., and Caten, C. E., eds. (1987). "Populations of Plant Pathogens: Their Dynamics and Genetics." Blackwell, Oxford.

Wood, D. M., and del Moral, R. (1987). Mechanisms of early primary succession in subalpine habitats on Mount St. Helens. *Ecology* **68:**780–790.

Wootton, J. T. (1994). The nature and consequences of indirect effects in ecological communities. *Annu. Rev. Ecol. Syst.* **25:**443–466.

Wright, S. (1931). Evolution in Mendelian populations. *Genetics* **16:**97–159.

Wright, S. (1932). The roles of mutation, inbreeding, crossbreeding and selection in evolution. *Proc. Int. Congr. Genet., 6th, 1932,* Vol. **1,** pp. 356–366.

Wright, S. (1935). Evolution in populations in approximate equilibrium. *J. Genet.* **30:**257–266.

Wright, S. (1937). The distribution of gene frequencies in populations. *Science* **85:**504.

Wright, S. (1938). Size of population and breeding structure in relation to evolution. *Science* **87:**430–431.

Wright, S. (1940). Breeding structure of populations in relation to speciation. *Am. Nat.* **74:**232–248.

Wright, S. (1941). On the probability of fixation of reciprocal translocations. *Am. Nat.* **75:**513–522.

Wright, S. (1943). Isolation by distance. *Genetics* **28:**114–138.

Wright, S. (1951). The genetical structure of populations. *Ann. Eugen.* **15:**323–354.

Wright, S. (1952). The theoretical variance within and among subdivisions of a population that is in a steady state. *Genetics* **37:**312–321.

Wright, S. (1969). "Evolution and the Genetics of Populations," Vol. 2. University of Chicago Press, Chicago.

Wright, S. (1977). "Evolution and the Genetics of Populations," Vol. 3. University of Chicago Press, Chicago.

Wright, S. (1978). "Evolution and the Genetics of Populations," Vol. 4. University of Chicago Press, Chicago.

Wu, J., Vankat, J. L., and Barlas, Y. (1993). Effects of patch connectivity and arrangement of animal metapopulation dynamics: A simulation study. *Ecol. Modell.* **65:**221–254.

Wynne-Edwards, V. C. (1962). "Animal Dispersion in Relation to Social Behaviour." Oliver & Boyd, Edinburgh.

Wynne-Edwards, V. C. (1971). Intergroup selection in the evolution of social systems. *In* "Group Selection" (G. C. Williams, ed.), pp. 93–104. Aldine-Atherton, Chicago.

Yeaton, R. I., and Bond, W. J. (1991). Competition between two shrub species dispersal differences and fire promote coexistence. *Am. Nat.* **138**:328–341.

Yoccoz, N. G. (1994). Deduction and inference in population biology: The role of models. The case-study of small mammals and their cycles. Ph.D. Thesis, Université Claude Bernard, Villeurbanne, France.

Yoccoz, N. G., Steen, H., Ims, R. A., and Stenseth, N. C. (1993). Estimating demographic parameters and the population size: An updated methodological survey. *In* "The Biology of Lemmings" (N. C. Stenseth, and R. A. Ims, eds.), pp. 565–587. Academic Press, London.

Yoccoz, N. G., and Lambin, X. (1996). In preparation.

Young, S., and Goldman, E. A. (1944). "The Wolves of North America." American Wildlife Institute, Washington, DC.

Index